The Probabilistic Revolution

The Probabilistic Revolution
Volume 1: Ideas in History

edited by Lorenz Krüger, Lorraine J. Daston, and Michael
Heidelberger

A Bradford Book
The MIT Press
Cambridge, Massachusetts
London, England

A substantial part of the research work for this publication was supported by the Stiftung Volkswagenwerk, Hannover, Germany, and by the Zentrum für interdisziplinäre Forschung, Universität Bielefeld, Federal Republic of Germany.
Gedruckt mit Unterstützung der Universität Bielefeld, Bundesrepublik Deutschland.

First MIT Press paperback edition, 1990

This book was set in Times New Roman by Asco Trade Typesetting Ltd., Hong Kong.

Library of Congress Cataloging-in-Publication Data

The probabilistic revolution.

"A Bradford book."
Includes bibliographies and indexes.
Contents: v. 1. Ideas in history/edited by Lorenz Krüger, Lorraine J. Daston, and Michael Heidelberger—v. 2. Ideas in the sciences/edited by Lorenz Krüger, Gerd Gigerenzer, and Mary S. Morgan.
1. Probabilities—History. 2. Science—History. 3. Social sciences—History. I. Krüger, Lorenz.
QA273.A4P76   1987   509   86-17972

ISBN 978-0-262-11118-8 (v. 1) (hc : alk. paper)   ISBN 978-0-262-61062-9 (v.1) (pb)
ISBN 978-0-262-11119-5 (v.2) (hc : alk. paper)   ISBN 978-0-262-61063-6 (v.2) (pb)

Contents of Volume 1

I REVOLUTION

II CONCEPTS

presented, and their impact on three statistical practitioners
(William Farr in England, and Dr. Louis-Adolphe Bertillon and
his son Jacques in France) are examined.

Urbanization, industrialization, and the rise of a market economy
promoted growing interest in numbers in the early decades of the
nineteenth century. Social statistics became the empirical basis of
social policy in Britain through the interplay of sanitary statistics
and social reform in the 1830s.

German social scientists and reformers were prominent among
those writers of the late nineteenth century who denied that
statistical regularities were laws of the average man. They came
to see statistical method as applying best to communities of
genuinely diverse individuals, governed only approximately by
mass regularities.

An institutional history of the Prussian Statistical Bureau at its
apogee is used to illustrate fundamental differences between
Western liberal and atomistic conceptions of probability and those
of a conservative and holistic German ideology.

The idea of probabilistic causation as expressed by the founders of
quantum mechanics should be regarded as continuous with earlier
discussions of "psychical," "qualitative," and "statistical"
causality among psychologists and social theorists. The latter
ideas, in turn, should be understood within the general context of
holistic social-political thinking designated "moderate liberal."

Contents of Volume 2

List of Contributors to Volumes 1 and 2

Contributors' names are marked with asterisks as follows: one asterisk, Fellows of the Zentrum für interdisziplinäre Forschung, Universität Bielefeld, 1982–1983; two asterisks, Fellows of the Zentrum für interdisziplinäre Forschung, supported by the Stiftung Volkswagenwerk; three asterisks, Fellows of the Zentrum für interdisziplinäre Forschung, 1982–1983, supported by the Alexander von Humboldt-Stiftung.

John Beatty*
Department of Ecology and
Behavioral Biology
University of Minnesota
Minneapolis, Minnesota

Marie-Noëlle Bourguet
Départment d'Histoire
Université de Reims
Reims, France

Nancy Cartwright*
Department of Philosophy
Stanford University
Stanford, California

I. Bernard Cohen*
Department of the History of Science
Harvard University
Cambridge, Massachusetts

William Coleman**
Department of the History of Science
University of Wisconsin
Madison, Wisconsin

Kurt Danziger
Department of Psychology
York University
Downsview, Ontario, Canada

Lorraine J. Daston*
Department of History
Brandeis University
Waltham, Massachusetts

Gerd Gigerenzer**
Fachgruppe Psychologie
Universität Konstanz
Constance, Federal Republic of
Germany

Ian Hacking*
Institute for History and Philosophy
of Science and Technology
University of Toronto
Toronto, Ontario, Canada

Michael Heidelberger*
Philosophisches Seminar
Universität Göttingen
Göttingen, Federal Republic of
Germany

M. J. S. Hodge
Division of History and Philosophy of
Science
The University of Leeds
Leeds, England

Robert A. Horváth*
Josef Attila Tudományegyetem
Allames Jogtudomńnyi Karának
Statisztikai Tanszéke
Szeged, Hungary

Gérard Jorland**
Centre de Recherches Historiques
Ecole des Hautes Etudes en Sciences
Sociales
Paris, France

Andreas Kamlah**
Fachbereich für Kultur- und
Geowissenschaften
Universität Osnabrück
Osnabrück, Federal Republic of
Germany

Eberhard Knobloch
Fachbereich Mathematik
Technische Universität Berlin
Berlin (West), Germany

Lorenz Krüger*
Philosophisches Seminar
Universität Göttingen
Göttingen, Federal Republic of
Germany

Thomas S. Kuhn
Department of Linguistics and
Philosophy
Massachusetts Institute of Technology
Cambridge, Massachusetts

Bernd-Olaf Küppers
Max-Planck-Institut für
biophysikalische Chemie
Göttingen, Federal Republic of
Germany

Bernard-Pierre Lécuyer
Institut d'Histoire du Temps Présent
Paris, France

Claude Ménard
Sciences Economiques
Université de Paris I, Panthéon-
Sorbonne
Paris, France

Karl H. Metz*
Historisches Institut
Ludwig-Maximilians-Universität
Munich, Federal Republic of
Germany

Mary S. Morgan**
Department of Economics
University of York
York, England

David J. Murray**
Department of Psychology
Queen's University
Kingston, Ontario, Canada

Anthony Oberschall**
Department of Sociology
University of North Carolina
Chapel Hill, North Carolina

Jan von Plato*
Department of Philosophy
University of Helsinki
Helsinki, Finland

Theodore M. Porter*
Corcoran Department of History
University of Virginia
Charlottesville, Virginia

Ivo Schneider*
Institut für Geschichte der
Naturwissenschaften
Ludwig-Maximilians-Universität
Munich, Federal Republic of
Germany

Stephen M. Stigler
Department of Statistics
The University of Chicago
Chicago, Illinois

Zeno G. Swijtink***
Department of Philosophy
State University of New York at
Buffalo
Buffalo, New York

John R. G. Turner
Department of Genetics
University of Leeds
Leeds, England

M. Norton Wise*
Department of History
University of California
Los Angeles, California

Preface to Volumes 1 and 2

Lorenz Krüger

During the academic year 1982–1983 an international, interdisciplinary group of twenty-one scholars gathered in the Federal Republic of Germany under the auspices of the Zentrum für interdisziplinäre Forschung (ZiF) of the University of Bielefeld and the Stiftung Volkswagenwerk to study the "Probabilistic Revolution." We used that somewhat tendentious shorthand to encompass the web of changes that made probability a part of philosophy, scientific theories and practice, social policy, and daily life between circa 1800 and 1950. We worked and lived together for almost a full year, and from time to time invited other colleagues to join us for conferences and seminars on related themes. In all, the project spanned some four years, being over two years in the planning, with a preliminary conference in September 1981, and followed up by an intensive week-long reunion of the group members in August 1984 to discuss final drafts and to organize their publication.

The prehistory of the project began quite a few years earlier when, in 1974, I had the privilege of participating in a research seminar given by Thomas Kuhn on the development of statistical physics. The idea of assembling an interdisciplinary group to study the rise of probability in the sciences occurred to me as a result of a conference of the International Union of the History and Philosophy of Science, held at Pisa in 1978, to which I was invited by Jaakko Hintikka. My understanding of the depth and the breadth of the problem grew in conversations with colleagues whom I met there, among them Ian Hacking and Nancy Cartwright. In 1980, when the project had been approved by the ZiF, Michael Heidelberger joined it. His active part in conceiving and guiding it throughout, as well as the assistance of Rosemarie Rheinwald during 1982–1983, were essential for its execution.

These two volumes represent the fruit of the year the research group spent together, not only of the individual essays conceived then, but also of the innumerable—indeed, well nigh constant—discussions, over the seminar table and in the laundry room, gathered in plenary session and in twos and threes. Thus, the present work represents more than an anthology of far-flung authors and topics, or the proceedings of a conference. Rather, it records a sort of scholarly experiment in loose-knit but sustained collaboration over disciplinary and national boundaries.

We are grateful to the institutions, and especially to the individuals who are the human face of such institutions, that made this somewhat utopian project both possible and pleasant. The Stiftung Volkswagenwerk generously supported a subgroup of seven members particularly concerned with the impact of the Probabilistic Revolution on man and society. The Zentrum für interdisziplinäre Forschung provided occasion, funds, research facilities, a friendly, helpful staff, and a sylvan setting in the Teutoburger Wald. Families of foreign group members abandoned home and hearth for the year and applied themselves to learning German; families of native group members and their colleagues were hospitable beyond reckoning.

We thank all most heartily.

The Probabilistic Revolution

Introduction to Volume 1

Lorraine J. Daston

All of the contributors to these volumes asked roughly the same question: How did the mathematical theory of probability find a domain of applications? For almost a full year, we pursued likely candidates jointly and individually: the nineteenth-century bureaucratic mania for statistics; the practices of insurers; the arguments of philosophers; the measurements of experimenters; the theories of physicists. How did probabilistic ideas eventually unseat the ideal of determinism shared by almost all European thinkers in 1800—if it did? What were the various meanings of probability, and how did they open up (or close off) possibilities for applying probabilities to, say, the crime rate of Belgium or the motions of gas molecules? If there was a "Probabilistic Revolution" that made chance a meaningful part of our conceptual vocabulary, what were the forces of counterrevolution that retained the forms of probability without the substance? What was the connection between probability and new standards of precision, argument, and objectivity?

We pursued these questions in common, but we pursued them as intellectual historians, social historians, philosophers, economists, sociologists, physicists, psychologists; as Germans, Canadians, Frenchmen, Americans, Hungarians and Britons, Finns and Dutch. We posed similar problems, we developed a shared vocabulary in which to talk about them, and we learned from one another, but we did not always convince one another. The present collection of partial answers to these questions is still indelibly stamped with the distinctions of discipline, national tradition, and sheer personal predilection. We do not offer a collective answer or a unified approach: our arguments as well as our agreements are present in these chapters, in both choice and treatment of subjects. The interested reader will find a philosophical analysis of the arguments for and against probability check-and-jowl by an account of the extraordinary career of Dr. Ernst Engel, *echt* Prussian bureaucrat; a full-dress history of the St. Petersburg Paradox alongside reflections on the contrasts between British and German statistical thinking in the nineteenth century.

Given this diversity, it has been no easy task to subdivide this collection into volumes and parts. The scheme we have chosen attends both to period and to filiation of subject matter within that period.

Volume 1 is devoted to the pre-twentieth-century development of probabilistic and statistical thinking and its applications. History is no respecter of dates, and some of the articles reach as far back as the seventeenth century, and others stretch into the twentieth, but the focus is on nineteenth-century events. The history of probability and statistics during this period is also no respecter of latter-day disciplinary boundaries: key concepts and techniques like the normal distribution migrate from mathematics to astronomy to sociology to physics to philosophy on the strength of analogies of substance, as well as form. Central figures like the mathematician Pierre-Simon Laplace or the social statistician Adolphe Quetelet make their entrances in quite varied contexts, and ancient philosophical problems like that of free will, seen in a new statistical light, surface both in heated debates over social reform and in the origins of mathematical psychophysics. Hence, in this volume, disciplinary subdivisions have given way to more general designations that seek to keep the historical connections intact. Volume 2, in contrast, deals largely with twentieth-century developments in which disciplinary boundaries are not so permeable, and specialization is the rule.

The four parts of this volume are entitled "Revolution," "Concepts," "Uncertainty," and "Society." Chapters in the latter three parts treat specific historical episodes in the history of philosophy, mathematics, and the natural and social sciences; the chapters in the first part take a more synoptic view that compares the spread of probabilistic and statistical ideas and methods to other important intellectual transformations.

The three chapters in the "Revolution" part look for underlying patterns in the history of science and relate these to the particular case of the history of probability and statistics. These chapters are historically informed, but aim to set forth a framework for understanding that history taken as a whole, rather than to explore any part of that history in depth. They sharpen theses about continuity and change in science both by philosophical analysis and by historical example and counterexample, and are intended as contributions to the ongoing discussion about these issues among historians, philosophers, and sociologists of science, as well as synthetic reflections on the history of probability and statistics.

They are emphatically *not* meant to provide a general vocabulary or an exclusive interpretative lens for the other chapters in these volumes. These latter are intended as part of the evidence against which such interpretations are to be tested, not necessarily as evidence *for* these interpretations. (And the plural "interpretations" must be stressed here, for our authors sometimes disagree among themselves.) However, these three chapters do serve as a general introduction to both volumes, for they assess the significance of the transformation the other chapters address in its specifics, and point out connections both among these specific aspects of the rise of probabilistic thinking, and also between this and other revolutions in theory and practice of comparable magnitude.

The "Concepts" part deals with the evolution of philosophical and mathematical ideas about uncertainty, chance, error, variation, freedom, and necessity. These chapters examine the interaction of mathematical probability theory with the applications that gave it impetus, and with the interpretations that gave it applications. Broadly speaking, these chapters all address the question of "What does probability mean?" in a historical vein, and collectively reveal how wide the range of answers was, from the epistemic views championed by Laplace to the radical indeterminism of Fechner. Probability theory is remarkable both then and now for its intimate relations with philosophical issues like determinism and induction on the one hand, and with concrete applications like error theory and statistical mechanics on the other, and this part brings that double conceptual dependence to the fore. Special domains of application like error theory virtually sustain mathematical probability for long periods; new philosophical positions about issues like determinism or the meaning of probabilities open up new areas of applications— in Fechner's psychophysics or the statistical turn of late nineteenth-century probability—and close off others—Laplace's law of succession or the probability of judgments. Probability theory in the nineteenth century was never very far from philosophy nor from practice.

Uncertainty is our constant companion in practice: we mostly dwell, as Locke says, in the twilight of probabilities. The third part deals with three important examples of attempts to master uncertainty: actuarial mathematics, measurement, and statistical inference. The context in which these methods emerge are quite varied—the feverish London insurance market, the venerable tradition of astronomical observations, the fledgling social sciences—but each case centers upon the issue of the stability of the phenomena. Do people die in regular proportions at regular intervals? How should we match our variable data to the invariable world? Is human judgment a source of errors or of corrections? The recurring themes are the inexorable substitution of standardized techniques for individual judgments, and the trade-off between real variability and human error. Not only do these methods for dealing with uncertainty transmute unpredictability, indecision, and variability into regularity, consensus, and uniformity; they also create new standards for objectivity, right reasoning, and rational risk-taking that have transformed both the natural and social orders since 1800.

The contributions to the final part on "Society" examine the efforts of bureaucrats, social theorists, and reformers to translate human experience into numbers. These efforts depend crucially upon the concepts and methods discussed in parts II and III, but contrasting cultural and professional milieux stamp them: British, French, and German officials count different people for different reasons; so do physicians and sociologists. The categories of the Prussian and French statistical surveys map the significant divisions of those societies for us—and tell us much about what social ills most worried the governments that sponsored them. To simplify: the French counted criminals, the British paupers, and the Prussians foreigners. And behind the question "Who counted what and why?" lay the anterior debate over whether to count at all, as both the French shift from ethnographic description to quantitative tallies and the German reaction to Quetelet illustrate. Statistics was variously a diagnostic, a body of observations, a set of techniques, and a social science for its nineteenth-century practitioners, and its relationships with mathematical probability theory ranged from remote to intimate. The statisticians often knew very little statistics; just as often the probabilists were equally ignorant of the mechanics of censuses and the tabulation of mortality tables. Yet the two groups challenged each other, the distributions of the probabilists inspiring the social theory of l'homme moyen and that social theory in turn provoking German probabilists to refine the mathematical language of distributions. Eventually, statistics and probability merged in theory, application, and practice, a story continued in volume 2.

One final note to the reader of this volume and its companion: These chapters were conceived as independent contributions, but they were not conceived independently. That is, they were written so that each could be read singly, at will, but they are also linked by common themes, overlapping subject matter, and a year's worth of discussion, argument, and shared perplexity. We forged no consensus, but we did find unsuspected connections, create new controversies, and pose similar questions. Because the connections, controversies, and questions were as often as not many-cornered, no simple linear arrangement of the individual essays can truly chart them. Nonetheless, we have tried to indicate by proximity in the table of contents particularly close affinities, and hope the reader will take advantage of this device to read at least the immediate neighbors of the chapter that drew him to these volumes in the first place.

I REVOLUTION

1 What Are Scientific Revolutions?

Thomas S. Kuhn

"What Are Scientific Revolutions?" attempts to refine and clarify the distinction between normal and revolutionary scientific development. After an introductory presentation of the issue, most of the chapter is devoted to the presention of three examples of revolutionary change: the transition from an Aristotelian to a Newtonian understanding of motion, from the contact to the chemical theory of the Voltaic cell, and from Planck's to the now familiar derivation of the law of black-body radiation.* A concluding section epitomizes three features common to the examples. All are locally holistic in that they require a number of interrelated changes of theory to be made at once; only at the price of incoherence could these changes have occurred one step at a time. All require changes in the way some set of interdefined scientific terms attached to nature, in the taxonomy provided by scientific language itself. And all also involved changes in something very like metaphor, in the scientist's acquired sense of what objects or events are like each other and of which differ.

It is now almost twenty years since I first distinguished what I took to be two types of scientific development, normal and revolutionary.[1] Most successful scientific research results in change of the first sort, and its nature is well captured by a standard image: normal science is what produces the bricks that scientific research is forever adding to the growing stockpile of scientific knowledge. That cumulative conception of scientific development is familiar, and it has guided the elaboration of a considerable methodological literature. Both it and its methodological by-products apply to a great deal of significant scientific work. But scientific development also displays a noncumulative mode, and the episodes that exhibit it provide unique clues to a central aspect of scientific knowledge. Returning to a long-standing concern, I shall therefore here attempt to isolate several such clues, first by describing three examples of revolutionary change and then by briefly discussing three characteristics which they all share. Doubtless revolutionary changes share other characteristics as well, but these three provide a sufficient basis for the more theoretical analyses on which I am currently engaged, and on which I shall be drawing somewhat cryptically when concluding this paper.

Before turning to a first extended example, let me try—for those not previously familiar with my vocabulary—to suggest what it is an example of. Revolutionary change is defined in part by its difference from normal change, and normal change is, as already indicated, the sort that results in growth, accretion, cumulative addition to what was known before. Scientific laws, for example, are usually products of this normal process: Boyle's law will illustrate what is involved. Its

---

* The three examples that constitute the bulk of this chapter were developed in this form for the first of three lectures delivered at the University of Notre Dame during November 1981 in the series Perspectives in Philosophy. In very nearly their present frame, but under the title "From Revolutions to Salient Features," they were read to the Third Annual Conference of the Cognitive Science Society in August 1981.

discoverers had previously possessed the concepts of gas pressure and volume as well as the instruments required to determine their magnitudes. The discovery that, for a given gas sample, the product of pressure and volume was a constant at constant temperature simply added to the knowledge of the way these antecedently understood[2] variables behave. The overwhelming majority of scientific advance is of this normal cumulative sort, but I shall not multiply examples.

Revolutionary changes are different and far more problematic. They involve discoveries that cannot be accommodated within the concepts in use before they were made. In order to make or to assimilate such a discovery one must alter the way one thinks about and describes some range of natural phenomena. The discovery (in cases like these "invention" may be a better word) of Newton's Second Law of motion is of this sort. The concepts of force and mass deployed in that law differed from those in use before the law was introduced, and the law itself was essential to their definition. A second, fuller, but more simplistic example is provided by the transition from Ptolemaic to Copernican astronomy. Before it occurred, the sun and moon were planets, the earth was not. After it, the earth was a planet, like Mars and Jupiter; the sun was a star; and the moon was a new sort of body, a satellite. Changes of that sort were not simply corrections of individual mistakes embedded in the Ptolemaic system. Like the transition to Newton's laws of motion, they involved not only changes in laws of nature but also changes in the criteria by which some terms in those laws attached to nature. These criteria, furthermore, were in part dependent upon the theory with which they were introduced.

When referential changes of this sort accompany change of law or theory, scientific development cannot be quite cumulative. One cannot get from the old to the new simply by an addition to what was already known. Nor can one quite describe the new in the vocabulary of the old or vice versa. Consider the compound sentence, "In the Ptolemaic system planets revolve about the earth; in the Copernican they revolve about the sun." Strictly construed, that sentence is incoherent. The first occurrence of the term "planet" is Ptolemaic, the second Copernican, and the two attach to nature differently. For no univocal reading of the term "planet" is the compound sentence true.

No example so schematic can more than hint at what is involved in revolutionary change. I therefore turn at once to some fuller examples, beginning with the one that, a generation ago, introduced me to revolutionary change, the transition from Aristotelian to Newtonian physics. Only a small part of it, centering on problems of motion and mechanics, can be considered here, and even about it I shall be schematic. In addition, my account will invert historical order and describe, not what Aristotelian natural philosophers required to reach Newtonian concepts, but what I, raised a Newtonian, required to reach those of Aristotelian natural philosophy. The route I traveled backward with the aid of written texts was, I shall simply assert, nearly enough the same one that earlier scientists had traveled forward with no text but nature to guide them.

I first read some of Aristotle's physical writings in the summer of 1947, at which time I was a graduate student of physics trying to prepare a case study on the development of mechanics for a course in science for nonscientists. Not surprisingly, I approached Aristotle's texts with the Newtonian mechanics I had previously read clearly in mind. The question I hoped to answer was how much mechanics Aristotle had known, how much he had left for people like Galileo and Newton to discover. Given that formulation, I rapidly discovered that Aristotle had known almost no mechanics at all. Everything was left for his successors, mostly those of the sixteenth and seventeenth centuries. That conclusion was standard, and it might in principle have been right. But I found it bothersome because, as I was reading him, Aristotle appeared not only ignorant of mechanics, but a dreadfully bad physical scientist as well. About motion, in particular, his writings seemed to me full of egregious errors, both of logic and of observation.

These conclusions were unlikely. Aristotle, after all, had been the much admired codifier of ancient logic. For almost two millennia after his death, his work played the same role in logic that Euclid's played in geometry. In addition, Aristotle had often proved an extraordinarily acute naturalistic observer. In biology, especially, his descriptive writings provided models that were central in the sixteenth and seventeenth centuries to the emergence of the modern biological tradition. How could his characteristic talents have deserted him so systematically when he turned to the study of motion and mechanics? Equally, if his talents had so deserted him, why had his writings in physics been taken so seriously for so many centuries after his death? Those questions troubled me. I could easily believe that Aristotle had stumbled, but not that, on entering physics, he had totally collapsed. Might not the fault be mine rather than Aristotle's, I asked myself. Perhaps his words had not always meant to him and his contemporaries quite what they meant to me and mine.

Feeling that way, I continued to puzzle over the text, and my suspicions ultimately proved well-founded. I was sitting at my desk with the text of Aristotle's *Physics* open in front of me and with a four-colored pencil in my hand. Looking up, I gazed abstractedly out the window of my room—the visual image is one I still retain. Suddenly the fragments in my head sorted themselves out in a new way, and fell into place together. My jaw dropped, for all at once Aristotle seemed a very good physicist indeed, but of a sort I'd never dreamed possible. Now I could understand why he had said what he'd said, and what his authority had been. Statements that had previously seemed egregious mistakes, now seemed at worst near misses within a powerful and generally successful tradition. That sort of experience—the pieces suddenly sorting themselves out and coming together in a new way—is the first general characteristic of revolutionary change that I shall be singling out after further consideration of examples. Though scientific revolutions leave much piecemeal mopping up to do, the central change cannot be experienced piecemeal, one step at a time. Instead, it involves some relatively sudden and unstructured transformation in which some part of the flux of experience sorts itself out differently and displays patterns that were not visible before.

To make all this more concrete let me now illustrate some of what was involved in my discovery of a way of reading Aristotelian physics, one that made the texts make sense. A first illustration will be familiar to many. When the term "motion" occurs in Aristotelian physics, it refers to change in general, not just to the change of position of a physical body. Change of position, the exclusive subject of mechanics for Galileo and Newton, is one of a number of subcategories of motion for Aristotle. Others include growth (the transformation of an acorn to an oak), alterations of intensity (the heating of an iron bar), and a number of more general qualitative changes (the transition from sickness to health). As a result, though Aristotle recognizes that the various subcategories are not alike in *all* respects, the basic characteristics relevant to the recognition and analysis of motion must apply to changes of all sorts. In some sense that is not merely metaphorical; all varieties of change are seen as like each other, as constituting a single natural family.[3]

A second aspect of Aristotle's physics—harder to recognize and even more important—is the centrality of qualities to its conceptual structure. By that I do not mean simply that it aims to explain quality and change of quality, for other sorts of physics have done that. Rather I have in mind that Aristotelian physics inverts the ontological hierarchy of matter and quality that has been standard since the middle of the seventeenth century. In Newtonian physics a body is constituted of particles of matter, and its qualities are a consequence of the way those particles are arranged, move, and interact. In Aristotle's physics, on the other hand, matter is very nearly dispensable. It is a neutral substrate, present wherever a body could be—which means wherever there's space or place. A particular body, a substance, exists in whatever place this neutral substrate, a sort of sponge, is sufficiently impregnated with qualities like heat, wetness, color, and so on to give it individual identity. Change occurs by changing qualities, not matter, by removing some qualities from some given matter and replacing them with others. There are even some implicit conservation laws that the qualities must apparently obey.[4]

Aristotle's physics displays other similarly general aspects, some of great importance. But I shall work toward the points that concern me from these two, picking up one other well-known one in passing. What I want now to begin to suggest is that, as one recognizes these and other aspects of Aristotle's viewpoint, they begin to fit together, to lend each other mutual support, and thus to make a sort of sense collectively that they individually lack. In my original experience of breaking into Aristotle's text, the new pieces I have been describing and the sense of their coherent fit actually emerged together.

Begin from the notion of a qualitative physics that has just been sketched. When one analyzes a particular object by specifying the qualities that have been imposed on omnipresent neutral matter, one of the qualities that must be specified is the object's position, or, in Aristotle's terminology, its place. Position is thus, like wetness or hotness, a quality of the object, one that changes as the object moves or is moved. Local motion (motion *tout court* in Newton's sense) is therefore change-of-quality or change-of-state for Aristotle, rather than being itself a state as it is for Newton. But it is precisely seeing motion as change-of-quality that permits its assimilation to all other sorts of change—acorn to oak or sickness to health, for examples. That assimilation is the aspect of Aristotle's physics from which I began, and I could equally well have traveled the route in the other direction. The conception of motion-as-change and the conception of a qualitative physics prove deeply interdependent, almost equivalent notions, and that is a first example of the fitting or the locking together of parts.

If that much is clear, however, then another aspect of Aristotle's physics—one that regularly seems ridiculous in isolation—begins to make sense as well. Most changes of quality, especially in the organic realm, are asymmetric, at least when left to themselves. An acorn naturally develops into an oak, not vice versa. A sick man often grows healthy by himself, but an external agent is needed, or believed to be needed, to make him sick. One set of qualities, one end point of change, represents a body's natural state, the one that it realizes voluntarily and thereafter rests. The same asymmetry should be characteristic of local motion, change of position, and indeed it is. The quality that a stone or other heavy body strives to realize is position at the center of the universe; the natural position of fire is at the periphery. That is why stones fall toward the center until blocked by an obstacle and why fire flies to the heavens. They are realizing their natural properties just as the acorn does through its growth. Another initially strange part of Aristotelian doctrine begins to fall into place.

One could continue for some time in this manner, locking individual bits of Aristotelian physics into place in the whole. But I shall instead conclude this first example with a last illustration, Aristotle's doctrine about the vacuum or void. It displays with particular clarity the way in which a number of theses that appear arbitrary in isolation lend each other mutual authority and support. Aristotle states that a void is impossible: his underlying position is that the notion itself is incoherent. By now it should be apparent how that might be so. If position is a quality, and if qualities cannot exist separate from matter, then there must be matter wherever there's position, wherever body might be. But that is to say that there must be matter everywhere in space: the void, space without matter, acquires the status of, say, a square circle.[5]

That argument has force, but its premise seems arbitrary. Aristotle need not, one supposes, have conceived position as a quality. Perhaps, but we have already noted that that conception underlies his view of motion as change-of-state, and other aspects of his physics depend on it as well. If there could be a void, then the Aristotelian universe or cosmos could not be finite. It is just because matter and space are coextensive that space can end where matter ends, at the outermost sphere beyond which there is nothing at all, neither space nor matter. That doctrine, too, may seem dispensable. But expanding the stellar sphere to infinity would make problems for astronomy, since that sphere's rotations carry the stars about the earth. Another, more central, difficulty arises earlier. In an infinite universe there is no center—any point is as much the center as any other—and there is thus no natural position at which stones and other heavy bodies realize their natural quality. Or, to put the point in another way, one that Aristotle actually uses, in a void a body could not be aware of the location of its natural place. It is just by being in contact with all positions in the universe through a cha:n of intervening matter that a body is able to find its way to the place where its natural qualities are fully realized. The presence of matter is what provides space with structure.[6] Thus, both Aristotle's theory of natural local motion and ancient geocentric astronomy are threatened by an attack on Aristotle's doctrine of the void. There is no way to "correct" Aristotle's views about the void without reconstructing much of the rest of his physics.

Those remarks, though both simplified and incomplete, should sufficiently illustrate the way in which Aristotelian physics cuts up and describes the phenomenal world. Also, and more important, they should indicate how the pieces of that description lock together to form an integral whole, one that had to be broken and reformed on the road to Newtonian mechanics. Rather than extend them further, I shall therefore proceed at once to a second example, returning to the beginning of the nineteenth century for the purpose. The year 1800 is notable, among other things, for Volta's discovery of the electric battery. That discovery was announced in a letter to Sir Joseph Banks, President of the Royal Society.[7] It was intended for publication and was accompanied by the illustration reproduced here as figure 1. For a modern audience there is something odd about it, though the oddity is seldom noticed, even by historians. Looking at any one of the so-called "piles" (of coins) in the lower two-thirds of the diagram, one sees, reading upward from the bottom right, a piece of zinc, Z, then a piece of silver, A, then a piece of wet blotting paper, then a second piece of zinc, and so on. The cycle zinc, silver, wet blotting paper is repeated an integral number of times, eight in Volta's original illustration. Now suppose that, instead of having all this spelled out, you had been asked simply to look at the diagram, then to put it aside and reproduce it from memory. Almost certainly, those of you who know even the most elementary physics would have drawn zinc (or silver), followed by wet blotting paper, followed by silver (or zinc). In a battery, as we all know, the liquid belongs between the two different metals.

If one recognizes this difficulty and puzzles over it with the aid of Volta's texts, one is likely to realize suddenly that for Volta and his followers, the unit cell consists of the two pieces of metal in contact. The source of power is the metallic interface, the bimetallic junction that Volta had previously found to be the source of an electrical tension, what we would call a voltage. The role of the liquid then is simply to connect one unit cell to the next without generating a contact potential, which would neutralize the initial effect. Pursuing Volta's text still further, one realizes that he is assimilating his new discovery to electrostatics. The bimetallic junction is a condenser or Leyden jar, but one that charges itself. The pile of coins is, then, a linked assemblage or "battery" of charged Leyden jars, and that is where, by specialization from the group to its members, the term "battery" comes from in its application to electricity. For confirmation, look at the top part of Volta's diagram, which illustrates an arrangement he called "the crown of cups." This time the resemblance to diagrams in elementary modern textbooks is striking, but there is again an oddity. Why do the cups at the two ends of the diagram contain only one piece of metal? Why does Volta include two half-cells? The answer is the same as before. For Volta the cups are not cells but simply containers for the liquids that connect cells. The cells themselves are the bimetallic horseshoe strips. The apparently unoccupied positions in the outermost cups are what we would think of as binding posts. In Volta's diagram there are no half-cells.

What Are Scientific Revolutions?   13

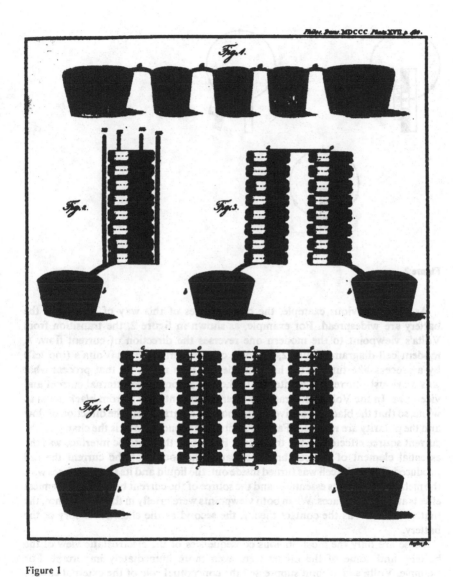

Figure 1

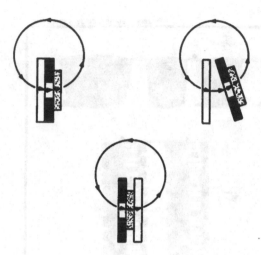

Figure 2

As in the previous example, the consequences of this way of looking at the battery are widespread. For example, as shown in figure 2, the transition from Volta's viewpoint to the modern one reverses the direction of current flow. A modern cell-diagram (figure 2, bottom) can be derived from Volta's (top left) by a process like turning the latter inside out (top right). In that process what was previously current flow internal to the cell becomes the external current and vice versa. In the Voltaic diagram the external current flow is from black metal to white, so that the black is positive. In the modern diagram both the direction of flow and the polarity are reversed. Far more important conceptually is the change in the current source effected by the transition. For Volta the metallic interface was the essential element of the cell and necessarily the source of the current the cell produced. When the cell was turned inside out, the liquid and its two interfaces with the metals provided its essentials, and the source of the current became the chemical effects at these interfaces. When both viewpoints were briefly in the field at once, the first was known as the contact theory, the second as the chemical theory of the battery.

Those are only the most obvious consequences of the electrostatic view of the battery, and some of the others were even more immediately important. For example, Volta's viewpoint suppressed the conceptual role of the external circuit. What we would think of as an external circuit is simply a discharge path like the short circuit to ground that discharges a Leyden jar. As a result, early battery diagrams do not show an external circuit unless some special effect, like electrolysis or heating a wire, is occurring there, and then, very often the battery is not shown. Not until the 1840s do modern cell-diagrams begin to appear regularly in books on electricity. When they do, either the external circuit or explicit points for its attachment appears with them.[8] Examples are shown in figures 3 and 4.

Figure 3

Figure 4

Finally, the electrostatic view of the battery leads to a concept of electrical resistance very different from the one now standard. There is an electrostatic concept of resistance, or there was in this period. For an insulating material of given cross section, resistance was measured by the shortest length the material could have without breaking down or leaking—ceasing to insulate—when subjected to a given voltage. For a conducting material of given cross section, it was measured by the shortest length the material could have without melting when connected across a given voltage. It is possible to measure resistance conceived in this way, but the results are not compatible with Ohm's law. To get those results one must conceive the battery and circuit on a more hydrostatic model. Resistance must become like the frictional resistance to the flow of water in pipes. The assimilation of Ohm's law required a noncumulative change of that sort, and that is part of what made his law so difficult for many people to accept. It has for some time provided a standard example of an important discovery that was initially rejected or ignored.

At this point I end my second example and proceed at once to a third, this one both more modern and more technical then its predecessors. Substantively, it is controversial, involving a new version, not yet everywhere accepted, of the origins of the quantum theory.[9] Its subject is Max Planck's work on the so-called black-body problem, and its structure may usefully be anticipated as follows. Planck first solved the black-body problem in 1900 using a classical method developed by the

Figure 5

Austrian physicist Ludwig Boltzmann. Six years later a small but crucial error was found in his derivation, and one of its central elements had to be reconceived. When that was done Planck's solution did work, but it then also broke radically with tradition. Ultimately that break spread through and caused the reconstruction of a good deal of physics.

Begin with Boltzmann, who had considered the behavior of a gas, conceived as a collection of many tiny molecules, moving rapidly about within a container, and colliding frequently both with each other and with the container's walls. From previous works of others, Boltzmann knew the average velocity of the molecules (more precisely, the average of the square of their velocity). But, many of the molecules were, of course, moving much more slowly than the average, others much faster. Boltzmann wanted to know what proportion of them were moving at, say, 1/2 the average velocity, what proportion at 4/3 the average, and so on. Neither that question nor the answer he found to it was new. But Boltzmann reached the answer by a new route, from probability theory, and that route was fundamental for Planck, since whose work it has been standard.

Only one aspect of Boltzmann's method is of present concern. He considered the total kinetic energy E of the molecules. Then, to permit the introduction of probability theory, he mentally subdivided that energy into little cells or elements of size ε, as in figure 5. Next, he imagined distributing the molecules at random among those cells, drawing numbered slips from an urn to specify the assignment of each molecule and then excluding all distributions with total energy different from E. For example, if the first molecule were assigned to the last cell (energy E), then the only acceptable distribution would be the one that assigned all other molecules to the first cell (energy 0). Clearly, that particular distribution is a most improbable one. It is far more likely that most molecules will have appreciable energy, and by probability theory one can discover the most probable distribution of all. Boltzmann showed how to do so, and his result was the same as the one he and others had previously gotten by more problematic means.

That way of solving the problem was invented in 1877, and twenty-three years later, at the end of 1900, Max Planck applied it to an apparently rather different problem, black-body radiation. Physically the problem is to explain the way in which the color of a heated body changes with temperature. Think, for example, of the radiation from an iron bar, which, as the temperature increases, first gives off heat (infrared radiation), then glows dull red, and then gradually becomes a brilliant white. To analyze that situation Planck imagined a container or cavity filled with radiation, that is, with light, heat, radio waves, and so on. In addition, he supposed that the cavity contained a lot of what he called "resonators" (think of them as tiny electrical tuning forks, each sensitive to radiation at one frequency, not at others). These resonators absorb energy from the radiation, and Planck's question was: How does the energy picked up by each resonator depend on its frequency? What is the frequency distribution of the energy over the resonators?

Conceived in that way, Planck's problem was very close to Boltzmann's, and Planck applied Boltzmann's probabilistic techniques to it. Roughly speaking, he used probability theory to find the proportion of resonators that fell in each of the various cells, just as Boltzmann had found the proportion of molecules. His answer fit experimental results better than any other then or since known, but there turned out to be one unexpected difference between his problem and Boltzmann's. For Boltzmann's, the cell size ε could have many different values without changing the result. Though permissible values were bounded, could not be too large or too small, an infinity of satisfactory values was available in between. Planck's problem proved different: other aspects of physics determined ε, the cell size. It could have only a single value given by the famous formula $\varepsilon = h\nu$, where ν is the resonator frequency and h is the universal constant subsequently known by Planck's name. Planck was, of course, puzzled about the reason for the restriction on cell size, though he had a strong hunch about it, one he attempted to develop. But, excepting that residual puzzle, he had solved his problem, and his approach remained very close to Boltzmann's. In particular, the presently crucial point, in both solutions the division of the total energy E into cells of size ε was a mental division made for statistical purposes. The molecules and resonators could lie anywhere along the line and were governed by all the standard laws of classical physics.

The rest of this story is very quickly told. The work just described was done at the end of 1900. Six years later, in the middle of 1906, two other physicists argued that Planck's result could not be gained in Planck's way. One small but absolutely crucial alteration of the argument was required. The resonators could not be permitted to lie anywhere on the continuous energy line but only at the divisions between cells. A resonator might, that is, have energy $0, \varepsilon, 2\varepsilon, 3\varepsilon, \ldots$, and so on, but not $(1/3)\varepsilon$, $(4/5)\varepsilon$, etc. When a resonator changed energy it did not do so continuously but by discontinuous jumps of size ε or a multiple of ε.

After those alterations, Planck's argument was both radically different and very much the same. Mathematically it was virtually unchanged, with the result that it has been standard for years to read Planck's 1900 paper as presenting the subsequent modern argument. But physically, the entities to which the derivation refers are very different. In particular, the element ε has gone from a mental division of the total energy to a separable physical energy atom, of which each resonator may have 0, 1, 2, 3, or some other number. Figure 6 tries to capture that change in a way that

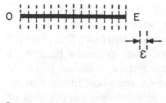

Figure 6

suggests its resemblance to the inside-out battery of my last example. Once again the transformation is subtle, difficult to see. But also once again, the change is consequential. Already the resonator has been transformed from a familiar sort of entity governed by standard classical laws to a strange creature the very existence of which is incompatible with traditional ways of doing physics. As most of you know, changes of the same sort continued for another twenty years as similar nonclassical phenomena were found in other parts of the field.

Those later changes, I shall not attempt to follow, but instead conclude this example, my last, by pointing to one other sort of change that occurred near its start. In discussing the earlier examples, I pointed out that revolutions were accompanied by changes in the way in which terms like "motion" or "cell" attached to nature. In this example there was actually a change in the words themselves, one that highlights those features of the physical situation that the revolution had made prominent. When Planck around 1909 was at last persuaded that discontinuity had come to stay he switched to a vocabulary that has been standard since. Previously he had ordinarily referred to the cell-size ε as the energy "element." Now, in 1909, he began regularly to speak instead of the energy "quantum," for "quantum," as used in German physics, was a separable element, an atomlike entity that could exist by itself. While ε had been merely the size of a mental subdivision, it had not been a quantum but an element. Also in 1909, Planck abandoned the acoustic analogy. The entities he had introduced as "resonators" now became "oscillators," the latter a neutral term that refers to any entity that simply vibrates regularly back and forth. By contrast, "resonator" refers in the first instance to an acoustic entity or, by extension, to a vibrator that responds gradually to stimulation, swelling and diminishing with the applied stimulus. For one who believed that energy changes discontinuously "resonator" was not an appropriate term, and Planck gave it up in and after 1909.

That vocabulary change concludes my third example. Rather than give others, I shall conclude this discussion by asking what characteristics of revolutionary change are displayed by the examples at hand. Answers will fall under three headings, and I shall be relatively brief about each. The extended discussion they require, I am not quite ready to provide.

A first set of shared characteristics was mentioned near the start of this paper. Revolutionary changes are somehow holistic. They cannot, that is, be made piece-meal, one step at a time, and they thus contrast with normal or cumulative changes like, for example, the discovery of Boyle's law. In normal change, one simply revises or adds a single generalization, all others remaining the same. In revolutionary change one must either live with incoherence or else revise a number of interrelated generalizations together. If these same changes were introduced one at a time, there would be no intermediate resting place. Only the initial and final sets of general-izations provide a coherent account of nature. Even in my last example, the most nearly cumulative of the three, one cannot simply change the description of the energy element ε. One must also change one's notion of what it is to be a resonator, for resonators, in any normal sense of the term, cannot behave as these do. Simultaneously, to permit the new behavior, one must change, or try to, laws of mechanics and of electromagnetic theory. Again, in the second example, one cannot simply change one's mind about the order of elements in a battery cell. The direction of the current, the role of the external circuit, the concept of electrical resistance, and so on, must also be changed. Or still again, in the case of Aristotelian physics, one cannot simply discover that a vacuum is possible or that motion is a state, not a change-of-state. An integrated picture of several aspects of nature has to be changed at the same time.

A second characteristic of these examples is closely related. It is the one I have in the past described as meaning change and which I have here been describing, somewhat more specifically, as change in the way words and phrases attach to nature, change in the way their referents are determined. Even that version is, however, somewhat too general. As recent studies of reference have emphasized, anything one knows about the referents of a term may be of use in attaching that term to nature. A newly discovered property of electricity, of radiation, or of the effects of force on motion may thereafter be called upon (usually with others) to determine the presence of electricity, radiation, or force and thus to pick out the referents of the corresponding term. Such discoveries need not be and usually are not revolutionary. Normal science, too, alters the way in which terms attach to nature. What characterizes revolutions is not, therefore, simply change in the way referents are determined but change of a still more restricted sort.

How best to characterize that restricted sort of change is among the problems that currently occupy me, and I have no full solution. But roughly speaking, the distinctive character of revolutionary change in language is that it alters not only the criteria by which terms attach to nature but also, massively, the set of objects or situations to which those terms attach. What had been paradigmatic examples of motion for Aristotle—acorn to oak or sickness to health—were not motions at all for Newton. In the transition, a natural family ceased to be natural; its members were redistributed among preexisting sets; and only one of them continued to bear the old name. Or again, what had been the unit cell of Volta's battery was no longer the referent of any term forty years after his invention was made. Though Volta's successors still dealt with metals, liquids, and the flow of charge, the units of their analyses were different and differently interrelated.

What characterizes revolutions is, thus, change in several of the taxonomic categories prerequisite to scientific descriptions and generalizations. That change, furthermore, is an adjustment not only of criteria relevant to categorization, but also of the way in which given objects and situations are distributed among preexisting categories. Since such redistribution always involves more than one category and since those categories are interdefined, this sort of alteration is necessarily holistic. That holism, furthermore, is rooted in the nature of language, for the criteria relevant to categorization are *ipso facto* the criteria that attach the names of those categories to the world. Language is a coinage with two faces, one looking outward to the world, the other inward to the world's reflection in the referential structure of the language.

Look now at the last of the three characteristics shared by my three examples. It has been the most difficult of the three for me to see, but now seems the most obvious and probably the most consequential. Even more than the others, it should repay further exploration. All of my examples have involved a central change of model, metaphor, or analogy—a change in one's sense of what is similar to what, and of what is different. Sometimes, as in the Aristotle example, the similarity is internal to the subject matter. Thus, for Aristotelians, motion was a special case of change, so that the falling stone was *like* the growing oak, or *like* the person recovering from illness. That is the pattern of similarities that constitutes these phenomena a natural family, that places them in the same taxonomic category, and that had to be replaced in the development of Newtonian physics. Elsewhere the similarity is external. Thus, Planck's resonators were *like* Boltzmann's molecules, or Volta's battery cells were *like* Leyden jars, and resistance was *like* electrostatic leakage. In these cases, too, the old pattern of similarities had to be discarded and replaced before or during the process of change.

All these cases display interrelated features familiar to students of metaphor. In each case two objects or situations are juxtaposed and said to be the same or similar. (An even slightly more extended discussion would have also to consider examples of dissimilarity, for they, too, are often important in establishing a taxonomy.) Furthermore, whatever their origin—a separate issue with which I am not presently concerned—the primary function of all these juxtapositions is to transmit and maintain a taxonomy. The juxtaposed items are exhibited to a previously uninitiated audience by someone who can already recognize their similarity, and who urges that audience to learn to do the same. If the exhibit succeeds, the new initiates emerge with an acquired list of features salient to the required similarity relation— with a feature-space, that is, within which the previously juxtaposed items are durably clustered together as examples of the same thing and are simultaneously separated from objects or situations with which they might otherwise have been confused. Thus, the education of an Aristotelian associates the flight of an arrow with a falling stone and both with the growth of an oak and the return to health. All are thereafter changes of state; their end points and the elapsed time of transition are their salient features. Seen in that way, motion cannot be relative and must be in a category distinct from rest, which is a state. Similarly, on that view, an infinite motion, because it lacks an end point, becomes a contradiction in terms.

The metaphorlike juxtapositions that change at times of scientific revolution are thus central to the process by which scientific and other language is acquired. Only after that acquisition or learning process has passed a certain point can the practice of science even begin. Scientific practice always involves the production and the explanation of generalizations about nature; those activities presuppose a language with some minimal richness; and the acquisition of such a language brings knowledge of nature with it. When the exhibit of examples is part of the process of learning terms like "motion," "cell," or "energy element," what is acquired is knowledge of language and of the world together. On the one hand, the student learns what these terms mean, what features are relevant to attaching them to nature, what things cannot be said of them on pain of self-contradiction, and so on. On the other hand, the student learns what categories of things populate the world, what their salient features are, and something about the behavior that is and is not permitted to them. In much of language learning these two sorts of knowledge—knowledge of words and knowledge of nature—are acquired together, not really two sorts of knowledge at all, but two faces of the single coinage that a language provides.

The reappearance of the double-faced character of scientific language provides an appropriate terminus for this paper. If I am right, the central characteristic of scientific revolutions is that they alter the knowledge of nature that is intrinsic to the language itself and that is thus prior to anything quite describable as description or generalization, scientific or everyday. To make the void or an infinite linear motion part of science required observation reports that could only be formulated by altering the language with which nature was described. Until those changes had occurred, language itself resisted the invention and introduction of the sought after new theories. The same resistance by language is, I take it, the reason for Planck's switch from "element" and "resonator" to "quantum" and "oscillator." Violation or distortion of a previously unproblematic scientific language is the touchstone for revolutionary change.

Notes

1. Thomas S. Kuhn, *The Structure of Scientific Revolutions*, 2nd ed., rev. (Chicago: University of Chicago Press, 1969). The book was first published in 1962.

2. The phrase "antecedently understood" was introduced by C. G. Hempel, who shows that it will serve many of the same purposes as "observational" in discussions involving the distinction between observational and theoretical terms (cf., particularly, his *Aspects of Scientific Explanation* (New York: Free Press, 1965), pp. 208ff.). I borrow the phrase because the notion of an antecedently understood term is intrinsically developmental or historical, and its use within logical empiricism points to important areas of overlap between that traditional approach to philosophy of science and the more recent historical approach. In particular, the often elegant apparatus developed by logical empiricists for discussions of concept formation and of the definition of theoretical terms can be transferred as a whole to the historical approach and used to analyze the formation of new concepts and the definition of new terms, both of which usually take place in intimate association with the introduction of a new theory. A more systematic way of preserving an important part of the observation/theoretical distinction by embedding it in a developmental approach has been developed by Joseph D. Sneed (*The Logical Structure of Mathematical Physics* (Dordrecht: Reidel, 1971), pp. 1–64,

249–307). Wolfgang Stegmüller has clarified and extended Sneed's approach by positing a hierarchy of theoretical terms, each level introduced within a particular historical theory (*The Structure and Dynamics of Theories* (New York: Springer, 1976), pp. 40 - 67, 196–231). The resulting picture of linguistic strata shows intriguing parallels to the one discussed by Michel Foucault in *The Archeology of Knowledge*, trans. A. M. Sheridan Smith (New York: Pantheon, 1972).

3. For all of this see Aristotle's *Physics*, Book V, Chapters 1–2 (224a21–226b16). Note that Aristotle does have a concept of change that is broader than that of motion. Motion is change of substance, change from something to something (225a1). But change also includes coming to be and passing away, i.e., change from nothing to something and from something to nothing (225a34–225b9), and these are not motions.

4. Compare Aristotle's *Physics*, Book I, and especially his *On Generation and Corruption*, Book II, Chapters 1–4.

5. There is an ingredient missing from my sketch of this argument: Aristotle's doctrine of place, developed in the *Physics*, Book IV, just before his discussion of the vacuum. Place, for Aristotle, is always the place of body or, more precisely, the interior surface of the containing or surrounding body (212a2–7). Turning to his next topic, Aristotle says, "Since the void (if there is any) must be conceived as place in which there might be body but is not, it is clear that, so conceived, the void cannot exist at all, either as inseparable or separable" (214a16–20). (I quote from the Loeb Classical Library translation by Philip H. Wickstead and Francis M. Cornford, a version that, on this difficult aspect of the *Physics*, seems to me clearer than most, both in text and commentary.) That it is not merely a mistake to substitute "position" for "place" in a sketch of the argument is indicated by the last part of the next paragraph of my text.

6. For this and closely related arguments see Aristotle, *Physics*, Book IV, Chapter 8 (especially 214b27–215a24).

7. Alessandro Volta, "On the Electricity Excited by the mere Contact of Conducting Substances of Different Kinds," *Philosophical Transactions*, 90 (1800), 403–431. On this subject see, T. M. Brown, "The Electric Current in Early Nineteenth-Century French Physics," *Historical Studies in the Physical Sciences*, 1 (1969), 61–103.

8. The illustrations are from A. de la Rive, *Traité d'électricité théorique et appliquée*, Vol. 2 (Paris: J. B. Bailière, 1856), pp. 600, 656. Structurally similar but schematic diagrams appear in Faraday's experimental researches from the early 1830s. My choice of the 1840s as the period when such diagrams became standard results from a casual survey of electricity texts lying ready to hand. A more systematic study would, in any case, have had to distinguish between British, French, and German responses to the chemical theory of the battery.

9. For the full version with supporting evidence see my *Black-Body Theory and the Quantum Discontinuity, 1894–1912* (Oxford and New York: Clarendon and Oxford University Presses, 1978).

2 Scientific Revolutions, Revolutions in Science, and a Probabilistic Revolution 1800–1930

I. Bernard Cohen[1]

Change in science occurs in at least three ways: by revolution, evolution, and emergence. Historians, philosophers, and sociologists of science have paid most attention in recent years to revolutions, but both the definition and the criteria for such revolutions remain controversial. The very word 'revolution' has considerably altered its meaning from a cycle in the astronomical sense to a violent rupture in the social and political order; the first uses of the word to describe epoch-making changes in the sciences surface in the late seventeenth century. 'Historical revolutions' should be distinguished from 'historians' revolutions': the former are events for which there exists objective and unambiguous historical evidence, whereas the revolutionary character of the latter arises primarily in the subjective judgment of one or more historians or historically minded scientists, philosophers, or sociologists. The revolutions of Lavoisier and Darwin are examples of the first type; that of Copernicus, the latter. Four criteria help us decide when a historical revolution has truly occurred in a science: the testimony of scientists and nonscientists active at the time; the impact of the alleged revolution on treatises, textbooks, and other essential documents; the judgment of competent historians; and the general opinion of scientists. Measured by these criteria, did a probabilistic revolution occur in the period circa 1800–1930? Probability theory itself shows little sign of such a transformation, but there is plenty of evidence of all four sorts for a historical revolution in the applications of probability to the social and natural sciences, as well as to the areas of medicine and public health.

Two popular models of scientific change are evolution and revolution.[2] The difference between them is extreme. One invokes the notion of a relatively slow and gradual process, consisting of a succession of small steps, perhaps occasionally punctuated by a greater one, while the other connotes a sudden change of a radical kind, a violent fragmentation of a system of concepts and theories that is followed by the introduction of something wholly new. The net effects of both may be equally profound even though they differ fundamentally in their time scales: there is an obvious analogy with events in the biological and political realms. Hence one of the problems in discussing any revolutionary set of changes in scientific thought or practice is the somewhat subjective decision whether the extent of the time scale implies that the process of change was a revolution or an evolution.

Historians, it must be acknowledged, are not in universal agreement concerning the nature of, and even the existence of, scientific revolutions or revolutions in science.[3] For example, modern science dates from the middle or late sixteenth century and is some four hundred years old. A three-hundred-year period of development in modern science might seem too long to be a revolution and hence more properly an evolution. And yet our foremost historian of the Scientific Revolution has set its boundaries at "1500–1800." There is no example of a political revolution that historians believe to have lasted this long. Furthermore, historians and philosophers of science do not agree on what constitutes or defines a revolution in science; they do not have an objective test for the occurrence of such a revolution. In the extreme, there has even been a minatory conclusion that it is at best confusing and at worst misleading (if not downright wrong) to apply to science the term 'revolution', which is said to belong to the discourse concerning political and social events and not to the analysis of the development of science.[4] The study of political and social revolutions, however, raises questions of its own, since there is no consensus on every event or series of events that may be considered a revolution.[5] Why, then, it has been asked, need we introduce into history and philosophy of science a 'foreign' term that has fundamental problems in its use in its own domains? Thus it may be seen that any serious discussion of an alleged probabilistic revolution in the nineteenth century poses very serious questions at the outset.

An additional historiographic problem is that most historians and philosophers of science believe that discussions of 'revolution' in science may be anachronistic to the degree that they are attempts to force events of the past into a twentieth-century mold of thinking in which revolution plays a decisive role. Here there are two major issues. One is related to the complex and compound event (or series of events) in the sixteenth and seventeenth centuries that produced our modern science, a tremendous change in the basis of our knowledge of nature or of the world of observation and experience, and in the modes of seeking and validating that knowledge. While there has been an awareness during at least three centuries that what we nowadays call '*the* Scientific Revolution' affected all of science and the whole knowledge industry, it was only a little more than a century ago that historians and philosophers began to regard this set of changes as a revolution. The other issue has to do with radical changes that occur in the individual sciences, or in branches of science. Of more limited scope, these revolutions are often linked with the name of a particular scientist. Familiar examples are the revolutions associated with the names of Newton, Lavoisier, Darwin, Einstein, and Freud. These were seen to be revolutions in their own times. My concern here is with this latter type—revolutions in the sciences—and not with such a singular event as '*the* Scientific Revolution'.

Thomas S. Kuhn and others have suggested that in addition to the Scientific Revolution in which modern science was born in the sixteenth and seventeenth centuries, there was a second Scientific Revolution, which occurred in the nineteenth century.[6] This was a time when there were notable changes in the structure of science. An effect of this second revolution was to reject the ideal of a scientist being universally learned in all branches of natural knowledge. From then on, it has been agreed that a scientist needs to be a specialist, concentrating on the narrow aspects of his subject.[7] The theme of the second Scientific Revolution has been developed in a daring and brilliant way by Ian Hacking in the chapter he has contributed to this volume; he suggests a link between a conceptual second Scientific Revolution and an institutional second Scientific Revolution. I myself, following Hacking's lead, have explored the possibility of yet other Scientific Revolutions:[8] perhaps a third, which occurred at the end of the nineteenth century, marked by the rise of new institutional structures such as industrial laboratories and university centers for scientific research and advanced training leading to higher degrees, and the tremendous enlargement of the scientific profession. Another possibility is a fourth Scientific Revolution in the twentieth century, of which the most distinguishing feature may be the growth of 'big science', with tremendous financial support by government. One of the aspects of this fourth Scientific Revolution has been the rise of group research, the coordination of effort on a high level in which the role of the single individual is uncertain.

How are we to consider a 'Probabilistic Revolution 1800–1930'? Is it, perhaps, a 'Scientific Revolution'? Is it a 'revolution in science'? Or, does it fail to fall neatly into either category? The very name suggests the problem of a time dimension: thirteen decades. We have seen that historians entertain the idea that *the* Scientific Revolution possibly had a duration of three centuries; it has been suggested by Stephen Brush that the second Scientific Revolution occurred during the period 1800–1950. If the Probabilistic Revolution was of this sort, then a one-hundred-and-thirty-year span would not be too extreme. But if this was a 'revolution in science', then it would be the only one with which I am familiar to have lasted almost a century and a half—an event without any parallel in the records of history. Accordingly, such a probabilistic revolution would have to have been either a continuing, or 'permanent', revolution that required about a century and a half to accomplish its goals or a series of individual and separate revolutions that, only when taken together as an ensemble, constitute the probabilistic revolution.

In order to illuminate the problem of such a probabilistic revolution, I have drawn on my own research on revolutions, which includes a series of tests for the possible occurrence of revolutions in science.[9] In this research my goal has been to make clear a distinction between a 'historical revolution' and a 'historians' revolution'. The former is an event for which there exists objective and unambiguous historical evidence, whereas the revolutionary character of the latter arises primarily in the subjective judgment of one or more historians or historically minded scientists, philosophers, or sociologists. For example, both Lavoisier and Darwin conceived that their new sciences were (or would eventually constitute) revolutions and both of them said so in print—a judgment that was shared by many of their contemporaries. This is a historical fact, fully supported by unimpeachable evidence, by full historical documentation. From the perspective of today, we—as critical historians or philosophers of science—may or may not agree with these evaluations of the past, but we cannot deny the historical fact of such evidence.

It is quite different for the Copernican revolution. There is no historical evidence that either Copernicus or his contemporaries ever conceived that the set of ideas and principles that today we call 'Copernican'[10] constituted a revolution. It is a historical fact that such an evaluation was not made until much later, apparently for the first time only some two centuries later, as a judgment by historians.[11] There is in any event, an additional and contrasting difference in the reaction of contemporaries to Copernican science and to Darwinian science or the science of Lavoisier. For it is a historical fact that at the time of Copernicus, and even at the time of Galileo and Kepler, it was not yet the custom to think of science in terms of revolution nor to apply this term to the development of the sciences. Hence we see

why we cannot profitably explore the nature of contemporaneous opinions concerning revolutions in science unless we first clarify some aspects of the history and
applications of the concept of revolution.

As an aside it may be noted that thinking about revolutionary changes in science
does not have to be stated by using the actual word 'revolution'. A near-synonym
with an ancient history going back to classical times is 'mutatio' (as in the expression 'mutatio rerum'). The revolutionary or near-revolutionary quality of the
science of the late sixteenth century and of the seventeenth century is indicated by
the occurrence of the word 'new' in the titles of books and articles, e.g., Bacon's
Novum Organum, Tartaglia's *New Science*, Galileo's *Two New Sciences*, Kepler's
New Astronomy, Boyle's *New Experiments*, and Newton's "New Theory of Light
and Colors." But, to a large degree, a full and careful reading of the radical authors
of this period shows that although they were consciously rejecting traditional
methods and accepted science, and although they were also openly aware of the
novelty in their own work, they nevertheless tended—to a greater degree than one
would have imagined—to write and to think of making improvements, of returning
to the kind of science espoused by certain ancient scientists and thinkers, and of
introducing changes in harmony with universally accepted ideals and standards.
Galileo, to take but one example, tells the Aristotelian character in his dialogue on
the world systems that if he will follow Galileo's way of critical observation and
experiment he will "think more Aristotelically" than he does in his current scholastic mode. Again and again Galileo either lauds or cites with approval both Plato and
Archimedes. Although in cumulative net effect, the results of these scientists seem
to us *in retrospect* to partake of revolution, our concept of revolutions in science
would have seemed somewhat alien to them and to men of their time. It is perhaps noteworthy that the recognition of revolution in something like our present
sense was simultaneous with the introduction and the beginning of a general use of
the term itself. This aspect of the subject of revolutions provides yet another datum
to the debate on the question of the existence of concepts independent of their
having names.

The word 'revolution', as its etymological roots suggest, did not originally signify
a radical change, accompanied by violence, but rather a rolling back, a return to
some antecedent condition. The primary signification of this term is that of a
constancy within change, as in the familiar use to describe the revolution of a planet
in its orbit, a rolling around to some point where it had been once before. This sense
is also preserved in the use of the term in mathematics, where a solid of revolution is
a figure determined by the revolution (or rotation)[12] of a plain figure about an axis.
For instance, if a circle is turned through $360°$ about an axis that is a diameter of a
circle, the ensuing figure will be a sphere; but if the axis about which the circle is
turned is located outside of the circle, the result will be a torus or doughnut. During
the Latin Middle Ages, the term 'revolution' became widely used in its astronomical
context, and associatively came to have an astrological significance. Through
astrology this term was introduced into considerations of affairs of the state and
even of human destinies—all believed to be controlled by the revolutions of the
stars or planets in their spheres.

The concept of revolution underwent a dramatic transformation whereby there was a shift from the idea of a return or a cycle to that of a radical change and the introduction of something new. One of the factors in this alteration of meaning is that throughout many centuries of human thought, the most obvious mode of radical improvement was to return to some better antecedent condition. In the Western tradition, this better precedent state has been primarily symbolized by Adam and Eve in the Garden of Eden before the Fall of Man. There is also the common image of a Golden Age in the past. The notion of producing a better world by regaining this primitive paradise is expressed in such a phrase as to build 'a new Jerusalem'. There has been a constant tendency, in other words, to see one's own age as poorer in many qualities than the ages of the past. Isaac Newton, in one of his manuscripts on the interpretation of Scripture, observed, "'Tis the nature of man to admire least what he is most acquainted with: and this makes us always think our own times the worst." And in this regard he observed further, "Men are not sainted till their vices be forgotten." [13] We may agree with Newton about so-called 'good old days', that "if we have a kindness for any age we have to deify it when it is old enough."

Of course, each generation resents being told by its elders that in many respects past ages were superior to its own. Yet is it not certain that children of the past have always been thought to have been better mannered than those of the present? Who can possibly doubt that most foods tasted better before being 'preserved', 'colored', and 'improved' by adulteration with chemicals? And is it not beyond question that older modes of leisurely travel had advantages we can never recapture when packed like sardines in a wide-bodied plane? So we may understand why political and social radicals of the sixteenth and the seventeenth centuries, and even many of those of the eighteenth, saw themselves introducing or advocating a radical improvement of the world by returning to some primitive or at least antecedent state. Thus the revolutionary program of the Levellers, as put forth in their *Manifestoes* of 1649, was described as a renewal of the "[Voluntary] Community [which existed] amongst the primitive Christians." Even the apparently forward-looking French Revolution, an event often said to mark the real beginning of 'modern times', took as its symbols the Phrygian Liberty Cap (presented in ancient Greek times to slaves on the occasion of their manumission) and the 'Fasces' (or bundle of sticks) of ancient Rome. Often, radical reformers advocate a program of return to conditions that existed a long time ago. In such cases the achievement of that goal would not only be a return in the original etymological sense of the term revolution, but would also constitute the establishment of a real novelty in relation to anything known in recent memory. In short, such an example shows how both the original and the present sense of political or social revolution could be implied in a single program or goal or event.

Yet another basis for the association of the concept of revolution with the introduction of something new has already been mentioned: the fact that throughout the centuries of the Middle Ages and Renaissance, and well into the seventeenth century, there was a widespread and very common belief that the affairs of man were controlled or influenced by the revolutions of the heavenly bodies in their spheres. Thus any important event would be astrologically associated with a revolution. In a late medieval chronicle, we have an instance of a discussion of a "revolution made by the citizens of Siena,"[14] apparently an allusion to the fact that this was a revolution somewhat independent of the stars and their influences, one that was the result primarily of human activity. This example shows how the concept of a great event conditioned by an astronomical or astrological revolution began to change into the notion that the same effect of revolutions could be produced in independence of the revolutions of the heavens.

A common image of the Middle Ages and Renaissance that links the idea of revolution to important human and social events is 'Fortune's Wheel'. The 'rota di fortuna' has always been (and still is) a major card in the tarocchi (tarot) cards used to tell the fortunes of men and women and of the state. The notion that the affairs of men and women and the destinies of the state might be determined by the turning of Fortune's Wheel gives an obvious link between great events and revolution. Furthermore, the designation of revolution was given to great events that were associated with the revolution of the great wheel of time, events that thus punctuated the passage of time. Time was clearly associated with revolution in the visible symbol of the hand of a clock on towers of the Middle Ages and Renaissance. The primary symbols of time were the revolution (or the apparent daily revolution) of the sun, producing the day, and of the moon, producing the month, and the annual motion of the sun in revolution, the year.

In the Renaissance and in the seventeenth and the eighteenth centuries, the concept of revolution did not necessarily mean a strictly cyclical set of events that repeated themselves exactly. The term was used also to signify a kind of ebb and flow, as in the motion of the tides. Thus a common use of the concept of revolution was to designate the rise and fall of empires or of cultures.

Historians agree that for the above reasons, and others, the word revolution began to be associated with events of change, particularly the production of something new in human affairs—new in the sense of being fundamentally different from the present. But we must also keep in mind that revolutions of the kind with which we have become familiar in the eighteenth, nineteenth, and twentieth centuries did not really occur prior to the seventeenth century. There were, of course, revolts, dynastic changes effected by violence, and other events of change; but in the late Middle Ages and during the Renaissance it would be difficult to find examples of thorough and radical social change or alterations of the political system that would be like the events that we today would call revolutions.

Historically, the first event to be recognized as a revolution in anything approaching the modern sense was the Glorious Revolution that occurred in England in 1688.[15] Of course, by today's standards—that is, after the French Revolution and the Russian Revolution of 1917—the Glorious Revolution must seem very mild, perhaps not even meriting the name of revolution at all. But in its days it was conceived as a very radical kind of event. David Hume, a conservative, and Joseph Priestley, a radical, were one in agreeing that this was a remarkable revolution in human affairs. Both Hume and Priestley were aware that the Glorious Revolution not only guaranteed the Protestant succession to the British monarchy, but established the principle that (at least to some major degree) the monarch ruled by and with the consent of the governed, rather than ruling by divine right. Of course, this set of events partook to some degree of a return, since many Englishmen held that there had been, in the first instance, a return to the established Protestant succession, which had suffered an interruption by the presence of Catholic monarchs. Furthermore the Revolution's settlement seemed a reaffirmation of traditional rights of Englishmen, rather than the establishing of something wholly new. In any event, the Glorious Revolution was very quickly recognized as both a radical change and a return; it was soon discussed as a revolution and it served to spread abroad the notion that radical changes may really occur and that they might be called revolutions. During the ensuing century, that is, up to the time of the American and French Revolutions, the Glorious Revolution was a symbolic event that announced both the old and the new sense of the term revolution and indicated that even a return to antecedent conditions could be considered the establishment of a new set of political arrangements.

The spread of this dual sense of revolution—following the Glorious Revolution—has been traced for us, particularly in France, in novels and plays.[16] As the new sense of revolution came into general consciousness, it did not take long for this same concept of radical change, the introduction of something wholly new, to be discerned in the recent events that had taken place in the sciences. Although in many ways Descartes was the most revolutionary figure in the sciences of the seventeenth century, and was also the scientist who most clearly articulated the idea that he himself had produced something radically new, the term revolution was apparently not applied in the sciences until a generation later. And this did not occur in the realm of the physical or biological sciences but of mathematics.

At the end of the seventeenth century, Fontenelle, the permanent secretary of the Paris Academy of Sciences, introduced the term and concept of revolution to indicate the very radical changes that had occurred in mathematics as a result of the invention of the calculus by Newton and Leibniz. He showed how the whole nature of mathematics had become altered. For instance, the very beginners of mathematics could now easily solve problems that before the invention of the calculus had required years and years of preparation and the acquisition of special skills. Obviously, great mathematicians could now advance rapidly to the frontiers of the subject and needed no longer to be occupied with the solution of the old thorny problems. Above all, Fontenelle pointed out, the new algorithms wholly revolutionized the teaching of the subject. Then, soon after Newton's death, Clairaut (in 1747) used this term in relation to Newton's *Principia*, stating expressly that Newton's *Principia* marked the "epoch of a great revolution in physics," not only using the term revolution in a simple and direct way in the new sense, but coupling it with the term epoch, in the sense of an event that begins a new era.

As the eighteenth century wore on, three scientists announced that the work on which they were engaged would produce a revolution. Two of them, the English electrician Robert Symmer and the French political leader and scientist Jean-Paul

Marat, predicted revolutions that never occurred. It was quite different in the case of Lavoisier, the chief architect of the successful Chemical Revolution. This announcing of an impending revolution became a fulfilled prophecy.[17]

Many of the writers in the eighteenth century who used the term revolution, both in the realm of human and social affairs and in science, wrote about revolutions in the dual sense of the introduction of something radically new and a return to some antecedent state. In the works of Voltaire and Rousseau, for example, every occurrence of the term 'révolution' has to be carefully examined in context in order to determine which of the two senses the author had in mind. In some instances, it is almost impossible to tell whether the author intended the old or the new meaning of the word. But the thinkers of the nineteenth century, writing after the French Revolution, tended almost universally to use the term revolution in much the same sense in which we would use it today in the realm of social and political affairs. That is, they meant by revolution the process of establishing some new system, usually by violence. Only in astronomy and mathematics did the old sense of revolution continue.

Accordingly, when we seek for evidence that 'a revolution in science' has occurred (meaning the creation of a radically new theory or system) by examining contemporaneous opinion, we must take account of the fact that this term itself was not generally used in this way prior to the end of the seventeenth century. Furthermore, in analyzing texts or documents of the eighteenth century, one has to examine each occurrence of the word revolution with care in order to see whether the sense is that of a return, some kind of cyclical event, or the production of something that had never existed before. By the time of the nineteenth century, however, there is no such problem, since the term revolution has come unambiguously to mean the kind of change symbolized in political and social affairs by the French Revolution. In short, when we look for contemporaneous evidence that an event in science was or was not a revolution, we have a different problem for science before 1800 or so and the science that was created afterward. In the case of our Probabilistic Revolution, the time frame spans the years from 1800 to 1930, so that no difficulty arises from the fact that the term revolution in its usual modern sense is of recent origin.

I referred earlier to the fact that my research on revolutions has dealt to some degree with evidence for the occurrence of revolutions in science. Here there is a pair of related questions. The first is, What is a revolution? And the second is, How can we tell whether or not a revolution has occurred? At first glance, it may seem that these are not wholly distinct questions. Yet there is a possibility of developing a working test for the occurrence of revolutions in science, even without a clear-cut definition.

All discussions of these topics have been greatly influenced by Thomas Kuhn's seminal work on revolutions (1962) with his characterization of a revolution in science as a shift in 'paradigms' (to use his original language) that arises when a series of 'anomalies' has produced a 'crisis'.[18] In attempting to use this important characterization in formulating a test for revolutions we face a triple problem: that of making precise the three notions of anomaly, crisis, and paradigm. Furthermore, some confusion arises because there are certain kinds of revolutions in science that do not exactly fit Kuhn's schema.

Finding the answer to the definitional question is not of historical significance. But we must be aware of the historical fact that during the four centuries or so in which modern science has existed, scientists and critical observers of science have tended to see certain events as revolutions. These include conceptual changes of a fundamental kind, radical alterations in the standard or accepted norm of explanation, new postulates or axioms, new forms of acceptable knowledge, and new theories that embrace some or all of these features and others. The Newtonian revolution entailed the radical concept of a gravitational force of attraction and achieved the goal of expressing and developing the principles of natural philosophy in mathematical terms. The Cartesian revolution was posited on the 'mechanical philosophy'—the goal of explaining all phenomena in terms of matter and motion. The Darwinian theory of evolution denied the fixity of species and introduced a science that, while causal, did not permit the prediction of single events. Relativity not only sounded the death knell of absolute time and space, but radically altered the apparently simple concept of simultaneity. The Harveyan revolution set forth the idea of a continuous circulation of the blood from the heart out through the arteries, and back into the heart through the veins; it rejected the ancient and well-established doctrine that blood merely ebbs and flows in the veins, that it is continually being generated in the liver. In each of these cases, an event occurred in science that has been (and is now) generally called a revolution. This is a historical fact, one that is quite independent of our personal predilections, whether we like the word 'revolution' or not, whether we are or are not able to produce a definition that fits all these examples and others.

It is on the basis of such historical evidence that a series of four tests may be devised that can be universally applied to all major scientific events that have occurred during the past four centuries. The basis of these tests is purely historical and factual. The first of these tests is the testimony of witnesses: the judgments of scientists and nonscientists active during that time. Among such witnesses I would include philosophers, political scientists, people active in political affairs, social scientists, journalists, literary figures, and even educated laymen. Both Newton and Leibniz were still alive and working on the further development of the calculus which they had developed when Fontenelle recorded his contemporaneous impression that their creation had produced a revolution in mathematics. It was less than a decade after Newton's death that Clairaut held Newton's *Principia* to have launched an era of revolution. Lavoisier himself referred to his radical reform of chemistry as a revolution, and so did his contemporaries on both sides of the Channel. There is no want of evidence among Darwin's contemporaries that they agreed with his valuation that the theory of evolution had implications of a revolution in biology. To many earth scientists of the 1920s and 1930s it was obvious that Wegener's ideas concerning the motion of continents, continental drift, were revolution-making, even before continental drift had changed its status from a mere revolution on paper to a revolution in science.[19] All these revolutions pass the first test—the testimony of contemporary witnesses.

We may observe, incidentally, that in three of the foregoing examples the scientist chiefly responsible for each revolution (Lavoisier, Darwin, Wegener) said expressly that his own work would create a revolution. This concurrence gives added strength to the testimony of other witnesses. But it is obvious that one must not make too much of the lack of such special testimony, since most scientists are usually too modest or too restrained by the conventions of the scientific enterprise to make such judgments about their own creations. On the other hand, I would not put much trust in a later historical judgment that a revolution in science had occurred if there were no witnesses to testify to the event. (This would be the case for a possible Mendelian or Babbagian revolution in science *in the nineteenth century*.)

A second test is a critical examination of the documentary history of the subject in which the revolution is said to have occurred. For example, a study of the treatises and textbooks of astronomy written between 1543 and 1609 do not show the adoption of Copernicus's ideas or methods. Such a study shows rather that neither in the practice of astronomers nor in the principles they adopted was there a radical or even significant change from the situation prior to the publication of Copernicus's book. Hence this important second test indicates the nonexistence, during the years from 1543 to 1609, of a Copernican revolution in astronomy.[20]

By contrast, the literature of mathematics of the early decades of the eighteenth century shows unambiguously that there had been a fundamental alteration in this subject. That is, the treatises, research articles, and textbooks are largely written in terms of the new calculus (whether the Leibnizian or the Newtonian algorithm); and this is true also of works combining applications of mathematics to physics and astronomy. Thus, this test gives confirmatory evidence to Fontenelle's statement that the invention of the calculus was the epoch of a revolution in mathematics.

Similarly, we have evidence of a Newtonian revolution if we compare and contrast post-1687 mathematical astronomy (and its new strong component of gravitational celestial mechanics) with astronomy before the *Principia*. For example, prior to the publication of the *Principia* in 1687, the problem of the moon's motion consisted entirely of attempts to fit various combinations of circles and curves to the data of observation, much in the manner traditionally associated with the use of Ptolemaic epicycles. After the *Principia*, the ideal for explaining the motion of the moon was put on the basis of the mutual gravitational action of the earth and moon, plus the perturbing effect of the sun.

This second test contains some element of subjective judgment concerning the degree of restructuring. That is, there is not any complete objectivity in estimating whether the degree of restructuring is of sufficient magnitude to constitute a revolution in science. But this test is absolutely conclusive in a negative judgment, in cases where no evidence of a fundamental change can be found in the major writings of a particular science. In almost all cases of acknowledged revolutions in science, however, the evidence of this second test tends to be overwhelmingly positive (as for the calculus or for the Darwinian revolution) or at least strongly confirmatory. A confluence of positive results of these first two tests gives us a powerful indication of the occurrence of a revolution in science. We may see such a concurrence in the case of Lavoisier's Chemical Revolution, the Darwinian revolution, and the revolution in earth science arising from the acceptance of Wegener's notion of continental drift.

This second test is, of course, absolutely essential for the occurrence of a revolution; some critical scholars might even rate it of higher importance than the first test. This test is not merely the crucial step in verifying whether a revolution has occurred in cases where there is no contemporaneous evidence for the revolution; it also serves as a direct check on statements by a scientist and his followers who were convinced that they had produced a revolution and said so explicitly. Here it may be that the comparison and contrast of the state of a science shows no dramatic change as a result of the alleged revolution. This is certainly the case for the revolution in optics claimed to have been effected by J.-P. Marat. An additional possibility, of course, is that research might uncover an episode in science for which hitherto there is no record of an evaluation as a revolution, but that did radically and fundamentally alter the state of a science. If there should be a real example of such a revolution, I do not know of it. This second test is of primary importance in actual practice in verifying the assumptions of historians. Thus, as we have just seen, this test shows conclusively that no revolution occurred in astronomy after the publication of Copernicus's magnum opus.

A third test for a revolution in science is the judgment of competent historians, notably historians of science and historians of philosophy.[21] Here I would include not only historians of the present and recent past, but also historians of the eighteenth and nineteenth centuries and even some of those of the seventeenth century. There is certainly no want of historians or historically minded scholars (philosophers, sociologists, and other social scientists) who have testified during the last three centuries for the occurrence of a Newtonian revolution. For at least a century, there has been agreement among historians about a Darwinian revolution. The conjunction of affirmative answers to all three of these tests leads to an overwhelming conviction that these events were revolutions.

But there are some episodes that are generally considered by historians to be revolutions for which there is no contemporaneous body of opinion that this was the case. A primary instance, to which I have already referred, is the Copernican revolution so-called. This is a particularly interesting case because the Copernican revolution is taken by many students of the subject as the paradigmatic example of a revolution in science. I have mentioned that the evidence of contemporaneous witnesses and the astronomical texts of the day do not support the opinion of historians, a sign that this revolution must be a fiction invented and perpetrated after the fact. The existence of a discrepancy between the testimony of ancient witnesses and the contents of texts, on the one hand, and the opinions of later historians, on the other, should have warned historians to treat this alleged revolution with skepticism. A close analysis of the events in this case shows how the error arose, in a conflation of Copernican astronomy with that of Galileo and Kepler a half century or more after the publication of Copernicus's *De revolutionibus* in 1543. It is nevertheless a fact of history that for at least two centuries, historians and scientists have believed that there had been a Copernican revolution. Such judgments, made long after the events, must always be examined critically, and this is especially the case when the people making the judgments lived before present-day standards of historical evidence.

When we attempt to apply the first and the third criteria to the development of probability in the nineteenth century, we encounter a real problem in that there does not seem to be a body of clear-cut statements of revolution—either by contemporaneous observers or participants or by later historians. I do not know of any simple and direct declarations of revolution that match the critical observations of Lavoisier and his contemporaries about the Chemical Revolution, or those of Fontenelle and Clairaut about Newton's revolution, or those of Darwin and his successors about the impact of evolution. Nor is there a consensus concerning revolution among historians of probability and statistics. But this situation may be to some degree only an index of our ignorance, an indication that the history of probability and statistics is a much neglected area of study. We do not at present know for sure whether there were some contemporaneous witnesses who wrote in terms of revolution concerning statistics and probability, but whose writings on these topics are not generally known! The paucity of serious historical research in the development of statistics and probability may well have the result that neither the first nor the third criterion for a revolution may be fully applicable to this case. But there is a fourth and final test that may be applied to the statistical revolution: the general opinion of working scientists today. In this case, the physical, biological, and social scientists of the twentieth century are almost universally aware that the establishment of a statistically based physics (radioactivity and quantum physics), biology (especially genetics), evaluation of experimental data, and social science have constituted so sharp a break with the past that no term of lesser magnitude than revolution should be used to characterize this mutation. There is, of course, also the basic fact of the creation of the discipline of statistics itself.

In this fourth test of revolutions, I may appear to be giving undue importance to the living scientific tradition, to the mythology that is part of the accepted heritage of practicing scientists. Here we are literally on a frontier of our understanding of science, of scientists, and of the scientific enterprise as a social institution. Myths certainly play a significant and as yet not adequately appreciated role in science, one that is in some ways analogous to that of myths in society at large. Myths current today about heroic figures of science and their revolutions do not of course of and by themselves constitute primary historical evidence concerning past events, but they do alert historians to major episodes that are considered to have had a formative significance in the development of science. The general beliefs of scientists about their past thus reinforce the kind of evidence supplied by the other three tests.

Of course, this fourth test is not wholly independent of the third. No one could seriously doubt that scientists are often influenced by historians just as historians tend to be influenced by scientists. And it is the case that both may be under the spell of a long tradition, as in the case of the Chemical Revolution. Even a tradition that is based ultimately on error may strongly affect the thinking of later historians and scientists, as we may see by the example of a Copernican revolution.

For me, the testimony of contemporaneous witnesses weighs very strongly.[22] Unlike later judgments, which are reflections less on the events of the revolution than on the revolution's long-term effects (that is, on the postrevolutionary history of the science), these evaluations provide a direct insight into what was going on. To me there is a real significance in the fact that Charles Darwin not only believed his new ideas would create a revolution but actually said so in the conclusion of his *Origin of Species* in 1859. In fact, his prediction of a "considerable revolution in natural history" is a rare if not unique instance when a scientist was so bold as to make such a declaration in print in the major publication announcing his discovery.[23]

Oftentimes there is background evidence related to a revolution in science, evidence that illuminates the thinking of an era and that may be of more than ordinary value in providing insight into the occurrence of a revolution in science. Let us take as a case in point the annual report of the President of the Linnean Society of London for 1858. Recall that this was the year in which there was a public presentation to the society of Darwin's and Wallace's first joint communication on the evolution of species by natural selection, a set of reports that were published in the society's journal. In his annual report, the president said that the past year had not been noted for one of the revolutions that change the face of a science. We are not to assume from this statement that he was merely insensitive to the revolutionary implications of evolution. A study of his report in detail[24] shows that he was a firm believer that revolutions regularly occur in science. Furthermore, he made evident his conviction that the time was now ripe for a significant revolution to occur in the life sciences. Hence his statement proves primarily that the great Darwinian revolution was not produced merely by the enunciation and publication of bold ideas concerning evolution and natural selection. For a revolution to occur, there was need of the careful and complete body of documentary evidence and the worked-up theory that Darwin produced in his book a year later. Darwinian revolution was not produced by the mere statement of radical ideas but by an interplay between an overwhelming mass of factual data and theoretical inferences on a high level.

At the beginning of this chapter I referred to two models of scientific change: evolution and revolution. Both represent a transition from one set of shared beliefs (methods, basic concepts, definitions, laws, theories) to another, a process that Kuhn has characterized dramatically as a shift in 'paradigm'.[25] Typical changes of this sort are the Newtonian rejection of the Aristotelian and Cartesian systems of the world in favor of a universe regulated by central forces and universal gravity, and the introduction of Harveyan physiology based on the circulation of the blood as a substitute for the Galenic physiology. In each of these cases one kind of science (or paradigm) is replaced by another during a process rapid enough to constitute a revolution. But the shift from one theory, kind of science, or paradigm to another is also characteristic of the process of evolution in science, which bears a similarity to biological evolution to the extent that the latter implies a long-term process in the gradual change from one living form of species to another.

During the nineteenth century and in the twentieth many scientists and historians or philosophers of science argued for an evolutionary model of scientific change, one that would embody some aspect or variant of the features of Darwinian theory. For example, Ernst Mach supposed that there was a "struggle for existence" among scientific ideas in which there was a "survival of the fittest." Ludwig Boltzmann also wrote of evolution as a factor in the growth of science. Some scientists even argued that it was time to discard revolution from the vocabulary of science in favor of evolution.[26] These two models of scientific change differ in more than mere time-scale. Revolution, in its post-1789 signification, suggests not merely abruptness in change but rupture, an act of violence on the plane of intellectual creativity. An illustration of this feature of revolutionary science is provided by a letter from Charles Darwin to Joseph Hooker in January 1844, when he was "almost convinced" that "species are not ... immutable"; to admit this, he wrote to Hooker, was "like confessing a murder." We may agree with the late Walter F. (Susan Faye) Cannon that it was a violent revolutionary act, that Darwin was indeed contemplating murder, a form of "intellectual parricide" that would destroy what "Lyell had stood for with his uniformitarian principle of eternal stability."[27]

Evolution, on the other hand, usually conveys the image of a cumulative process, an accretion of changes that produce their effect incrementally and collectively rather than individually.[28] But the net change may be just as profound and far-reaching (i.e., the magnitude of the change may be just as 'revolutionary') when produced by evolution as when caused by revolution. And it may be further observed that only slight variants are required in order to adapt the four tests of revolution to what may legitimately be called 'revolutionary change by an evolutionary process'. It may also be observed that these tests can be applied to revolutionary changes in domains other than science.

In mathematics and the sciences evolution and revolution are not the only ways in which something new may appear. For instance, if we look at the development of theories of probability during the seventeenth century, in which this subject first began to take form, we would not find that there was a conscious revolt by Pascal, Fermat, Huygens, and Bernoulli against an established form of mathematics and its rejection in favor of something else. Since a revolution is always a process in which the reigning order is overthrown and a new one established, a revolution by necessity is an act *against* something—the state, the social or economic system, or the received philosophy or accepted systems of science and mathematics. The rise of the new subject of probability was not in this sense a revolt or a revolution. Nor, strictly speaking, was it a case of evolution. As Darwin made clear in his *Origin of Species*, his theory of evolution considers a process in which natural selection occurs among individuals with heritable characters that vary from one to another. This could account for the evolution of one species from an existing species but was not directly relevant to the very different questions of the origin of life itself or of the coming-into-being of the primitive ancestor from which all species derived in common descent.

The origins of probability theory may exemplify a third mode of scientific advance in which a new field develops—neither evolution nor revolution. This process is neither a long-term or gradual type of change from one kind of science or mathematics to another, nor the sudden rejection of an existing kind of science or mathematics in favor of another; it therefore cannot strictly be categorized as either 'evolution' or 'revolution'. The coming-into-being of probability may, however, be perfectly described by using the term introduced into the literature of probability history by Ian Hacking: an 'emergence'.[29] This emergence of the subject of probability carried the subject to a high level of maturity with the work of Laplace in the early years of the nineteenth century.[30] By this time, probability was well enough developed that there could be a revolution in the sense in which we use this term in referring to a Newtonian revolution, a Chemical Revolution, or a Darwinian revolution. The century and a third from 1800 to 1930 was not, however, marked by a continuous revolution in the technical aspects of probability. There were certain real changes that occurred and that might be considered individual revolutions in science. That is, probability theory and its applications in statistics had developed to such an extent by the 1800s that thereafter there could occur certain radical transformations or rejections of what had become accepted modes of work within the areas of probability and statistics. For instance, one such revolutionary event would be the example given by Ian Hacking,[31] in which the work of Jerzy Neyman and E. S. Pearson "transformed Fisher's study of inference into the confidence approach which is understood in terms of 'inductive behaviour'."[32] Such an event, no doubt, would satisfy all four tests for revolution that I have outlined above. Another would be "the Fisherian revolution," to use the term introduced by one of R. A. Fisher's biographers.[33]

I believe, however, that there is an aspect to revolution in science that applies to the case of probability and statistics and that may allow us to discuss a revolution in this area in a somewhat different way. I have in mind what can be called a revolution by application rather than a revolution in science. I believe that a convincing case can be made that the introduction of quantitative statistical reasoning produced a set of very radical changes in thinking and methods in the natural and exact sciences and in the social sciences in the nineteenth and twentieth centuries. This revolution appeared first in the domain of the social sciences and most notably in the work of Adolphe Quetelet in what was known as 'moral statistics'. Above all, there was a radical and fundamental transformation in thinking in the areas of medicine and public health as a result of a similar introduction of considerations of statistics and probability. This would not be a 'probabilistic revolution' in the traditional sense of a revolution in the techniques and subject matter of probability so much as it would be a revolution deriving from the introduction of the methods and considerations of probability into the areas of social thought and analysis and of medicine and public health.

We may see signs aplenty of this revolution. A most competent witness was Sir John Herschel, whose famous statement about Quetelet's findings and their implications was published in the *Edinburgh Review* in July 1850:

Men began to hear with surprise, not unmingled with some vague hope of ultimate
benefit, that not only births, deaths, and marriages, but the decisions of tribunals,
the results of popular elections, the influence of punishments in checking crime—
the comparative value of medical remedies, and different modes of treatment of
diseases—the probable limits of error in numerical results in every department
of physical inquiry—the detection of causes physical, social, and moral,—nay,
even the weight of evidence, and the validity of logical argument—might come
to be surveyed with that lynx-eyed scrutiny of a dispassionate analysis, which, if
not at once leading to the discovery of positive truth, would at least secure the
detection and proscription of many mischievous and besetting fallacies.

Quetelet's statistics had been awesome in their conclusions. If, for example, there
was a regularity in the number of crimes committed (and even in their varieties),
then the conclusion was inescapable that to some major degree society was to blame
and not the criminal.

One way of gauging whether the new statistical analysis of society was sufficiently
profound to be reckoned a revolution is to take cognizance of the intensity of
reactions against the new statistical way of thinking. Two opponents of statistically
based science or knowledge were Auguste Comte and John Stuart Mill. Comte, in
his *Course of Positive Philosophy*, scorned "the empty claim of a great number of
geometers to render social studies positive through a chimerical subordination to
the illusory mathematical theory of chances." Comte regretted that James Bernoulli
and particularly Condorcet had sought to apply probability and statistics to social
theory (or to sociology). There was, he wrote,

no excuse for Laplace's sterile reproduction of such a philosophical aberration at
a time when the general state of human reason was already beginning to permit
the discernment of the true fundamental spirit of sound political philosophy, so
well prepared, as I have shown, by the labors of Montesquieu and of Condorcet
himself, and, in addition, powerfully stimulated by the radical convulsion of
society. With even greater reason we cannot by any means condone the current
prolongation of this absurd illusion among the inferior imitators who, without
adding anything fundamental to the subject, limit themselves to repeating, with
ponderous algebraic verbiage, the obsolete expression of these empty claims,
thereby grossly abusing the credit that has up to now been so justly extended
to the true mathematical spirit. Far from indicating, as it did a century ago,
a premature instinct for the indispensable renovation of social studies, this
aberration constitutes today, in my view, only the involuntary decisive evidence
of a profound philosophical impotence, usually combined, in addition, with a
sort of algebraic mania now too familiar to the general crowd of geometers,
and perhaps also stimulated sometimes by the desire, so common in our days,
to create for oneself at little expense a certain reputation, ephemeral but
productive, of great political consequence. Would it be possible, indeed, to
imagine a conception more radically irrational than that which consists in giving
to the whole of social science, as philosophical basis or principal means of final
elaboration, an alleged mathematical theory in which, habitually taking signs for
ideas, in accordance with the usual character of purely metaphysical speculations,
we try to subject to calculation the necessarily sophistic notion of numerical

probability, which leads directly to giving our own real ignorance as the natural measure of the degree of likelihood of our various opinions?[34]

Comte's opposition to statistics and probability was very likely based on his conviction that "all sciences aim at prevision" (that is, precise prediction), as he wrote in an essay of 1822 on "reorganizing society."[35] To this end, "the laws established by the observation of phenomena" should enable the scientist to foretell the succession of phenomena. It follows that "observation of the past should reveal the future in politics as it has done in astronomy, physics, chemistry, and physiology." Comte expanded this theme in the fourth volume of his *Course of Positive Philosophy*, where he argued (p. 230; repr. p. 109) that social phenomena are "inevitably subject to true natural laws, regularly entailing rational prevision." What Comte had in mind was the simple causal predictions found in classical rational mechanics—which he saw as the antithesis of the 'inexact' predictions of statistics and probability.

In John Stuart Mill's *System of Logic*, described as his most important or "principal philosophical work," a stand is taken against statistical arguments or improper use of probability in science or social science. According to Mill, "It would indeed require strong evidence to persuade any rational person that by a system of operations upon numbers, our ignorance can be coined into science." Mill added that it was "doubtless this strange pretension" that "has driven a profound thinker, M. Comte, into the contrary extreme of rejecting [this doctrine] altogether," despite the fact that it "receives daily verification from the practice of insurance, and from a great mass of other positive experience."[36] This statement, like some other similar ones in the first edition (1843) of the *Logic*, was eliminated in the second and later editions; but no reader can escape the obvious conclusion that Mill took a rather unfavorable view of the foundations of probability and the usefulness of applying it.[37] Mill left no doubt concerning his position when he said that "misapplications of the calculus of probabilities" have made it "the real opprobrium of mathematics."[38]

Many scientists, as well as philosophers, either came out directly against the use of probability and statistics in science or voiced strong doubts concerning their use. As late as 1890, in the second edition of his *Properties of Matter*, Peter Guthrie Tait could still take an antistatistical posture and write of the remaining difficulties in the kinetic theory of gases as "greatly enhanced by an apparently unwarranted application of the *Theory of Probabilities*, on which the statistical method is based." Tait was concerned about the introduction into physical science of nondeterministic or probabilistic reasoning. The pioneering work in this area had been done in thermodynamics (kinetic theory of gases) and eventually produced statistical mechanics. The two early major architects of this revolutionizing of physics were J. Clerk Maxwell and Ludwig Boltzmann. Both, we now know, had been influenced by the radical work of Quetelet, providing an interesting example of the way in which the natural or exact sciences may be influenced by the social sciences.[39]

A more continual and outspoken critic of the use of statistics and probability in science was Claude Bernard, often called the founder of modern experimental physiology. In his *Introduction to the Study of Experimental Medicine*, Bernard said simply that he could not possibly understand "how we can teach practical and exact science on the basis of statistics." The use of statistics, he believed, must necessarily "bring to birth only conjectural sciences," and "can never produce active experimental sciences, i.e., sciences which regulate phenomena according to definite laws." Furthermore, he argued, "By statistics, we get a conjecture of greater or less probability about a given case, but never any certainty, never any absolute determinism." Since "facts are never identical," statistics can serve only as "an empirical enumeration of observations." Hence if medicine were based on statistics, it could "never be anything but a conjectural science; only by basing itself on experimental determinism can it become a true science, i.e., a sure science." Here Bernard was expressing the difference between what he denominated the point of view of "so-called observing physicians" and that of "experimental physicians." Bernard saw experimental science leading to a strict determinism that he and other physiologists believed to be incompatible with probabilistic or statistical considerations.[40]

In an address read to the Congress of Arts and Sciences at the Universal Exposition in St. Louis in 1904, the philosophically minded theoretical physicist Ludwig Boltzmann discussed briefly the applications of statistics to science and to social science. Defending the "theorems of statistical mechanics," which he claimed to be as valid "as all well-founded mathematical theorems," he noted that there was a difficulty in other applications of statistics, for instance, in assuming "the equal probability of elementary errors." Alluding to the broadening application of statistics to "animated beings, ... human society, ... sociology, etc., and not merely ... mechanical particles," he called attention to "difficulties of principle" that arise from basing such studies on the theory of probability. This subject, he said, "is as exact as any other branch of mathematics if the concept of equal probabilities, which cannot be deduced from the other fundamental notions, is assumed."

In the twentieth century, we have witnessed a true revolution in the physical sciences resulting from the substitution of probability and statistics for the old Newtonian simple causality of assigned cause and effect. A similar movement in biology has centered on genetics and evolution. There is ample testimony to this revolution, beginning in the social sciences in the mid-nineteenth century and reaching a culmination in the physics of radioactivity and quantum mechanics in the first third of the twentieth century. By 1914, in a book entitled *Chance*, a nontechnical exposition of probability and statistics "in the various branches of scientific knowledge," the French mathematician Emile Borel indicated that we have been present, "almost without being aware of it, at a genuine scientific revolution" (p. ii). This revolution, a revolution in application, easily passes all four tests for revolution that I have presented above.[41] Hence, even if the decades 1800–1930 do not show a single revolution in the domain of probability, they provide evidence of a *probabilizing revolution*, that is, of a true revolution of fantastic consequences attendant on the introduction of probability and statistics into areas that have undergone revolutionary changes as a result.

Notes

1. Some of the ancillary topics presented in this chapter are developed more fully in my *Revolution in Science* (Cambridge, MA: The Belknap Press of Harvard University Press, 1985).

2. A third model, emergence, will be discussed below.

3. I make a distinction between a revolution in science, an event that changes a branch or a part of science (e.g., the Chemical Revolution or the Darwinian Revolution), and a scientific revolution, a connected series of events that affects the whole of science (e.g., *the* Scientific Revolution of the sixteenth and seventeenth centuries).

4. Stephen Toulmin, *Human Understanding* (Princeton: Princeton University Press, 1972).

5. Even though there are legitimate grounds for doubting whether the concept of revolution is generally useful in understanding any change—whether political, social, economic, artistic, intellectual, religious, or scientific—the fact remains that in political history there are great classic events (the Glorious Revolution, the French Revolution, the Russian Revolution of 1917) that have been and are universally referred to as revolutions. They have had both an existential and symbolic influence on later thought and action; they embody special qualities that are connoted by the commonly accepted name of revolution.

Peter Calvert has wisely advocated that we "retain the term 'revolution' itself as a political term covering all forms of violent change of government or regime originating internally," as "a simple recognition of the fact that it is the meaning most widely used in the modern world." See his *Revolution* (London: Pall Mall Press; New York: Praeger Publishers, 1970), p. 141.

6. The concept of a 'second Scientific Revolution' was formally introduced into the literature of the history of science by T. S. Kuhn at a symposium sponsored by the Joint Social Science Research Council-National Research Council Committee on the History of Science in 1959, published in *Isis* in 1961. Kuhn particularly directed our attention to "an important change in the character of research in many of the physical sciences" that occurred some time between 1800 and 1850, "particularly in the cluster of research fields known as physics." This change, "the mathematization of Baconian physical science," is said by Kuhn to be "one facet of a second scientific revolution." Kuhn stresses the fact that various "changes affected all the sciences in much the same way." Accordingly, some other factors are needed in order to "explain the characteristics that differentiate the newly mathematized sciences of the nineteenth century from other sciences of the same period." See p. 190.

7. Roger Hahn would see the signs of a "'second' scientific revolution in the early nineteenth century" in the enormous increase in science professionals and the institutions that support them. Hahn's concept of the second Revolution thus differs notably from Kuhn's. For Hahn, whose views appear in his celebrated study *The Anatomy of a Scientific Institution: The Paris Academy of Sciences, 1666-1803* (Berkeley: University of California Press, 1971), pp. 275ff., the second scientific revolution is "the crucial social transformation that ushered science into its more mature state and, like the first revolution in the seventeenth century, cut across national boundaries." In this presentation, Hahn focuses on the institutional changes that were features of the revolution: "the eclipse of the generalized learned society and the rise of more specialized institutions" and "the concurrent establishment of professional standards for individual scientific disciplines." This second Scientific Revolution was accompanied by the rise of universities and research institutes, and particularly the cultivation of "professionalized science" in "institutions of higher learning." It was an age when "specialized laboratories" were replacing the "academies that had dominated the scene since the middle of the seventeenth century." Hahn also directs our attention to the enormous increase in the size of the scientific community—a factor of bulk that, by itself, "required institutional differentiation." He finds that the growth of professionalization was a necessary consequence of an "increased technicality of disciplinary problems" within each of the sciences, as well as

"experimental requirements peculiar to each subject." Hahn would link the rise of specialization to a "narrowing of the gap between science and its direct applications."

Some other historians who have discussed a second Scientific Revolution are Hugh Kearney (1964), Everett Mendelsohn (1966), Stephen Brush (1982), and Enrico Bellone (1976, 1980). For details see my *Revolution in Science* (1985), Chapter 6.

8. See note 1.

9. See note 1.

10. J. L. E. Dreyer, in his *History of the Planetary System from Thales to Kepler* (Cambridge: at the University Press, 1906), p. 344, explains simply and boldly that "Copernicus did not produce what is now-a-days meant by 'the Copernican system'."

11. E.g., by the eighteenth-century historians, J.-E. Montucla and J.-S. Bailly.

12. Until well into the eighteenth century scientists did not always make our current distinction between rotation and revolution.

13. Yahuda Collection, National Library, Jerusalem.

14. For details concerning the previous and following paragraphs, and this episode, see my *Revolution in Science* (cited in note 1), Chapter 4.

15. Many would consider the Reformation to have been a revolution predating the Glorious Revolution, even though not strictly a revolution in the social order or the political system. The consideration of the Reformation as a revolution is discussed in historical perspective in my *Revolution in Science*, §4.3.

16. See Jean Marie Goulemot, *Discours, révolutions et histoire* (Paris: Union Générale d'Éditions, 1975).

17. For documentation concerning these early statements of revolution and of self-declared revolution, see my *Revolution in Science*.

18. T. S. Kuhn, *The Structure of Scientific Revolutions* (Chicago: The University of Chicago Press, 1962; rev. ed., 1970).

19. In my *Revolution in Science*, I define four basic stages of revolution in science. The last two are the public stages: revolution on paper, revolution in science. The stage of a 'revolution on paper' occurs whenever a revolutionary new theory, concept, or method is published, that is, made available for other scientists to study, use, or reject. It is only when a sufficient number of working scientists accept and make use of the new theory, concept, or method that the 'revolution on paper' becomes transformed into a 'revolution in science'.

20. Here I am concerned only with a possible revolution in astronomical science. I do not consider the possibility that there may have been a revolution in philosophical cosmology.

21. I am fully aware of the apparently grossly subjective quality of the adjective 'competent'. But I have in mind only a distinction between a scholar and a popularizer—without at all going into the question of whether a scholarly historian may uncritically accept vulgar errors and traditional points of view and may also make mistakes.

22. As does the primary study—comparison and contrast—of the state of a science before and after an alleged revolution. One needs the other.

23. Research over some two decades has not uncovered a second example of this kind. Hermann Minkowski spoke of a great revolution when he read a paper introducing his four-dimensional space-time reformulation of Einstein's theory of relativity, but as Peter Galison has shown (*Historical Studies in the Physical Sciences*, *10* (1979), 85-121), he toned this down in the public version. Lavoisier did refer in print to his revolution, but not in the major announcement of his Chemical Revolution; see my forthcoming article on this subject in *Ambix*.

24. See my *Revolution in Science*, pp. 286ff.

25. Kuhn's original concept of paradigm—expounded in his *The Structure of Scientific Revolutions* (Chicago: The University of Chicago Press, 1962, 1970)—has been modified to the extent of replacement; see his *The Essential Tension* (Chicago: The University of Chicago Press, 1977).

26. See my *Revolution in Science*, Chapter 18, §18.2, §26.1.

27. Walter F. Cannon, "The Bases of Darwin's Achievement: A Revaluation," *Victorian Studies*, 5 (1961), 117; Sir Gavin de Beer, *Charles Darwin: Evolution by Natural Selection* (Garden City, NY: Doubleday, 1964), p. 135.

28. I do not wish to discuss here the varieties of non-Darwinian evolution, such as the possibility of 'punctuated equilibria' (as proposed by Stephen J. Gould and Niles Eldredge: 1972, 1977), a model applied to the progress of science by Garland Allen ("Morphology and Twentieth-Century Biology: A Response," *Journal of the History of Biology*, 14 (1981), 159–176; see my *Revolution in Science*, p. 387).

In this context it must be kept in mind that nineteenth- and twentieth-century scientists tended to use the term 'evolution' in two distinct senses to characterize the development of the sciences. One was rather strictly Darwinian, with its overtones of a "struggle for existence," or "fight for life," the "survival of the fittest"—as if the growth of science might, to some degree, be like the proverbial "nature red in tooth and claw." We have seen an example of this predatory view of the evolution of science in the writings of Ernst Mach. But for most scientists and historians, evolution in the context of the advance of science has not had so specific a Darwinian signification. Rather, they have tended to equate the evolution of science with mere gradualism, with progress by a cumulative effect of many lesser steps. In short, these scientists and historians—their number includes such diverse figures as Claude Bernard, J. B. Conant, Pierre Duhem, Albert Einstein, R. A. Millikan, Lord Rutherford, George Sarton—have tended to use evolution merely to denote a gradual or incremental model of scientific growth in opposition to the violent model of sudden revolution.

29. Ian Hacking, *The Emergence of Probability* (London, New York: Cambridge University Press, 1975).

A careful analysis of the historical process of emergence as an alternate mode of scientific change would take us far afield from the main topics of my discussion. Here we have merely to take cognizance of the fact that there are more than two ways—evolution and revolution—by which a science can change and by which a new science (or branch of science) can come into being. I am fully aware of the seemingly endless and fruitless debates among biologists and philosophers over evolutionary emergence and I do not wish to start a similar debate in the historical realm. I have used the term 'emergence' only because Ian Hacking had already introduced it in more or less the context I had in mind, and definitely not for its cognate biological overtones.

30. I am not here concerned with the essentially nonfruitful problem of whether Laplace's work was the end of an 'emergence' or part of an 'evolution' of a subject that by his day had become sufficiently well established to admit of evolution.

31. See his chapter in this volume.

32. Hacking has suggested other examples of 'revolutions in science' occurring in probability theory (though he does not explicitly use this term of mine). They include Galton's "discovery of correlation and the theory of regression towards mediocrity" in relation to "patterns of hereditary genius" and Quetelet's recognition (or "mistaken claim") "that distributions of a property across a biological species (including humanity) are just *as if* they were a distribution of errors around an objectively existing mean."

33. Norman T. Gridgeman in *Dictionary of Scientific Biography* (1972), Vol. 5, p. 9a.

34. Auguste Comte, *Cours de philosophie positive*, 3rd ed., Vol. 4 (Paris: J. B. Baillière et Fils, 1809), pp. 366–368; reprinted in *Physique sociale: Cours de philosophie positive, leçons 46 à 60*, ed. Jean-Paul Enthoven (Paris: Hermann, 1975), pp. 168–169. The English version by Harriet Martineau tones down Comte's French by omitting, for example, many strong adjectives.

35. See Ronald Fletcher, *The Crisis of Industrial Civilization: The Early Essays of August Comte* (London: Heinemann Educational Books, 1974), p. 167.

36. John Stuart Mill, *A System of Logic, Ratiocinative and Inductive, Being a Connected View of the Principles of Evidence and the Methods of Scientific Investigation*, ed. J. M. Robson, *Collected Works of John Stuart Mill*, Vols. 7 and 8 (Toronto: University of Toronto Press; London: Routledge & Kegan Paul, 1973–1974), p. 1142.

37. Ibid., pp. 1140–1153.

38. Ibid., p. 538. Lorraine J. Daston reminds me that "Mill's major objections were to applications still regarded with grave suspicion by many probabilists, viz., the probability of judgments and testimonies (which assumes constant individual probabilities for truth telling and correct decision making, the independence of individual juror's decisions, etc.), and the 'law of succession' (which uses Bayes's theorem in a controversial fashion, assuming uniform prior probability distributions in a state of ignorance)."

39. See Charles C. Gillispie, "Intellectual Factors in the Background of Analysis by Probabilities," pp. 431–453 of Alistair C. Crombie, ed., *Scientific Change* (London: Heinemann Educational Books; New York: Basic Books, 1963); also Theodore M. Porter, "A Statistical Survey of Gases: Maxwell's Social Physics," *Historical Studies in the Physical Sciences, 12* (1981), 77–116.

40. Claude Bernard, *An Introduction to the Study of Experimental Medicine*, trans. Henry Copley Greene (New York: The Macmillan Company, 1927); reprint 1957, introduction by I. Bernard Cohen (New York: Dover Publications), pp. 138–139.

41. I have not explored here the actual ways in which a probabilizing revolution altered the sciences of physics and biology and the social sciences, a topic introduced in Ian Hacking's chapter and others. Some discussions of special interest are R. A. Fisher, "Statistics," pp. 31–55 of A. E. Heath, ed., *Scientific Thought in the Twentieth Century* (New York: Frederick Ungar, 1953); M. Kac, "The Emergence of Statistical Thought in Exact Sciences," pp. 433–444 of Jerzy Neyman, ed., *The Heritage of Copernicus: Theories More "Pleasing to the Mind"* (Cambridge, MA: The MIT Press, 1974); K. Mendelssohn, "Probability Enters Physics," pp. 45–67 of A. C. Crombie, ed., *Turning Points in Physics* (Amsterdam: North-Holland; New York: Interscience Publishers, 1959); and especially Hans Reichenbach, *Selected Writings 1909–1953*, eds. Maria Reichenbach and Robert S. Cohen, 2 Vols., trans. Elizabeth Hughes Schneewind, section 35, "The Problem of Causality in Physics [1931]" (Dordrecht, Boston, London: D. Reidel, 1978).

3 Was There a Probabilistic Revolution 1800–1930?

Ian Hacking

There was no such probabilistic revolution in the sense of Kuhn's Structure, *although between 1800 and 1930 there were many small* Structure-*like revolutions in probability ideas. Nor was there a revolution in probability, in Cohen's sense. But there is another idea of scientific revolution, deriving from the scientific revolution of the seventeenth century, and from Kuhn's mention (not in* Structure*) of a second scientific revolution around 1800. This idea is analyzed in terms of institutional innovation. It is argued that there is an event, 1800–1930, connected with probability concepts, and which fits this analysis. Hence there is a sense in which there was a probabilistic revolution during this period. Moreover, this event can be broken into four segments, each displaying a dominant conceptual shift, and collectively making for the larger revolution. Note that even if one witholds the word "revolution" from events connected with probability 1800–1930, the associated radical changes in concepts and practice would, in common speech, be correctly called "revolutionary." It is our task to understand the nature and extent of these revolutionary changes.*

The research reported in these volumes began with a simple observation: today our vision of the world is permeated by probability, while in 1800 it was not. Probability is the great philosophical success story of the period. Other philosophical ideas have waxed and waned and sometimes grown again, but probability has been monotone. It has waxed and waxed, shone and shone. It has been a success in metaphysics, epistemology, and pragmatics, to mention but three of the classic philosophical fields. In metaphysics, partly thanks to physics, we witness an incredible mutation, for the whole structure of causality seems to be overthrown. In 1800 the world was deemed to be governed by stern necessity and universal laws. Shortly after 1930 it became virtually certain that at bottom our world is run at best by laws of chance. In epistemology, largely thanks to logicians and statisticians, it is now a commonplace that much of our learning from experience, and much of our foundations for knowledge, are to be represented by probabilistic models. Laplace may have said as much by 1800, but the present conviction derives from the tradition established by F. P. Ramsey's paper of 1926.[1] Finally, in pragmatics, risk analysis and decision theory are regarded as at least essential window dressing for any practical action of moment to the public. Nor need we be highbrow philosophers to see that something has happened: nowadays people seem incapable of talking about sport, war, acid rain, medicine, and even sex without bringing in a few probabilities or chances or at least statistics. More statistics are cited during American prime time television than acts of violence are portrayed on the screen.

In casual speech it would be quite right to describe the transition to our present state as revolutionary, but nothing much would hang on that. The philosopher and the historian would then go on to analyze what happened, to try to understand the conceptual and practical changes, to speculate about their causes, and to contemplate how this past has shaped our present and perhaps determines our future. But the word "revolution" is a bogey here, a word that's weighty. It is not so clear whether it is simply overweight, a euphemism for being fat, or whether the weight is all muscle. It is mostly thought to be muscle, so that it was useful rhetoric for us to say that we were studying the "probabilistic revolution." That made our work sound truly important. But it also invited skepticism. Indeed I recall that at the opening session of our research group the Rektor of Bielefeld University wondered, in effect, if we were not just using a fashionable word.

The word "revolution" is more than fashionable in the history and philosophy of science. It has some precise senses, but I say senses, for the word is not unequivocal. T. S. Kuhn's *The Structure of Scientific Revolutions* ushered in a whole new field of enquiry over twenty years ago, and it is fitting that his paper should open this volume. One is equally glad that I. B. Cohen is the next contributor, for his new book *Revolution in Science* will be the classic source work for many years to come. But the projects of the two books are fundamentally different, and I would say that they have substantially different notions of revolution. Thus "revolution in science" is not to be thought of as a stylistic variant of "scientific revolution." I would argue, although not here, we ought not even to see the two notions of revolution as in competition for providing the "true" history of science.[2] They are different kinds of conceptual telescope, for observing phenomena that only loosely relate to each other. Nevertheless, the works of these two men sufficiently define for us ideas of revolution in terms of which we may seriously pose our question, "Was there a Probabilistic Revolution 1800–1930?"

In his paper, and in more detail in his book, Cohen provides criteria for being a revolution in science. Do the kinds of transformations that I mention in my opening paragraph satisfy his criteria? Cohen has already answered: no. At most we may talk of a revolution in the application of probability ideas.

What of revolution in the sense of Kuhn's *Structure*? Do the events I mention illustrate a revolution that has Kuhn's structure? Once again the answer is, no. I carefully rehearsed, or rather labored, the reasons for saying this in an earlier version of this this paper, but I cannot imagine that anyone who has read *Structure* will disagree.[3]

It is, however, worthwhile to consider Kuhn's modest revisions and changes of emphases, as represented in the opening chapter of the present volume. He argues for three characteristics of scientific revolutions; they are not exactly criteria by which we recognize revolutions, but they are significant clues to what Kuhn will count as a revolution.

The first is a "shift in the way things fit together." After describing his own experience in finally coming to understand Aristotle's *Physics*, he writes, "That sort of experience—the pieces suddenly sorting themselves out and coming together in a new way—is the first general characteristic of revolutionary change that I shall be singling out.... Though revolutions leave much piecemeal mopping up to do, the central change cannot be experienced piecemeal, one step at a time."[4] This is his present version of what, in earlier writing, he used to compare to the Gestalt-switches of 1930s psychology. His second characteristic is a replacement for his earlier talk of meaning-change and incommensurability. It is now expressed in terms of taxonomy: "What characterizes revolutions is, thus, change in several of the taxonomic categories prerequisite to scientific description and generalizations. That change, furthermore, is an adjustment not only of criteria relevant to categorization, but also of the way in which given objects and situations are distributed among preexisting categories. Since such redistribution always involves more than one category and since those categories are interdefined, this sort of alteration is necessarily holistic."[5]

His third characteristic is "a central change of model, metaphor, or analogy—a change in one's sense of what is similar to what, and of what is different."[6] Thus in his example of the battery, Volta took the cells of the battery to be like Leyden jars, and resistance to be like electrostatic leakage. In that little revolution after 1800, these patterns of similarities were replaced by others.

Now suppose that we can speak of an event, a long event (but history is full of long events) that corresponds to that transformation in the status of probability that is manifest between 1800 and 1930. Does this event display Kuhn's three characteristics? Obviously not. I conclude: *There is no reason to suppose that there is an event spanning 1800–1930, connected with probability, and which exhibits Kuhn's structure.*

Not surprisingly, there are many events within the period 1800–1930 that are *prima facie* candidates for being revolutions according to Kuhn's *present* account. I say "present" because certain features of *Structure* seem not to be emphasized in his present paper. I think especially of the period of crisis that precedes revolution in *Structure*. I shall list a few events concerning probability, and which seem to me to display Kuhn's three characteristics. None has the feature of being preceded by a crisis. My examples are all from my own field of expertise, the development of statistics. A comparable list might be devised for other fields, although I suspect that if we are to consider the whole period 1800–1930, it is statistics that will give us the most regular sequence of Kuhn-structured minirevolutions.

1. In the 1820s a vast range of human phenomena, especially those connected with deviance, were of a sudden conceived of as lawlike: as subject to the calculus of probabilities. (The most important single stimulus for this event was the publication, under the direction of Fourier, of *Recherches statistiques sur la ville de Paris et le département de la Seine*, 1823.)[7]

2. Quetelet's recognition (we would now call it his mistaken claim) that distributions of a property across a biological species (including humanity) are just as if they were a distribution of errors around an objectively existing mean. Quetelet really hit on this after reading a summary of the chest diameters of 5,738 soldiers from Highland regiments. He was seized by this insight in 1844.

3. Galton's discovery of correlation and the theory of regression toward mediocrity, which he used as the explanation of the patterns of hereditary genius that he claimed to observe. He was taking shelter from a rainstorm during a country outing: "There the idea flashed across me, and I forgot everything for a moment in my great delight."[8] That happened in 1888, and mathematical statistics never looked back.

4. The work of E. S. Pearson and Jerzy Neyman, which transformed Fisher's study of statistical inference into the confidence approach, which is understood not in terms of inference, but rather in terms of "inductive behavior." We can just squeeze this into our time span, for the original paper was published in 1928, although the mature presentation came in 1933.[9] (To tell the truth, I regret that convention binds us to decades and centuries. My periodization would be 1823–1936, the latter being the year of von Neumann's no-hidden-variables theorem for the quantum theory.)

Each of my four examples exhibits, to some degree, the characteristics noted by Kuhn. In each case we do have a central change in model, metaphor, or analogy. That is least marked in the case of Galton, yet he perhaps best exhibits the sudden and sharp Gestalt-switch that so interests Kuhn. Things do fit together in new ways. My examples are weakest on Kuhn's second characteristic, of changing taxonomies, but I think Kuhn himself would agree that "taxonomy," understood literally, is only a part of what he is trying to get at with his second character. Not all revolutions of the sort that concern him are mere changes in classificatory organization.

I have no doubt that there are many examples of what one might call neo-Kuhnian revolutions within the history of probability and its applications. My (1)–(4) only scratch at a heap of examples from statistics, and other disciplines and applications will provide similar heaps. Perhaps we ought to have been studying "the Probabilistic *Revolutions* 1823–1936"? Well, that would be one thing to study, or rather, a collection of many different things. There might, however, be another concept of revolution that could fill our bill. The best candidate is provided by Kuhn himself, although not in *Structure*.

In his remarkable paper on "The Function of Measurement in Modern Physical Science" he writes, "Sometime between 1800 and 1850 there was an important change in the character of research in many of the physical sciences, particularly in the cluster of fields known as physics. That change is what makes me call the mathematization of Baconian science a second scientific revolution."[10] The alleged change was a remarkable one, if Kuhn is correct. To exaggerate what he himself writes, there was an official mythology in the Baconian sciences that the world was written in the language of mathematics. However, one did not take this seriously except in the non-Baconian sciences that clustered around celestial mechanics. Although there was a lot of rhetoric about counting, our present vogue for measuring and enumeration came into being, suggests Kuhn, between 1800 and 1850, say. That was the time in which the present-day main scientific professions came into being—physicist, biologist, chemist. They were among the new measurers. By the end of the nineteenth century the official philosophy was that you understood something only when you could measure it.

I quote and explain this passage of Kuhn's for two reasons. One is that I think the sudden burgeoning of measurement is connected with a possible "probability revolution." The other is that it indicates a sense of "revolution" entirely different from that found in Kuhn's *Structure*. How can Kuhn, who writes of myriad revolutions involving groups of 100 or so research workers, casually speak of a "second" scientific revolution? Manifestly he has in mind something like the so-called scientific revolution of the seventeenth century, which, once again, is not the sort of revolution of which *Structure* treats. Do we here have a clue to a concept of revolution that might fit the idea of a probabilistic revolution? I shall answer with a cautious yes.

Cohen's note 7 and chapter 6 of *Revolutions in Science* provide a little history of the idea of a "second scientific revolution." The authors whom he cites have different candidates for what revolution is "second."[11] I shall not enter that fray, for I am more concerned with the very notion of revolution that is being deployed. My strategy is as follows. Suppose that there is a coherent idea of revolution, of which the scientific revolution of the seventeenth century is the prime example. Suppose we take Kuhn's second scientific revolution as a candidate for another example. Can we find features common to them that will serve as markers of the idea of revolution in question? If so, is there an event 1800–1930, connected with probability, and which also exhibits these features?

We could think of fancy names for the kind of revolution that Kuhn suggests, but let us simply speak of *big revolutions*. One of their features is that they are inter-disciplinary. Many of us still take Herbert Butterfield's *The Origins of Modern Science* as our primer on "the" scientific revolution, and he writes of Harvey, of Boyle, of Bacon, of chemists. Likewise Kuhn's second scientific revolution involves the very inauguration of our modern disciplines. Perhaps "interdisciplinary" is anachronistic. We see a group of different disciplines only because they have been forged by the revolution itself. We might suggest that big revolutions are in part predisciplinary.

At any rate a fundamental feature of big revolutions, as shown by our two present exemplars, is that they are embedded in, pervade, and transform a wide range of cultural practices and institutions. I would emphasize the last word and suggest the following rule of thumb: *Don't look for a big revolution until you find new kinds of institution that epitomize the new directions created by the revolution.*

In the seventeenth-century case we have at once the Royal Society of London, and all the national academies that followed in suit. London, Paris, Berlin, Petersburg, and the rest had a family of interconnected roles, such as status, meeting, and communication of information. Many other organizations had those features earlier. A specific mark of the academy was that it published its transactions. Although there had been a few earlier sporadic attempts to publish science communications, the academies created a new kind of thing: the scientific periodical.

What about Kuhn's second scientific revolution? Were there new institutions? Certainly. It is not insular to attend to the special case of Great Britain. In the period 1800–1850 that Kuhn singles out, the European continent was rife with revolution, and we might attribute new institutions to political circumstances. Great Britain, however, preserved the outward forms of government, so new institutions cannot simply be attributed to violent political change. There occurred, in 1830, the foundation of a remarkable new institution, the British Association for the Advancement of Science, usually known as the B.A. It was founded partly because it seemed to many younger men that the Royal Society had become moribund. More important, it was founded because there was new work to do: measure. In contrasting the Royal Society and the B.A., we ought not to use evaluative criteria, saying that the B.A. was more open, more honest, more talented, or less stuffy. We should use formal criteria. One simple criterion is the place of meeting. The Royal Society, as its name implied, was of London. Unlike the national academies, the B.A. met annually in different towns, often taking advantage of the enormously active "provincial" scientific societies of Manchester, Birmingham, and the like. By 1908 the B.A. went as far afield as Winnipeg. I rather like the thought of the cream of British science (and it was the cream—everyone went) steaming across the Atlantic and up the St. Lawrence and then taking the Canadian Pacific Railway across the continent. Now *that* was an excursion that puts modern-day junketing to shame. (And Millikan took the train up from Chicago; on the way back, the story goes, he had figured out how to do the oil drop experiment properly.)

The B.A. is the model upon which most national professional associations work. They have an annual (or whatever) convention in successive towns in which their interests are being actively pursued. Moreover, each professional association has its own house journal, just as did the B.A. from the start. Thus the B.A. was not merely another society falling under an established rubric. It was making up the rules of a game that we still play. But in one respect it was importantly different from the modern professional society. It was interdisciplinary or, as I have suggested, almost predisciplinary.

How does my "institutional" rule of thumb work for probability? Rather well. During the 1830s numerous statistical societies were founded. Never before had there been people who formed clubs simply to talk about the numbers of this or that, in general. There was a new topic of conversation, not numbers of people, or the size of the national debt, or the number of men under arms, but just—numbers of anything. It may be protested that this is only an instance of the B.A. phenomenon. Thanks to a small power play by a number of individuals, including Quetelet, who was on a visit, the B.A. was conned into admitting a Section F—a statistical section—in 1833. But it is important that it was a power play, involving even a little duplicity or at least politicking, for it was the already formed statisticians who wanted a niche in the national scientific association.

We may note another kind of institution, rather later: the statistical congress. The first international statistical congress was held in Brussels in 1853. At that time international congresses on scientific matters were virtually unknown. The statistical congresses were a bridge between political congresses (The Congress of Vienna, etc.) and scientific congresses. This is because their membership was a mix of bureaucrats and what may loosely be called scientists. The political congresses were run by civil servants for the benefits of their masters. But in Brussels, in 1853, civil servants met at an international site for their own benefit, and hobnobbed with more purely scientific individuals. They drew up the ground rules for the international congress, taken from the political sphere, and these were in turn passed over to purely scientific congresses. The ground rules drawn up by the statistical congresses are still the ground rules of congresses supported by UNESCO and the like.

Just like the B.A., the international congress is something new in kind that serves as a model for future generations. This adds to the *prima facie* case that we may have something like a probabilistic big revolution.

It is important not to restrict the idea of an institution to fairly academic matters. Far more important than statistical clubs and congresses were the various national statistical bureaux. There was many a national office for collecting numbers before the nineteenth century. The chief concern was taxation and recruitment. The numbers were all kept secret, the privy property of the king's ministers. An entirely new institution emerged in the nineteenth century, an institution in the service of the state, but which made public a very great many summaries of its data. Each country did this in its own way. England and Wales was typically an ill-coordinated jumble of internally rather excellent offices, such as the Board of Trade or the office of the Registrar General. In France each ministry produced its own annual volumes— thus we find compendia from Justice, Education, and the like—but the ministries themselves took care to coordinate their reports. In Prussia there is the ideal of a central statistical establishment. That was commenced in 1810, but achieved true centrality only in 1860. The Prussian Statistical Bureau shortly after then became the model for all national bureaux. I describe it in some detail later in this volume.

Of course a new institution does not a revolution make. It is only a symptom that I elevate into a necessary condition. Undoubtedly and big scientific revolution must be connected with substantial social change. Notoriously the scientific revolution of the seventeenth century went hand in hand with the emergence of bourgeois society. Similarly Kuhn's second scientific revolution is intimately connected with the rise of capitalism and with the industrial revolution (however that be periodized). Arguably the thrust toward measurement arose from industrial problems. We know that the science of thermodynamics grew out of the early steam engine. A fundamental concept was the "duty" of an engine, a measure of its efficiency. The mine manager needed to know whether it was better to use a steam engine or draught horses to haul tin up from his mine. So he had to measure.

Do we find comparable social events that would accompany our suggested probabilistic big revolution 1800–1930? All too many, alas. In connection with examples already given, I would emphasize the philanthropic utilitarianism that worked hand in hand with the new statistics. The sanitary movement is an example. It is no small example. It increased life expectancy more than anything in history. The marvels of modern medicine produce modest increases in life expectancy that are peanuts compared to that coconut of an increase provided by the sanitary movement and its band of well-meaning statisticians. Most people like myself, whose ancestors lived in an urban slum sometime in the nineteenth century, would simply not *exist*, had there not been the sanitary movement. Our ancestors would have died before breeding.

In the end, a big revolution must go along with a change in our attitude to the world. Even as Namieresque a historian as Herbert Butterfield wrote that "the" scientific revolution was accompanied by a change in our sense of the "texture" of the world, in a different "feel" for the world. It would have to be argued that in the course of Kuhn's second scientific revolution, people came to regard much of nature as essentially numerical and measurable. Such a case is rather easily made. The same is true for a possible probabilistic big revolution 1800–1930. I mentioned some of the symptoms in my opening paragraph. The world is no longer deterministic. It is pervaded by lawlike chance. We cannot regard an action as rational unless it computes the probabilities. Beliefs are accompanied by probabilities. This is an ongoing process. When I was a boy I inanely wanted to be a weatherman. The textbooks said that a sound meteorologist should never say that a thunderstorm is probable or that fine weather is likely. Such an utterance would be a tautology, or a pleonasm, for of course all forecasting is only probable. The weatherman should simply predict thunder or fine weather. Now all is changed. When was the last time you heard a weather forecast not tricked out in probabilities?

I have sketched some similarities between a probability "event" 1800–1930, and the first and second scientific revolutions. But nagging doubts remain. Perhaps the most important is the most obvious. Cohen ironically remarks that if there were such a probabilistic revolution, it would be the longest revolution in history. We do not take seriously the official doctrine that Mexico has been in a state of permanent revolution since 1910. My own examples of institutions that bolster up the claim to a probabilistic revolution occur at specific times: a bureau founded in 1810, and reformed in 1860; statistical clubs springing up in the 1830s; the congress period of the 1850s and '60s. These give no credence to a revolution occupying 130 years. Revolutions, including political ones, may, however, occur in stages. We may restore some credibility to the doctrine of revolution if we think of articulated revolution, in which we break down the impression of overwhelming long-run change into a series of segments.

There is no need to imagine that there is a unique segmentation of the period 1800–1930 from the point of view of probability. People segment the industrial revolution in dozens of different ways, according to their interests. So I can at most offer my articulation of the stages, deriving from my research problems.

I have two complementary labels for what is of most interest to me: *the taming of chance* and *the erosion of determinism*. The latter is what first strikes a philosopher. In 1800 we are in the deterministic world so aptly characterized by Laplace. By 1936 we are firmly in a world that is ultimately indeterministic. But the former label, the taming of chance, stands for the deeper thought, because determinism was eroded precisely by making chance manageable, intelligible, existent, and governed by laws of probability. Chance, which, for Hume, was "nothing real," was, for von Neumann, perhaps the only reality. How did this transformation occur? I find it convenient for purposes of analysis to say that it occurred in four stages, as follows. I do not mind the fact that my stages overlap. Notes refer to essays in which I say more about one of the segments.

1. *The avalanche of printed numbers* (1820–1840).   The practice of printing in general increased linearly after the Napoleonic wars, although the slope was steep. But the practice of printing numbers increased exponentially during the same period. This new fetishism for public numbers is not unconnected with Kuhn's second scientific revolution and the development of measurement. But it is most

striking not in Kuhn's sphere of the physical sciences, but in what were to become the social sciences. For the first time it was possible to perceive (seeming) regularities in facts about human behavior, and to model them by probabilistic laws. This was itself a revolutionary change.[12]

2. *Faith in the regularity of the numbers* (1835–1875). Once the numbers were there, probability laws could be investigated. But at this stage, determinism was not in trouble. Quetelet in particular saved the day by proposing that the biological and social phenomena were primarily distributed like the Gaussian law of errors. That in turn is derived as a limiting case of the binomial distribution, and it was thought that one could explain the binomial distribution as the result of interactions among small independent deterministic causes. The mathematical establishment of this claim occupied many of the better mathematicians, culminating, perhaps, in Poincaré much later. At this stage probabilistic laws were not seen as a threat to determinism. On the contrary, there was much hue and cry about statistical determinism that seemed to preclude free will![13]

3. *The autonomy of statistical law* (1875–?). During the second stage it was an essential part of the doctrine of chances that there was always an underlying causal structure. It was the task of the analyst to find that structure. In this third stage people become indifferent to that. I call something autonomous if it can be used to explain something else, without itself having to be reduced. Thus, for example, Galton explained the apparent fact that outstanding parents (in height, intelligence, or whatever) had less outstanding offspring by the fact that regression toward mediocrity is a logical consequence of the "fact" that the trait in question is normally distributed. Reduction of this normal distribution to an underlying causal structure is simply irrelevant to this explanation. Galton's statistical laws are autonomous. At this stage, not coincidentally, we find the first serious tests of statistical hypotheses in the modern sense of testing. That is due to Lexis, in 1875, and testing procedures develop apace. By the 1890s we find the first serious philosophical statement of modern indeterminism. The author was the cantankerous C. S. Peirce, and at first hardly anyone took him very seriously. For my part, I would fill in the "?" in my dating scheme by 1897, the year that Durkheim published *Suicide*, but I grant that my choice is eccentric.[14].

4. *Possible to actual indeterminism* (1892–1936). Here, for the first time (in my opinion) physics becomes central to the taming of chance and to the erosion of determinism. I do not underestimate the statistical mechanics of Maxwell and Boltzmann, but Einstein unwittingly did much to make their mechanics real, rather than instrumental, when he and others in 1905 derived the Brownian movement from propositions of statistical mechanics. More important, we have the quantum theory evolving, a story more familiar than my earlier stories that tend to focus on dusty bureaucrats.

Justification of this periodization requires a book, not a page. You will notice that my four examples of *Structure*-style scientific revolutions, given earlier, fit the above four segments. One is drawn from each segment. I think this is no accident. I think that the microrevolutions will all be found to be internal to one or the other of the segments, and will not cross segments. If this is true, it adds to the objectivity of my articulation.

But do the segments add up to a revolution? If they do, it is certainly not to what Cohen calls a revolution in application. There was *also* a revolution in application, but my segments are changes in ontology, in our vision of the world in which we live. That is perhaps what matters to the essays in these two volumes. We need not go on further quibbling about whether we have a scientific revolution in the sense of *Structure*, or in the sense of Kuhn's opening chapter here, or a revolution in science, in the sense of Cohen above, or something like Kuhn's second scientific revolution. What is clear, beyond all scholasticism, is this. The taming of chance and the erosion of determinism constitute one of the most revolutionary changes in the history of the human mind. I use the word "revolutionary" not as a scholar but as a speaker of common English. If *that* change is not revolutionary, nothing is. That is the real justification for talk of a Probabilistic Revolution 1800–1930.

Notes

1. F. P. Ramsey, "Truth and Probability," in *The Foundations of Mathematics and Other Logical Essays by Frank Plumpton Ramsey*, ed. R. B. Braithwaite (London: Routledge and Kegan Paul, 1931), pp. 158–198. All the central probability ideas were revived at about the same time, in roughly the same place; see my "The Theory of Probable Inference: Neyman, Peirce and Braithwaite," in *Science, Belief and Behaviour*, ed. D. H. Mellor (Cambridge: Cambridge University Press, 1980), p. 142.

2. Ian Hacking, review of I. Bernard Cohen, *Revolution in Science* (Cambridge, MA/London: Belknap Press, 1985), in *New York Review of Books*, February 27, 1986, and cf. October 9, 1986, pp. 56–57.

3. Ian Hacking, "Was There a Probabilistic Revolution, 1800–1930?" in *Probability since 1800*, ed. M. Heidelberger et al. (Bielefeld: Universität Bielefeld and B. Kleine Verlag, 1983), pp. 489–493. This earlier version also contains methodological discussions omitted from the present paper.

4. Thomas S. Kuhn, "What Are Scientific Revolutions?" p. 9.

5. Kuhn, "What Are ...," p. 20.

6. Kuhn, "What Are ...," p. 20.

7. Adolphe Quetelet, "Sur l'appréciation des documents statistiques et en particulier sur l'appréciation des moyennes," *Bulletin de la Commission Centrale de Statistique*, 2 (1845), 258–262.

8. Francis Galton, *Memories of My Life* (London: 1908), p. 300.

9. J. Neyman and E. S. Pearson, "On the Use and Interpretation of Certain Test Criteria for Purposes of Statistical Inference," *Biometrika*, A20 (1928), 175–240 and 263–294; and "On the Problem of the Most Efficient Tests of Statistical Hypotheses," *Philosophical Transactions of the Royal Society*, A231 (1933), 289–337.

10. Thomas S. Kuhn, *The Essential Tension* (Chicago: University of Chicago Press, 1977), p. 220. (The paper in question was given in 1960 and published in *Isis* in 1961, before *The Structure of Scientific Revolutions* was published (1962).)

11. I would add to Cohen's list an essay that more explicitly takes off from Kuhn himself: Kenneth L. Caneva, "From Galvanism to Electrodynamics: The Transformation of German Physics and Its Social Context," *Historical Studies in the Physical Sciences*, (1978), 63–159.

12. Ian Hacking, "Biopower and the Avalanche of Printed Numbers," *Humanities in Society*, 5 (1982), 279–295.

13. Ian Hacking, "Nineteenth Century Cracks in the Concept of Determinism," *Journal of the History of Ideas*, 44 (1983), 455–475.

14. Ian Hacking, "The Autonomy of Statistical Law," in *Scientific Explanation and Understanding*, ed. Nicholas Rescher (Lanham/London: University Press of America, 1983), pp. 3–19.

II CONCEPTS

4 The Slow Rise of Probabilism: Philosophical Arguments in the Nineteenth Century

Lorenz Krüger

Probabilism is taken to denote the view that statistical laws may be fundamental in scientific explanation. The remarkable time lag between the rise of statistical practice and the recognition of probabilism demands an explanation. It is argued that philosophical problems and arguments, whether developed or employed by philosophers or by scientists, will help to make such an explanation intelligible. Three themes are analyzed: (1) Determinism acquired a particularly stable position around 1800 due to the fusion of traditional epistemology and recent successes of mechanical science. (2) The notoriously unclear notions of causation proved to be a rich source of deterministic accounts of statistical regularities. (3) The regular coordination of instances falling under a statistical law is hard to understand. The analysis of these themes is illustrated by historical examples from Kant and Laplace to Maxwell.

1 Introduction: Philosophical Aspects of the Rise of Probabilism

This essay is not meant to tell a story about the actual development of probabilistic thinking or statistical practice. Even if successful, it can contribute to such a story only indirectly. I hope it will do at least the latter by collecting and connecting philosophical arguments that occur at different stages of the transition from a deterministic to a probabilistic attitude in the natural and social sciences. This transition, we may assume, has been brought about by many different factors belonging to various kinds, all of which a full explanation would have to take into account. Yet, since the transition is an intellectual transformation, arguments will be among the causal factors shaping it; and since the transformation is of extraordinary depth, these arguments are of a kind that is usually called 'philosophical', whether philosophers or scientists happen to be their authors or users. To review, to interpret, and to examine such arguments may therefore help—indeed may be indispensable—to make the transformation intelligible. Moreover, I take them to be causally efficient, first of all for the simple reason that they are employed, and reacted to, by the actors on the historical scene, but also, second, because the history of the sciences would not be the history of a rational enterprise if reasons did not figure as causal factors in it.

To say all this, however, is less than claiming that the causal connections of reasons are eventually made out in the following essay itself. Except for a few direct quotations of one important author by another, the material offered here should rather be viewed as a preparatory collection of concepts and arguments that displays a certain inherent logical structure. My hope is that this or a similar conceptual pattern will be useful heuristic tool for working out a fuller and more concrete history of probabilism without getting lost in too many contingent and unilluminating details. But even the conceptual preparation will have to be extended to include other disciplines, especially chemistry and physiology, which lie beyond the scope and page limit of this chapter.

I want to mention another feature that is specific to the history of probabilistic thinking and that speaks in favor of the influence of philosophical ideas and arguments: although, in the course of the nineteenth century, statistical practice and a corresponding amount of probability theory spread rapidly through various disciplines, most scientists and philosophers remained opposed to taking probability as fundamental or irreducible, or according it an explanatory function. Indeed, that function appeared to imply the reality of possibilities or of indeterminateness; hence to recognize it seemed to involve too high an ontological price. At any rate, there is a time lag between the widespread use of statistics and probability on the one hand and the adoption of a probabilistic view of, or attitude toward, reality (or parts of reality) on the other. This delay seems to me to indicate the fundamental, or, if you like, philosophical, character of the intellectual change discussed in these volumes.

The time lag also indicates the fact that most of the arguments in question were directed *against probabilistic views*, or intended to *support determinism* in the face of statistical appearances. The philosophical history of probability is, in large stretches, intellectually conservative. This circumstance turns determinism into an *explanandum* rather than a mere starting point of the story: in order to understand the slow rise of probabilism one has to inquire into the sources of determinism, too.

2 The Attractions of Determinism

At least three sources of determinism deserve to be listed: (a) the success of mechanics, (b) the ideal of knowledge, or of scientific explanation and understanding, and (c) the desire for domination over nature.

2.1 Newtonian Mechanics in Laplacian Perspective

The overwhelming success of Newtonian mechanics induces the belief in the universal validity of mechanistic philosophy, whose core is the explanation of phenomena in terms of a mechanism in which the whole succession of motions is determined by the state of the moving system at any particular time (example: ideally precise clockwork). But what is the argument, if any, that could be advanced either to produce or to back up this belief? It is as crude as it was effective: (i) A number of phenomenal changes have been explained successfully in terms of matter in motion, where matter is characterized by one intrinsic magnitude, mass, and the motion explained by forces that, in turn, were correlated uniquely with an observable configuration of matter. (ii) At least one important subset of these phenomena has proved accessible to quantitative determination without known limits of precision and, within the actual limits of precision, is predictable over time. (iii) Therefore, we are entitled to assume that all phenomena are, at least in principle, no less accessible to such prediction, even though it may be difficult to work it out in practice.

According to this argument the belief in determinism reduces to the belief in the truth of a theory that has some important and impressive applications, first of all in planetary astronomy. The latter belief, in turn, can, as far as *this* argument goes, only be based on the empirical success of the theory. This success, however, is limited in actual precision and in its actual domain of application. This weakness of the argument is not just the weakness of every empirical generalization; the present generalization jumps the heterogeneities of the phenomena and it extrapolates to the ideal limit of infinite precision, which we must, as a matter of principle, consider as unattainable. And, of course, unnoticeably small uncertainties in the initial conditions of a mechanism may entail large differences in the phenomena after a sufficiently long time.

How could the empirical argument for determinism ever have been persuasive, nevertheless? A possible answer seems to be this: true, the success of Newtonian physics was confined to a limited area, but there it surpassed everything the students of nature had ever hoped for. Was this a "mere obligingness on the part of the solar system," which provided a model that not only satisfied the mechanical theory perfectly but also happened to display a stable history throughout recorded time, i.e., a history in which no impredictable event—say, the explosion of a planet—occurred? If so, "the high success of Newton's astronomy was in one way an intellectual disaster: it produced an illusion [i.e, the belief in determinism] from which we tend still to suffer."[1] But the psychological impact of Newton's successful alliance with the obliging heavens cannot be expected to be the only source of determinism, not even for an astronomer like Laplace, the master of nineteenth-century necessitarians.

2.2 The Merger of Epistemology and Science: Kant and Laplace

Since antiquity the ideal of knowledge has been tinged by determinism, though in an epistemological sense of that term. Knowledge, as opposed to opinion, was thought to require that we can be certain about the object of our knowledge, which implies that we know of the object that it cannot be otherwise than it actually is.[2] We cannot know this, in turn, without knowing the why,[3] i.e., a sufficient reason. For Leibniz the principle of sufficient reason was the "determining" principle of all actual existence.[4] Moreover, only someone who knows a sufficient reason of a thing can explain this thing or, what was often taken to be equivalent, can fully understand it.[5] Kant interpreted Leibniz's principle as that of causality,[6] which means that he restricted the legitimate use of the principle to the domain of experience, and hence the admissible reasons to antecedents in time, or causes. In this way the explication of the concept of knowledge and the most successful form of empirical science began to merge. Furthermore, the Aristotelian condition of knowledge was not only further specified as a condition for the possibility of *empirical* knowledge; the latter condition, according to the basic idea of Kant's transcendental idealism, was simultaneously conceived of as a necessary condition of the (existence of) *objects* of experience.[7] This identification of the conditions of the possibility of experience and the conditions of the possibility of the objects of experience achieved two things at the same time: it conferred upon reality the conceptual "necessity" of epistemology, and, conversely, it elevated the predictability of events that was actually found in science to the status of a universal truth about the world.

As a consequence of this philosophical *tour de force* the problem of possible limits of the applicability of mechanics disappeared. Kant believed that the empirical science of external nature *in general* could be founded on the *a priori* construction of the concept of "the movable in space." [8] Given this belief, it must have been all the more plausible for Kant to view determinism as a mere corollary of his causal version of the principle of sufficient reason: "All that happens is hypothetically necessary," that is, necessary, given certain antecedents. Another version of this proposition which Kant adds at once is, "Nothing happens through blind chance (*in mundo non datur casus*)." [9] Chance is excluded not only from the domain of scientific concepts but also from the world of events; laws of chance or probability theory cannot possibly refer to reality.

Does this view contradict the existence of the contingent? Does Kant imply that everything that exists is necessary? Of course not. He claims the following proposition to be analytic: ". . . every contingent existence has a cause." [10] In other words: Kant identifies the concepts 'contingently existing' and 'causally determined' with each other. The argument he adduces in this connection is worth noting, since it might be used outside the framework of transcendental theory as well: If the concept 'contingent' is to have empirical meaning, that is, if we want to present actual instances of the contingent, i.e., nonnecessary phenomena, we must always refer to change and not only to the mere possibility of *thinking* the opposite. For instance, if a moving body comes to rest, it is evident that the opposite properties 'moving' and 'at rest' are *both* empirically possible. But this in itself does not yet show that the movement was contingent when it occurred. In order to know *that*, we must know that the body could have been at rest during the time when it was actually moving. Now, this knowledge can only be derived from an analysis of the conditions of movement and rest, including the laws of motion. With the help of these laws we may indeed conclude that, at a time when the body was actually moving, it would have been at rest if certain conditions, e.g., forces or positions of other bodies, had been different. "And in this way," Kant concludes, "one recognizes the contingency (*Zufälligkeit*) from the fact that a thing can only exist as an effect of a cause; hence, if a thing is assumed to be contingent, it is analytic to say that is has a cause." [11] This subtle non sequitur dissolves the contingent into the necessary, of course not into the absolutely necessary but the hypothetically necessary. The existence of the latter is compatible with the contingency of the empirical world as a whole.

Kant's theory impressed many scientists and philosophers in the nineteenth and still in the twentieth centuries. Even where his transcendental idealism was not accepted or not even properly understood, the fusion of the rationalistic epistemological tradition with Newtonian empirical science greatly strengthened the case of determinism. Laplace is an instance of such a fusion, if only on a more intuitive basis. The first section of his *Essai philosophique sur les probabilités* [12] starts with the sentence, "All events, even those which, by their insignificance, seem not to follow from the great laws of nature, follow from them just as necessarily as the revolutions of the sun." Here our first source of determinism is clearly operative. A few sentences later, however, the second source is also exploited: "The present events and those that precede them have a connection that is based on the evident principle that a thing cannot begin to be without a cause producing it. This axiom, known under the name of *principle of sufficient reason*, extends even to actions that one judges to be indifferent." The paradigms of such actions are, of course, the deeds of free agents. But even these are determined by motives. After this brief side-glance at an alleged counter-example Laplace draws his well-known conclusion: "We must therefore view the present state of the universe as the effect of its preceding state and as the cause of the state that will follow." [13] The subsequent sentence introduces his famous demon.

How does Laplace relate the two sources of determinism? Unlike Kant, he does not derive their identity from a philosophical theory of knowledge. Instead he restricts himself to a few vague hints. In astronomy, the human mind offers a feeble image of the omniscient demon, or: it is a specific tendency of the human race, one that ranks man above animals, to strive after the cognitive ideal represented by the mechanical intelligence of the demon. Passages like these suggest that epistemological reasons for determinism are fundamental and empirical successes only additional confirmations. But a page or so further on it is again an astronomical event, viz., Clairaut's corrected prediction of the return of Halley's comet, which spurs the sweeping conclusion, "The regularity that astronomy shows us in the motion of the comets takes place, without any doubt, in all phenomena." Without *any* doubt? In *all* phenomena? Laplace seems to have been serious. He explained at some length how we can conceive psychology as "a continuation of the visible physiology" that is already known. As the object of the extrapolated physiology he introduced the *sensorium*, the seat of our thoughts. He modeled it, roughly speaking, after a system of resonators coupled to the environment. Its vibrations should than be subjected to the laws of dynamics, which assumption he thought confirmed by experience, so that the analogies we discover between material objects and intellectual phenomena turn out, at bottom, to be identities. [14]

How did the eminent scientist Laplace persuade himself by such reasoning? It may help our understanding to observe his motives. Moreover, these will point the way to the third source of determinism. The example of Clairaut's calculation and the dynamical theory of the *sensorium* have a common background in Laplace's desire to forestall ignorance, error, prejudice, and superstition. Earlier appearances of Halley's comet had been taken to announce imminent disaster; only the prediction of its path from general laws could break these childish apprehensions. The *sensorium* theory is introduced in a section entitled "On the Illusions in the Estimation of Probabilities." Laplace quickly broadens this subject to include all sorts of false beliefs and unfounded fears. He sees them all connected with physiological causes, but for this very reason also mechanically destructible by actually performing contrary thoughts and actions. [15] Thus, the program of the Enlightenment, Laplace thought, was to be carried out by recognizing and utilizing the laws of the physical and intellectual worlds. [16] On this foundation only can natural judgments of probability be explained and due corrections be judged *and* produced.

2.3   The Practical Importance of Determinism

At least since Francis Bacon the knowledge of laws was considered as a means of dominating nature. This view is natural enough for an experimentally based discipline, which from its beginning was dependent upon technology and vice versa. It is naturally extended to nonexperimental disciplines (e.g., social science) if the latter are modeled upon the former. Even if intervention or production should not be possible (e.g., in astronomy), prediction seems the indispensable minimum of a useful science; otherwise the available knowledge would not even permit us to adjust ourselves to the circumstances. It is sometimes said that the one and only goal of science is the ascertainment of laws; but that is, of course, an exaggeration. It is even true that all practical applications involve some location or other in space and time, if they are not completely restricted to a single such location; but particular instances are only tractable to the extent to which they are determined by previous circumstances, and only usable to the extent in which they determine future events. In short: knowledge-based action can only be secure if it employs deterministic laws—a practical non sequitur, to be classed with the theoretical ones we have already encountered, again subtle enough not to have been obvious. But even if it were, one could still search for laws to improve one's adjustments and actions. To assume insurmountable limits to such improvement, therefore, tends to frustrate a central human interest that, once the possibility of its satisfaction had become plausible, was unlikely to be easily abandoned. *Methodological* determinism was and will remain attractive.

The relevance of this point to the case of probabilism may be illustrated by the debate about the single case. Two authors who disagreed fundamentally in their views concerning probability in most other respects, De Morgan and Venn, nevertheless both subscribed to the assertion that the single case can be an object of (future-related) knowledge only if it is lawfully determined.[17] Limits of determination are then limits of knowledge, hence unacceptable to those who seek knowledge for the sake of enlightened action.

3   Statistical Regularities and Causation

According to the above account, determinism has the peculiar feature, not atypical for philosophical rationalism, of merging conceptual and empirical elements into a unified whole, as if they were not different in nature. This strategy of the determinist appears to have significantly influenced the thinking about observed regularities or uniformities. To a determinist all regularities must naturally appear as expressions of the lawlike character of reality, hence as something to be explained within the framework of a deterministic theory. Regularities that are only statistical are then an intricate problem. But it was difficult to recognize the true nature of this problem, or even to see a problem at all. I shall attempt to illustrate this point by analyzing the connection that was established between lawlike statistical distributions and causation. It would be useful to investigate the use of probability for the analysis of causes in general. But I shall restrict myself to the explanation of stable averages in terms of constant (or regularly varying) and accidental causes, since this explanation played an important role for the reconciliation of statistical practice with deterministic theory, and hence also for the time lag between that practice and the rise of probabilism.

A well-known passage from Laplace may be taken as a starting point.[18] Laplace discusses an urn filled with black and white balls from which a series of drawings with replacement are made. With respect to this example he states Bernoulli's theorem, and then proceeds immediately to what he calls "a consequence" of it and wants to consider as a "general law": "... the ratios of the effects of nature are very nearly constant if these effects are considered in large number." After a few examples from nature and social life he notes a second "consequence": "... in a series of events continued indefinitely the actions of the regular and constant causes must, in the long run, get the better of that of the irregular causes." The application of this consequence to the example of the urn immediately discloses a problem that did not even cross Laplace's mind, though it was clearly seen by others somewhat later: Each drawing, hence also the entire series of drawings, is fully determined by the motion of the system, the shaking of the urn and the movement of the hand in drawing and replacing. These movements themselves may be conceived to be fully determined by certain forces, e.g., gravity, elasticity, friction, electrical impulses in the brain of the person who performs the drawings, etc. But none of these dynamical elements shows up in the effects—apart from the fact that every so often a ball appears from the urn and disappears again into it. The ratio of the two effects 'white ball drawn' and 'black ball drawn', however, cannot be attributed to any dynamical cause operative in the process. The constant "cause" that brings about this ratio is, of course, the ratio of white to black balls in the urn. This is clearly a real condition of the dynamical process, which will therefore figure in a complete description of it, but we may call it a "structural" condition, since it is not a dynamical part of the history of the system. Laplace's omniscient demon would be able to predict the outcome of each drawing, hence also the correct ratio of the white to black drawings, but should he care to inform us about his calculations, he would not find any occasion to mention the numerical composition of the urn as an element from which he deduces anything. This is no surprise because there simply is not any *dynamical* connection between the composition of the urn and the actual drawings. We conclude therefore that if the composition is a cause at all, it is a cause in a sense that is very different from the sense the dynamical determinist has in mind. He has not yet discovered the incoherence between his astronomical realization of the idea of determination on the one hand and the actual nature of some of his favorite phenomenal correlations on the other. One of the reasons, then, for the slow rise of probabilism is to be sought in the fact that this incoherence was not perceived.

It would be interesting to investigate why it was not easily perceived, but I cannot do this within the scope of this paper. I should expect to find at least three sources of the confusion: first, the general tendency to turn limited scientific findings into metaphysical truths about the world; second, the amalgamation of epistemological and scientific versions of determination mentioned above and found in leading figures like Kant and Laplace; and third, the notoriously obscure nature of causation. A fourth source may have been the lack of discrimination between determination in limited systems and Laplace's determinism concerning the whole universe, a point to be discussed below.

Laplace's doctrine of constant and irregular causes was to become standard for a long time. It commanded a central position in Quetelet's social physics,[19] and from there it infiltrated the writings of most nineteenth-century social scientists and historians concerned with statistics.[20] Rather than giving a survey of this story, I want to stress two further features of this approach. First, it needs a piece replacing the missing dynamical link between the constant cause and its effect. Laplace finds it in what we may call *the principle of the development of probabilities*. He speaks of "the development of the respective probabilities of simple events" and claims that "they must present themselves more frequently when they are more probable,"[21] where the probability is of course to be measured independently in terms of equipossibility of alternatives with respect to which we are in a state of equal undecidedness ("également indécis").[22] From this principle Laplace deduces his interpretation of Bernoulli's limit theorem, an interpretation that was later attacked by empiricist critics who would not accept the rationalistic identification of conceptual alternatives with real possibilities, or of the development of possibilities with causation. The principle reappears, however, in a number of later authors.[23]

What is perhaps a more remarkable case is a version of the principle used by Boltzmann in order to circumvent a dynamical derivation of the actual motion of a gas toward thermal equilibrium. At first, he believed he had given "an exact proof" of the law of entropy on purely mechanical grounds (his famous H-theorem of 1872);[24] but he later learned from Loschmidt's reversibility objection that he had to refer explicitly to probabilistic arguments. In a paper of 1877 he used the mechanical theory for no more than the selection of the appropriate sample space, i.e., phase space, for which he then did his probabilistic calculations. The calculations are based on the assumption that the different phases, or microstates, play the role of Laplace's equipossible simple events, so that complex events, i.e., given macrostates, are more or less probable according to the greater or smaller number of microstates by which they can be realized. Once having made the transition from mechanics to probability theory, he advances his analogue to Laplace's development of probabilities in the following form: "In most cases the initial state will be very improbable; the system will pass from this through ever more probable states, reaching finally the most probable state, ..."[25] This assertion seems intuitively no less evident than Laplace's reading of Bernoulli's theorem. But there is, of course, no a priori reason why a mechanical system could not remain in improbable states or even reach such states from more probable ones; Loschmidt's objection had just shown that, on purely mechanical grounds, this was perfectly possible. In the late nineteenth century an eminent physicist like Boltzmann was still under the spell of the (con)fusion of mechanical determinism and conceptual evidence, which was so characteristic for the classical view of probability.[26]

Another point related to Laplace's doctrine of constant and irregular causes is this: As soon as the mechanical concretization of causation is disregarded, the constant "cause" may be advanced from the status of a structural condition to that of a normal cause. The outcome of drawings from an urn can be influenced either by changing the "mechanics" of mixing and grasping the balls or by changing the composition of the urn. Now, in the most important application of the doctrine, i.e., in social statistics, there is not much point in distinguishing mechanical from structural causes; they are both factors that influence the phenomena and may be subjected to manipulation. In this sense, then, they are on a par. Indeed, actual manipulation via structural causes, e.g., legislation, will appear more important, and hence be the dominant, if not the only, object of interest for the social scientist.[27] But someone who is after the real dependence of social phenomena on structural causes will not notice that he does not have any (quasi-)mechanical models of this dependence; he is unlikely to realize that those models need not be deterministic, let alone to think that they may turn out to be conceptually incoherent with deterministic theories about individual and collective life histories. How could a social scientist have anticipated a problem of Boltzmann's type?

At least one requirement had to be fulfilled first: the discovery of the conceptual disparity between causal and probabilistic analyses. Steps in this direction were made by some well-known critics of Laplace with empiricist leanings in philosophy. The next section sketches their contributions; it will also illustrate that this critical line of argument could hardly be successful short of inventing an entirely new paradigm for probability theory: the frequency notion of probability.

4   Distinguishing Statistical from Causal Investigations

4.1   Laplace Criticized: Jakob Friedrich Fries

An early critic of Laplacian views of probability was Jakob Friedrich Fries. In his philosophy he was a Kantian but criticized the rationalistic mistakes of Kant's own way of establishing the critical philosophy. Thus he was prepared to spot the problem discussed above when, late in his life, he approached the subject of probability.

A concise version of one major point is found in the *Selbstrezension* of his *Versuch einer Kritik der Principien der Wahrscheinlichkeitsrechnung* of 1842.[28] He stresses that induction involves nonnumerical "philosophical probability," which must not be confused with "mathematical probability," as was done by Condorcet, Laplace, Lacroix, and again recently by Poisson. Mathematical induction based on Bernoulli's theorem or Poisson's generalization of it must not be used in the discovery of natural laws from observation. Why not? As a striking example of the confusion of his predecessors Fries presents Laplace's rule of succession, according to which the probability of the sun's rising p more times after it rose m times in the past is equal to $(m + 1)/(m + p + 1)$. If the revolutions of the sun obey a law of nature, the formula should be valid for an arbitrarily large p, so that for any given m the probability of the law would be zero. Hence, if we are entitled to make inductions from a finite stock of observations to general laws at all, mathematical probability cannot be their foundation. But the paradigm case of Bernoulli's theorem, that of games of chance, warrants an even stronger conclusion: "This entire method of calculation of the mean subjective probability a posteriori is always based on the background assumption that, within its scope, no necessary laws of nature are valid, but rather that an open range still remains possible for the alternation of unknown equipossible cases."[29] In other words: probability considerations and causal determination are complementary; the former are only admissible where the latter has not (yet) succeeded. Lawlike connections must be absent or destroyed, e.g., by mixing the cards or shaking the die, in order to make room for valid applications of probability theory. Fries appears to have seen that statistical regularities, far from being explainable in terms of causes, are not even compatible with the possibility of a complete regulation of the relevant events by determining laws.

Nevertheless, Fries has no claim to having grasped the full depth of the problem. If he talks about a remaining open range ("Spielraum") for alternative cases, he does not mean to imply indeterminism. The outcome of rolling a die is "already fully decided in nature"; we are only ignorant about it.[30] Although we cannot determine the additional conditions of the single case, we may be able to partition the sphere of the possible quantitatively because we can "count how many different kinds of them [i.e., of determining conditions] are possible and how frequently, in general, one kind will occur in proportion to another. In this way we can divide the extension of our knowledge into a certain number of equally possible cases, correspondingly calculate what must happen in the majority of the cases, and direct our surmises accordingly."[31] The reappearance of equipossible cases makes it difficult to interpret counting as an empirical activity, the more so since Fries, immediately after the quoted passage, embarks on an extensive discussion of games with their typical a priori probabilities. But then the phrase that something "*must* happen in the majority of cases" comes dangerously close to Laplace's principle of the development of probabilities.

4.2 The Discovery of a New Explanandum: Robert Leslie Ellis

A straightforward critique of this principle and of Laplace's interpretation of Bernoulli's theorem was offered by Ellis[32] almost simultaneously with Fries's book in a paper read in 1842. Like Fries, he rejects Bernoulli's theorem as a foundation of inferences to natural laws. But unlike Fries, not being under the spell of Kant, he thinks that the theory of probability is compatible only with "sensational philosophy." Counting for him is, therefore, an *empirical* activity. Now, if this activity is essential in any determination of probabilities, it becomes conceptually true that what is more probable will occur more frequently in the long run. Bernoulli's theorem is therefore unnecessary, which is all for the better, since it cannot perform the job Laplace had assigned to it, viz., in Ellis's words, to serve as "a demonstration of a general law of nature," i.e., what Laplace meant by the law of constant ratios of effects in the long run. By removing the alleged foundation of such a law Ellis restores its character as an explanandum: why is it that we count stable relative frequencies under certain conditions?

In the 1842 paper Ellis does not have much more to offer than the traditional reply to this question. It is twofold; in Ellis's words:[33] (i) We base our expectations of stable frequencies "not on the fortuitous circumstances of each trial, but on those which are permanent"; *prima facie* this looks like the usual reference to constant causes. (ii) Our judgments, in such cases, "involve the fundamental axiom that, on [sic] the long run, the action of fortuitous causes disappears"; this phrase refers to the compensation problem of causes to be discussed in a subsequent section. What is to be noted at this point, however, is the fact that Ellis later attempts to reinterpret (i) in *noncausal* terms and that he connects (i) with (ii) in a unified account of the possibility of statistical regularities. Moreover, it was already clear to him in 1842 that (i) and (ii) are not consequences of probability theory, e.g., to be read off from Bernoulli's theorem, but presuppositions of any meaningful application of the theory to reality.

Whereas Fries, having made the distinction between statistical and causal analyses very clearly, did not have any philosophical view to back it up, Ellis developed a novel and peculiar account of it. In 1854 he returns to what he now calls "the fundamental principle of the theory of probabilities."[34] The factual content has, of course, not changed; the application of probability rests on the existence of stable frequencies in the long run. But the causal account of this fact is replaced by an account in terms of genus and species: In order to do statistics we need "similar trials"; e.g., we refer to a certain "genus of phenomena to which the different results are subordinated as distinct species."[35] Contingency is located at the level of individuals; it disappears gradually as we overcome the narrow limitations of the single case, which is determined by circumstances extraneous to the idea of the genus and its species. Here the causal talk still lingers but the principle of probability theory has been completely cleansed of it. Ellis says that "the fundamental principle of the theory of probabilities may be regarded as included in the following statement;—'The conception of a genus implies that of numerical relations among the species subordinated to it.'"[36] He adds at once that we have to conceive this statement as telling us something about the nature of reality if probability theory is to be a "real science." Without going into the reasons Ellis offers for his realism (which would lead us too far afield), we may note that he confronts "the idea," i.e., the idea of the generic structure of reality, with "the fact," i.e., something that, on our dark earth, can only be a partial realization of the idea. Causes figure therefore as no more than disturbances of the Platonic structure. This structure of genera and species, instead of a causal or dynamical structure of reality, is responsible for the existence of statistical regularities.

The last statement, however, cannot be upheld without a serious qualification. Since there are disturbing causes on the level of the individual, how can the genus be preserved in spite of them? We need not, says Ellis, conceive the genus after the model of a game of chance with strictly independent trials. The progressive alteration of the "permanent circumstances" (Ellis's term for what used to be called "constant causes"), which we observe even in rolling dice, may be such "as to restore the balance which the result of past trials has disturbed."[37] So far we have paid but little attention to such "balancing tendencies"; now we ought, he thinks, to develop this branch of investigation, e.g., in climatology.

So, there is a *causal* problem after all, indeed no other than the familiar one of the mutual compensation of accidental causes. But it is put in an interesting new framework: that of a dynamical theory of stable structures or equilibria. *A new kind of explanandum appears* on the scene, an explanandum the classical determinists had not been aware of. The dazzling success of mechanics had made it almost impossible, or so it appears, to see the deep dissimilarity between two complementary tasks of the science and philosophy of nature: that of explaining (and predicting) events on the one hand, and that of explaining states or kinds of matter or things on the other.[38] Since it was particularly difficult, given the history of classical physical science, to recognize this dissimilarity, the need for novel conceptual elements that did not fit deterministic patterns was not easily perceived, be it thermodynamics, the theory of atoms, or social science.

4.3   Natural Kinds as an Empirical Fact: John Venn

John Venn, who was the first to work out Ellis's frequency view of probability into a full-blown philosophical system, took a further step in the direction Ellis had taken. He once again took issue with Laplace's principle of the development of probabilities. He grants that the principle may give empirically correct results in the special case of games, if only thanks to further conditions (e.g., mixing), but he calls it "unmeaning" when applied to facts of nature in a more general form.[39] In cases that are less artificial than the ideal coin or die, the complementary conditions or "agencies" are clearly on a par with the properties of the object under investigation; and here he thinks, e.g., of male versus female births, or of acting human beings. It would, therefore, be completely arbitrary to single out some of the many antecedents that produce the statistical regularity, and assign them to the objects as "objective probabilities." But it is only by doing this that a necessary condition for the application of Laplace's principle of development can be satisfied. This principle, or equivalently (Laplace's interpretation of) Bernoulli's theorem, was therefore diagnosed by Venn as "one of the last remaining relics of Realism, which after being banished elsewhere still manages to linger in the remote province of Probability." One may doubt whether realism could be banished so easily; but even if not, Venn was certainly right to see "the inveterate tendency to objectify our conceptions" at work in traditional probability theory.

But Venn not only did recognize the conceptual confusion that I ascribed to rationalism and that he ascribed to realism; he drew a further consequence of considerable import from his empiricist approach to probabilities:

> On the theory adopted in this work we simply postulate ignorance of the details, but it is not regarded as of any importance on what sort of grounds this ignorance is based. It may be that knowledge is out of the question from the nature of the case, the causative link, so to say, being missing. It may be that such links are known to exist, but that either we cannot ascertain them, or should find it troublesome to do so. It is the fact of this ignorance that makes us appeal to the theory of Probability, the grounds of it are of no importance.[40]

This is less than an admission, or a confession, of indeterminism, but at least it is a plea for agnosticism in matters of universal causation. Venn does not want any longer to invoke the traditional argument of the mutual compensation of causes in the long run. The gap that was already seen between causal and probabilistic accounts by Fries and Ellis is systematically emphasized by Venn; he groups probability with logic or induction, since "it is simply a body of rules for drawing inferences."[41] How then does he describe the relationship between this inferential discipline and reality? The existence of series of observations that "combine individual irregularity with aggregate regularity" is the presupposition of its being used.[42] Conversely: "The mere regularity of the observed statistics ... seems to me scarcely to have any connexion with causation ... but to lead to an entirely distinct class of conclusions. It is in this way that the fact is ascertained ... that almost all the properties of natural classes of objects preserve a general uniformity amidst individual variations."[43]

This clear rejection of a causal account of statistical regularities is already to be found in the first edition of *The Logic of Chance* of 1866; it might well have been inspired by Darwin's revolutionary thoughts on the nature of natural kinds.[44] Venn himself, however, acknowledged another ancestry. In the second edition of 1876 (as well as in the third edition of 1888), he enlarged the expository first chapter to include an explicit reference to Ellis. The few sentences he quoted in full culminate in Ellis's "fundamental principle of the theory of probabilities" quoted above. Though, for special reasons, he preferred the term 'series' to the term 'genus', Venn considered his view of probability to be substantially similar to that of Ellis. In opposition to Ellis, however, he disliked all metaphysics, especially a realistic interpretation of probabilistic statements; hence agnosticism in matters of causation replaces Ellis's speculation on "balancing tendencies" built into the nature of things. For Venn the theory of probability is a branch of logic, not a "real science." This attitude allowed him to reject unwarranted causal imputations more clearly than ever before, but it also prevented him from recognizing a novel and most important explanandum of statistical science: the existence of stable equilibria or invariant kinds amidst individual variations.

5   The Mutual Compensation of Accidental Causes

The deterministic account of statistical regularities was built on two complementary ideas: (i) the causal efficacy of structural conditions, e.g., of the mixture of white and black balls in the urn, of the legal constitution of a society, or of the macroscopic properties of gas, *and* (ii) the mutual compensation of accidental causes, e.g., of the contingencies determining the single draw, of the individual biography or of the irregularities of the walls of the vessel that contains the gas. In the last section I have reviewed some critical analyses of (i). The discussion will now be completed by an analysis of (ii). For the deterministic explanation of statistical phenomena is inadequate without (ii), so that a successful criticism of (ii) may open another door for probabilism. Conversely, as became clear with Ellis, a causal theory of equilibrium structures may possibly serve to buttress determinism in the face of probabilistic appearances.

5.1   Crossing Causal Lines: Antoine Augustin Cournot

It may be useful to start out from an ambiguity of determinism that was unresolved in the classical period and, as far as I am aware, never explicitly discussed as a major problem in the nineteenth century. Kant's principle of causality reads, "Everything that happens (begins to be) presupposes something upon which it follows in accordance with a rule."[45] This version of determinism is most naturally interpreted in terms of well-defined kinds of events and specific laws of nature applicable to a particular kind. Experiments are especially suitable instantiations of the principle. The formula 'same cause, same effect' suggests a similar view. Let us call it 'local determinism'. To it we may contrast Laplace's 'global determinism' of successive states of the universe. As opposed to the former, the latter is purely speculative.[46] How then could it receive so much credit? Presumably because some paradigm cases of local determinism deserve credit, and the extrapolation of local to global dimensions seemed innocuous enough, indeed in view of the ever present threat of external disturbances of localized systems even attractive. The attraction may have veiled the non sequitur, which is nicely expressed in the following words: "It is one thing to hold that in a clear-cut situation—an astronomical or a well-contrived experimental one designed to discover laws—'the result' should be determined, and quite another to say that in the hurlyburly of many crossing contingencies whatever happens next must be determined; ...."[47] Even if local determinism is true, global determinism may be false. There is a somewhat stronger version of this statement that points to its relevance for the problem that statistical regularities pose, even if Kant's local version of determinism is unexceptionally true: Although local determinism is *globally* true, global determinism may be false. That is, we may be able to discover a difference in the antecedents *whenever* their consequences turn out to differ, but we may still be unable (in principle) to predict everything, simply because there is no way to recognize in advance *all* circumstances that are relevant for a certain event. There may be coincidences of causally fully determined occurrences, after all. True, a demon who, by definition, knows *all* circumstances in the universe at a particular time can foresee even these coincidences, but only trivially so; the coincidences are, *ex hypothesi*, not themselves instances of a rule connecting them with other circumstances of a certain type.[48]

The empirical, if not metaphysical, possibility of precisely this state of affairs supports the attempt at reconciling statistical regularities with causal determinism. Cournot adopted this approach in a particularly clear way.[49] After an initial exposition of the familiar combinatorial techniques of probability, he tries to connect the mathematics with empirical reality in a chapter entitled *Du hazard, de la possibilité et impossibilité physiques*. At first, he states that "every event has its cause"; this he calls "an absolute and necessary rule" (§39). But since there are many mutually independent causal chains, chance phenomena can exist as coincidences of the criss-crossing causal chains (§40). His model is the genealogical tree (§41).

Independence and coincidence of causal chains are objective features of the world. Hence Cournot finds fault with the traditional notion that Laplace's omniscient being would have no use for the concepts of chance and probability; it would be superior to us only by the perfection with which it could determine the amount of chance in the world (§45). Thus, the being would be able to predict everything, yet it would predict the coincidences or chance events with the help of a kind of knowledge we humans are necessarily lacking.

So far so good. It is perhaps clear how causality and chance are made compatible. It is less clear how chance and probability could be related. Single coincidences in themselves do not have a measure. In answering this question Cournot falls back on the traditional combination of conceptually disparate elements. His first step consists in an analysis of the "physically impossible." When a sphere is set in motion by the impact of another body, it is physically impossible that the sphere emerges from the collision without a rotational component of its motion. Why? "Because there is no reason why the accidental combination of mutually independent phenomena or causes by which alone the physically impossible event would be produced should occur rather than infinitely many alternative combinations" (§43). This is the old argument from indifference, for which the objective independence is no more than a *necessary* condition. Apparently Cournot got this wrong. At any rate, what he says is that the general and abstract notion of the independence of causes and of the infinite number of possible combinations offers the reason of physical impossibility, and that there is no need of empirical concepts of matter as provided by our senses (§43). This is classical a priori speculation.

Cournot's next step is to apply the notion of the physically impossible to a Bernoulli series. From the preceding argument he concludes that an event whose mathematical probability is infinitely small is physically impossible. Now, since, according to Bernoulli's theorem, the mathematical probability of a deviation of relative frequencies from the corresponding probabilities becomes infinitely small in the long run, it becomes physically impossible to obtain such a deviation in the long run. In this way, Cournot thinks, probability theory as a whole can be attached to the notion of physical impossibility. One further assumption, however, in addition to that of indifference is required, viz., that "the same uncertain events can recur arbitrarily often under the influence of mutually independent and accidentally combined causes" (§44). Now, the repetition of coincidences that appear as "the same events" requires a certain regular structure of the net of criss-crossing causal chains, i.e., an order of nature that has nothing to do with the causal order. Here again Cournot resorts to a familiar device. He specializes the idea of the coincidence of causal chains to that of the coincidence of a constant cause with variable causes. In his discussion, however, it is obvious in a way in which it had not been before that the causal explanation of empirical probabilities requires the idea of a *causally unfounded coordination of causes*.

Thus, we have reached a result similar to that of the last section: we found there that statistical regularities have little to do with causation; instead they raise a problem of order in classes of similar cases. This problem can now be recognized more distinctly as a problem of coordination. It is by no means clear, however, whether the things to be coordinated will turn out to be causes. This latter assumption prevails until about 1870. Venn abandoned it in 1866; parallel moves around that time will be discussed below. In the remainder of this section I shall pursue the coordination problem through its causal phase, at least in rough outline.

5.2 A New Law: Adolphe Quetelet

Since the elements of a statistical aggregate are to be conceived as mutually independent, the coordination problem appears to be unavoidable. It was announced as early as 1795 in Laplace's account of Bernoulli's result, in which he says that the variable causes "produce effects alternately favorable and contrary to the regular succession of events," and that they "destroy each other."[50] It enters error theory and is made explicit by Hagen and Bessel in their proofs of the law of errors in the 1830s. It is still alive 120 years after its birth when Timerding presents the law of large numbers as a still unexplained empirical fact and reviews the long succession of unsuccessful attempts at dealing with it. He calls the explanandum "statistical compensation" ("statistischer Ausgleich"[51]).

Before the discovery of statistical physics around 1870 the scientifically important statistical regularities were those of social science, so that Laplace's ideas acquired their relevance predominantly in this field. Quetelet, the doyen of the new empirical social science, fully adopted the explanation of stable averages in terms of constant and accidental causes, including the claim of the mutual compensation of the latter in the long run.[52] But he introduced two major features into this account that are both relevant to the theme of probabilism and both related to his much debated conceptual innovation, the "mean man": (1) He blurred the clear distinction between the levels of constant causes characterizing the statistical whole, and accidental causes characterizing the single instance. (2) He replaced the compensation problem by a new principle. Step (1) helped him to take step (2).

Quetelet's statistical ensembles were located at the level of society. Moreover, he was a determined antireductionist theoretician of society.[53] Excluding what is accidental, he wanted to view man "as a fraction of the species."[54] Nevertheless, he modeled society in terms of a quasi-individual, his ominous "mean man." True, he assured his reader at once that this creature is "a fictive being" and that he will be concerned not with a theory of man but only with "the facts and the phenomena concerning him" (p. 21). Presumably he was thinking of *social* facts and phenomena here. Yet, later in the book he starts talking in terms of properties like "penchant au crime," "tendance au marriage," etc. There are a number of reasons that make it almost irresistible to interpret these expressions as referring to *dispositions of individuals*, or at least as including such a reference in their meaning. More important, it seems almost built into the task of accounting for statistical regularities that some such reference should be involved: (i) It is artificial—and Quetelet's readers prove the point—to suppress an individual-related connotation of the terms; we all have the experience of inclinations, or of natural drives, etc. (ii) Even if the properties of the mean man should be functions of the state of society at large, they can hardly be functions *only* of that state; rather they would be expected to be combined results of social conditions and individual dispositions. It is the actions of individuals, after all, that are the objects of moral statistics. (iii) Quetelet himself entertained the idea that different societies, and the same society at different times, statistically scatter around what one may call a universal anthropological mean (pp. 23, 25).[55] (iv) If it is assumed that the constant conditions are inherent tendencies not in individuals but in society, the question arises where to locate them. Society as the structured ensemble of many individuals may be a natural carrier of statistical properties as such, e.g., an average rate of marriages per year; but is this also true of some further feature that could be looked upon as the cause of this rate? In this respect the case of society differs from that of, say, the die. The die is an ordinary object to which we can ascribe ordinary properties. Moreover, it even has an identifiable constitution—its symmetry, which corresponds to a probabilistic disposition, the equipossibility of its faces. To ascribe a "constitution" to society in a similar way is problematical.[56]

In sum, even in a nonreductionist social theory some of the so-called constant causes of social phenomena may plausibly be located at the level of individuals. Now, once they are located at this level, it becomes inevitable to conceive their nature in a new way, as was already indicated in the preceding paragraph by the term 'disposition' as well as by Quetelet's own terms 'penchant', 'tendency', and the like. In order to make sense in the context of social statistics these terms must not only denote the property of an individual to react under fully determined conditions in a unique and fully predetermined manner; if understood in this way, Quetelet's tendencies would not even match their descriptive, let alone their explanatory, task. Rather the dispositions in question are to be taken as multivalued properties that are to be characterized by continuous distributions of numerical values. If so, they cannot be ascribed to actual individuals, at least not as properties characterizing any of them at a well-defined time. Quetelet's solution is to ascribe them to his mean man. Seen in this way, this concept, unhappy as it may be, can be recognized as more than an accidental invention.

Nevertheless, it does not seem to be able to achieve more than an expression of certain statistical facts in an unfamiliar language and to serve as a reminder of the anthropological ingredients of social states. For Quetelet this situation changed completely when he found empirically that some properties of human individuals were normally distributed about a mean.[57] The analogy to the law of errors (which was familiar to Quetelet from his astronomical work) suggested a surprising interpretation: it was as if nature had aimed at an ideal value but only obtained somewhat disturbed results. The carrier of the ideal values could then be interpreted as the type nature had aimed at. In this case the mean (or mean man) turned out to be "l'homme type." The distribution appeared as a law of nature that serves to preserve the species.[58] In general, he thought, a normal distribution indicates a "proper mean," i.e., a meaningful average value that characterizes a natural type or kind, whereas an arbitrary arithmetical average is devoid of any scientific significance.[59] Since the deviation of exemplars from the mean was ascribed to accidental causes, Quetelet called this law of the conservation of the type "the law of accidental causes."[60] He devoted an entire book to the task of showing that "the law of accidental causes is a general law that applies to individuals as well as peoples and that rules our moral and intellectual qualities no less than our physical qualities," indeed "that dominates our universe and seems destined to spread life trough it."[61]

It is worth noting that Quetelet, the supporter of a Laplacian social science, arrives at the same conception as the critics of Laplace: the order of things according to natural kinds. Moreover, he makes a further step, which one may view as an alternative to the one Ellis made: he not only notes the existence of natural classes as a *fact*, as Venn did, he elevates it to the status of a *law* that governs the universe. This move contrasts with Ellis's view, which culminates in a presumably causal model of actual compensation processes within a natural class of phenomena, whereas in Quetelet the compensation problem disappears behind the weighty term 'law'. He did not bother to explain in dynamical terms why such a law should hold; he seems to have been overwhelmed by the discovery of order at the very heart of apparent disorder.[62]

Now, if the law is not only phenomenal or secondary but fundamental, it may seem to express a basic probabilistic feature of the world: that it is the nature of individuals, as far as they belong to a natural kind, to display accidental variations about a mean in accordance with a law. Indeed, Quetelet says that the law "confers an infinite variety to everything that breathes, without impairing its principles of preservation."[63] This sounds as if he intended to show that individuality and a lawlike order are compatible. Individual characters appear to be accidental; yet, if these contingencies were not regulated by law, the species could not survive. The combination of both features, however, amounts to a reality structured by probabilistic laws. This reading would accord with Quetelet's treatment of free will, which he reckons among the accidental causes of social phenomena.

5.3 The Hierarchy of Laws: Henry Thomas Buckle

In contradiction to the foregoing remarks the reader will find Quetelet not only less explicit but hopelessly ambiguous on the question of freedom, contingency, and determinism. The regulation of the accidental features including free will appears, at other places, conceived in terms of a rigid coordination of all instances of a statistical ensemble: "It is society that prepares the crime . . . . The culprit is nothing but the instrument of its execution . . . . The wretched man who carries his head to the scaffold is . . . an expiatory victim of society."[64] It is difficult to construe this passage as *not* saying that the apparently contingent single instance is nevertheless determined by its connection to the general conditions. The phenomena of social and biological life suggested ideas like 'accidental variations', 'law of accidental causes', and the like; but probabilism needed more momentum in order to become acceptable in science. In a time steeped in the ideal of mechanistic determinism, the compensation problem was bound to be brought under something like Cournot's superposition of independent causal lines.

This very idea, reformulated in terms of laws, is vaguely present in the background of the last quotation from Quetelet. At least, this is how he was understood by his most influential reader, Henry Thomas Buckle, and through him by many others. Buckle was interested in establishing solid foundations for scientific historiography, which in his view implied the total regulation of all events by more or less general laws. In the first chapter of his major work[65] he therefore tried to refute the most critical source of contingency, human free will, by an argument from social statistics. In this context he not only quotes the first two sentences of the above quotation from Quetelet, but also develops the implications of the view that is expressed by them. In the discussion of suicide he comes to say that "... the individual felon only carries into effect what is a necessary consequence of preceding circumstances. In a given state of society, a certain number of persons must put an end to their own life. This is the general law; and the special question as to *who* shall commit the crime depends of course upon special laws; which, however, in their total action must obey the large social law to which they are all subordinate...."[66] Perhaps Buckle was not aware of it, but this statement develops, at least *in nuce*, the dynamical substitute for Laplace's development of probabilities. The link between the general law, i.e., the statistical uniformity, and the individual action is not some intuitive a priori assumption but a mediating hierarchy of more or less special laws whose superposition inevitably produces the action in question. It is clear that the serious dynamical determinist must hold some such view. The idea of crossing causal lines, which we found in Cournot, is to be complemented by the idea of the correlation of the relevant laws, which is brought about by their subordination to a more general law.

Clearly, this rough sketch is no solution to the problem, since nothing is said (and perhaps nothing can be said) about the particular arrangement of laws that is needed in order to achieve the desired result. There are well-known cases of the subordination of one law to another, e.g., of the law of free fall to the law of gravitation. The general law in combination with particular boundary conditions allows one to subsume the special law under it, and to derive the constant involved in the latter from the constant characterizing the former. But it remains mysterious how we should conceive of a law that regulates the correlation of several laws. The analogue in the case of gravitation would be a law telling us something about the relative frequencies of different conditions leading to different gravitational accelerations. This is a peculiar and unknown kind of law.

The compensation problem is now hidden in the assumption of special laws that, if they exist at all, are certainly not of the familiar dynamical kind, and presumably not derivable from them either. Buckle himself recognized that the subordinate laws may "*disturb* the normal activity" of higher laws.[67] But if *that* is possible, the relation between the higher and lower laws in itself is not such that the higher laws govern the facts, as it were, through the administration of their subordinate servants. Why then should the disturbances of the higher laws level out?

The deterministic explanation of statistical regularities seemed obvious to many nineteenth-century scientists and philosophers. And yet it failed. No plausible account of structural causation was discovered, and the compensation problem of accidental causes remained unsolved. Nevertheless, it must have been almost impossible to recognize this failure in so complex a subject matter as human society. It is perhaps not surprising that further conceptual clarifications were suggested by physics rather than by sociology or history.[68]

6 The Emergence of Probabilism: James Clerk Maxwell

Maxwell is generally recognized to be the founder of statistical physics; he invented a new theoretical method, that of statistical ensembles, which was later developed to its present form by Gibbs. What had been a descriptive tool in social science and the treatment of data, statistics, was transformed by Maxwell into an explanatory device figuring in the microtheoretical account of observable phenomena like heat flow, diffusion, etc.[69] This transformation was prepared and accompanied in Maxwell's thought by far-reaching epistemological and ontological reflections that, in conjunction with the new kind of physical theory, led him finally to a probabilistic and indeterministic philosophy. It is not to be expected that Maxwell had ready solutions for the standard problems of statistical laws, such as a new concept of structural causation or an answer to the compensation problem for accidental causes; but through his work both problems began to appear in a very different light than they had done before.[70] What Maxwell understood much better and much earlier than most other philosophers and scientists was that a conception of the lawful order of things was needed that was much more flexible than the classical deterministic framework constructed by Kant and Laplace.[71]

6.1 Maxwell on Analogies: The Plurality of Laws

At the same time that Buckle conceived the hierarchy of laws in order to defend determinism and bar the way to free will, Maxwell pondered on the hierarchy of laws as well, but in a very different spirit.[72] His aim was to develop a philosophical framework that could accommodate human intellect and will as well as inanimate matter or organic life. To him it seemed unpromising to reduce all laws to those of motion, or of thought, for that matter. The resemblances between laws of different classes, he argued, were not "identities" (as Laplace had thought[73]) but only "analogies."[74] This characterization makes it possible to modify the view of the law-governed world in two ways: (i) not all laws need be of the same kind, e.g., not all need necessitate the phenomena for which they are laws, and (ii) the monolithic hierarchy of laws may be replaced by a cooperative plurality of laws. Maxwell introduces both of these modifications together. The laws of physics are ruled by the maxim that might is right; their instances are characterized by the *strength* of driving forces inherent in matter. The laws of biology do not drive but only direct; the forces they describe are *molds* for motion rather than sources of motion. The instances of these laws are natural types characterized by *health*. Further laws regulate animal emotions, which tend to promote the enjoyment of life. Superior to all three kinds of laws are the laws of thought; their superiority, however, is not due to strength or to the capability of reproducing a type but consists in their being right and true.[75] Their supremacy, as Maxwell quotes De Morgan approvingly fourteen years later,[76] does not preserve them from violation, so that the analogy to natural laws will not help us to understand the nature of their authority. Finally, there are specifically moral laws, which can direct "distinct acts of the will," but share their influence with "physical necessity," "organic exitement," and "attractions of pleasure or the pressure of constraint activity."

Though the different kinds of laws are dissimilar with respect to their subject matter and manner of operation, they all share the common property of presenting themselves as "a continuation of the analogy of Cause." [77] From this point of view the merger of a priori epistemological principles and mechanical theory, discussed in section 2.2, must appear as an unholy alliance. Neither the fundamental laws of motion nor the laws of thought are fit to explain what Maxwell saw to be a real phenomenon: "the existence of one set of laws of which *the supremacy is necessary, but to the operation contingent*." [78] What is important to note here is the combination of the ideas of law and cause with the ideas of contingency and indeterminacy of the single instance. Maxwell admits that these thoughts are "somewhat diffuse," even "confused on the subject of moral law" (which is the focus of his interest in this paper), but he insists that "if we are going to study the constitution of the individual mental man, and draw all our arguments from the laws of society on the one hand, or those of the nervous tissue on the other, we may chance to convert useful helps into Wills-of-the-wisp." [79] This statement in favor of theoretical pluralism is diametrically opposed to what Buckle said in his famous book a year later.

6.2 Two Kinds of Knowledge

At this point in Maxwell's career the flexible plurality of laws and causes did not yet have any connection with the problem of statistical regularities. I concur fully with Theodore Porter, who argues that the sources of Maxwell's indeterminism on the one hand and of his statistical physics on the other are different. The first is his devotion to religion and his belief in free will, the second the encounter with the kinetic theory of heat and matter, especially the work of Robert Clausius.[80] But fourteen years later the rivers issuing from these two sources had met and mixed. Not only did the statistical character of knowledge serve as a defense against the refutation of free will, the indispensability of a plurality of laws and causes also helped to form a new conception of statistical laws. When Maxwell wrote again on free will in 1873 he did two things at once: (i) he placed his plea for the possibility of free will within the framework of statistical physics, and (ii) argued for a fundamentally new conception of physical investigations on the grounds that "if the molecular theory of the constitution of bodies is true, *all* our knowledge of matter is of the statistical kind." [81]

This statement acquires its full weight only in conjunction with Maxwell's elaboration of the contrast between "two kinds of knowledge ... the Dynamical and Statistical," the second belonging "to a different department of knowledge from the domain of exact science." It is worth noting that Maxwell tied *all* philosophical implications of the progress of physical science he discussed at some length to a single new development: molecular science. Of this "the most important effect," he thought, was again the very distinction between dynamical and statistical knowledge.[82] Since the avowed ideal still remains exact science and the dynamical study of nature as "the only perfect method in principle,"[83] its unavailability for physical foundations becomes crucial for the outlook of science in general. As a matter of principle, the traditional foundations of deterministic mechanical science are shaken. But if so, the familiar hierarchy of the disciplines loses its meaning: whereas Laplace, Buckle, and Quetelet thought to base social science on, if not reduce it to, physics, Maxwell feels free to base the new physics on the usual procedures of the statisticians[84] and to cite the three mentioned figures as authorities for it.[85] The irony of this citation resides in the literally and figuratively revolutionary exchange of the roles of physics and social science, a move that was prepared by Maxwell's earlier thoughts on analogies. It consists further in the subtle but profound reinterpretation of statistical averages and the implications of their existence.

Why do statistical laws not belong to exact science? Because, if they cannot be construed as reducible in principle to more basic dynamical laws any longer, they destroy the unbroken chain or net of determinations and predictabilities in principle. Maxwell's well-known analysis of the second law of thermodynamics with the help of his demon makes this point.[86] Moreover, Maxwell recognizes further features of the new statistical kind of knowledge in physics. He flags the most notable at once: the reversibility of physical processes in time is lost; in the theory of heat, for instance, we can only infer the future, and in the study of mental processes generally only the past. Maxwell goes on to ask what might be the reason for so fundamental a difference. It is in *this* context that he introduces another idea that for him was connected with the problem of free will: the existence of branching points in the dynamical path of a sufficiently complex mechanical system.[87] Now, if there are such branching points in any complex system, some systems may still be predictable, whereas others are not. The bias of the determinist was to pay attention only to the former. Maxwell recommends overcoming "the prejudice of determinism" by studying "the singularities and instabilities, rather than the continuities and stabilities of things."[88]

Furthermore Maxwell assumes that "in phenomena of higher complexity there will be a far greater number of singularities, near which the axiom of like causes producing like effects ceases to be true." His examples are systems involving a triggering mechanism whose size and energy (almost) vanishes with respect to the total size and energy of the system: "the little spark which kindles the great forest" or "the little gemmule which makes us philosophers or idiots."[89] In all such cases, the energy of the trigger may be "infinitesimally small" in proportion to the energy released in the subsequent process. Maxwell is careful to withhold judgment on whether or not an ideally precise knowledge of the microscopic conditions would bring things back to full determination and predictability. What he stresses, however, is the purely *metaphysical* character of the doctrine that from the *same* antecedents follow the *same* consequents. In our world this doctrine is to be replaced by the useful *physical* axiom that from *like* antecedents follow *like* consequents. The point now is that this so-called axiom turns out to be sometimes true, e.g., in astronomy or elementary dynamics, and sometimes false, namely, in complex and instable systems. Whether or not it holds may be a question that cannot sensibly be asked except under the general presupposition that exact sameness of circumstances either never occurs or is never accessible to human observers. But since this presupposition is undeniably and universally true, the question aims at a *real physical difference between stable and unstable systems.*

A comment on the term 'real physical difference' will be helpful at this point. As far as I am aware, Maxwell consistently and cautiously restricts himself to the weaker epistemological claim: he does not worry about whether precisely the same circumstances ever *actually recur*; for him it is enough to know that precise sameness can never be *ascertained*. He repeatedly emphasizes that our limited human capabilities of discernment are the decisive reason for the move from dynamics to statistics. In a famous passage[90] he goes so far as to say that "the idea of dissipation of energy depends on the extent of our knowledge," and "... confusion, like the correlative term order, is not a property of material things in themselves, but only in relation to the mind which perceives them." One may be tempted to take such sentences as an early version of current information theoretic notions of entropy. I think we must resist this temptation. The concluding sentences of the article return to the unversal condition of *our* science: that *we* are beings who can lay hold on some forms of energy whereas others elude our grasp. This statement amounts to naming a transcendental condition of knowledge (in Kant's sense of the term); if so, it is irrelevant whether we are able consistently to *conceive* another kind of knowledge that is more in line with traditional philosophical ideals. In his essay on necessity versus free will,[91] Maxwell starts out by claiming that metaphysics depends on the available physics; and at the end of the paper he warns us to assume "that the physical science of the future is a mere magnified image of that of the past." Now, what could more literally violate this warning than to anticipate the demonlike knowledge of precise dynamical paths of microobjects, as if they were planets traveling around the sun? It seems to me mistaken to construe Maxwell's probabilism as subjective, if this term is to imply the expectation that we might meaningfully, if perhaps forever hypothetically, develop a full dynamical account of nature following the pattern of the good old days of classical mechanics. If his view is subjective, then it is so in a totally different sense in which we admit that our knowledge of the world will be inevitably formed and informed by our specific cognitive constitution as well as our specific position in nature.

6.3   The Proper Objects of Statistics

Given that the difference between stable and unstable systems *is* real for all human intents and purposes, the particular class of systems that are at the same time sufficiently complex, hence rich in branching points, *and* nevertheless stable (within sufficiently precise limits and for sufficiently long periods) will be the proper subject matter of statistical science. But then a new type of physical interaction or at least of collective properties of a system must be postulated. Ellis's problem of natural kinds reappears in a generalized form. It now comprises, e.g., all thermally normal macroscopic bodies, as well as natural organisms or even entire species. Whatever a still nonexistent theory of such systems will turn out to be, it will transform the ideas of a mutual compensation of accidental causes and of the efficacy of constant causes into a new conception of collective action of many parts to be subject to new kinds of laws. At any rate, Maxwell's analogical view of nature leaves room for this conception.

In the very paper in which Maxwell, following a hint by Boltzmann, explicitly formulated the concept of what was later known as statistical mechanics[92] he also made clear that the new method required as its physical foundation a special property to be satisfied by any system coming within the compass of the new statistics: the ergodic property of passing through *every* phase that is consistent with the total energy of the system. Since this property amounts to requiring that no other constant of the motion besides energy must exist, its possibility, as Maxwell knew, cannot be explained on purely mechanical grounds. He therefore invoked external causes, the irregular surface of the vessel containing the substance under consideration, in order to introduce the necessary "disturbances" beyond the mechanical level that lead to jumps of the system from one undisturbed path to another, or in other words, to the branching of paths. We have learned that Maxwell's ergodic assumption in the above cited form is simply false for closed systems; but it was never claimed for a closed system by Maxwell himself. A fairer criticism would be the following: Maxwell himself had argued that "the so-called disturbance is a mere figment of the mind, not a fact of nature, and that in natural action there is no disturbance."[93] Maxwell could therefore be accused of not having lived up to this remark, so far as his theory of 1878 went. Presumably he would have granted the point immediately. Hence Maxwell cannot be said to have offered the solution of the problem of stable statistics, a problem still largely open today, but he can be praised for having grasped the task, and for having pointed to the need for turning the *coordination* problem of causes into the problem of finding the laws governing the mutual *interaction* of the parts of complex systems with numerous internal instabilities or branching points.[94]

According to this idea the preservation of equilibrium or of the type is not construed in the fashion of Quetelet (and later Galton) as a law of the universe that, if interpreted as a fundamental feature of reality, would express a *tendency* of nature to realize her products with a specific distribution of values around an ideal mean. In Maxwell's case it does not make sense to ascribe to each of the molecules an inherent dispositional property of tending toward the most probable velocity, or the like. One may perhaps say that a stable thermal equilibrium is characterized by a certain average kinetic energy of the molecules. But the *average* kinetic energy is *not* the *actual* kinetic energy. In this *prima facie* inconspicuous difference lurks the entire probabilistic revolution. The former property of the molecules is in fact a property of the whole body that has no meaning for its elementary parts. To be ergodic is such a property as well. From this it should be clear that Maxwell's probabilism, i.e., the probabilism of classical physics, is possibly profoundly distinct in nature from quantum probabilism. In the domain of the latter it is at least a serious option to construe the probabilities as propensities inherent in each single elementary part of a macroscopic collective governed by a statistical law—e.g., to ascribe a certain probability of disintegration to each radium atom separately. As a matter of fact, we know that the interaction with the environment, including the other atoms of the sample, is irrelevant to the decay rate.

7 Conclusion

The preceding selection of arguments from Kant and nineteenth-century sources reveals at least three distinct sorts of intellectual reasons for the slow rise of probabilism: (1) the fusion of an ancient epistemological tradition concerning the nature of knowledge with more recent successes of mechanical science; (2) the inscrutable and profuse depths of causal thinking; and (3) the difficulty of understanding the nature of statistical laws.

Reason (1) helped to entrench deterministic beliefs. Seen in the light of section 2 of this paper, determinism appears as a new version of modern philosophical rationalism of Cartesian descent wrapped in progressive physical science. Reason (2) is a notorious and persistent source of philosophical confusion. In this capacity it served well, and for quite some time, as a protective belt for the hard core of determinism. It is instructive, however, to observe the ambivalence of this intellectual weapon; the use to which it is put depends on more general philosophical views, which tend to tinge the notion of cause and law. Empiricist leanings in philosophy helped to uncover the discontinuity separating statistical from causal investigations. Being phenomenalistic in outlook, however, the empiricist arguments proved too weak to break the spell of truly explanatory theoretical views.

Reason (3) for the slow rise of probabilism can obviously only be overcome by transforming deeply rooted views about the nature of cause and law, indeed of scientific knowledge in general. Provided that a theoretical explanation of statistical statements and methods is sought at all, the required transformation of scientific thinking appears to involve a double ontological price; two pieces of traditional science will have to be abandoned: (a) reductionism and (b) determinism. The rejection of (a) is suggested by the fact that the properties that can explain the regular behavior of a statistical aggregate cannot meaningfully be ascribed to the single parts—they characterize the aggregate as a whole, or at least sizable portions of it. This feature is at variance with the analytical spirit of previous science according to which the behavior of any system ought to be explained in terms of its composition and the behavior of its elementary parts. Theoretical pluralism had not thus far been an official part of modern science. The rejection of (b) is suggested as a consequence. If the apparent contingency of a single instance is explained by placing it within a whole whose relevant properties *cannot* be reduced to properties of the parts, even the hypothetical assumption of a complete determination of the instance in question will appear doubtful, because incompatible with a novel explanatory scheme, indeed the only scheme that eventually does the work of explaining the phenomena. Saving they do not need, since they exist.

The obstacles impeding the rise of probabilism are not rooted in the facts of everyday life and of science. Workable statistics abound, and probability has always been the great guide of life. The difficulties arise out of our attempts to understand and to explain. If probabilism means to accord probability an explanatory fuction, it could not be realized short of a revolution in thought, a rethinking of the relationship of human reason and factual contingency.

Notes

1. G. E. M. Anscombe, *Causality and Determination* (Cambridge: University Press, 1971), p. 20.

2. See Aristotle, *Analytica Posteriora*, A 2, 71b 10–12, as a locus classicus.

3. Ibid.; cf. *Metaphysics*, A 1, 980a24ff.

4. E.g., G. W. Leibniz, *Principes de la nature et de la grâce fondés en raison*, §7: "... il faut s'élever à la Métaphysique, en nous servant du Grand principe ... qui porte, *que rien ne se fait sans raison suffisante*, c'est-à-dire, que rien n'arrive sans qu'il serait possible ... de rendre une Raison qui suffise pour déterminer pourquoi il en est ainsi, et non pas autrement." It is worth noting the use of the term 'déterminer', which is not yet restricted to temporal antecedents of events.

5. Thus, H. von Helmholtz, for instance, says that the theoretical scientist seeks to understand ("begreifen") by identifying the sufficient cause: *Über die Erhaltung der Kraft*, 1847, reprinted as No. 1 of Ostwald's Klassiker der exacten Wissenschaften (Leipzig: Engelmann, 1889), Introduction, p. 4.

6. I. Kant, *Kritik der reinen Vernunft*, A 200ff. = B246.

7. Ibid., A 158 = B197: "... die Bedingungen der *Möglichkeit der Erfahrung* überhaupt sind zugleich die Bedingungen der *Möglichkeit der Gegenstände der Erfahrung* ..."; emphasis added.

8. Kant, *Metaphysische Anfangsgründe der Naturwissenschaft*, 1786.

9. *Kritik der reinen Vernunft*, A 228 = B280.

10. Ibid., B289: "Alles zufällig Existierende hat eine Ursache."

11. Ibid., B291.

12. S. P. de Laplace, *Essai philosophique sur les Probabilités* (1895), first published as an introduction to the 2nd edition of his *Théorie analytique des Probabilités* (1814); *Oeuvres Complètes*, (Paris: Gauthier-Villars), vol. VII, 1886.

13. So far all quotations are taken from the very first pages of the *Essai: Oeuvres* VII, pp. VI–VIII.

14. Ibid., pp. CXXIII and CXXXVIIff.

15. Ibid., p. CXXIII.

16. Ibid., p. CXXXVIII. The formula of the two worlds seems to qualify Laplace as a kind of dualist, but the other formula of the identities on the very same page, indeed his entire *sensorium* theory, renders this interpretation very doubtful. Therefore, I don't agree with Ian Hacking and Cassirer that the idea of a unified determinism embracing body and mind was born only in the 1870s (see I. Hacking, "Nineteenth Century Cracks in the Concept of Determinism," in *Probability and Conceptual Change in Scientific Thought*, ed. M. Heidelberger and L. Krüger, Bielefeld University: Report Science Studies No. 22, 1982, pp. 5–34, reprinted in *Journal of the History of Ideas* 44 (1983), 455–475, esp. pp. 456 and 463). Even Kant was not a dualist in the required sense; for his dualism is *not* that of mind and body, but of *noumena* and *phaenomena*.

17. John Venn, *The Logic of Chance* (London/Cambridge: Macmillan, 1866), XIV.§17; similarly 4th ed. (identical with 3rd ed., 1888) (New York: Chelsea, 1962), X.§3.

18. Laplace, *Essai philosophique* (note 12), pp. XLVIIff.

19. To mention one passage among many: A. Quetelet, *Sur l'homme et le développement de ses facultés* (Paris: Bachelier, 1835), vol. I, pp. 13ff. This reads almost like a quotation from Laplace.

20. I may mention Drobisch as an example, because he illustrates the doctrine with the Laplacian model of the urn and explicityly calls its composition the constant cause of the ratio of black to white drawings: M. W. Drobisch, *Die moralische Statistik und die menschliche Willensfreiheit* (Leipzig: Voss, 1867); pp. 8-11.

21. Laplace (note 12), p. XLVII.

22. Ibid., p. VIII.

23. E.g., Poisson, Windelband, Überweg, Trendelenburg. For references and critical discussion see T. E. Timerding, *Die Analyse des Zufalls* (Braunschweig: Vieweg, 1915), pp. 44-45, 55-57.

24. L. Boltzmann, "Weitere Studien über das Wärmegleichgewicht unter Gasmolekülen" (1872), quoted from Boltzmann, *Wissenschaftliche Abhandlungen*, ed. F. Hasenöhrl (Leipzig 1909, repr. New York 1968), vol. I, pp. 316-402; see esp. p. 345.

25. L. Boltzmann, "Über die Beziehung zwischen dem zweiten Hauptsatze der mechanischen Wärmetheorie und der Wahrscheinlichkeitsrechnung respektive den Sätzen über das Wärmegleichgewicht" (1877), in Boltzmann, *Abhandlungen* (see note 24), vol. II, pp. 164-223; quotation from p. 165.

26. Boltzmann himself discovered somewhat later that probability theory as such would not even give him the motion of a system toward the most probable state. Precisely if all microstates are equiprobable, the system will have to spend an equal amount of time in each of them in the long run; i.e., it will have to return to improbable states ever again. Movements from more probable to less probable states will be no less frequent than those from less probable to more probable states. The H-curve, or the entropy, will be symmetric with respect to time, so that the validity of the second law of thermodynamics can only be based on assumptions about the initial states of the system to which the law is applied (see Boltzmann, *Vorlesungen über Gastheorie*, vol. 2 (Leipzig, 1898), §87, for a particularly clear statement). These results are extremely unsatisfactory in view of the reality of irreversible mixing processes. The connection of physical processes and probability cannot be of the simple type envisaged in Laplace-like principles of the development of probabilities.

27. A. Quetelet expresses this interest clearly enough: e.g., *Sur l'homme* (see note 19), p. 10.

28. Jakob Friedrich Fries, *Versuch einer Kritik der Principien der Wahrscheinlichkeitsrechnung* (Braunschweig: Vieweg, 1842); "Selbstrezension" of this book, *Neue Jenaische Allgemeine Literaturzeitung*, I (1842), 258-260; both reprinted in Fries, *Sämtliche Schriften*, ed. G. König and L. Geldsetzer, vol. 14 (Aalen: Scientia, 1974).

29. "Selbstrezension" (see preceding note), p. 3 in the pagination of the reprint.

30. Fries, *Versuch einer Kritik* (note 28); Einleitung, §VIII, p. 23 (in the original pagination).

31. Ibid., §VI, p. 19.

32. Robert Leslie Ellis, "On the Foundations of the Theory of Probabilities," *Transactions of the Cambridge Philosophical Society*, vol. VIII, part I, 1844; the entire vol. VIII appeared in 1849.

33. Ellis, "Foundations" (note 32), p. 3.

34. Robert L. Ellis, "Remarks on the Fundamental Principle of the Theory of Probabilities," *Transactions of the Cambridge Philosophical Society*, vol. IX, part VI, pp. 605-607; the entire volume appeared in 1856.

35. Ellis, "Remarks" (note 34), p. 605.

36. Ellis, "Remarks" (note 34), p. 606.

37. Ellis, "Remarks" (note 34), p. 607.

38. Equilibria or stable structures, e.g., the chemical elements, were not recognized as specific explananda by the mechanical determinists. Prominent instances of this are Kant's chapter on dynamics in his *Metaphysical Foundations of Natural Science* (1786), or Hermann Helmholtz' Introduction to his *Über die Erhaltung der Kraft* (Berlin: Reimer, 1847). Helmholtz assumes immutable forces or qualities to be the ultimate causes of all phenomena. Neither he nor Kant seem to have noticed that these forces, being permanent structural conditions of all motion, are not of the same kind as the conditions from which a given event is to be derived in accordance with deterministic rules or laws. Nor did they ask for an explanation of the specific nature and multiplicity of such forces. It was only Maxwell who became puzzled by the immutability and exact likeness of all atoms or molecules of a given kind; he noted it as a problem that lay beyond the reach of mechanical theory.

39. John Venn, *The Logic of Chance* (note 17). The relevant passage, also for the following text, is Chap. II, §§8-11, in the 1st ed., and Chap. IV, §§14-17, in the 3rd ed., which show only minor deviations from each other.

40. Ibid., 3rd ed., X.3; a similar passage is found in the 1st ed., XIV.17.

41. Ibid., 3rd., X.7.

42. Ibid., I.3, both 1st and 3rd eds.

43. Ibid., 1st ed., XIV.23. The 2nd and 3rd eds. preserve the theme of natural kinds and emphasize their indispensability for the application of probability: II.3.

44. I have not come across any evidence for this (plausible) surmise; Darwin is not mentioned in *The Logic of Chance*.

45. I. Kant, *Kritik der reinen Vernunft*, A 189.

46. It is sometimes said that also local determinism is metaphysical (e.g., by Maxwell—see section 6 below). But then it is assumed that the sameness of causes refers implicitly to a cosmically extended set of circumstances, so that it becomes a non empirical notion, and the causal principle empirically useless. Even so the principle may be taken to assert that if two effects differ, there is a further condition differentiating their causes; and this existential proposition is no less and no more unempirical than other existence claims. Moreover, within limits of accuracy, fully predictable systems are empirically known.

47. Anscombe, *Causality* (note 1), p. 22. The passage quoted in the text continues as follows: "... or to say that the generation of forces ... is always determined in advance of the generating procedure; or to say that there is always a law of composition, of such a kind that the combined effect of a set of forces is determined in every situation." Thus extended, the objection makes clear why *a unified theory*, like mechanics, must seem attractive, if not indispensable, to the determinist. For a lesser price a solid foundation for his creed can hardly be had. Seen from this angle, determinism entails reductionism.

48. It may, therefore, be misleading to say that global determinism could still be *false* if local determinism is globally true. Laplace had probably nothing else in mind other than this kind of additive knowledge of all causal chains. As a matter of metaphysical principle there is no non sequitur, once local determinism has been granted globally (a concession, though, that itself does not follow from the known success of science). Rather, what I want to point out is the strictly metaphysical character of global determinism. Under the empirically meaningful

interpretation given in the text, global determinism does not follow any more from the global, i.e., empirically unrestricted, truth of local determinism.

49. A. A. Cournot, *Exposition de la théorie des chances et des probabilités* (Paris: Hachette, 1843). References in the text are by sections of this book.

50. Laplace, *Essai philosophique* (note 12), p. XLVII.

51. Timerding, *Analyse des Zufalls* (note 23), pp. 35ff., esp. 42ff.

52. Quetelet, *Sur l'homme* (note 19), pp. 10ff.; for the idea of compensation see esp. p. 12.

53. Cf. Lorraine Daston, "Rational Individuals versus Laws of Society: From Probability to Statistics"; Theodore Porter, "Lawless Society: Social Science and the Reinterpretation of Statistics in Germany, 1850–1880"; both in this volume.

54. Quetelet: *Sur l'homme* (note 19), pp. 4–5; cf. p. 6. References in the following text refer to this work.

55. True, Quetelet calls this anthropological mean "the *social system*" (*Sur l'homme*, note 19, pp. 23, 25; the emphasis is mine); but he also speaks of "the laws of nature that concern man" and raises the problem of whether we humans can upset the stability of society by our ability to change these laws (p. 17). He deems this problem "mysterious"; but if it is to be a problem at all, it presupposes that there are, at bottom, universal properties of humans that are the source of even the statistical laws. In a paragraph added to this section in the 2nd ed. of 1869 Quetelet goes so far as to *define* the mean man as "representing our whole species and, furthermore, possessing all properties of other men to an average degree"—admittedly a very obscure formula but obviously an anthropological one.

56. Hence Venn argued that a "constitution" corresponding to the facts of social statistics is unknown; *The Logic of Chance* (note 17), 3rd ed., IV.§§15–17.

57. The discovery is dicussed for the first time in Adolphe Quetelet, "Sur l'appréciation des documents statistiques, et en particulier sur l'appréciation des moyennes," *Bulletin de la Commission Centrale de Statistique, Bruxelles*, 2 (1844); for a full account the reader is referred to Theodore M. Porter, *The Rise of Statistical Thinking* (Princeton: Princeton University Press, 1986); cf. also Porter, "A Statistical Survey of Gases: Maxwell's Social Physics," *Historical Studies in the Physical Sciences* 12 (1982), 77–116 (on pp. 91–93).

58. Quetelet, *Sur la théorie des probabilités appliquée aux sciences morales et politiques— Lettres au Duc des Saxe-Cobourg et Gotha* (Bruxelles, 1846), p. 137.

59. The distinction is developed, e.g., in Quetelet, *Sur la Théorie* (note 58), second part.

60. Quetelet, "Sur la statistique morale et les principes qui doivent en former la base," *Mémoires de l'Académie Royale des Sciences et Belles-lettres de la Belgique*, vol. XXI; separately published (Bruxelles: Hayez, 1848), p. 10.

61. Quetelet: *Du système social et des lois qui le régissent* (Paris, 1848), pp. IX and 16. The law of errors was already seen as an element of universal order somewhat earlier, e.g., by Fourier; cf, Porter, "A Statistical Survey" (note 57), pp. 90–91.

62. Cf. Porter, "A Statistical Survey" (note 57), pp. 92–93. Later criticism, specially by statisticians and economists, was directed against the lawlike character of the illegal law. Porter describes this line of argument in detail in this volume.

63. Quetelet, *Du système social* (note 61), p. 16.

64. Quetelet, *Sur l'homme* (note 19), vol. 2, p. 341.

65. Henry Thomas Buckle, *History of Civilisation in England*, vol. 1 (London, 1857).

66. Ibid., Chap. 1, pp. 21–22.

67. Ibid., p. 27.

68. Biology is at least as important as physics, if not more so; but for lack of competence and space I shall not even attempt to include it in the present essay. The reader is referred to part V of volume 2 of this work.

69. The history of this transformation, which is, at the same time, a history of transfer from social science to physics, has been carefully studied. Important sources are Charles C. Gillispie, "Intellectual Factors in the Background of Analysis by Probabilities," in A. C. Crombie (ed.), *Scientific Change* (New York: 1963), pp. 431–453; Elisabeth W. Garber, "Aspects of the Introduction of Probability into Physics", *Centaurus*, 17 (1972), 11–39; Porter, "A Statistical Survey" (note 57). The exclusive attention to Maxwell in my present paper is not meant to belittle Boltzmann's great contribution to statistical physics. It only reflects the fact that Boltzmann never trusted statistical as opposed to dynamical principles (which led him to most important discoveries, notably the Boltzmann equation), *and* that he was far from conceiving revolutionary changes in science that would lead beyond deterministic mechanics.

70. The immediate effect of Maxwell's new attitude was perhaps not great. But from the perspective of 1915 he could be recognized as the one authority in matters of probabilism: Timerding, *Die Analyse des Zufalls* (note 23), Chap. 4, esp. p. 43.

71. To label the framework of nineteenth-century science with Kantian and Laplacian is certainly one-sided and biased; there is much more on the market of ideas—e.g., Hegel and German Naturphilosophie, or derivations thereof in science, such as in the work of Fechner, analyzed by Michael Heidelberger in this volume. But for a leading scientist like Maxwell there was obviously no doubt about what was to count as a full and ultimately satisfactory account of nature: "the dynamical method of study ... is the only perfect method in principle," even for the study of man (Maxwell, "Does the Progress of Physical Science Tend to Give Any Advantage to the Opinion of Necessity (or Determinism) over That of the Contingency of Events and the Freedom of the Will?" in Lewis Campbell/William Garnett, *The Life of James Clerk Maxwell* (London: Macmillan, 1882, repr. New York/London: Johnson, 1969), pp. 434–444: quote on p. 439). A well-known contemporary mathematician went even further: whereas Maxwell despaired of a dynamical theory of the structure of matter, William K. Clifford envisaged a time "when the structure and motions inside of a molecule will be so well known that a future Kant or Laplace will be able to make an hypothesis about the history and generation of matter" ("Atoms," in *Lectures and Essays* (London, 1879), p. 133; I owe this quote to Silvan Schweber). Our two determinist heroes were still symbols of scientific hope in the 1870s; the astronomical theory of the generation of our solar system still served as a model for the generation of the atom.

72. James Clerk Maxwell, "Are There Real Analogies in Nature?" dated February 1856, in Campbell/Garnett, *Maxwell* (note 71), pp. 235–244.

73. See section 2.2, esp. note 14 and the corresponding text.

74. Maxwell, *Analogies* (note 72), p. 241.

75. Ibid., pp. 239–240.

76. J. C. Maxwell, "Address to the Mathematical and Physical Sections of the British Association," given on September 15, 1870, in *The Scientific Papers of James Clerk Maxwell*, ed. W. D. Niven, 2 vols. (Cambridge: 1890; repr. New York: Dover, 1965), vol. 2, pp. 215–229 (cf. esp. p. 229).

77. *Analogies*, (note 72), p. 240.

78. Ibid., p. 241; the emphasis is mine.

79. Ibid., p. 243.

80. Porter, *A statistical survey* (note 57).

81. Maxwell: "Does the Progress" (note 71), p. 439; emphasis added.

82. Ibid., pp. 438 and 440.

83. Cf. note 71. See also J. C. Maxwell, "On the Dynamical Evidence of the Molecular Constitution of Bodies," lecture Feb. 1874, in *Scientific Papers* (note 76), vol. 2, pp. 418-438 (cf. esp. pp. 418ff).

84. J. C. Maxwell, "Molecules" (a lecture delivered in 1873), in *Scientific Papers* (note 76), vol. 2 (see p. 373).

85. Maxwell, "Does the Progress" (note 71), pp. 438-439; the reference to Quetelet is only implicit but unmistakable.

86. The history of Maxwell's demon is pursued in Edward Daub, "Maxwell's Demon," *Studies in History and Philosophy of Science*, 9 (1976), 213-227. For a concise statement of Maxwell's epistemological conclusions see Maxwell, "Tait's Thermodynamics," in *Scientific Papers*, vol. 2 (note 76), pp. 660-671 (esp. pp. 669ff.), or the conclusion of his *Theory of Heat* (London: Longmans, 1871).

87. Also in his published writings Maxwell connects the action of the soul in the body with singularities of dynamical theory; see his review of a book by B. Stewart and P. G. Tait, "Paradoxical Philosophy," in *Scientific Papers* (note 76), vol. 2, pp. 756-762. His references are to Stewart, Saint-Venant, and Boussinesq. For further references see Hacking, "Nineteenth Century Cracks" (note 16), pp. 16ff.

88. "Does the Progress" (note 71), pp. 440 and 444.

89. Ibid., p. 443.

90. J. C. Maxwell, "Diffusion," article written in 1878 for the Encyclopedia Britannica, in *Scientific Papers* (note 76), pp. 625-646 (on p. 646).

91. Maxwell, "Does the Progress" (note 71).

92. J. C. Maxwell, "On Boltzmann's Theorem on the Average Distribution of Energy in a System of Material Points," 1878, in *Scientific Papers* (note 76) pp. 713-741 (see esp. pp. 713-715 for the following).

93. J. C. Maxwell, "Address" (note 76), p. 229.

94. I leave entirely open what the ultimate source of instabilities Maxwell had in mind will prove to be. Perhaps he thought that the continuum is essential, a view advocated in M. Norton Wise, "The Maxwell Literature and British Dynamical Theory," *Historical Studies in the Physical Sciences*, 13 (1982), 175-205 (esp. p. 189 in connection with p. 200).

5 The Decline of the Laplacian Theory of Probability: A Study of Stumpf, von Kries, and Meinong

Andreas Kamlah

Laplace's paradigm dominated the theory of probability for about a century and was replaced by other types of exposition (partly frequentist) in our century. I try to show that a gestalt switch took place between the old and new paradigms that was not necessitated by any new empirical facts but rather was due to a shift in the general epistemological attitude toward nature, and to a different way of dealing with statistical data. I first show that Laplace's theory can deal with certain anomalies arising in its application if it is augmented by auxiliary hypotheses, and that it was not abandoned because of its supposed untenability, but rather because other accounts seemed to be less complicated.

Laplacian probabilities are as well suited for a logical as for a dualist interpretation. I discuss the logical interpretation of C. Stumpf, and the dualist interpretation of A. Meinong, and the physical interpretation of J. von Kries.

Though these interpretations deal perfectly well with many statistical phenomena, the frequency account of probability ultimately becomes more important in probability textbooks. A new way of looking at the facts becomes popular. The gestalt switch has occurred.

1 Have There Been Gestalt Switches in the Probabilistic Revolution?

This volume is dedicated to the exploration of the probabilistic revolution. This label has been coined to denote the change in the natural and social sciences due to the introduction of probabilistic concepts into their theories. The outlook of physics, biology, psychology, and sociology were changed considerably by the application of probability. This important scientific change also led to a different interpretation of the probability concept itself, and to its recognition within epistemology as an important tool of scientific enquiry.

Even if all these historical facts must be admitted as true, the question of whether the concept of probability led to a scientific revolution is still undecided. What is a scientific revolution? I do not want to explain this concept here in detail. Kuhn's article in this volume may help the reader to understand it to the extent that our context requires. Let us simply assume that we must consider the overthrow of the old paradigm and the gestalt switch as essential for a scientific revolution.

In many cases the gestalt switches do not change anything and are nothing but a reinterpretation of the same data already explained by the old theory. Thus Copernicus was not forced by the facts to give up the Ptolemaic theory of planetary motion. Before Galileo and Kepler the heliocentric theory was nothing more than a reformulation of the geocentric account that was somewhat simpler than but not strikingly superior to Ptolemy's theory. In some other cases the old theory is confronted with new data that it cannot account for without serious revisions, and the new theory contains the explanations of the old data and the new ones at the

My thanks go to Ted Porter for his help with style and quotations.

same time. Such a case is the change from the caloric theory of heat to the kinetic gas theory. I shall deal in this paper with the first case, a gestalt switch that does not involve any changes in the efficiency and power of the theory—but rather one that removes some odd auxiliary hypotheses.

There is a good indicator for gestalt switches of both kinds that is easily applied by the historian of science. The historian who has become bilingual and talks the languages of the old and of the new paradigms may be struck by the extent to which later scientists misunderstand the older ones, or by claims in both paradigms that are held to be self-evident or trivial, and that are mutually inconsistent.

What was the so called *probabilistic revolution*? The process so labeled is a complex one embracing half a dozen sciences and, above all, the theory of probability. There might have been gestalt switches in any of these fields—so we might be left with a whole bundle of loosely connected revolutions. But we ought also to look for the old paradigms that were overthrown: what was replaced by probabilistic methods and probabilistic theoretical ideas? In some sciences, such as medicine and sociology, there was resistance to the extensive use of probability. But was this resistance reactionary or was it due to the success of competing nonprobabilistic methods? In some other sciences probabilistic methods might have been the first effective ones, and the same may be the case for probabilistic models. In such cases nothing was there to be overthrown. Thus it is not immediately obvious that the adoption of probabilistic thinking in the sciences involved the main features of a scientific revolution in any of these fields. Nevertheless we have not yet found an argument against revolutionary changes or gestalt switches in the history of applied probability. We simply have to look for such phenomena.

One such gestalt switch, suggested by a distorted account of the old paradigm, involves the theory of probability itself. It is the change from the paradigm of Laplace to modern frequentist foundations of applied probability. Laplace's *Théorie analytique des probabilités* dominated the discipline for about a century. It was the model for nearly all textbooks that appeared in that time, as was Euclid's *Elements* for two thousand years in geometry or Newton's *Principia* for three hundred years in celestial mechanics.

Even if one century is a short time compared with the lifetime of the *Elements*, it is long enough to qualify Laplace's work as a paradigm. And I think that there are good reasons for the long reign of the classical theory of probability. It worked very well and provided a firm basis for the solution of many problems in the application of probability. It was compatible with a personalist, a dualist, and a logical interpretation of probability. It was superior to an exclusively objectivist interpretation, since it included a version of Bayes's principle that the objectivist had to reject without being able at that time to replace it by something equivalent. All these virtues of Laplace's theory will be discussed in the next section in more detail.

Nevertheless many writers in the twentieth century have shaken their heads over the classical theory and wondered how it was possible that such a stupid theory could be accepted for such a long time. There are several unjustified objections raised by, for example, von Mises and Reichenbach. One objection is directed against the law of large numbers as it is expressed by the theorems of Bernoulli,

Poisson, and Chebyshev. Von Mises writes, "The mathematical deductions of Bernoulli, Poisson, and Tschebysheff, based as they are on a definition of probability which has nothing to do with the frequency of occurrence of events in a sequence of observations, cannot be used for predictions relative to the results of such sequences. They have therefore, no connexion whatsoever with the general empirical rule formulated by Poisson in the introduction to his book." [1] This claim agrees with von Mises's interpretation of "equiprobable case": "The phrase 'equally likely cases' is exactly synonymous with 'equally probable cases'." The critique shows that von Mises no longer understood expressions like "There is no reason to believe that one of these cases should more tend to arrive than the others (*doit arriver plutôt que les autres*)," in short, to say that one event is more to be expected than another one. He is not willing to see in the classical definition any real content: "... we may say that, unless we consider the classical definition of probability to be a vicious circle, this definition means the reduction of all distributions to the simpler case of uniform distributions." [2]

Another objection is that the so-called a posteriori probabilities cannot be expressed by the definition as quotients of the numbers of favorable cases divided by the numbers of possible cases. What are the equipossible cases for the biased die, or for male and female birth? Both Reichenbach[3] and von Mises[4] consider the classical theory unable to account for such phenomena. Until recently, I too considered this objection decisive. But we shall soon see that the position of the Laplacians is not so hopeless as it seems at first sight from the modern standpoint. It might be wrong after all, but that cannot be decided so easily as Reichenbach and von Mises thought. If they are right, it is completely unimaginable that three generations of mathematicians, philosophers, and scientists were quite happy with Laplace's account of a posteriori probabilities.

These indications of the existence of a gestalt switch in the transition from Laplacian to frequentist probability may be sufficient for the moment. The aim of this paper is to contribute to a detailed analysis of this transition using material from German philosophy of nature. If the analysis is successful, we shall come to the conclusion that there is at least one tiny scientific revolution in the historical process that we call with more or less justification "the probabilistic revolution." There may be other ones as well.

To stress the gestalt switch even more and to clarify that people looked in a quite different way at the same probabilistic phenomena I want to quote an author of the *ancien régime*, C. Sigwart, a logician who feels indebted to Lotze's "reformation" of logic. I do not choose Sigwart because he was eminent. Rather, he is interesting for us as a typical representative of the old style of thinking. Sigwart says that the fact that all sides of a die appear nearly equally often *veils* the purely subjective character of probability.

Sigwart writes that probability is nothing but a measure for the degree of expectation and continues, "This purely subjective character of the theory of probabilities is often concealed by the fact that the illustrations chosen contain further knowledge; ... we know further ... from comparing our observations of a series of throws, ... that in a long series of throws the particular throws will appear on approximately equal number of times. . . ." [5] Relative frequencies are known to Sigwart but considered to be not very illuminating for the study of the nature of probability. He admits that a "homogeneous variability of causes" leads to the appearance of all equipossible cases with equal relative frequency in the long run. But that is simply not what probability is about. The concept of probability is more general than that of relative frequencies and also applicable to cases where such long-run predictions cannot be made.

For the physicist M. Smoluchowski writing three decades later it is just the other way round. Relative frequency is now *the* concept that is identified with probability: "In contrast to this, exact natural science is concerned . . . with objective or 'mathematical' probability, that is, with the relative frequency of the occurrence of certain random events. It thus employs this . . . highly ambiguous concept of probability in a very limited sense . . . which . . . permits an exact mathematical treatment. In this respect probability is like many other expressions, such as force, work, energy, and heat, which the physicist comprehends in an essentially different way than is customary in everyday life." [6]

The same attitude can also be found in H. Reichenbach's thesis not much earlier. It is for him self-evident that probability is a physical quantity: "In modern insurance companies we are presented with institutions whose success is only possible because statistical laws govern what actually occurs; and each shareholder in an insurance company demonstrates through the purchase of his shares that he has complete confidence in these laws." And after some remarks about modern statistical physics he continues, "All this authorizes us to surmise that the laws of probability are instances of objective laws of natural events, whose validity must be susceptible of philosophical demonstration." [7]

Insurance companies and stockholders, however, are no inventions of the twentieth century. They are about as old as probability theory, and the older companies behaved in the same way as the "modern" ones. So all theorists of probability knew about their existence, and many of them were not led by this fact to the conclusions drawn by Reichenbach. How can different conclusions be drawn in the face of the same premises? I think that such an observation indicates a gestalt switch that, in our case, involves a transformation in the concept of probability itself and in the foundation of its axioms.

It seems clear that Reichenbach suspects all previous writers on probability of being subjectivists, and of failing to see that probability is a physical quantity. He lives in a world that is quite different from that of the past century. It is puzzling to observe this discrepancy. Is it due exclusively to a new approach to the theory of probability or are there additional factors responsible for this rejection of the classical theory?

We have now seen the switch. It is not a change in any substantial belief about the world, but in the attitude toward reality and in the paradigmatic examples for the theory. The label "probability" is used for that function of theory, which is considered to be central for it and to be the most fundamental one. This function after the switch is different from the one before it.

Is this all that we can expect? Are there only separate changes in a bundle of different sciences, among which the replacement of Laplacian probability by relative frequency or related concepts is particularly significant? Or is there still a common feature of all these changes that may justify the label "probabilistic revolution"?

Let us look at the last quotations from Reichenbach: Reichenbach talks of "Wahrscheinlichkeitsgesetze" (statistical laws) that govern real processes and that are objective laws of nature. It is hard for us today to grasp that the concept of statistical law was a revolutionary innovation in science. But for many philosophers and scientists of the nineteenth century statistical laws were inconceivable. All statistical regularities predict future events with some probability and are consistent with the occurrence of the events and with their nonoccurrence as well. H. Lotze writes in his *Logik* referring to the observation of statistical regularities, "... and so it is very unsafe to characterize the results of such observations as *laws* of what happens.... A law ... is a hypothetical judgment and enunciates the necessary validity of a consequent provided the antecedent be valid." [8]

The last quotation makes clear that Lotze does not classify statistical regularities as lawlike statements. He does not have the same criteria for lawlike statements as we have today. Such criteria are an important part of the *framework* that we use for the interpretation of nature and society, and many scientific revolutions involve changes in the criteria of lawlikeness. It would be interesting to study how far *the* probabilistic revolution was essentially a replacement of an old criterion for laws by a new one; but this must be the subject of another paper. The paper of L. Krüger in this volume offers some insights on this question.

Thus it may be the case that there is indeed a general probabilistic revolution besides the local revolution in theory of probability. This, if it has taken place, was a change in the concept of natural law.

2 Interpretations of Probability

In philosophical discussions of probability the different interpretations of this term have played a leading role for more than 150 years. The discussion of these interpretations, however, suffers from the fluctuating meanings of words like "subjective," "objective," "logical," "personal," etc. The imprecision of these terms is in principle not different from the vagueness of other philosophical labels for "isms" of a different kind. Thus frequently such terms denote partly or essentially identical viewpoints. There exists no real difference between the views of two people if we can translate one view into the other. Thus a philosopher might claim to be a subjectivist toward probability, defining probability as the degree of justified belief. He might then develop a theory of determination of such beliefs that is clearly translatable into what would ordinarily be considered a frequentist view. Such an author should not be classified as a subjectivist, even if the introduction of his book on probability strongly suggests this label. An epistemic way of talking about methods, however, might be replaced by an ontic one in many issues of philosophy of science, and is by itself in no way significant. We may talk, for instance, about explanations in a context in which we also could talk about causes. Carnap calls the first way of talking the "formal mode of speech" and the second the "material mode of speech." [9]

I would like to exemplify this point using an example taken from the actual history of probability. In contemporary textbooks probabilities are functions of one or two arguments $p(A)$ or $p(B/A)$. In books of the nineteenth century, however, we find just the letter p as a symbol for probabilities. The use of this incomplete notation led to a lot of puzzles and difficulties. We know from the history of physics that incomplete notations frequently suggest views that would not have appeared if better symbols had been available. A famous example is Galileo's way of conceiving $a = b/c$ as $a \sim b$, $a \sim 1/c$, which does not show whether there is a third quantity $d \neq 1$ with $a = d \cdot (b/c)$ or not.[10] In the same way an argument of the probability function may be overlooked by the use of the symbol "p" for probability without any further specification. Today we tend to use at least a two-place function for probabilities. Frequentists read $r = p(B/A)$ as "Given the experimental condition A, the outcome B has the probability r." Adherents of logical probabilities would read this expression as "Given the total evidence $\hat{A}$, the probability that the proposition $\hat{B}$ is true has the value r." In a real-life situation the condition A and the total evidence $\hat{A}$ on the one hand and the outcome B and the proposition $\hat{B}$ on the other may correspond. Let $\hat{A}$ be what a person knows about his age and the state of his health, and let $\hat{B}$ be his death at a certain age of his life; then the subjective probability $r = \hat{p}(\hat{B}/\hat{A})$ may account for his contract with an insurance company. For the latter the probability $r = p(B/A)$ is the relevant fact, the frequency of people who know that they are in the state A and are in the state B, when they later die at a certain age. The actual determination of $r = p(B/A)$ and $r = \hat{p}(\hat{B}/\hat{A})$ may be exactly the same. So what is the difference? The insurance company uses the material mode of speech while the customer uses the formal mode. John Stuart Mill thought that there was such a difference and turned from a "frequentist" in the first edition of his *Logic*[11] into a "subjectivist" in the second. He simply wrote in the first edition from the standpoint of the insurance company and in the second from that of its customers. He does not go into the details of probability theory in his book. Therefore it is impossible to see whether there is a real change in his view after all.

Mill's change in view about probability might have been due to the fact that he did not see that the probability function was the same in both cases. Frequentist probability is a relative frequency of outcomes in a certain reference class. As acting persons, however, we need probabilities for future events without any recourse to a reference class, single-case probabilities. These do not exist when we assume that determinism is true. (The frequentist may consider the event e as an element of the class A of all events that happen under the same statistically relevant conditions. If determinism is true, this class A is just $\{e\}$, a one-element set containing only e. The probability function will then have only the values 1 and 0.) Subjective probability theory does yield single-case probabilities. Thus it seems at first sight that by switching from frequentist to personal probability we replace a two-place function by a one-place function. So it may be the case that in fact Mill just switched from an ontic to an epistemic account, from the material mode of speech to a formal mode of speech.

Another example of the lack of clarity about the arguments of the probability function is a paper written by A. Fick.[12] This author noticed that there is some structural similarity between a probability and an implication. That B under the condition A has the probability r seems to be something like a weakened implication. And indeed as we know today $p(B/A) = 1$ is not much different from $A \Rightarrow B$, if we read it as "If A, then necessarily B." Fick concludes that the probability function must have $A \Rightarrow B$ as its argument, that is, $p(A \Rightarrow B) = r$ instead of $p(B/A) = r$. That Fick's proposal is clearly wrong is not difficult to see by performing some operations in modal logic or the propositional calculus. Neither a necessary implication nor a material implication will make $p(A \Rightarrow B/T)$ equivalent to $p(A/B)$ (T is here the tautology). We are thus warned not to believe the claims of mathematicians and philosophers about their own interpretations. Insufficient notations and the correspondence between the material mode of speech and the formal mode are reasons for a different outlook on what are actually identical interpretations of probability. There are indeed a lot of cases in which a probability $p(B/A)$ can be read in both ways. I think that this is the reason for what I. Hacking calls the Janus-faced nature of probability.[13] If we have a fair die the physical features of which are sufficiently well known, its probability of 1/6 of yielding a 6 after having been thrown is such a case. It may be correctly looked at as both a relative frequency and as the degree of justified belief.

But there are also cases of a different kind, and that is why it makes sense after all to talk about different interpretations. I want here to give a classification of views about probability that "really make a difference," since their different probability functions are determined in different ways. I try to avoid the labels "objective" and "subjective" here, and to use as crucial criterion for the interpretation of probability the method of its derivation. We may distinguish[14]

Logical probabilities: These are functions of pairs of propositions A and B, $r = p_1(A/B)$. There is either one unique function, such that $r = p_1(B/A)$ is an analytical statement—in which case we deal with *unique logical probabilities*—or there are many of them, which might be distinguished by additional parameters $\gamma_1, \ldots, \gamma_n$: $r = p_1(B/A; \gamma_1, \ldots, \gamma_n)$—in which case the probabilities are *partly conventional*.

Personal probabilities: These functions are determined by testing the person π at the time t: $r = p_1(A; \pi, t)$. The argument A of the probability function is a proposition.

Physical probabilities: These are determined approximately by determining relative frequencies in (long) series of chance events: $r = p_1(B/A)$. For physical probabilities, the arguments of the probability function are sets of events. These arguments may be replaced by propositions that describe these events, but this would not turn physical probabilities into logical ones, since $r = p_1(B/A)$ is not held to be an analytical sentence.

If we ask which of these probabilities are objective and which are subjective, we shall not get a clear answer in any case. Logical probabilities may be interpreted both as subjective and as objective functions. A logical probability is simply calculated from the propositions A and B (and in the partly conventional case also from some parameters $\gamma_1, \ldots, \gamma_n$). If these propositions are looked at as states of affairs, the coefficient r is an objective fact. If they are looked at as thoughts, the coefficient $r = p_1(B/A)$ is subjective. It is clear that personal probabilities can only be considered subjective, and that physical probabilities must be objective.

The different views about probability might be connected by certain principles. There might, for instance, be a principle that any person who knows a frequency will bet in a way that his wagers are in agreement with the determined physical probabilities. This principle would connect personal with physical probabilities. As far as I know, this principle has been tested by psychological experiments and found to be wrong.

There might also be a principle connecting logical with physical probabilities: $p_{ph}(B/A) = p_1(B/A \wedge T)$, where T is a physical theory, from which the physical probability $p_{ph}(B/A)$ may be derived as a logical probability. Such a principle is assumed by both J. von Kries and by C. Stumpf, as we shall see later.

Also logical and personal probabilities might be connected by such a principle. Let $K(\pi, t)$ be the knowledge of a person π at the time t. Then the following relation may be valid:

$$p_{per}(A; \pi, t) = p_1(A/K(\pi, t)).$$

And finally we may relate personal and physical or logical and physical probabilities by the principle that in a series of events of the same kind A the probability p_1 or p_{per} that a future event $e \in A$ will also be $e \in B$ will converge to a certain limit p_{ph} with an increasing number of observations made of the series.

Thus we may think of several different interpretations of probability, which assume one or two of the probabilities p_{per}, p_1, and p_{ph} as reasonable concepts, and subscribe to one or some of the aforementioned principles that connect these concepts.

3   The Main Anomaly in Laplace's Theory

If we want to understand the vigor of Laplace's paradigm, we should immediately study what has been considered its Achilles heel, the distinction between a priori and a posteriori probabilities. Laplace and all his followers[15] first introduce a priori probabilities by the classical definition. Later we are taught that there is a second kind of probabilities, which we can approximately determine by the application of Bayes's principle. This principle enables us to compute from evidence the probability of the hypothesis that the probability of, say, a male birth is larger than 0.5 and smaller than 0.52. We are thus led to probabilities for which it is more or less probable that they have values within certain intervals. These new probabilities, which can themselves be subject to probabilistic treatment, seem to be something completely different from the a priori probabilities. How can they evolve from the classical definition of the number of favorable cases divided by the number of possible cases? This seems to be possible in some special cases, for instance, when we are dealing with urns, but not in general, in cases like radioactive decay, birth rates,

or genetic mutations. The fundamental difference between the two kinds of probabilities is revealed by the exposition common to most old-fashioned textbooks,
which use urns as examples for the application of Bayes's principle.[16] Usually the
following exercise is treated: Let there be an urn containing white and black balls in
an unknown proportion. A ball is drawn and replaced again m times. Of these balls
n prove to be white and $(m - n)$ to be black. If M is the total number of balls in the
urn, what is the probability in the light of the given evidence that the urn contains
N white and $M - N$ black balls?

In this example we are not asking for the probability of an objective probability
but rather for that of the physical state of the urn. From this physical state a
probability distribution may be derived via the classical definition of probability,
but the hypothesis itself, the probability that we want to estimate, is not a probabilistic statement. Therefore we have no problem with the classical definition for a
posteriori probabilities. The familiar exposition of Bayes's principle is already
found in Laplace's *Essai philosophique*.[17] But in the original text of his *Théorie
analytique des probabilités*[18] he does not apply this simplified treatment. He speaks
instead in an abstract way about causes and their probability:[19]

The probability of most simple events is unknown: considered *a priori* it would
seem to be capable of assuming all values between zero and unity; but, if a result
composed of numerous such events has already been observed, the way in which
they appear renders some of these values more probable than the others. Thus,
in proportion to the increase of simple events, compounded into an observed
result, their true possibility makes itself more and more known (*leur vraie
possibilité se fait de plus en plus connaître*), and it becomes more and more
probable that it will fall within limits that, becoming narrower without limit,
must end by coinciding with one another if the number of simple events becomes
infinite. To determine the law according to which this objective probability is
revealed, we shall call it x.... If one considers the different values of x as so
many causes of this result, then the probability of x will be ....

Then Laplace writes down the well-known Bayes formula for the probability of
causes.

The crucial phrase is here "...Si l'on considère les différentes valeurs de x comme
autant de causes de ce résultat...." (If one considers the different values of x as so
many causes of this result.) Laplace just assumes that there will be in any case a
cause, which corresponds to the "vraie possibilité" of the observed series of events.
If we ask that where causes are, we have to answer—so far as we are not dealing with
special cases like the selection of urns—that they can only be described as "causes
for the event to occur with probability x," which is not very different from the
famous "dormative potency" of opium in Molière's comedy *Le malade imaginaire*.

Here is the point of departure for two different interpretations of Laplace's
theory. *Either* we admit the existence of causes that are characterized by nothing
more than their possible effects. Then we accept a propensity theory of probability,
and conclude that Laplace's definition of probability does not hold for the general
case, since it cannot be applied to most a posteriori probabilities. *Or* we postulate

that in any case the observed probabilities could be calculated from a detailed description of the physical system using Laplace's definition if this system were known to us in all its details. Under this assumption we may claim that Bayes's principle is correctly derived for all possible applications from Laplace's definition, and is not introduced as an additional principle into the theory. We can preserve the uniformity of the theory of probability and hold that there is only one concept of probability that is subjective in cases of insufficient knowledge about the causes and objective in the case of sufficient knowledge.

If we adopt the first position, we are not committed to any metaphysical assumption. The second position, however, rests on the following a priori principle:

"Any observed long-run frequency can be calculated from Laplace's definition $p = n/N$ in an unambiguous way if the features of the physical systems in question are sufficiently well known."

The first position might have been very common in the nineteenth century. The second standpoint can be attributed to C. Stumpf. We could also consider this theory as the starting point of J. von Kries's reflections on probability. A Meinong was influenced by both philosophers and tried to find a synthesis of their conceptions.

4 German Philosopher Psychologists

There had been some discussion of the foundations of probability concurrently in Wilhelmian Germany and Austria. With the exception of Fries, Central European philosophers before that time made no effort to analyze the concept of probability.[20] The theory of probability was imported from France, and all textbooks of the first half of the century were translations from French originals.[21] But natural science underwent rapid changes in Germany after about 1840, and one of its most remarkable results was the rise of modern physiology, sensory physiology, and somewhat later, psychology. While physiology was always a discipline within the faculty of medicine, psychology arose as a branch of philosophy and was not institutionally separated from it in German universities before the Nazi time.[22] Philosophers and some physiologists focused their attention on problems of psychology because they were dissatisfied with the metaphysical speculations of German idealism, and thought the success of contemporary natural science offered a promise of constructive reform. This resulted in a desire for empirical work, which motivated some philosophers to study the human mind, a subject that had belonged to their domain since antiquity. The investigation of the mind included, besides such subjects as emotions, perceptions, and volitions, concept formation and logical inference. Thus probability, too, became a subject of psychology. While psychologists of our day use probabilistic methods as an instrument for research, for nineteenth-century psychologists it was a subject of direct investigation. Besides W. Wundt and C. Stumpf, A. Meinong and K. Marbe were psychologists, and in some ways G. Th. Fechner, who created a discipline called "Kollektivmasslehre," was also. A. Fick and J. von Kries were physiologists. R. H. Lotze, F. A. Lange, W. Windelband, and C. Sigwart have to be classified as philosophers or logicians, as we use these labels today.[23] For physiologists the approach to probability came from sense physiology, and from the problem of measurability of psychological quantities. According to the Weber-Fechner law, there is a certain quantitative correlation between physical stimuli and subjectively perceived quantities. J. von Kries rejected this law and claimed that no psychological intensities are capable of measurement.[24] Thus he was automatically led to the question whether this is also true for the degree of expectation, i.e., for probability, and so he had to deal with this subject.

It would be a misunderstanding however, to conclude from these remarks that the philosopher psychologists investigated the way in which people come to have personal probabilities. While some of them defended psychologism, others soon assumed a domain of logic that is intersubjective, i.e., independent of the state of mind of individuals. They considered the objectivity of logic to be an important insight. Stumpf, von Kries, and Meinong agreed that probability cannot be personal probability, but is at least either logical or physical.

The trend toward psychology exemplified by many Wilhelmian philosophers and their Austrian contemporaries and their tendency to discuss philosophical problems in psychological terms may be compared with the tendency of analytical philosophy (in the broadest sense) to reduce philosophical problems to linguistic ones. Both approaches have led to important philosophical insights. This is also true of the interpretation of probability.

I shall discuss only authors from the German scene. From A. Fick (1883) to A. Meinong (1915) these authors were barely influenced by ideas coming from England or America. They all had read Laplace, Cournot, and other French authors whose works existed in German translations, including Bertrand and Poincaré. But G. Boole and J. Venn were hardly noticed, and selected quotations from Venn were given to characterize him as a quite ridiculous philosopher who believed that relative frequencies will converge after a finite number of chance events.[25] The only English philosopher writing on probability who seemed to be well known was J. S. Mill, whose *Logic* had been translated into German.[26] Thus the German philosophical scene was quite independent of the English, and only after World War I, when the Anglo-Saxon and Central European contexts of discussion merged together, did philosophy of probability become an international issue.

5   Carl Stumpf's Logical Interpretation

I want to start with the theory of C. Stumpf, even though it appeared some years later than that of von Kries.[27] The reason is that Stumpf's interpretation is nearer to Laplace's original account, and we can understand von Kries's book better if we have studied Stumpf's article first.

From the two different ways to read Laplace—the dualist one and the logical one—Stumpf adopted the second, which is the simpler one, since it involves only one concept of probability. He started with Laplace's definition and tried to correct it. Laplace defines "probability" by "equal possibility," and therefore has a circularity in his definition, Stumpf thought, since possibility is but another word for probability. Accordingly, Stumpf was at pains to point out that probability rests on nothing but lack of knowledge. If I absolutely do not know if A_1 or A_2 will be the case, then both are equally probable. It is important, however, that I am quite clear about my knowledge B. $p(A_1/B)$ and $p(A_2/B)$ are functions of both A_1 or A_2 and of B, which describes what I know. This function is a logical relation between two propositions and one number. Thus we can say more accurately that A_1 and A_2 are equally probable under the condition B:

$$p(A_1/B) = p(A_2/B)$$

if B does not give any reason for a preference of A_1 over A_2, or vice versa. We are now prepared for a definition of Stumpf's probability that is better than his own:[28]

Let there be n mutually exclusive Propositions $C_1, \ldots, C_n$. Let A be the disjunction

$$A \leftrightarrow C_1 \vee C_2 \vee \cdots \vee C_m$$

and B the disjunction

$$B \leftrightarrow C_1 \vee C_2 \vee \cdots \vee C_n.$$

Then $p(A/B) = m/n$ if we cannot conclude from B anything that would let us know more about C_i than about C_j.

It looks as if we could blame Stumpf for the same circularity as Laplace. What does it mean "to know anything" or "nothing" about a certain event? If I know more than nothing about A and less than that A is true, I shall know some facts that indicate that A is more or less probable. Thus there may be an underlying concept of "gradual knowledge" that is as yet unanalyzed.

Stumpf tried to escape this difficulty. But his solution is entirely insufficient if it is not completed by additional criteria. He says that B has to be such that we *cannot know anything* about C_i and C_j on grounds of B. But then we would also know nothing about $C_i \vee C_{i+1}$ being more likely than C_{i+2}, and thus have no basis for any quantitative estimate of A. This objection leads to the von Kries-Bertrand paradox, which will be discussed in the next section.

The logical character of Stumpf's concept of probability implies that time plays no role in probability. Lotze and W. Wundt had the view that only future states of affairs could be more or less probable.[29]

Stumpf refers to some examples that show that we frequently talk about past events that were very improbable under the given circumstances though they certainly took place. Somebody might, for instance, have obtained 100 consecutive 1s with a die.

As has already been pointed out, logical probabilities $p(A/B)$ can be both subjective and objective. In the first case B is the knowledge of a person and in the second case B describes an objective fact not dependent on anybody's knowledge about it. Thus B might in the objective case be a physical description of a die containing all relevant details, and of the circumstances under which the die-throwing experiment takes place.

If one wants now to consider all physical probabilities as objective logical ones, one has to assume that for any of them such a proposition B exists.

For this purpose one has to postulate that all kinds of chances can be explained by a mechanism analogous to that of an urn. Let us look at an example: the ratio of male to female births.

We know from geneticists that the sex of a baby is determined at the moment when a spermatozoon enters an egg (ovum). There are two kinds of spermatozoa. Some produce boys and some girls. Now we can look at the sperm as an urn from which the egg takes a spermatozoon instead of a ball. If we can show that in the sperm are 18 boy-producing spermatozoa per 17 girl-producing spermatozoa, and if the chances of boy-producing and girl-producing spermatozoa invading the egg are equal, and if there is also no difference in the chances of boy embryos and girl embryos surviving, we can reduce the probability that a newborn baby will be a boy to Laplace's classical definition.

Stumpf seems to believe that for any a posteriori probability there is an explanation of this kind: "To posit (*hypothetisch setzen*) probabilities can only mean this: to posit certain real conditions that, if they were to hold, would provide the basis for ascribing just this probability to the event—e.g., ratios in a mixture of 1 : 1, 2 : 1, 3 : 1, and so on—on the basis of which one can ascribe probabilities of 1/2, 2/3, 3/4, and so on for the drawing of a white ball from the urn. The implication of this conclusion is thus that real conditions are analogous to the drawing of balls from an urn in which they are mixed approximately in proportion to the observed distribution."[30]

But what shall we do, if we have no idea what these real hypothetically assumed circumstances are like, as is mostly the case when we apply Bayes's principle? We know that there has to be an unknown hypothesis from which a certain value for the probability—which has been determined approximately—may be obtained by application of Laplace's definition. But how does this help us? Stumpf has to concede for the case of possibility of male birth: "The difficulty lies in showing how our knowledge of the real conditions has really increased or changed. For it could be said that, at bottom, we know absolutely nothing about the determination of sex, either before or after."[31]

In spite of the fact that we do not know the explanation of the chance $p = 106/206$ for male birth, we know at least something that is relevant to such an explanation. But how can we know a chance or probability if we do not know the facts in terms of which it is defined? Stumpf answers that we know at least that from 206 equipossible cases there are 106 favorable cases, no matter in which explanatory model these equipossible cases appear. The proportion $p(\text{male})/p(\text{female}) = 106/100$ is[32]

a numerical ratio of equally possible cases, namely, of 106 favorable against 100 unfavorable (or the reverse), among the 206 equally possible combinations of circumstances. We must suppose these 206 possible and 106 favorable combination to be distinguishable in themselves, although we are unable to designate them more precisely. We must conclude that the totality of conditions—whatever they may be—that leads directly in each case to the determination of sex in humans permits under the conditions that have produced these statistical results approximately 206 possible combinations, of which 100 fall under the concept "female," 106 under the concept "male" (i.e., give rise to the female or male sex), and that we find ourselves disjunctively (regarding any one in relation to all the others) in complete ignorance about each of these 206 possible combinations of circumstances, even if we are able to specify adequately the nature of these circumstances.

Thus Stumpf's definition of chance is the following: "A has the chance r under the condition B, ch$(A/B) = r$, means that there are in a (not necessarily known) theory M possible mutually exclusive sets of circumstances consistent with B, from which N sets are also consistent with A. There is no reason to expect the existence of any of these sets more than the existence of any other of them. Finally $r = N/M$."

One gets the impression that this definition is very intricate and clumsy. It refers to a theory that is true but may be unknown. Is such a definition logically admissible? I think there cannot be any objections of a purely logical character. It is quite possible to use unknown objects or entities for the definition of something that can be observed.

But even if we cannot refute Stumpf by logical arguments, we still have the impression that he is trying to save a paradigm that suffers from serious anomalies. His account is quite as complicated as Ptolemy's system of spheres, and one awaits a Kepler who will cut through the Gordian knot and will give a drastically simplified picture of the situation. Laplace's theory in Stumpf's interpretation is waiting for the gestalt switch.

6 Some Problems That Arise for C. Stumpf's Interpretation of Probability

A logical interpretation of probability has to face some important objections, which are seen as devastating by many philosophers of science. I want to discuss two of them here:

1. The *von Kries-Bertrand paradox*, which says that for a continuum of possible cases the classical probability cannot be uniquely determined.

2. The *loaded-coin objection*. This argument says that for a series of coin-throwing experiments the classical definition of probability leads to the prediction that a relative frequency of about .5 for heads is almost certain, while the coin may in fact be loaded.

The von Kries-Bertrand paradox was first discussed by von Kries and later in Bertrand's textbook.[33]

I do not want to discuss who discovered the "paradox" here. I think that anybody who deals with continua of possible cases should pose the question how a division into equipossible parts of such a continuum could be possible at all.

Von Kries discusses the paradox as follows: If one does not know anything about the size of the different countries in the world, one may assume that the probability that Denmark will be hit by a meteorite in the next 24 hours is the same as the probability for Mexico or the North Sea. But one could also divide the surface of the earth using criteria of a different kind. One could divide it into five continents and three oceans and would thus obtain a probability for a meteorite hitting Europe different from the probability obtained by application of the first approach.

For von Kries this "paradox" was one of the most important arguments for replacing Laplace's definition of probability by a more specific one, which I want to discuss later in a section on von Kries.

Stumpf's answer to the objection, however, is completely unsatisfactory.[34] He believes that the paradox is only apparent and is due to the fact that we pretend not to know more about the countries of the earth than their existence. It is simply not possible to abstract from a knowledge that one in fact has. If one considers a certain event to be more or less probable, one does this in the face of all the facts one knows. We are simply unable to abstract from the totality of evidence available to us when we try to judge intuitively about probabilities.

It may be that Stumpf is influenced here by an idea that is very important. Kant said that we can imagine a space without any things in it. Stumpf answers that one should really make this experiment, and that one would then notice that all representations of a space are representations of things in it that all have sensible qualities: colors, temperature, odor, etc.[35] Thus an experiment of mental representation should contain all details that have to be present in an imagined real case. We cannot discuss the intuitive plausibility of a thesis evaluating incomplete representations.

This applies to probability, too. Any realistically imagined situation will contain so much information that we can decide which is the set of really equally possible cases. Let us instead of the earth imagine a plane divided into five sections. A ball is thrown onto the plane, and if it hits a section it will drop into a bag corresponding to the section. We no longer ask, "What is the probability for section number 5 to be hit?" but rather, "What is the probability for the ball to be found in bag number 5?" Thus the sections of the plane are replaced by bags that can more easily be considered as equivalent goals of the ball than the planes.[36]

Stumpf's argument seems to have a certain appeal. But can we really say that it makes a difference whether we collect the balls in bags attached to the five sections or in two bags, one coordinated to section 1 and 2, and the other one to 3, 4, and 5? According to Stumpf, in the first case the probability for the ball to hit section 1 or 2 is $p = .4$ and in the second case it is $p = .5$. The values of probability are different but the situations are different, too.

I think that Stumpf's answer is unsatisfactory. As long as the procedure for the determination of the equally possible cases is not completely described, he simply cannot show that in any case there will be a unique result. If nothing else could be said in favor of a reasonable determination of equally possible cases, the conception of logical probability should be abolished altogether. There is, however, still a way out. In many cases a certain choice of equivalent alternatives seems to be more plausible than any other. In some other cases we may feel uncertain about it. But we could also determine the probability function partly by convention, as has been proposed by Poincaré.[37]

In many calculations we may start with more or less arbitrarily chosen probability functions and end up with a practically unique result. Poincaré discusses two types of such calculations, the derivation of the probability for gambling mechanisms, like rouge et noir (which plays also an important role in von Kries's theory[38]), and the application of the Bayes formula on the basis of a sufficient amount of evidence.[39] In the latter case the arbitrariness of the initial choice of a priori probabilities for the "causes" will average out if knowledge about sufficiently many "effects" is evaluated. These considerations show that von Kries's and Bertrand's paradox is an obstacle for the logical interpretation of probability but not sufficient for its refutation, even if Stumpf does not get the right answer.

We now come to Stumpf's second problem, the loaded coin objection. It runs as follows: We assume that the probability a tossed coin will yield heads is $p = 1/2$, the same its probability to yield tails. We do not know whether the coin is loaded or not, but we know that these tosses do not influence each other. Therefore we have to assume that the probability for the jth toss to yield heads is $p = 1/2$, independent of the preceding tosses. From these assumptions, it follows by Bernoulli's theorem that for 1,000 tosses it is more probable than the contrary that more than 450 and less than 550 tosses yield heads.[40] How can we deduce from a lack of knowledge such a specific conclusion? How can we deduce knowledge from ignorance?

Laplace would have responded to this objection that we clearly had made a mistake. Any toss is causally independent of the preceding toss, that is true. But that does not mean that we will expect a given result with equal intensity irrespective of the tosses made before. We can see this easily if we regard Laplace's formulation of the multiplication axiom. If there are on the one hand two events that are independent of each other ("indépendant l'un de l'autre"), their joint probability is the product of the probabilities of each of them. On the other hand if they are connected in such a way that the supposition of the first influences the expectation of the second, the relation is

$$p(E_1 \wedge E_2) = p(E_1) \cdot p(E_2/E_1).$$

Laplace uses the formulation, "If the simple events are arranged among themselves in such a way that the supposition of the occurrence of the first influences the probability of the occurrence of the second, ..."[41] This clearly shows that he is here using "probabilité" in the sense of "degree of expectation." For he does not say that the *existence* of the first event influences the probability of the second, but rather its "supposition," which is only possible for subjective probabilities. In the case of tosses of coins with unknown long-term behavior, the assumption that any long-run frequency is equally probable will yield the following probabilities ("h" for head, "t" for tail, "ht" for head-tail, etc.):

$p(h) = \frac{1}{2};$ $p(t) = \frac{1}{2};$ $p(hh) = \frac{1}{3};$ $p(ht) = p(th) = \frac{1}{6};$ $p(tt) = \frac{1}{3};$

$p(hhh) = p(ttt) = \frac{1}{4};$ $p(hht) = p(hth) = p(thh) = \frac{1}{12};$

$p(tth) = p(tht) = p(htt) = \frac{1}{12};$ etc.

These results are obtained by application of the Bayes formula. In a series of n tosses there will be equal probabilities for obtaining n heads, $n-1$ heads, ..., 2 heads, 1 head, and for obtaining only tails. It follows that $p(t/h) = 1/3$, and $p(t/t) = 2/3$, which clearly shows that independence does not hold.

For C. Stumpf and for A. Meinong the *loaded coin objection* poses a difficult problem. Stumpf formulates it in terms of urns filled with black and white balls. He asks, How can it make a difference if we *know* that an urn contains 50% black and 50% white balls or if the proportion of black and white balls is completely unknown? In the first case it is reasonable to predict about 500 ± 50 from 1,000 drawings of white balls with probability more than 1/2. In the second case one has to be mad if one strongly expects such a result. In both cases, however, we start with a probability $p = 1/2$ for drawing a single white ball and use the same rules from the calculus of probability. Stumpf writes, "I must confess that this difficulty tormented me for some time, for it seems to close every door." [42]

It is strange that Stumpf does not simply make use of the solution just mentioned that Laplace would have given, though some pages earlier he seems to be aware of it. He knows that for Laplace any proportion of black to white balls is equally probable (see p. 64). Evidently he did not completely understand the mathematical consequences of Laplace's assumption. The main reason for Stumpf's difficulties is his thesis, emphasized frequently in his paper, that probability rests on nothing else than lack of knowledge: "Only ... so long and so far as we are in ignorance, so long and so far does a probability take place...." [43] Therefore he does not see that the choice of a probability function can also rest on some knowledge about the world, and that only in such a case can we predict relative frequencies with a high degree of confidence. His solution, therefore, is also rather strange. He calls it "simple enough". [44]

Let us imagine instead of one urn in case B that there should be as many urns as drawings, e.g., 20,000, and that each time the drawing will be made from another urn. We are able to know that in each urn w or b (white or black) or a mixture of both types is present, but we lack the least basis to suspect any particular design or other constant causes that could have produced a certain similarity in the filling of the urns. In this case, even common sense does not run counter to the expectation that the final result of the drawings will yield, with great probability, an approximately equal distribution of w and b.

Stumpf seems to evade the difficulty by discussing another example. That was, according to a joke, the strategy of Khrushchev, who, when confronted with the high American standard of living during a visit to the United States, objected, "You Americans suppress the colored people." One cannot solve the difficulties arising in one situation just by dealing with another one. And above all Stumpf does not see that for a large number of drawings it makes no difference if one draws one ball from each of 20,000 urns or if the content of these urns is thrown together into one big urn, and if one draws 20,000 times from the compound urn (replacing the balls after each drawing). What counts here is the average proportion of white and black balls in the urns.

Thus Stumpf's interpretation of probability remains unsatisfactory after all. But that is not due to the theory of logical probability by itself. It only shows that *lack of knowledge cannot be the only basis* for assigning values to a probability function. There are some probability functions that are well adapted to the real world, and some that are badly adapted to it. We may start with any of them, if it does not prevent learning from experience. The logical theory of probability was later much improved by R. Carnap, and many objections that still apply to Stumpf's account can be rejected on the basis of Carnap's theory.[45] Thus we may say that logical Laplacian probability is still alive, defended by a minority of scientists, and was never refuted. The gestalt switch of the majority is not a result of brute facts, neither mathematical nor empirical ones.

7 Johannes von Kries and Alexius Meinong

As was already pointed out in the last section, von Kries's and Bertrand's paradox does not prevent the application of probability if we assume with Poincaré that we may start with a whole class of probability functions from which we choose one by convention for our calculations. The functions belonging to this class have to satisfy some general criteria. Thus they have to vary rather smoothly with any continuous parameter and they must not be zero for any proposition that is not a contradiction. We may thus interpret Laplace's theory either in terms of partly conventional logical probabilities alone, or additionally in terms of physical probabilities. The latter dualist standpoint may be the more convenient one. It allows for statements like

$$p_l(p_{ph}(B) = r/A) = s,$$

referring to the logical probability of physical probabilities.

A monistic interpretation of probability, however, may be considered as philosophically more satisfactory than a dualistic one. In this interpretation we have to replace physical probabilities by logical ones $p(B/A \wedge T)$, where A formulates some boundary conditions and T is a scientific theory from which these probabilities, the values of p for given A and B, may be calculated. The theory T itself does not contain any propositions about probability. The function $p(B/A \wedge T)$ is thus a logical but objective probability function, since A and T do not denote the knowledge of some person but the objective facts that are causes of the physical event B. There is no path, however, leading by logical deduction from a nonprobabilistic theory—say, about mechanical systems, their forces, etc.—to probabilities. We need at least some general a priori principles for such a derivation. This situation is well known from statistical mechanics. We cannot derive probability distributions from mechanics alone. We need additional premises like Liouville's theorem, the ergodic hypothesis, or temporal invariance assumptions. J. von Kries's account of probabilities can be understood as a logical interpretation for objective probabilities $p(B/A \wedge T)$ of this kind. He does not try to define probability by the method used to determine them empirically, as frequentists do, nor in terms of decisions or bets, as personalists do, but he does it rather by the way they are explained in natural science.

I have described von Kries's theory in some detail in another paper,[46] and therefore I want to deal with it here only very briefly. Von Kries starts with the theory of games of chance like roulette, dice, etc. All these games make use of mechanical systems that produce events with certain fixed probabilities. This fact is usually explained in the following way:

Let $s(t)$ be the state of the mechanical system S, the chance apparatus, at time t. $s(t)$ is a vector of parameters $\alpha_1, \ldots, \alpha_n$ that describe this state. Then we may assume a probability distribution

$$dp_0 = q(s)\, d\alpha_1 \cdots d\alpha_n$$

for the probability of the system S at time t_0 to be in the state s. dp_0 may, for example, be the probability that the croupier starts the roulette wheel at t_0 with angular velocity between w and $w + dw$. The distribution dp_0 is arbitrary. It should, however, be a sufficiently smooth function. The theory of chance designs now shows that the resulting probabilities $p(B_i)$ at time t_1 for the outcomes B_i are largely independent of the choice of dp_0 as long as this density is smooth. Therefore these probabilities are no longer subject to the von Kries-Bertrand paradox. We may derive them from the laws of mechanics and the boundary conditions, using as the only probabilistic presupposition the smoothness of dp_0. This result is used by von Kries for the definition of probability $p(B_i)$. There is a unique procedure for obtaining them from the theory T and the boundary conditions A that completely fixes the probability $p(B_i/A \wedge T)$ as a mathematical function.

It is clear that von Kries's definition is without any real value if it applies only to games of chance. But he believes that it is quite generally valid, and that any probability that appears in science with a definite value under certain conditions may be derived in this way. And indeed this seems to be true, at least for statistical mechanics. The mechanisms that are responsible for diffusion, Brownian motion, and the like are very similar to the roulette or to a dice-throwing machine.

But even if it is true that probabilities can be calculated in this way, we still do not know what they have to do with experience. Like most of the German writers on probability, von Kries points out that any empirical result is consistent with any value of an objective probability $p(B_i/A \wedge T)$.[47] It is indeed possible to throw a series of 1,000 aces with a fair die. But we would not expect that such an event really would happen. Therefore the relation between calculated probabilities and empirical results is a bit more complicated. The connection is established by the "Princip der Spielräume"! The principle says—if we translate it into the language of the preceding brief exposition of von Kries's theory—that B_i and B_j are to be expected equally if the calculated values $p(B_i/A \wedge T)$ and $p(B_j/A \wedge T)$ are equal. From this principle we may therefore conclude that B_i is almost certain if $p(B_i/A \wedge T)$ is nearly equal to 1. The "Princip der Spielräume" is held to be a priori by von Kries.[48] This leads to the following question: If the probabilities $P(B_i/A \wedge T)$ are obtained by completely a priori considerations from B_i, A, T, and from the "Princip der Spielräume," what can we do if they do not agree with observed relative frequencies? Shall we conclude that what in fact happens is very improbable? Von Kries answers that in such a case either A or T may be false propositions, and we should look for other explanations A' or T'. Thus a good theoretical explanation of observed relative frequencies should not make them improbable. In this respect von Kries's objective logical probabilities agree with experience.

Before ending this very brief account of von Kries's important ideas, I want to add two points in order to prevent misunderstandings. As has already been mentioned, von Kries was led to the study of probability by his wish to defend the thesis that psychological quantities are not measurable. Since probability is measurable, it could not be psychological according to his thesis. Von Kries was convinced that his claim was confirmed by his findings about probability. That means that he does not accept personal probabilities or logical probabilities $p(B/A)$ where A is the knowledge of a person.

On the other hand, I think that von Kries's rejection of subjective probabilities is not essential to his account of the objective ones. His theory could equally well be adopted by a monist like Stumpf as by a dualist like W. Lexis or A. Meinong. The next point to be made here is that the full impact of von Kries's ideas was not understood by most of his readers. Thus neither W. Stumpf nor A. Meinong really grasped von Kries's fundamental points. This is partly due to von Kries's rather involved style, which is not atypical for his time. Furthermore, von Kries did not work out his ideas in mathematical terms but preferred a verbal account, and only today in light of some later theories of statistical mechanics is it easy to understand what von Kries was talking about. We are now able to see why his *Principien der Wahrscheinlichkeitslehre* was the most intelligent and sophisticated book on probability in Germany before World War I.

The last of the three philosopher psychologists or philosopher physiologists considered is Alexius Meinong. He was the same age as von Kries and five years younger then C. Stumpf. Like the latter, he was strongly influenced by F. Brentano. Meinong has become well known because of the attention paid to him by B. Russell, who wrote a review of Meinong's most important book, *Über Annahmen*.[49] Russell's critique of Meinong's theory of meaning was one of the landmarks of the incipient analytical philosophy. This may be one of the reasons why Meinong has become classical in some respect.

Meinong's account of probability is inseparably connected with his philosophy in general. Therefore it is hardly possible to give a short outline of it without reviewing his semantics and ontology. If one neglects Meinong's philosophical background and gives only a simplified and reduced exposition, one might say that Meinong's analysis rests upon the work of von Kries and Stumpf. He is a dualist and accepts both logical and physical probabilities. The first ones are held to be probabilities proper: "Whoever wishes to give or receive information about a higher or lower probability (of an event) obviously relies on the greater on lesser strength of the suppositions associated (with the idea of it)—not, of course, suppositions that he just happens to have in his mind, but rather justified ones, that is—as we know—the self-evident suppositions." [50] And he was convinced that the "unsubjective use of the word probability (*Wahrscheinlichkeit*) remains more foreign to our linguistic instinct than the subjective one." [51] But that does not mean that there are no physical probabilities, which he calls "possibilities" (*Möglichkeiten*), as was already done by Laplace, Poisson, and Cournot. Meinong points out that possibilities are not directly found in the physical world. They belong to the domain of the "unsubjective"; they are independent of our individual states of mind. Meinong, like most German philosophers, points out that probability is "nowhere to be met in the domain of the actual, even with the most attentive search." [52] Nobody has ever found an infinite series of relative frequencies converging to a certain limit. Every finite series is compatible with any value of possibility. (This argument has already been used by von Kries.) But on the other hand, if it is a fact that an apparatus or creature will react to a certain stimulus A with a certain probability p of a response B, this will be a characteristic feature of it, a disposition. Meinong analyzed the relation of physical probabilities to dispositional possibilities and necessities, i.e., modalities, and here he came to his most valuable results. [53]

In spite of these new and interesting insights, Meinong is still an exponent of the old Laplacian paradigm, and does not see any reason to accept frequentist or exclusively physical probabilities. If he is asking for the nature of probabilities, he does not look for a method of its determination or for an operational definition, but for the way in which probabilities are conceived in the human mind, and there probabilities will always be connected with expectations. He was looking for "descriptions of essence" (*Wesensbeschreibungen*), and though he knew all the modern ideas of the frequentists like J. Venn, he could not see how these people gave "descriptions of essence" for the concept of probability—which in fact they did not do, if the term "Wesensbeschreibung" is understood in Meinong's sense. He could not think of any other kind of rational reconstruction of a concept, however. Therefore he was also unable to accept any operational approach to probability, an approach that, in the hands of von Mises, Reichenbach, and others at about the same time, led to a new foundation of the theory of probability.

8 The New Foundation

In the first decades of our century the change in the theory of probability took place. Textbooks of the Laplacian type gradually became rarer and finally disappeared. One of the last ones in the German language was E. Czuber's *Wahrscheinlich-keitslehre*, published in 1902 and 1908. But already in 1905 we have in H. Bruns's *Wahrscheinlichkeitsrechnung und Kollektivmasslehre* a different kind of textbook. The title points already to the roots of the new theory, to Fechner's *Kollektiv-*

masslehre, which was published 1897, nearly a decade after his death.[54] Since
M. Heidelberger has written extensively about Fechner in this volume, it is not
necessary to say much about "collectives." These are described by Fechner as
potentially infinite sets of objects with a chance distribution for some quantities or
features. Bruns had also taken notice of K. Pearson's papers, but these were less
important for him than Fechner's book.

Bruns's textbook was still a hybrid of the Laplacian and the frequentist approaches. The author was aware of this fact and sympathized with Hausdorff's
completely frequentist foundation for probability.[55] He considered the old exposition, however, as more convenient for undergraduate students. But in the crucial
points he had abandoned the Laplacian style of argument. He introduced the
theory of probability first as a purely formal theory, and then added an interpre-
tational rule, the "Satz von der gleichmäßigen Erschöpfung der Fälle" (theorem of
equal exhaustion of cases).[56] It says that in a series of events, equipossible cases
appear equally often in the long run. This principle defines equipossibility as an
empirical concept. Another essential point is the abolition of *Bayes's principle*. He
aknowledged a restricted application of Bayes's theorem to the case in which only
frequentist probabilities are involved.[57] But he no longer needed to move from
observations to estimated physical probabilities, as Poincaré still believed necessary.[58] Probabilities, he thought, are relative frequencies and, therefore, the a
priori probabilities of Bayes's formula for hypotheses become meaningless. One has
to apply other methods for estimating physical probabilities from observations. If
Bruns asks for the adequacy of a measured value for a physical quantity, he no
longer asks, "What *is the probability* that the mean value of my measurement is
equal to the true value of the quantity within certain given limits of error?" He
rather asks, "If I make measurements of the same kind, *how often in the long* run
will I get an estimated value that is equal to the true value of the quantity within
certain given limits of error?"[59] For the answer to the first question he would have
needed Bayes's formula involving subjective probabilities. For the answer to the
second question, which is quite different, he can dispense with subjective probabilities altogether.[60] When the axioms of probability theory were derived from the
relative frequency definition of probability, and when methods for estimating
physical quantities and probability distributions were found that rendered Bayes's
principle superfluous, the Laplacian theory of probability was dead. A completely
new approach, a new disciplinary matrix or paradigm—using Kuhn's expressions—had become dominant. The term "probability" was no longer identified
with subjective probability, which contrasts physical "possibility" or "chance"
to physical probability. People now tended to say that probability *is* a physical
quantity, while for the older generation it was quite clear that "Wahrscheinlichkeit"
(probability) was held by any native speaker of German to denote a degree of
expectation. Thus a theory of physical probability finally replaced the dualist or
logical interpretation of the theory of Laplace, and the new interpretation was
one that "really made a difference," since a very important statistical method,
namely, the application of Bayes's formula, was no longer justifiable within the
new framework.

But why did the new approach not come earlier? What were the reasons that finally caused the change? I cannot give a satisfactory answer to that question. There was no new fact that was discovered and was inconsistent with the old view. The gestalt switch was caused at least partly by external reasons. On the one hand the increasing investigation of probability distributions, i.e., of the "a posteriori probabilities," led to the impression that the "a priori probabilities" are only unimportant special cases of quantities that usually have to be determined by experience. Such a new attitude toward probability did not change much as long as one still believed subjective probabilities to be the necessary starting point for the determination of the objective ones. But the more probabilistic laws, such as Mendel's laws or the statistical laws of Brownian motion or radioactive decay, became known in natural science, the more it became clear that physical probabilities are physical quantities, while subjective probabilities are at best methodological tools—if they can be justified at all. Such a change in the general perspective, however, did not lead by itself to a new theory, no more than medieval criticism of Aristotle's physics could by itself create Newtonian mechanics. Some people, for example, O. Sterzinger, even believed that physical probabilities did not necessarily satisfy the axioms of Laplace.[61] Like any set of physical quantities, they can be related to each other by empirical laws that cannot be known beforehand, and that must be explored by empirical investigations. The derivation of the axioms from the relative frequency definition of probability, already given in informal terms by J. Venn, appeared quite late in the German mathematical literature. It is clear that the increasing number of known statistical regularities led to the insight that these really are natural laws, and supported a change of criteria of lawlikeness. At the end of section 1, I already pointed out that the old concept of a natural law did not allow for statistical laws, neither in physics nor elsewhere. The adoption of a new type of natural law may also have influenced the observed change in the concept of probability.

We may also think of external factors, like the increasing tendency to separate applied from pure mathematics, which characterizes the nineteenth century as a whole. For Kant mathematics still was essentially about the forms of intuition. Any kind of experience therefore had to follow the mathematical laws. Thus for Kant mathematics was essentially applied to physical or psychological reality. But in the nineteenth century analytical geometry, calculus, and many other branches of mathematics that dealt with numbers or functions were no longer considered as applying to something physical. The study of geometrical probabilities then showed that the theory of probability could also be considered as a branch of pure mathematics in terms of numbers, functions, and the like. Von Kries's and Bertrand's paradox in some way formulates the insight that something has to be added to the considerations of pure mathematics if one wants to talk of probabilities, and thus to use a concept of applied mathematics. These considerations have some similarity to Poincaré's discussion of the conventionality of geometrical concepts.

Besides that there is the fundamental conceptual change taking place at the end of the nineteenth century, which may be called the linguistic turn in science. The change from Laplacian to physical probability or from deterministic to probabilistic laws is logically independent of the linguistic turn. Psychologically, however, there is a connection, as we have seen in the case of A. Meinong. It was difficult for philosophers or mathematicians to reject subjective probability as long as they considered logic (including philosophy of science) as a branch of psychology and not primarily as an enterprise dealing with symbols and with language.

Thus the revolution in the theory of probability is a very complex phenomenon, which was influenced by a host of conceptual changes in science, and moreover by several external factors. I do not think that the changes mentioned in the general scientific attitude are by themselves satisfactory explanations for the gestalt switch in probability theory. But I hope I have contributed a bit to the understanding of the decline of Laplacian probability.

Notes

1. Richard von Mises, *Probability, Statistics, and Truth*, 2nd ed. (London: Allen and Unwin/ New York: Macmillan, 1957), p. 109.

2. Von Mises, *Probability*, p. 68.

3. Hans Reichenbach, *The Theory of Probability* (Berkeley/Los Angeles: University of California Press, 1949), p. 354.

4. Von Mises, *Probability*, pp. 69ff. (note 1).

5. Christoph Sigwart, *Logic* (New York: Macmillan, 1895), vol. 2, p. 224 (§85.8), *Logik* (Freiburg i.B.: Mohr, 1893), vol. 2, p. 315 (§85.8).

6. M. von Smoluchowski, "Über den Begriff des Zufalls und den Ursprung der Wahrscheinlichkeitsgesetze in der Physik," *Die Naturwissenschaften*, 6 (1918), 253–263, on p. 254.

7. Hans Reichenbach, *Der Begriff der Wahrscheinlichkeit für die mathematische Darstellung der Wirklichkeit* (Leipzig: Barth, 1915), p. 13. See also Henri Poincaré, *Science and Hypothesis* (New York: Dover 1952), p. 188.

8. Hermann Lotze, *Logik*, 2nd ed. (Leipzig: Hirzel, 1880), §287.

9. Rudolf Carnap, *Logische Syntax der Sprache* (Wien: Springer, 1934), p. 181, §64.

10. See E. J. Dijksterhuis, *Die Mechanisierung des Weltbildes* (Berlin/Göttingen/Heidelberg: Springer, 1956), §IV, 93, p. 381.

11. John Stuart Mill, *Collected Works* (Toronto: University of Toronto Press/London: Routledge & Kegan Paul, 1967–1974), vols. 7–8: *A System of Logic* (vols. 1 and 2), vol. 7, pp. 534–547, vol. 8, pp. 1140–1150.

12. Adolf Fick, *Philosophischer Versuch über die Wahrscheinlichkeiten* (Würzburg: Stahel'sche Universitäts-Buch- und Kunsthandlung, 1883), pp. 12–13.

13. Ian Hacking, *The Emergence of Probability* (Cambridge: Cambridge University Press, 1978), p. 12.

14. The classification given here is essentially identical with that of Leonard J. Savage, in *The Foundations of Statistics* (New York: Dover, 1972), p. 3.

15. See, for example, Emmanuel Czuber, *Wahrscheinlichkeitsrechnung und ihre Anwendung auf Fehlerausgleichsrechnung, Statistik und Lebensversicherung*, 2nd ed. (Leipzig/Berlin: Teubner, 1908).

16. Emmanuel Czuber, *Wahrscheinlichkeitsrechnung*, pp. 171–173 (note 15).

17. Pierre Simon Marquis de Laplace, *Essai philosophique sur les probabilités*, in *Oeuvres complètes*, vol. 7.2 (Paris: Gauthier-Villar, 1886).

18. Pierre Simon Marquis de Laplace, *Théorie analytique des probabilités*, *Oeuvres*, vol. 7.2 (note 17).

19. Pierre Simon Marquis de Laplace, *Oeuvres*, vol. 7.2, p. 370 (note 17).

20. Jakob Friedrich Fries, "Versuch einer Kritik der Prinzipien der Wahrscheinlichkeitsrechnung," in *Sämtliche Schriften*, vol. 14 (Aalen: Scientia, 1974). See p. 74 for Fries's adoption of the Bayes principle.

21. See Ivo Schneider, "Laplace and Thereafter: The Status of Probability Calculus in the Nineteenth Century," this volume.

22. This became clear to me when I checked all "Vorlesungsverzeichnisse" (catalogues) of German Universities from 1932.

23. Wilhelm Wundt, *Logik*, vol. 1: *Allgemeine Logik und Erkenntnistheorie*, 3rd ed. (Stuttgart: Enke, 1906), pp. 407–436; Carl Stumpf, "Über den Begriff der mathematischen Wahrscheinlichkeit," *Sitzungsberichte der philos.-philol. u. der historischen Classe der Königlich bayerischen Akademie der Wissenschaften zu München*, Jahrgang 1892, pp. 37-120; Alexius Meinong, *Über Möglichkeit und Wahrscheinlichkeit*, (Leipzig: Ambrosius Barth, 1915); Karl Marbe, *Die Gleichförmigkeit der Welt* (München: Beck, 1916); Gustav Theodor Fechner, *Kollektivmasslehre* (hrg. von Gottlob Friedrich Lipps) (Leipzig: 1897); Johann von Kries, *Die Principien der Wahrscheinlichkeitsrechnung* (Tübingen: Mohr 1st ed. 1886, 2nd ed. 1927); Hermann Lotze, *Logik* (note 8); Adolf Fick, *Versuch* (note 12); Friedrich Albert Lange, *Logische Studien* (Leipzig: Baedecker, 1894), chapter 5: "Das disjunktive Urteil und die Elemente der Wahrscheinlichkeitslehre," pp. 99-126; Wilhelm Windelband, *Die Lehren vom Zufall* (Berlin: Henschel, 1870); Christoph Sigwart, *Logik* (note 5), §85; Alexius Meinong, *Über Möglichkeit und Wahrscheinlichkeit* (Leipzig: Ambrosius Barth), reprinted as vol. 6 of *Gesamtausgabe* (Graz: Akademischer Verlag, 1972).

24. Johannes von Kries, *Principien*, p. 3 (note 23).

25. Johannes von Kries, *Principien*, p. 294 (note 23); Carl Stumpf, "Wahrscheinlichkeit," p. 81 (note 23); Alexius Meinong, *Möglichkeit*, pp. 2, 6, 9ff., 13 (note 23).

26. See note 11.

27. Carl Stumpf, "Wahrscheinlichkeit" (note 23), Johannes von Kries, *Principien* (note 23).

28. Carl Stumpf, "Wahrscheinlichkeit," p. 48 (note 23).

29. Hermann Lotze, *Logik* (note 8), pp. 414, 432ff., 434; Wilhelm Wundt, *Logik* (note 23), p. 393.

30. Carl Stumpf, "Wahrscheinlichkeit," p. 102 (note 23).

31. Carl Stumpf, "Wahrscheinlichkeit," p. 103 (note 23).

32. Carl Stumpf, "Wahrscheinlichkeit," p. 106 (note 23).

33. Johannes von Kries, *Principien* (note 23), pp. 8–9; Joseph L. F. Bertrand, *Calcul des probabilités* (Paris: 1888, 2nd ed.; Paris: 1907; repr. Bronx: Chelsea, 1978), pp. 4ff.

34. Stumpf, "Wahrscheinlichkeit," p. 50 (note 23).

35. Stumpf, *Über den psychologischen Ursprung der Raumvorstellung* (Leipzig: Hirzel, 1873), p. 20.

36. Stumpf, "Wahrscheinlichkeit," pp. 70ff. (note 23).

37. Henri Poincaré, *Science and Hypothesis* (New York: Dover 1952), pp. 203ff.

38. Andreas Kamlah, "Probability as a Quasi-Theoretical Concept: J. V. Kries' Sophisticated Account after a Century," *Erkenntnis* 19 (1983), 239-251 (see pp. 245ff).

39. Poincaré, *Science*, pp. 201-206 (note 37).

40. Stumpf, "Wahrscheinlichkeit," pp. 87ff. (note 23).

41. Laplace, *Theorie analytique*, p. 182 (note 17).

42. Stumpf, "Wahrscheinlichkeit," p. 88 (note 23).

43. Stumpf, "Wahrscheinlichkeit," pp. 62ff. (note 23).

44. Stumpf, "Wahrscheinlichkeit," pp. 88ff. (note 23).

45. Rudolf Carnap, *Logical Foundations of Probability* (Chicago: University of Chicago Press, 1950).

46. Kamlah, "Von Kries' account," pp. 243-247 (note 38).

47. Von Kries, *Principien* (note 23), p. 21. See also Meinong, *Möglichkeit* (note 23), pp. 2, 9.

48. Von Kries, *Principien* (note 23), pp. 157ff. See also Andreas Kamlah, "The Neo-Kantian Origin of Hans Reichenbach's Principle of Induction," in Nicholas Rescher (ed.), *The Heritage of Logical Positivism* (Lanham/New York/London: University Press of America, 1985), pp. 164ff.

49. Alexius Meinong, *Über Annahmen, Gesamtausgabe* (note 23), vol. 4, 1977; Bertrand Russell, "Meinong's Theory of Complexes and Assumptions", *Mind*, 13 (1904), 204-219, 336-354, 509-524.

50. Meinong, *Möglichkeit*, p. 489 (note 23).

51. Meinong, *Möglichkeit*, p. 305 (note 23).

52. Meinong, *Möglichkeit*, p. 2 (note 23).

53. Meinong, *Möglichkeit*, p. 261 (note 23).

54. Czuber, *Wahrscheinlichkeitslehre* (note 15); Heinrich Bruns, *Wahrscheinlichkeitsrechnung und Kollektivmasslehre* (Leipzig/Berlin: Teubner, 1906); Fechner, *Kollektivmasslehre* (note 23).

55. Felix Hausdorff, "Beiträge zur Wahrscheinlichkeitsrechnung," *Berichte der math.-phys. Klasse der Kgl. sächsischen Gesellschaft der Wissenschaften* (1901).

56. Bruns, *Wahrscheinlichkeitsrechnung*, p. 13 (note 54).

57. Bruns, *Wahrscheinlichkeitsrechnung*, p. 31 (note 54).

58. Poincaré, *Science*, p. 204 (note 37).

59. Bruns, *Wahrscheinlichkeitsrechnung*, pp. 236ff. (note 54).

60. John Venn, Charles Sanders Peirce, and Karl Pearson already had ideas of this kind. See J. Venn, *The Logic of Chance* (1st ed., London and Cambridge: Macmillan 1866), pp. 203-211; K. Pearson, *The Grammar of Science* (Gloucester, MA: Smith 1957), p. 146; Ch. S. Peirce, *The Philosophy of Peirce. Selected Writings* (ed. by Justus Buchler, London: Trench, Trubner & Co., 1940), pp. 327, 217

61. Othmar Sterzinger, *Zur Logik und Naturphilosophie der Wahrscheinlichkeitslehre* (Leipzig: 1911), p. 101.

6 Fechner's Indeterminism: From Freedom to Laws of Chance

Michael Heidelberger

Gustav Theodor Fechner (1801–1887), known as the founder of psychophysics, was the first universal indeterminist. He not only held that man is able to act without being determined by previous events but that nature in general possesses a certain amount of indeterminateness, which "truly depends on freedom" and not on "our ignorance of the conditions."

In the first section Fechner's psychophysical parallelism is sketched. The second section deals with his indeterministic thinking of 1849 in detail. There, he develops four different types of indetermination. Since present events are never just exact repetitions of earlier events, there is some amount of undetermined novelty that comes into the world with each new moment.

In the third section some aspects of Fechner's thinking are related to a wider context: There is late idealism criticizing the inability of Hegel's logic to account for freedom, individuality, and the organic realm. There is also a link to contemporary philosophy of history, epigenetic thinking, and antiessentialism.

The fourth section deals with Fechner's posthumously published Kollektivmasslehre *of 1897. This is the mathematical formulation of his earlier ideas on indeterminism.*

The fifth section shows that Fechner's theory led to the frequency theory of von Mises. The two postulates of von Mises, the existence of the limit and the excluded gambling system, are already present in Fechner's conception.

The conclusion stresses that the roots of twentieth-century indeterminism lie further back than is usually supposed.

Wir wissen mit weit mehr Deutlichkeit, dass unser Wille frei ist, als dass Alles, was geschieht, eine Ursache haben müsse. Könnte man also nicht einmal das Argument umkehren und sagen: Unsere Begriffe von Ursache und Wirkung müssen sehr unrichtig sein, weil unser Wille nicht frei sein könnte, wenn sie richtig wären?

Lichtenberg

1 Introduction, and Historical Setting

The science of the nineteenth century was ruled by mechanistic physics, and mechanistic physics is intimately connected with a deterministic conception of causality. Determinism was the subject of heated debates in which many philosophers, historians, and theologians developed their arguments and counterarguments. It seems, however, that philosophical claims cannot be held responsible for having raised indeterminism to a respectable and very often essential part of our contemporary scientific outlook. It is rather scientific practice itself that demanded and brought about the transition to indeterminism. Richard von Mises, to whom

I am grateful to Raine Daston, Marilyn Marshall, and Gad Freudenthal for their comments on an earlier version of this chapter.

we shall turn again at the end of this paper, put it like this: "The causal principle is not invariable and it shall submit to the requirements of physics."[1]

There was, however, at least one exception. The physicist and philosopher Gustav Theodor Fechner developed an indeterministic alternative by subordinating the causal principle to the requirements of our freedom as human beings. This was done without giving up the scientific outlook of his time and without taking recourse to an arbitrary suspension of causality. Unlike many other attempts with a similar goal, essential parts of Fechner's ideas survived and eventually became an integral component of the twentieth-century theory of probability.

In the following paper I shall discuss the origin and development of Fechner's indeterminism. The story begins with Fechner's solution to the mind-body problem, where he adopts a special variant of "identity-theory." This is the core of Fechner's thinking, both scientific and philosophical. The identity-theory requires that the freedom of the mind (as well as of any other spiritual element) becomes manifest also in the physical world. Fechner presented the solution of this problem in 1849 by developing four conceptions of objective indeterminism.[2] Several aspects of Fechner's thinking during this period can be linked to anti-Hegelian "late idealism," to philosophy of history, and to biological ideas of epigenetic thinking and antiessentialism.

In the later period of his life Fechner tried to provide a mathematical expression for his ideas. To this end, he follows the tradition of Quetelet's moral statistics and the mathematics of astronomical error theory. His posthumous *Kollektivmasslehre* of 1897 deals with the distributions formed by the chance variations of the elements of collective objects. He develops simple tests of homogeneity and insists on the objective reality of asymmetric distributions. Fechner's theory was to become a quarry from which several scientists were to gather new building blocks for their theories. Richard von Mises was the most eminent example in this respect. By consolidating and streamlining Fechner's ideas, he was to develop one of the most influential and important contributions to probability theory in our century, his relative frequency theory built on the notion of an objective randomness.

Who was this man Fechner who resisted mainstream determinism?[3] His life spanned almost the whole of the nineteenth century, from 1801 to 1887. At the age of 16 he began studying medicine at the University of Leipzig, where he would remain for the rest of his life. Medicine made him an atheist and a believer in mechanistic determinism. In 1820, however, he came across a copy of Lorenz Oken's *Naturphilosophie*, which shook these convictions and led him to read Schelling, Steffens, and other *Naturphilosophen*. This influence was, however, counterbalanced by Fechner's interest in physics. He started lecturing on this subject in 1824, receiving a chair in physics ten years later. In the meantime he supported himself by translating French science textbooks—in particular, those of Biot and Thénard. The revisions he made in successive editions were so substantial that these works eventually lost almost all similarity to their French originals, becoming virtually Fechner's own works. In 1827 Fechner received a scholarship to study with Biot, Thénard, and Ampère in Paris for three months. His own researches in physics led to almost 30 articles in physics journals and to several books, which mainly treat topics from the theory of electricity. Fechner's revisions were the chief channel for the reception of French mathematical science in Germany at this time, and the reform of German physics that resulted from it.

From 1840 to 1843, Fechner went through a deep crisis, both physical and mental, which led to his temporary blindness. He lost his physics chair, taken over by Wilhelm Weber, though he was able to remain a member of the university. From 1846 on he lectured again on topics that were to occupy him for the rest of his life: philosophy, the mind-body problem, psychophysics, aesthetics, topics from physiology, and anthropology. In 1851 his main work in philosophy appeared, the *Zend-Avesta*. His *Elemente der Psychophysik*, for which he is best known, came out in 1860.

William James characterized Fechner as "a typical *Gelehrter* of the old-fashioned German stripe. His means were always scanty, and his only extravagances could be in the way of thought, but they were gorgeous."[4] The main thrust of Fechner's "gorgeous" thinking occurred during a period of transition in German history that also left its mark on science. After the July revolution of 1830 and Hegel's and Goethe's deaths shortly thereafter, romanticism and idealism were to encounter a serious challenge from the new realist movement. As F.A. Lange was to put it, "The material interests developed, and, as in England, they allied themselves with the natural sciences against everything that seemed to divert man from his most immediate interests."[5] Under pressure were all those conceptions of science that saw scientific theories as possessing an intrinsic emotive and moral dimension. This led to an increasing separation of science from philosophy and the humanities. In an 1845 review of some works on plant chemistry, the botanist Matthias Jakob Schleiden, one of the founders of cell theory, expressed this idea in the following way: "The whole of recent physiology drives irresistibly toward the goal of separating sharply mind and body, to give to God that which belongs to God, and to give to Caesar that which belongs to Caesar; or to put it without metaphor, to acknowledge the mind in its freedom, but also at the same time to recognize that the body is under the complete constraint (*Gebundenheit*) of the mathematical laws of nature and that these laws can explain *all* material reactions."[6]

Fechner would have agreed completely with Schleiden's recognition of the universality of material explanation but he would vehemently have denied the premise. Mind and nature are not to be separated: they are the two sides of the same coin. If this idea already seemed to be losing touch with the general trend, it was to become completely outdated after the failure of the revolution of 1848. This date marks the complete separation of science and philosophy, of "mind and body" so to speak. The German states pushed forward their industrial revolution; the sciences withdrew into an increasing specialization and regarded philosophy in general as an outmoded enterprise. Schleiden himself realized in 1863, in his "*Über den Materialismus der neueren deutschen Naturwissenschaft, sein Wesen und seine Geschichte* (Leipzig: W. Engelmann), that the separation of *Geist* from *Körper* had led to a dangerous denial of mind and its freedom.

In the end, Fechner's philosophical work was considered as something of a relic from bygone times, admired perhaps for its beauty but not to be taken seriously.

This changed to some extent at the turn of the century when William James[7] and Friedrich Paulsen[8] incorporated some ideas of Fechner into their own systems.

2 Psychophysical Parallelism

Before one can understand Fechner's contribution to the probabilistic revolution, one has to develop his views on the relation between the physical and the psychical, the corporeal and the spiritual in more detail: these views are basic to his whole work, whether philosophical or scientific. Fechner adheres to a special kind of Neo-Spinozism, or "identity-theory," according to which the spiritual and the material are two aspects under which one and the same thing (*Grundwesen*, substance) can appear.[9] If the substance appears to itself, it is viewed as spiritual; if it appears to something (or someone) else, it appears as material.

Applied to human beings this means the following: If I appear to myself, I perceive psychical processes: "general feelings, sensory perceptions, imaginations, desires, etc."[10] I see myself as a psychic, spiritual being. If somebody else looks at me, I appear to him *prima facie* as a corporeal, physical being; he sees me as a material entity. Since I can view myself also from the perspective of the other (if I look, e.g., in the mirror, or at my foot), I have a double access to myself. And since I can also see other beings as persons, I have a double access to other people, too. Two quotations from Fechner, which could easily be multiplied, should suffice to illustrate this:

Body and mind or corpus and soul or material and ideal or physical and psychical ... are only different as a result of the point of view or aspect one takes but not according to their ultimate ground or essence .... The same essence has two aspects [*Seiten*], a spiritual, psychical one, insofar as it appears to itself, and a material, corporeal one, insofar as it is able to appear to something other than itself in another form. But body and mind or corpus and soul never cling to each other as would two essentially different substances."[11]

What from an inner point of view appears to you as spiritual if you yourself are this spirit [*Geist*], appears from an outer point of view as corporeal under-pinning. It makes a difference whether one thinks with one's brain or whether one looks into the brain of the thinker. There, completely different things appear; but the point of view is also different; there it is an inner, here it is an outer one.[12]

Starting from this "double aspect view" of substance Fechner draws several con-clusions: As a first consequence he derives a "parallelism of the spiritual and corporeal."[13]

This means that the physical and the psychical are strictly coordinated but that no causal relation whatsoever holds between them. The relations that instead exist Fechner characterizes as lawlike "functional relations" (*functionelle Beziehungen*) or as "interrelations" (*Wechselbezüglichkeit*).[14] The laws that formulate these relationships are therefore not of a causal nature. They are, to put it in modern terms, laws of coexistence rather than laws of interaction or succession. Psychophysics is the science that deals with the functional relations between body and mind. It is the "exact theory of the functional or dependency-relation between body and soul, more generally, between the corporeal and the spiritual, physical and psychical world."[15]

Causal laws of interaction can be formulated separately for the physical realm as well as for the psychical one, but not "across" these two realms. Since substances and their interrelations can be viewed in two different ways, either physically or psychically, each causal interaction of substances can equally be represented in two different ways. So Fechner is opposed to all interactionist mind-body theories, especially those that see free human actions as the outcome of an irregular interference of the mind with the material world.[16]

Consequently, Fechner sees himself at the same time as both a materialist and an idealist, "for the spiritual has to change in proportion to the modification of the corporeal in which it is expressed. It appears insofar as it is completely dependent on it, as a function of it; it could even be completely translated into it,"[17] and vice versa.

As a second consequence of the double aspect view Fechner holds that not only man has a consciousness but also animals, the plants, the earth, the stars, nature as a whole! If we say that something has a soul we mean that there is also an inner side to it, that it appears to itself in form of sensation, feeling, etc. We also mean that there is a unity of consciousness that puts these inner appearances into a joint interrelation. We experience this soul only in our own case; in other cases we can only believe it. But this belief is not without grounds and it can be justified scientifically.

Fechner's theory of the soul met opposition from all sides and so he was forced to put much care into his arguments. He was very conscious of the fact that with his doctrine of the soul he gave up strict empiricism, to which he otherwise ardently adhered. He had, however, a sophisticated argument for making the extension of theories to unobservable realms such as the psychical one a scientifically admissible (and even necessary) procedure. Very often, we are forced by our naturally limited standpoint as human beings to "complete" our direct experience with the help of inferences that go beyond the observable. These inferences come in two modes, inference by induction and inference by analogy. Both these inferences never lead to certain but only to probable conclusions.

Now, if one wants to infer the existence of a soul other than one's own, one has to employ analogy as the principal method. Whereas in induction you infer from the existence of an attribute in many things that it belongs to all things, in analogy you infer from the similarity of many known attributes of things that they are similar in all respects. Analogy is by no means a method invented only for Fechner's limited purpose; he sees it as an inference by which we complete our physical experience constantly, in everyday life as well as in science.[18] His most striking example from science is the existence of the atom: although nobody has ever seen an atom we have good grounds to believe in its reality.[19] We are allowed to treat such a probable, hypothetical conclusion as if it had the status of direct experience.

We even order our experience according to it if the following conditions are fulfilled:[20]

1. ... one must be able to picture it in the form of, and in consistency with, possible inner or outer experience;

2. it has to be inferred from the way actual experiences are related to each other;

3. its supposition—while completing our experience without leading to contradictions—should also not come into conflict with our practical interests, rather it should be useful and serviceable for them.

With this last condition, Fechner makes clear that the belief in the existence of souls other than one's own is not just an eccentric exercise in ontology of no consequence, but that it has a direct bearing on our conduct of life. This idea, of course, later became very familiar with pragmatism.[21] (Fechner's arguments also bear, by the way, a striking similarity to those that today's "scientific realists" adduce in favor of the reality of unobservable theoretical entities, except that today's realists rarely consider psychic attributes.[22])

The same completion of experience as in the case of the atoms is taking place in the psychical realm as well. Other people are not only physical entities to me but also animated beings. I arrive at this belief by inferring the existence of the inner mind of my fellow man from outer material signs. Other people are in so many respects similar to me that there is ample reason to believe that they are similar to me also in respect to their psychical qualities. The psychical, inner side of the plants, of nature as a whole, etc., is inferred from the outer experience at hand by the same method.

Fechner lists a great many observable attributes that must be present to permit one to infer the existence of a soul by analogy. There is good reason to think that a system of things has a soul if the following conditions are fulfilled: it forms a unified whole; it can be individuated and isolated from similar systems; it is self-developing in some way—it can bring about an unlimited variety of effects, partly determined by natural laws, partly unpredictably novel (with this condition we shall deal more fully later on); it is a teleological system, i.e., all its parts serve the purpose to maintain the system as a whole.[23]

Fechner's position eventually culminates in pantheism. Nature as a whole has a soul, and this soul is to be identified with God. The "material world of the appearences" is the "outer side of the divine existence," and the self-appearence of the world is God's spirit and the world's inner side.[24] Thus we can approach nature in a second way, besides that of science: One can try to find out by analogy how the sentiments, motives, and thoughts of a being would look to whom the whole material world appears in an inner way.

3 Indeterminism

We have seen that an animated being must be able to bring about unpredictable novel effects besides those determined by natural law. How can one make sense of the freedom of the spiritual in a material world determined by strict causal laws? How are the physicist Fechner and the philosopher Fechner reconciled to one another? Fechner gave his answer to this problem in two addresses of 1849, the first one being on "The Mathematical Treatment of Organic Forms and Processes" and the second one, "On the Causal Law."[25] Fechner gave his speeches at public sessions of the Royal Saxon Society of the Sciences at Leipzig as the vice-secretary of the society's class of mathematics and physics.[26] The first speech was given at the occasion of the king's birthday and the second one at the anniversary of Leibniz's death. The society made it a regular custom to celebrate this day.

Fechner went on to treat the same problem in many later writings. In the *Zend-Avesta* he reproduces his address on causality in almost identical form. Even thirty years later in his book on the *Tagesansicht gegenüber der Nachtansicht* he returns to the topic again.[27]

The overall goal of the two talks is to show that the idea of a partly indeterministic world can consistently be held within "exact research," and that one can even argue for the reality of this idea although there is no empirical procedure for us to decide the issue. Thus, for the first time, the idea that the world has some amount of objectively existing "*Indetermination*" or "*Unbestimmtheit*" (indeterminacy), which "really depends . . . on freedom," as Fechner puts it, and which is not just the result "of our ignorance of the conditions,"[28] is put into practice.

My interpretation of Fechner here is that he implicitly develops four different types of indetermination. Before going into any detail I shall simply list these types in my own words:

type 1: indetermination due to the *imprecision* of (the empirically accessible attributes of) the scientific object as such;

type 2: indetermination due to *suspension of the causal law*;

type 3: indetermination due to the constant occurrence of intrinsically *novel initial conditions* in the course of the world's development;

type 4: indetermination due to a *limitation of predictability* lying in the objective processes themselves.

Fechner thinks that none of the types 1, 3, or 4 necessarily implies type 2. This is taken as a proof of his assumption that indetermination is consistent with science. Although indetermination of type 2 is rejected as implausible, its assumption cannot be falsified by our experience. Type 4, however, is more than just consistent with science. One can show, although not by experiment, that this type of indetermination is actually realized in our world. If this is true, we have good grounds for assuming that there is also some truth to the claim that type 1 and type 3 are realized in the world because type 4 is not independent of either of them.

In the first paper Fechner sets out to refute the proposition "that the organic circumstances [*Verhältnisse*] cannot be treated mathematically or at least that the mathematical treatment is less successful than in the inorganic case."[29] The arguments that Fechner gives as normally cited in support of this proposition are the following: (i) The spiritual element that one sees active in the formation and movement of the organic is marked by its freedom. This freedom influences the material side of the organic and imparts to it a free and indeterminable character. Mathematics, however, can only be applied to a realm governed by law, "which carries the mark of necessity with it." [30] Therefore, the organic cannot be treated mathematically. (ii) The formation, development, and movement of the organic is much too complex to receive exact mathematical treatment. The peculiar nature of the organic does not allow exact mathematical determination. (This second argument can be upheld even if one denies an element of freedom in the sense of the first argument.)

In response to the first argument Fechner does not set out to disprove the existence of a spiritual element as one might imagine a physicist would. Instead, he remains neutral and tries to show that the argument is a non sequitur: Even if there were an element of freedom in the organic, it would not follow that it is mathematically untraceable. To show this, Fechner invokes empirical observations, saying that everyone, whether he believes in a spiritual element or not, has to admit that the formation and movement of the organic follows *some* kind of order, is under the influence of *some* kind of general norm or "character." And this amount of lawlikeness is sufficient to allow the use of mathematics even in the organic realm, whether you are convinced that the special case shows an undeterminable, indeterminate, free character or not. This way of employing mathematics does not differ from its application to the unorganic realm, "in all the uncountable cases where the conditions of the events are only partly known. There we also find an indeterminacy; this indeterminacy, however, instead of abolishing the use of mathematics, gives it a special form, which includes the measure and the kind of the indeterminacy itself in a determinate manner." [31] As example for indeterminate phenomena that nevertheless show a general determinate character Fechner cites the fixed frequency of vibration of a string that can produce different sounds, the dimension and fixing of a flag that can move differently, and the general equation of a conic section that can have many different models.

The strategy taken by Fechner in order to disarm the second argument is similar to the one taken in the first argument: In order to safeguard the applicability of mathematics in science one is not forced to deny the existence of indeterminacy. If one is convinced that organic phenomena such as the physiognomy of a human face, the development of a chicken in the egg, and the circulation of blood in our body do not admit an exact representation by a finite mathematical formula, then, by the very same argument, one has to entertain the same conviction for the inorganic realm! In both cases we would have the "impossibility of a perfectly precise determination but [we have also the] possibility of continuously approximating it to whatever degree." [32] As examples for successful mathematical approximation of organic phenomena, Fechner cites the approximation of the form of the egg, the description of snail shells, and the determination of the curvature of the human eye. The difficulty that arises in the *practical* development of a useful approximation is not confined to the organic realm. Inorganic meteorological descriptions show the same difficulty, for example.

What has Fechner achieved by defending the applicability of mathematics to the organic realm in such a way? He has shown that the idea of perfect precision in science is an empirically unwarranted conception because it cannot be reached in a finite period of time. Therefore, from a strictly empirical point of view, science is consistent with the assumption that the scientific object as such in its special individual form is "incommensurable with mathematical determinability,"[33] and that this kind of indeterminacy is the result of the spiritual freedom of the world. The mathematics that one can use under this assumption does not conflict with the mathematics used under the opposite assumption, which says that the objects of science are completely determinate but that we are in ignorance of all the relevant working conditions.

There is, however, a price Fechner has to pay for this belief. He can no longer make a categorical distinction between a natural law that explains the phenomena (i.e., that tries to find "the unknown causes of the occurrence out of their known effects," as Helmholtz put it in 1847), and a purely mathematical description (a "general rule," as Helmholtz would say).[34] In the first argument, the opponent was characterized as someone who believes that mathematics can only be applied to realms that are governed by necessity. If Fechner shows the admissibility of mathematics in the biological sphere, does this mean that he wants to extend necessity in some way also to the organic realm, at least to its general character? In the Kantian understanding of science, mathematical treatment of nature means to construct its concepts in pure intuition, that is, to show the necessity and objectivity of its laws. Thus, the successful application of mathematics in science asserts that the causes one talks about as natural laws are necessarily connected to their effects, and that causes can therefore be distinguished from mere concomitant variations.

But Fechner draws the opposite conclusion: If the organic realm proves to be mathematically tractable even under the assumption that there is no necessity governing the single case, and if the inorganic realm can be interpreted the same way, then, in order to guarantee the objectivity of science, there is no need to suppose that its mathematical character asserts relationships of necessity in nature itself. Instead, "the purely mathematical is everywhere only in our imagination" (*das rein Mathematische liegt überall bloss in unserer Vorstellung*).[35] The idea of the fixity of the scientific object is the unwarranted projection of an imagined converging limit into reality.

In the second paper, Fechner attempts to clear away another barrier to indeterministic thinking, namely, the conception that the causal law only admits of determinism. He starts by posing the question about what possibly could be the highest, the most general law for events in the realm of nature and of spirit, which amounts to asking for the "supreme regulating principle of our inferences in the whole area of experience."[36] A law is more general than another if it either embraces more special laws or is valid for more phenomena. Therefore, the most general experiential law one could imagine has to cover all other laws and all events. The only law one can imagine to satisfy these criteria is the following: "Everywhere and at all times, insofar as the same conditions (*Umstände*) recur, the same effect (*Erfolg*) will recur; insofar as the conditions are not the same, the effect will not be the same."[37] One could be tempted to consider alternatively a law that indeed covers all empirically possible events but leaves open the "choice" either between different effects that could take place under the same conditions or between different conditions that could be responsible for one and the same effect. But this would mean that there are cases that "escape" the "grip" of generality, i.e., that would be lawfully undetermined. So, we can say that for Fechner a realm of phenomena is "lawfully determined" (*gesetzlich bestimmt*) if the conditions and their effects can be mapped one-to-one, i.e., if the causal chains do not allow for branching either backward or forward in time.

Fechner seems to adhere to what one could call a 'chemical' conception of causality: A cause is the coincidence of two (or more) conditions that leads to an effect. If hydrogen, chlorine, and an electric spark are brought together, they will produce hydrochloric acid, whose qualities are different from its ingredients. Or, to take a more biological illustration: the meeting of the egg with a sperm causes the existence of a new being.[38]

Fechner goes on to show that the lawful determination of our world in the sense of his law is not only conceivable but also realized. One could object that if one considers an event, one has to take into account the totality of all conditions preceding it. This totality, however, never repeats itself and therefore one cannot speak of a law at all since laws admit of repeatable applications. To this, Fechner replies that in every confirmation of a natural law we abstract the more from a condition the further away it is in space and time. We are allowed to apply the same method in the case of the causal law. Since neither the same conditions nor the same effects recur exactly, a law can only be confirmed in approximation. We find by experience that insofar as conditions are similar, their effects will be similar. So we have the right to claim the validity of the supreme law "insofar as experiences allow such a conclusion at all."[39]

Fechner takes great pains to show that the general law applies both to the inorganic as well as to the organic, to the material as well as spiritual realms. In addition, he makes it clear that the "highest law" is needed in order to show that the world as a whole has a spiritual side. The highest law binds together and dominates the most distant spaces and times, and it allows for relations among distant parts of the world. This enables us to identify events that are separated in time and space. Without the law the world would be fragmented; with it the world forms a unified whole.[40]

So far, Fechner's theory appears just as a (somewhat whimsical) variant of the typical nineteenth-century belief in causality, although more in the spirit of J. S. Mill than of Helmholtz. One can easily mistake him as someone who argues thereby for universal determinism in the same nineteenth-century spirit.[41] After having paid tribute to the causal law, however, Fechner uses all his force to show that the causal law is "compatible with every freedom and indetermination ... even the greatest one can think of."[42] The type of indetermination he has in mind one could call 'indetermination by novelty'. In the course of the world's development new initial conditions occur that also have not happened before. These conditions will then, according to the highest law, lead to an effect that has also not occurred before: "To be sure, the law says that insofar as the same conditions recur, the same effect shall recur, and insofar as they do not recur, the same effect shall not recur.

But its formulation neither implies something about the kind of the first effect ...
that follows any given set of conditions nor does it determine the occurrence of the
first conditions in any way. In this respect, it leaves everything free."[43] Every set of
conditions has some similarity with old conditions (i.e., with conditions that have
appeared before), and therefore their effect will not be completely free. Nevertheless, there is something new in every set of conditions, and therefore every event has
some kind of freedom:[44]

Indeed, according to our law the coming effects will only be preconditioned and
predetermined by past conditions insofar as these conditions are themselves
recurrences of old conditions. But this is never completely the case. Every
other space and every other time carries with it again new conditions or new
modifications of the old conditions. The conditions therefore call anew for new
or modified effects that are neither determined nor determinable by those
laws that are based upon other places and former times. Or at least, they are
determined insofar as the old survives in the new. But it never survives
completely. The world changes from place to place and it is continually
developing in time. ... Indeterminism can make its whole doctrine of freedom
depend on this circumstance without doing any damage to the universal
lawlikeness of nature and spirit.

Fechner uses two examples to illustrate his point. First he considers the creation of
man. Man's first occurrence was the result of special conditions that were intrinsically new and could not have been explained out of earlier conditions. The law of
man's occurrence "was not given before his first creation, yet it will now last until
the end of time."[45]

In the second example, Fechner applies his theory to human freedom, which he
takes as the capability to produce new and therefore undetermined effects out of
new inner or outer conditions. These effects cannot be calculated. Yet, since earlier
events are legislative (*regelgebend*) for later events, man with his character 'determines himself' more and more in the course of his life insofar as old conditions of
his former life recur. But human life is never a complete repetition of old conditions.
And therefore, "the freedom to determine oneself in future in this or that direction
never ceases completely."[46]

The determinist can now raise the objection that what appears to be new is in
reality only the modification or combination of old circumstances. The future is
only a function of the past. He could refer to the planetary system as a good example
for this: It is true that the position of the masses of the system at any given time will
never recur another time. "Nevertheless, its whole movement is determined by rules
that are completely based on the past."[47] And the sole conditions relevant in this
case are the masses, distances, velocities, and directions of the bodies. What can be
done in the case of the solar system can be done in every case.

Fechner's answer to this is, "It is not possible to decide this quarrel [between the
determinist and the indeterminist] experimentally."[48] Determinism is possibly true
but its truth is not guaranteed. In Lakatos's jargon we could say, determinism is
a research program but neither is it proven nor is it an a priori principle. As long

as determinism has not been 'completed', the indeterministic program remains equally possible and has the same initial plausibility as its opposite.

There are, however, scientific observations that give indeterminism some excess weight. If we continue to use Lakatos's expression, we could say that Fechner tries to show the degeneracy of the deterministic program. He refers to the results of his friend, the physicist Wilhelm Weber, who had found in 1846 that the force with which two electrical particles act on one another depends not only on their electric mass (= charge), distance, and relative velocity, but also on their acceleration. This implies, however, "that the general law governing the effect of the action of two particles on each other is changed through the admission of a third, then of a fourth particle, etc. It changes in a way that is the harder to account for, the more complex the system becomes: the connection that forms a whole has an influence that cannot be calculated out of the composition of [unrelated] particulars (*Einzelheiten*)."[49] Weber himself had pointed out that already Berzelius had a similar supposition and that he saw this effect as the result of catalytic forces.[50]

What has Fechner achieved by arguing for indeterminism by intrinsic novelty? First, he has shown that science is at least compatible with the supposition that events occur during the development of the world that show an objectively undetermined character. These undetermined events are uncaused. Since laws do not come into existence before their first manifestation and indeterministic events are the first manifestations of laws, it follows that the uncaused character of these indeterministic events does not imply a violation of causality at all. Second, Fechner has shown that human freedom is a naturalistic quality that can be accounted for in the realm of the appearances. One does not have to take recourse to noumenal spheres and things-in-themselves, as Kant did, in order to reconcile causal necessity with freedom's spontaneity. Third, by putting human freedom in a naturalistic way on a par with other indeterministic events Fechner resists the temptation of his contemporaries and successors to separate the human from the natural realm, to claim a categorical difference between *Natur* and *Geschichte*.

Fechner's paper could end here, but he prepares for another blow. He sets out to show that even if the determinist were right, he could not help acknowledging a last vestige of indeterminism in his system, namely, that there is a limit, in principle, to predictability. On the other hand, the indeterminist must equally acknowledge a last vestige of determinism in his system: the possibility of unlimited retrodiction. If this could be shown, the dispute between the determinist and the indeterminist would be settled (*geschlichtet*), although perhaps not finally decided, and a "compromise view" that combines determinism and indeterminism would have been found.

Fechner starts from the supposition that the capability of predicting circumstances of a given complexity depends on the complexity of the predictor himself. The higher the complexity of the predicted event, the higher the "degree of the development of the mind" (*Entwicklungsgrad des Geistes*)[51] of the predictor must be. So, if one supposes that the complexity of the world as a whole increases steadily with time—this goes without saying for Fechner—then the complexity of the predictor's mind must increase as well in order to 'keep up' with the world's development. (So must his material complexity also increase, since body and mind are a function of one another.) But this means that prediction always comes to a limit. Take any predictor of a finite complexity. There will always be a future state of the world that is more complex and hence unpredictable for this predictor. "A worm will never be able to predict how an ape will behave, an ape never a man, a man never god." [52] This applies even to the development of any individual man: one cannot predict of oneself how one will behave on a higher "level of formation" (*Bildungsstufe*).

One can see immediately that an asymmetry between prediction and retrodiction arises. In the same way as one can survey the motives of one's own actions better if one has reached a higher level of formation, so the predictor can retrodict any state of affairs less complex than his own without reaching a principal limitation. If one now presupposes that it is only indetermination that makes the world more complex, i.e., that the appearance of indetermination by intrinsic novelty is the necessary and sufficient condition for any increase in complexity, then unlimited retrodiction remains in principle possible in such an undetermined world, even for those events that were undetermined at the time of their occurrence. [53]

If there is an asymmetry between the past and the future in the way Fechner suggests, than it follows as an obvious corollary, often hinted at by Fechner, that the laws governing the universe on a higher level of development are not reducible to the ones on lower levels, even though these laws remain somehow contained in the higher stage.

Fechner immediately discusses the obvious objection of the determinist that at the least a predictor of *infinite* complexity would not reach a limit in his prediction. Therefore, our actual inability to predict is just a result of our finiteness and ignorance and does not prove an objectively real indetermination founded in the nature of things. The vestige of indeterminism would be an illusion.

In response, Fechner invokes a metaphor in order to convince us that even God is incapable of predicting the course of the world. Could a poet predict how a poem that he will create tomorrow will look? He cannot, because the very moment he is predicting it he has already created it. And this also applies to God. If we suppose that "the course of the world is the outer appearance of God's train of thoughts," then even God himself cannot predict the world's development, since He cannot predict his thoughts without actually having them. "The very moment where He sees ... [a new stage in the development of the world] for the first time, it also comes into existence for the first time." [54]

From this Fechner concludes that the objective character of the world and not our subjective condition makes its course unpredictable in principle: "What, by no knowable rule, no being can predict, not even God himself, what, at the time of occurrence cannot be derived sufficiently out of something earlier, this has to be regarded as something that is indeterminable in reality and in itself." [55]

The indeterminist can also raise an objection against the "compromise view." He might say that if the possibility to count back to any past event from the present state of the world really existed, then "the free will" of man would be something completely determined. For "in order for the reasons [of the free will] to be followed back to any desired degree they have to exist." [56]

Without disputing this conclusion Fechner asks the indeterminist whether he really wants to continue to talk of *Determination* or *Vorausbestimmung* (predetermination, predestination) if this exists "neither for a finite nor an infinite knowledge."[57] And even if we nevertheless decided to call the compromise view deterministic, all the ethical disadvantages the indeterminist normally sees in determinism would have disappeared.

If one supposes in addition that the law of God's ordering of the world is on the whole a good one, "which in the end determines evil persons through the effects of their evil deeds necessarily to the good, here or there,"[58] then the "compromise view" indeed makes the bad consequences of determinism and indeterminism disappear. Fechner concludes his paper with the affirmation that both strict indeterminism (which believes in indetermination by novelty) and the compromising view (which believes in objective unpredictability) do not come into conflict with the causal law and therefore also not with science. Both views, however, have the advantage that the scientist is not forced to give up his belief "that something factually undeterminable lies in the very circumstance that he also is allowed to call his freedom."[59] He is not forced to believe in those interactionist theories that presuppose a suspension of causality for man. Such a belief, although it cannot be empirically refuted, remains implausible and unnecessary for Fechner.

If we step back for a moment and survey the three types of indetermination Fechner finds acceptable (types 1, 3, and 4 according to my list), we see that he has very little to say about the question how these types are related to each other. The most plausible solution is, it seems to me, to assume that indetermination by imprecision and indetermination by novelty are equivalent, since they refer to one and the same kind of freedom. What then would the relation of indetermination by limited predictability be to the two other types of indetermination? At one point Fechner suggests that increasing complexity might be *explained* by the occurrence of novel initial conditions.[60] Since it is precisely the increasing complexity of the world that sets a limit to its predictability, one could say that the novelty of the conditions can also be used to explain the limit in predictability. So indetermination by novelty explains indetermination by limited predictability.

4 Background

What is the tradition in which Fechner expresses his views on indetermination and freedom? I would like to treat here the two most important contexts for his thinking. The first context is given by a special kind of turn Hegelian philosophy took after Hegel's death. The second context is given by the peculiar mixture that epigenetic thinking in biology and ideas from the philosophy of history formed in Germany in the first half of the nineteenth-century.

The criticism of Hegel's system that began after his death is characterized by the conviction that true reality cannot without loss be reduced to the conceptually intelligible. Logic and ontology refer to the form of being but do not reach reality itself.[61] The later Schelling had already put forward this idea in his *Das Wesen der menschlichen Freiheit* of 1809. The same thought was the point of departure for Feuerbach and the different variations of materialism (including the Marxist) that originated with him; for Kierkegaard,[62] as well as for the so-called 'late idealists' Christian Hermann Weisse, Immanuel Hermann Fichte, Hermann Lotze, and others. Late idealism in philosophy and scientific thinking is characterized by the conviction that in order to explain the phenomena of life and the living, of freedom and individuality in the world, one has to invoke some amount of indetermination. This indetermination should be formulated in naturalistic terms even if this conflicts with established science. In this assumption late idealism differs not only from other critiques of Hegel, but also from materialism and Neokantianism later in the century.

The most important figure of late idealism is the philosopher Ch. H. Weisse (1801–1866),[63] who became professor of philosophy in Leipzig in 1828. Weisse's position is often described as speculative theism. He was the closest friend of Fechner,[64] who was of the same age, and he was, as teacher and later as friend, the decisive influence on the physician and philosopher Lotze (1817–1881).[65] Fechner also had a close personal relation with Lotze.[66] As long as Lotze was still in Leipzig (before he went to Göttingen in 1844), the three formed the inner core of a circle of friends that met regularly once a week.

Weisse's relation to Hegel is complicated. On the one hand, Weisse acknowledges Hegel's logic and the method of his philosophizing (at least he does this at the beginning of his career). On the other hand, he denies the exclusive validity and relevance of Hegel's logic to reality. It is true that logic carries with it all the principles of the form of the being but not being itself. It is basic to being but does not exhaust it. In a characteristic phrase Weisse writes that he found himself "heavily attracted and cruelly repulsed at the same time [by Hegel's system]. The formal truth of Hegel's philosophy, the solidity of its method, and the desolate baldness of its results forced themselves upon my mind with equal evidence. They incited my mind, straining to the utmost, to look for the solution of the contradiction."[67]

This mixture of agreement and opposition led contemporaries to opposing judgments of Weisse's philosophy: partly Weisse was seen as a Hegelian, partly as Hegel's key opponent. He was called a "half-Hegelian" as well as a "pseudo-Hegelian."

What does Weisse criticize in Hegel? His major argument is that Hegel confuses form with content. If the form of thinking has some kind of necessity, it does not follow that the content of thought also hangs together by necessity. Concrete reality is not the product of a logically necessary development of ideas; there is something in it that transcends all necessity. The world is a "free creation" and not the realm of pure necessity. Although one could say that Hegel's philosophy "has explored the concept of freedom in a deeper way than any former philosophy and in a more complete way, nevertheless, according to this doctrine, freedom is never realized in reality, but individual man, as well as the spirit of nature and the *Weltgeist* on the whole are beings acting with a machinelike necessity."[68] Therefore in Hegel's philosophy the world seems to lie before us "in exhausted immobility," which allows only for a "dull, monotonous course of wheels and pendula ... as if the irresistable might of the abstract idea—whose conceptual determinations and categories wander about like heavily armed but tired soldiers on the battleground—has sucked all the refreshing living humors out of the world and has stripped off all the fragrant blossoms."[69] In order to provide concrete reality again with liveliness and fragrant blossoms Weisse wants to help "freedom [to its right] as a general predicate of nature in contrast to logical necessity."[70]

It is interesting to note that Weisse's criticism of Hegel's "anancasticism" (as Peirce later expressed it) has a definite aesthetic connotation. This seems to have been a general feature of the time. Remember how Goethe reacted to Baron d'Holbach's *Système de la nature* of 1770. This book functioned at the time as the paradigm of determinism, both in the moral as well as in the cosmological sense. Goethe did not say that Holbach was wrong or nonsensical. Instead he criticized the system as gray, miserable, deathlike, *triste*, empty, hollow, senile. A system with such qualities cannot do justice to the ornate character of nature and therefore cannot be taken seriously.[71] Weisse, Fechner, and Lotze would all agree with Goethe that science has to retain some emotional value in order to be true. Weisse had worked out a much praised aesthetics in 1830, and both Fechner and Lotze tried to keep this tradition alive in their writings.

The concept of freedom is thus loaded in Weisse's system with aesthetic connotations, and it is made the central aspect of life and living reality. As a consequence Weisse makes freedom the central concept of his philosophy. He asks again and again "how freedom and necessity are related in nature and in the world of the spirit." The answer to all these questions is "the center or the starting point for every philosophical investigation that has reached the level to which *Hegel* has raised philosophy," although "Hegelian philosophy proper lacks the consciousness of this great question."[72]

What is Weisse's concept of freedom? He locates freedom in the individual character of reality. "Individuality or particularity" cannot completely be cast in general terms; they are not "preserved (*aufgehoben*) in the generality of pure thought,"[73] in the Hegelian sense. Nature has an "excess of reality", an "incommensurability with the purely metaphysical concept ... which Hegel denies."[74]

This 'embedding of freedom into nature', as one could call it, leads to a high revaluation of individual variation and deviation. A distinction between "concrete reality and its conception" (*Begriff*) can only be made if one realizes[75]

that every aspect of fortuitousness (*Zufälligkeit*) of all the individuals that are comprised under a generic concept is not the external, *bad* negation, as one thought until now, but rather that it is the explicit appearance of the *immanent dialectical negativity* by which the natural creature and spiritual individual tears himself away from the generality of the concept and its rigid, life- and soulless necessity. And by doing this the individual assimilates the concept to an attribute of himself and reaches his freedom. Only through this process is freedom posed as existent and real, and it is admitted as being the other, equally essential and real, side of logical necessity.

Weisse has clearly seen that his conception of freedom in nature raises problems for epistemology. For Hegel, knowledge is the subordination of the concrete individual under general concepts. In Weisse's "system of freedom," as he himself calls his philosophy, "the realm of nature appears as a *free creation* that can, as such, be known not by dialectical method but *historically*. The historical way of knowing is related to the dialectical one as freedom is to necessity: it presupposes dialectics in the same way as true freedom presupposes necessity, but it is nevertheless more and higher."[76]

Weisse never really put his programmatic ideas into practice, and therefore he never seems to have been forced to confront them with any conception of causality in natural science. He later occupied himself mainly with theology. Nevertheless, his ideas had a deep influence on several people who worked at the borderline between science and philosophy, among them, as already mentioned, Fechner and Lotze.[77]

How could we now characterize Fechner's position vis-à-vis Weisse's? It appears as the attempt to reconstruct as much as possible of Weisse's basic ideas in the framework of a strict empiricism and of the physics of the day. On the one hand, Fechner wants to make room for indeterminism in the world of the empirical phenomena and for a naturalistic conception of freedom and individuality. On the other hand, he is opposed to any solution that would see freedom connected with a suspension of causality.[78] Materialism is the legitimate attitude of the scientist.[79] Therefore, Fechner is also opposed to any vitalism that hypostatizes the spiritual as a substance in the form of an independent *Lebenskraft* or any other form.

The second context for Fechner's indeterminism is given by the mesh of ideas derived from epigenetical thinking in biology on the one hand and from philosophy of history on the other. In the early nineteenth century Germany saw many attempts to mingle both domains. Owsei Temkin has shown that the idea of ontogenetic development served as the most important bridge between the two views.[80] With Caspar Friedrich Wolff's *Theoria generationis* of 1759, a slow change begins in embryology (and with it of the life sciences in general) that alters the theoretical outlook from preformationism to epigenesis. Epigenesis teaches that in the course of development of organic creatures something novel appears in such a way that there is no longer any similarity between the germ and the final product of the development. Preformationism on the other hand assumes that every part of the later organism is already there in the egg and gets 'unfolded' in the course of development.

Epigenesis harmonized with contemporary German philosophy of history. The most important names here are Herder, Schelling, and Hegel. The mingling of epigenesis with their philosophy of history led to a "three fold parallelism," as Temkin put it, "between (1) ontogeny and the ages of man, (2) successive creation of species, and (3) the history of mankind through successive civilizations."[81]

As my prime representative of the second context I choose Karl Friedrich Burdach (1776–1847), a physiologist of great versatility and cultivation. Burdach also spent some time in Leipzig from 1793 to 1798 and 1799 to 1811. Von Baer was Burdach's student before he went to Döllinger; later he became Burdach's colleague and close coworker.[82]

Burdach constantly reflected on the philosophical side of his subject. In 1842 he made the attempt to transpose the idea of epigenesis to nature as a whole and to ask if one could not give a more satisfactory account of nature's multiplicity and differentiation and of the development of individual variations by viewing the development of nature as an epigenesis. His solution reads like a first intuitive sketch of Fechner's ideas, and I find it highly plausible to assume that Burdach's paper inspired Fechner to write his address on the causal law.[83] In order to show the similarity between Burdach und Fechner I shall quote at length the beginning of Burdach's paper on *Die persönliche Besonderheit:*[84]

King Solomon says that nothing new happens under the sun; Leibnitz, who also was a King although only in the domain of philosophy, retorts that not a leaf grows that is perfectly identical to an already existing one. In spite of all contradiction, we have to agree with each of them according to his standpoint: with the former, if we survey the world in its generality, according to its eternal laws; with the latter, if our vision reaches deeper into the particular phenomena of the endless multiplicity. Nature with her forces and laws is eternally unchangeable: the elements are everywhere and always the same; resting on the same foundation there appear similar phenomena in every place and every time; and what has stirred the emotions of the man to whom the sun shone first will also stir the bosom of the one who will live to see the sun enlighten the globe for the last time. But—it is always the case that only something similar recurs and never something identical.

The majority of the elements makes possible a multiplicity of compounds, and through the coincidence of such connections arises an endless series of circumstances. The creative force of the world is infinite in its essence and therefore it is infinite in its manifestations: inexhaustible in her combinations, nature produces only novelties to all eternity; she does not tolerate uniformity, and she never allows a law to govern alone in naked rigidity, but tempers its severity through the interference of another law.

If one takes this idea of the infinite variation of nature that never allows for the recurrence of identical appearances, and combines it with Fechnerian causality, then one obtains the kernel of Fechner's conception of indetermination by novelty: The occurrence of intrinsically novel initial conditions is a new cause for a new effect; new causes are themselves uncaused without suspending any other existing law.

Burdach never gets tired of emphasizing the "immense abundance," the "un-bounded" and "immeasurable manifold" (*Mannigfaltigkeit*) of life. He sees the aim of life in individualization.[85] This leads to a rejection of typological (essentialistic) conceptions of the species. The human species is not a *Stereotypenausgabe* (a stereotype edition), but rather the "epitome of a collection of mobile letters whose different combinations allow for a manifold that cannot be exhausted in eons."[86] This clearly amounts to a population conception of the species, to use the term that Ernst Mayr has made popular.[87]

Burdach names three factors that produce individual variations: external circumstances, heredity, and what he calls the "power of the species": "It can only be the power of the *species* that ordains the individual with its peculiarity as it does with its general character. The law is something infinite whose exhaustion by single individuals is possible only to a limited degree. The type of the species can realize itself only through an infinite multiplicity of individuals." [88]

The laws that show the power of the species determine the "frequency" with which certain attributes, e.g., malformations, are found in the species. These laws can only be found in large numbers. [89] In this connection, Burdach refers often to Quetelet, e.g., when he discusses the relation of male to female births. [90] The species brings forth "a proportion of the sexes that suits [the] conception" of the species. [91] There is another phenomenon that indicates the power of the species; we call it today the regression to the mean: "the return from the unusual and abnormal to the normal, the giving in of the ... deviation to the course of the type that lies in the middle." [92]

This last remark shows that Burdach's population thinking allows for a last typological vestige. On the one hand, the concept of a species is not defined by an essential strain realized in every individual alike; rather it is an essential characteristic of every species that all its individuals vary and deviate from each other so that no two are identical. On the other hand there is something like a *type* of a species that, it seems, can be represented by the mean value of the variations, although Burdach does not explicitly claim as Quetelet does that this mean value is an archetype that underlies all actual individuals.

Fechner's thinking appears just as a logical consequence of Burdach's view. Fechner adds another kind of parallelism to the three we have cited from Temkin before. Now, the course of the universe as a whole in time (including the origin of life) is taken as an epigenetical process. So, not only organic creatures are true individuals, but every time and place carries with it some novelty that makes it an individual phenomenon. Thus, the course of the universe cannot be a determined process.

5 *Kollektivs*

If Fechner's variant of indeterminism had been without any consequences, one would be tempted to judge it as a curious and eccentric, or at best a premature, aberration from mainstream scientific thinking. In the following, I set out to show that being indeed premature it was mainly Fechner's work that led to the decline of the classical Laplacian theory of probability and the rise of frequency theory in our century. Determinism in the nineteenth century allowed for indetermination only as a result of human ignorance, while Fechner's theory led to the conception that probability theory is an empirical science of chance phenomena in nature.

Fechner did not rest content with a merely philosophical or metascientific formulation of his indeterminism. He tried to develop a mathematical language in which he could express the indeterminate character of individual phenomena and the lawlike character of their general aspects. This work resulted in his unfinished *Kollektivmasslehre*—a book of 483 pages posthumously edited and completed by Gottlob Friedrich Lipps by order of the Royal Saxon Academy of the Sciences in 1897.

In order to develop his new mathematics Fechner draws on two traditions; the moral statistics of the Belgian statistician and astronomer Adolphe Quetelet (1796–1874) and the error theory of the mathematical astronomers Gauss, Encke, Bessel, and Hauber.[93]

In his *Sur l'homme* of 1835 (which was renamed *Physique sociale* in the second edition of 1869), as well as in his subsequent work, Quetelet tried to establish an analogy between the Laplacean model of celestial mechanics and a physics of human society.[94] In the same way as the gravitational force acts as a constant cause keeping a planet in its orbit, so society is governed by constant causes as well. The constancy and regularity of moral phenomena show that constant causes are indeed operative in human society. Planets are also influenced by variable or disturbing causes that have the effect of small local perturbations. These perturbations obey the Gaussian law and leave the center of gravity unmoved. In human society, the differences that exist between individuals have to be explained in the same way. They are symmetrically distributed around the mean, which forms the maximum. One can neglect them if one considers large numbers of individuals. Thus Quetelet transformed the law of error into a law of variation.

We do not know when Fechner first became aquainted with Quetelet's work. He does not mention Quetelet in his writings of 1849, but one could very well imagine that he had Quetelet's theory in mind when, in the paper on the *organische Formen*, he developed the idea of the general norm or lawlike character of natural phenomena that can be formulated mathematically but does not imply any necessity for the individual case as such. It might very well be possible that he learned about Quetelet from his friend, the mathematician and philosopher Moritz Wilhelm Drobisch in Leipzig. Drobisch mentions Quetelet already in a book he wrote about Herbart's philosophy in 1834. In 1849 he wrote a review of a *mémoire* on moral statistics by Quetelet in which he dismisses the view that the regularity of moral phenomena (crime, suicide, marriage, ...) is incompatible with human freedom. This review aroused much attention, and Drobisch worked it into a book that appeared in 1867.[95]

Fechner started to work on probability theory and statistics shortly after 1852, at the latest. In the *Kollektivmasslehre* he reports that "(in the fifties) I got hold of the lists of ten Saxon lotteries from 1843 to 1852 from the authorities concerned with 32,000 to 34,000 numbers each. These are lists in which the winning numbers are reproduced in the random succession [*zufällige Folge*] in which they were drawn."[96] Fechner uses these lists as a "substitute for the probability urn ... with infinitely many white and black balls of equal number," in order to submit his derivations to an "empirical confirmation."[97]

Another use of statistics is made in a book of 1856 where Fechner deals with objections the botanist Matthias Jakob Schleiden had raised against his doctrine of the soul. There he tries to solve questions like the following: "Has the moon an influence on the weather? Has the moon an influence on the process of life, or of illness of men and of other earthly creatures?"[98] He discusses a vast number of data from scientific journals and finds out that the majority of scientists who had dealt with this question affirmed an influence of the moon on the weather, although they were not able to give any causes of this influence. The whole work reads like the attempt to reject the opinion that small effects like those of the variable causes can be neglected and that they level out in the long run.

Fechner used probabilistic methods next in his work on psychophysics from 1859 on, and later in his writings on aesthetics. In this context, the most important application (also known to Francis Galton[99]), is made in the *Elemente der Psychophysik* of 1860. There, in developing the so-called "method of right and wrong cases," the Gaussian law is applied to the fluctuations in judgment with a stimulus kept constant. The "measure of precision" h that appears in older formulations of the Gaussian law is used by Fechner in order to express the differential sensitivity. (In modern terms h is equal to $1/\sigma\sqrt{2}$.[100]) Fechner does not give any indication in his work on psychophysics that human freedom makes it necessary to formulate the fluctuation in judgment probabilistically. If there were a relation between probability and freedom, one would then have to interpret the fluctuations not as erroneous deviations from the true value, but as the free mode of reaction of the individual to a stimulus. It is clear, however, that the work in psychophysics grew out of Fechner's more philosophical and religious views. He explicitly affirms a connection of his *Psychophysik* with his *Zend-Avesta* and his *Über die Seelenfrage* from 1860.[101]

Between the *Psychophysik* and the *Kollektivmasslehre* Fechner wrote three articles that dealt with error theory. In one of these articles Fechner developed the notion of the *Centralwerth C* (literally: central value), i.e., the median.[102] He gave a full treatment of its properties and computation. In his *Vorschule der Aesthetik* of 1876 Fechner used the method of extreme ranks for subjective judgments.[103]

In his memorial speech on Fechner, Wilhelm Wundt, the founder of the first psychology laboratory in Leipzig, reported his complete surprise when he discovered Fechner's manuscript on the *Kollektivmasslehre* upon perusing Fechner's papers after his death. "Nobody knew of the existence of this work, neither Mrs. Fechner nor any of his friends and colleagues, although he had carried around the plan for this work for about twenty years and had probably worked almost a decade in realizing this project."[104]

Fechner sees the "main result" and the "main source" of his investigation in the "mutually controlling mathematical justification and empirical confirmation it gives to a generalization of the Gaussian law of chance variation.... As a result, the restriction of this law to symmetrical probability and relative small deviations on both sides from the arithmetical mean is lifted. As another result, lawful relations become visible that were hitherto unknown."[105] The generalized Gaussian law and the newly found lawful relations apply to the so-called *Kollektivgegenstand* (literally: collective object). Fechner defines this as "an object consisting of innumerable specimens that vary randomly (*nach Zufall variierende Exemplare*) that are embraced by a concept of species or genus."[106]

This is almost the concept of the collective as introduced by the statistician Gustav Rümelin, except that Fechner adds the crucial condition of random variation that Rümelin had excluded. In 1863 and 1874, Rümelin distinguished between particulars that are typical of their genus and individuals that do not allow a straightforward inference as to the nature of the genus as a whole. Heterogeneous individuals that have an attribute in common form a "collective concept."[107]

The specimens or examples of a collective object in Fechner's sense can differ in space or time so that one can distinguish between a spatial or temporal object. Since the *Kollektivmasslehre* is, literally, the "doctrine of the measure of collective objects," only measurable attributes are treated. Fechner only considers attributes with continuous arguments; he apparently does not see that one can also extend his doctrine to the discrete case. He makes, however, a distinction between one-dimensional and many-dimensional magnitudes (i.e., correlations between magnitudes) that could equally well be subject to his theory.[108]

Fechner's first example for a collective object is man. A collective object in the narrower sense would be humans of a certain sex, of a certain age, of a certain race. "It is essential to anthropology, zoology, and botany that they deal with collective objects since in these sciences one does not characterize single specimens but the entirety of them."[109] Other examples of collective objects include daily or yearly thermic and barometric values (as examples for temporal collective objects), sizes of recruits, measures of human skulls, weights of human inner organs, measures of ears of rye, amount of rainfall, sizes of books and visiting cards, and dimensions of paintings from several galleries.

As we have already seen, Fechner uses the word *Zufall* (chance) in his definition of the collective object. He made it absolutely clear that this concept was to be taken in its ontic sense, as signifying an objective and real phenomenon. The following quotation shows that Fechner was very well aware of the controversial nature of such a conception but that he tried to circumvent the problem:

Since our concept of a collective object implies the concept of a *chance* variation of the examples, a definition of chance and an explanation of its nature would be desirable. For the following investigation, however, it would hardly be useful to attempt a treatment from a philosophical point of view. Here, one has to be content with the factual point of view, which is of a negative more than of a positive character. A *chance* variation of the specimens, as I see it, is independent of any *arbitrariness* that arises in the measuring process as well as of any *law of nature* (*Naturgesetzlichkeit*) that interferes with the state of the values. It does not matter which of these variations plays a role in the determination of the objects; only variations that are independent of these changes happen by chance.

In other words: Take a collective object, measure the values of a certain attribute, and correct these values, first by taking account of the errors committed in the measuring process, and second by subtracting any special external influence that is active in this case. The variations that are left are objectively due to chance. The "laws" that govern these variations are the proper objects of the *Kollektivmasslehre*.

One can of course be convinced that these remaining variations are the effect of hidden causes that are hitherto unknown and could be found out by science. Yet, in echoing his insight of 1849 that the question of determinism and indeterminism cannot be empirically decided, Fechner retreats to his strict empiricist attitude, claiming the right to talk of chance as long as it is impossible to determine the precise value of an arbitrarily chosen example neither by deduction from theory nor by induction from experience. The continuation of the former quotation is a carefully formulated saving clause that avoids any explicit commitment:[111]

This view does not imply the denial that from a *general* point of view, chance does not exist, since according to the existing natural laws and under the existing conditions the value of each single specimen can be regarded as determined by necessity. But we speak of chance as long as we are neither capable of deriving the single values (*Einzelbestimmungen*) from such general laws nor of inferring them from the facts that are present. Insofar as this would be the case, chance would cease to be, and the applicability of the laws presented here would cease or would be disturbed.

The great reluctance with which Fechner addresses philosophical questions when addressing a scientific audience seems to have become his general attitude after about 1860, as Wundt notes. Having encountered so much opposition to his writings, both from the scientists and the philosophers, Fechner changed the "*tactics* of his procedure," without, however, giving up his overall goal. He now set out to give scientifically exact and refined confirmations of several anticipations and speculations of the *Zend-Avesta*, and at least when talking to scientists, he tried to avoid as much as possible the dreamlike and mystical jargon he had used in the *Zend-Avesta* and similar writings. Wundt especially stresses this in relation to Fechner's *Psychophysik*.[112] I think that the same is true for the *Kollektivmasslehre* as well, which has to be read as the proof of another part of Fechner's *Weltanschauung*. A proof, however, not for the vulgar but for the wise and learned who have eyes to see. Wundt notes repeatedly that Fechner avoided discussions of philosophical questions even when he was himself working and writing on philosophical subjects.[113]

What is the nature of the new laws put forward in the *Kollektivmasslehre?* They differ from the usual natural laws in several ways. First of all, they are laws *of chance*. "The subordination of chance under more general laws ... not having been considered before"[114] bestows philosophical interest upon the subject. The practical interest, however, Fechner estimates (at least in print) to be much higher than the philosophical one.

Second, the laws of collective objects are *distributions*; they show "how the specimens of a collective object are distributed according to measure and number. The term *distribution* signifies the determination by which the number of the specimens of a given collective object varies with the size"[115] of the attribute under consideration. Thus, the *Kollektivmasslehre* is a theory of frequency distributions.

Third, laws of collective objects never deal with single cases. Since the sizes of the specimens of a collective object are not subject to a natural law but instead depend on chance, "one cannot determine by a chance law how large this or that *single* specimen is, although [one can determine] the limits of values in which a given number of these specimens will lie with such and such a degree of probability."[116] This quality of the new laws is particularly relevant to anthropology, zoology, and botany.

Laws of collective objects are of a *probabilistic nature*. One has to note, however, that according to Fechner probability laws can come in two different modes. As laws of collective objects they *rule* these objects, whereas, especially in astronomy and physics, probability laws can also play an *incidental* role as "determinations of certainty" of a value. This role is much less essential than the former one.[117] In astronomy you infer the true value of a single object with probability theory from many repeated measurements. In the *Kollektivmasslehre*, however, "the specimens of a collective object, regardless of their deviation from the arithmetical mean, ... are equally real and true, and a consideration that prefers one to the other ... makes, of course, no sense."[118]

The goal of the *Kollektivmasslehre* is now to find the distribution laws for the collective objects and to "distinguish the different collective objects by characteristic constants that are to be derived from the circumstances of their distributions."[119] The first important step is to find out whether a given set of data represents a collective object or not. Certain formal requirements must be satisfied. One needs a large mass of data, "so many, that the ideal laws of chance, which can be claimed strictly speaking only for an infinite number, can be confirmed with an approximation that is sufficient for the desired degree of precision."[120] In addition to this, the collective object has to be "complete" (*vollzählig*), which means that one has not the right to "mutilate" a collective object, i.e., to leave out any outlier. A collective object must not consist of "disparate components;" i.e., the mixture of two collective objects does not form a collective object itself.[121] If a set of given data shows more than one maximum, i.e., if a "densest value" appears more than once, we have an indication for a mixed object.

The essential aspect that has to be fulfilled by a collection of data in order to be a collective object is chance variation. That means excluding specimens that "show a kind of dependence by natural law upon one another that steps out of the realm of the laws of chance."[122] In order to do this, Fechner develops very simple tests of homogeneity by which one can find out whether a collective object is governed by chance or whether the laws of chance are "disturbed" by laws of nature—and if so, by how much. First, one has to find a series of data that varies purely by chance (a random series as we would call it today) and compare it as a standard to the other series under consideration. A series of draws out of an urn with infinitely many numbered balls would be the ideal representation of a chance series. As a substitute for this Fechner takes the results of the Saxon lotteries already mentioned above. "If anywhere, then it is here that chance plays its pure role."[123] In comparing other series of data with the Saxon lottery one can measure the degree of "independence of succession" of the members of another chance series.

To this end, Fechner counts how often in a given series of data the successor of a number follows its predecessor in being odd or even and how often it changes in this respect. Fechner finds that for the Saxon lotteries these two numbers are about equal. In a series of meteorological data, however, taken in the order of their appearance during a day, we find that the successor of a number deviates from the mean more often in the same direction as its predecessor than in the opposite one.

Fechner takes this as a proof for the dependence of these values, a dependence "that surpasses the laws of chance."[124]

A second method consists in counting how often in the series a number is larger than its predecessor and how often it is smaller. Fechner again finds a very significant difference between lottery and weather in this respect, and therefore a second "proof that the increase and decrease in meteorological data from day to day does not follow pure laws of chance." In chapter XV Fechner tries to expand this method. There he wants to find out whether a set of data is fluctuating because of external circumstances or because of mere chance. His examples are births, deaths, and suicides in different seasons, male and female births, and the number of thunderstorms at different places. He tries to answer the question how much we should bet in a given case that a fluctuation is due to external circumstances and not to mere chance. When in this context he investigated the "pure" chance behavior of the lottery, he divided the numbers up in series of 3, 10, 50, and 100 numbers and examined their behavior. In order to do this, he counts the absolute value of the difference between the amount of even and odd numbers in these series and finds out by theory and in practice that these selected series show the same chance behavior. Later on Fechner tries to use similar ideas in estimating the mutual dependence of two statistical series, i.e., to develop a kind of correlation index.[125] In chapters XXIII and XIV Fechner develops rank methods similar to Kendall's tau.[126]

The basis for the investigation of collective objects is the so-called "distribution table." It shows how many specimens for one value have been found for the collective object. The values are ordered by size and each must differ from its predecessor by a constant interval that has to correspond to the accuracy in observation. These frequencies have to be corrected, however, before one can determine their mutual relations. First one has to get rid of the "inessential deviations." These deviations arise because one does not have an infinitely long distribution table. They can be detected and estimated by the decrease in deviation that appears if one successively increases the number of specimens of a collective object.[127] A further correction is necessary in order to account for the fact that one always arrives at discrete values because of finite observational precision.

Now one is in a position "to derive the *determining parts* or *elements* of the collective objects that allow for the characterization of the object and for its comparability with other objects in a quantitative way."[128] The most important elements already taken into account before Fechner were of course the total number of specimens m in the distribution table, their arithmetical mean A, the deviations Θ from A, and perhaps the extremes E. But Fechner finds this list highly unsatisfactory and expands it by several other elements that are not reducible to the former ones. The most important ones are the median C (*Centralwerth* or central value),[129] for which the number of positive and negative deviations is equal, and the "densest value" D (*der dichteste Werth*), for which the number of examples of a collective object is a maximum and is therefore the most probable value. A, C, and D are called the "main values" (*Hauptwerthe*), to which all other values are related. Other values include the "dividing value" R (*Scheidewerth*), which has an equal sum of values above and below, the "heaviest value" T (*schwerster Werth*), for which the product of value and number of the specimen is a maximum, and the "center of mass of the deviations" F (*Abweichungsschwerwerth*) for which the product of the frequency of the specimens and the deviations (from a main value) are a maximum.

Fechner considers the deviations Θ from the main values as German mathematical astronomers had since the early-nineteenth-century.[130] The "mean deviation" (*mittlere Abweichung*) is $\varepsilon = \sum \Theta / m$, the "probable deviation" w (*wahrscheinliche Abweichung*) is the median of the deviations, and the "quadratic mean deviation" (*quadratischer Mittelfehler*) is $q = \sqrt{\sum \Theta^2 / m}$, later called standard deviation after Pearson's work of 1894.

The main reason why Fechner introduces other "main values" besides the arithmetical mean lies in his conviction that asymmetric distributions are the normal case in nature. For symmetric distributions one need consider only the deviations from the arithmetical mean in order to calculate the most probable value. The "main values" do not deviate from each other in this case.

This conviction was quite revolutionary since it ran counter to the whole statistical approach of the nineteenth century. Quetelet and Galton both saw the normal distribution as the key to the science of man. It was not until Pearson's work in the 1890s that the normal distribution ceased to be preeminent. (Bessel, however, had already noted in 1838 that there are also asymmetric error distributions and that the normal distribution cannot therefore be a general law valid for all cases.[131]) For Quetelet, the asymmetric distribution of data was an indication that one has not yet collected enough data and that fortuitious influences (variable causes) have not yet leveled out and still mask the true value.[132] For Fechner, however, the asymmetry of a collective object is not the result of a temporary external influence that produces misleading results but a characteristic mark whose special form distinguishes one law of chance from another. "Nothing is less self-evident," writes Fechner, than to carry over the Gaussian law from observational errors to collective objects:[133]

There is indeed from the outset a big difference between deviations from the arithmetical mean, which are obtained by measuring a *single* object repeatedly because the measuring instruments or the senses lack precision and because there are fortuitious external disturbances; and deviations from the mean, which the *many* specimens of a collective object show because of reasons that lie in the nature of the objects themselves or in the external circumstances that influence them. It was therefore completely impossible to predict a priori whether nature in these deviations from the mean follows the law of the observational errors or not. One had to establish the law directly by examining the collective objects themselves.

The sizes of recruits, the first collective objects examined, initially seemed to follow the Gaussian law. This was misleading, however, as Fechner says, since other distributions were later found that were undoubtedly asymmetric. There is also a theoretical reason against the universality of the normal distribution noted by

Fechner: The probability for negative deviations from the mean can in theory never be zero. In reality, however, it becomes zero if the deviation is greater than the mean value itself.[134]

As the general law for collective objects Fechner develops a so-called "two-sided" or "double-columned" Gaussian law that treats the symmetric case as a limit. The deviations are counted from D and not from A as in the case of the normal distribution. The idea now is to treat the left and the right branches of the distribution curve *separately*, as if they stemmed from two distinct normal distributions that agree in D but differ in their measure of precision h. Fechner lists some simple theorems he has found on the relations between the "main values" of the two-sided distribution.[135]

In the last part of the *Kollektivmasslehre* Fechner gives detailed examples of empirical distributions: heights of recruits and students at different times and places, different measures taken from blades and ears of rye (he painstakingly lists the places and dates (1863) where and when he has taken samples of rye in the environs of Leipzig), dimensions of paintings (he investigated 10,558 paintings using information from catalogues of 22 galleries), and meteorological values (amounts of rainfall from Geneva in 1845). Apparently he also wanted to add a chapter on the asymmetry of error distributions. In elaborating this chapter the editor G. F. Lipps came to the conclusion that error series do not show such an asymmetry.

Let us step back and ask how Fechner's *Kollektivmasslehre* is related to his philosophical ideas about indetermination. The most obvious link one can establish is to the essay of 1849 on *Die mathematische Behandlung organischer Gestalten und Processe*. There, Fechner wanted a mathematics that allowed for the treatment of nature in its general aspects without implying necessity and strict determination of the single case. The *Kollektivmasslehre* is the detailed elaboration of this idea. Although the specimens of a collective object vary haphazardly, one can formulate laws for the collective object as such. Fechner nowhere admits a connection to his philosophy explicitly, but this can be explained by his general reluctance to disclose his philosophical motives in strictly scientific contexts. Even his change in jargon can be explained this way: in 1849 he spoke of indetermination and indeterminism in the way a philosopher would; later he spoke of the "laws of chance" with the most innocent face, as if he just meant "probability theory" by this term, in the same way as the French did with their expression *les lois du hasard*.

There are three contemporaries of Fechner who confirm a relation of the *Kollektivmasslehre* to the idea of the intrinsic imprecision of the scientific object as put forward in the *mathematische Behandlung* ... in 1849. These are Wilhelm Wundt, the philosopher-psychologist Oswald Külpe, and the Leipzig astronomer Heinrich Bruns. Wundt and Bruns at least had firsthand knowledge of Fechner's manuscripts.[136]

Unfortunately, Fechner does not give any hint as to how the ideas of his second article of 1849 are related to the *Kollektivmasslehre*. In particular, we do not know how the two other types of indetermination (by novelty and by limited prediction) bear on collective objects. The most likely relation, if Fechner still wanted one, could be established in this way: The constant occurrence of novel initial conditions leads to the chance variation of the attribute in a collective object; i.e., intrinsic novelty *explains* chance behavior. On the other hand, chance variation serves as an explanation for why there is factual unpredictability in nature—an explanation that is empirically much more satisfactory than the a priori argument from ignorance to which the determinist must take resort. One could very easily imagine that Fechner chose to continue the line of the *mathematische Behandlung* ... rather than of the *Causalgesetz* because he saw it as more likely to yield fruitful and empirically relevant new results. Novelty and unpredictability cannot be conclusively proved, but the search for workable criteria of irregularity and chance might prove successful.

6   From Fechner to von Mises

A word must be said about the reception of Fechner's ideas. For Fechner's papers of 1849 this can be done very quickly, for they seem to have been simply ignored. The first reference I have found is from 1890. In a study on the development of the problem of causality, Edmund Koenig interprets Fechner's *Causalgesetz* as the attempt to avoid a complete determination of mind by matter without postulating a violation of causal determination: "In other words, Fechner tries to represent the course of the world as a developmental process in which laws are not to be considered as rigid norms beforehand that necessitate the individual case, but rather as later descriptions of the factual behavior of the entities that are determined in their mode of action by totally different motives." [137] This view, though logically possible, is for Koenig of no account since one can nowhere demonstrate that an entity or being newly creates the law of its action by itself.

The general neglect of Fechner's ideas of 1849 is easily explained. Either they were superficially read as the mere repetition of typical nineteenth-century determinism of the Laplacean variety, which took natural laws as expressions of absolutely rigid and necessary causal relations,[138] or one dismissed them, as Koenig did, as a perhaps interesting but empirically unprovable theory.

The reception of the *Kollektivmasslehre* was different. In 1897, while the *Kollektivmasslehre* was still in press, the Leipzig astronomer Heinrich Bruns (1848–1919) gave a general solution to Fechner's problem of how to find a mathematical representation for frequency distributions. Taking up a suggestion from Bessel of 1838, Bruns gave an expansion of an arbitrary function into an infinite series whose terms are successive derivatives of the Gaussian distribution. This function plays a similar role for frequency distributions as the Fourier series does for the treatment of periodic processes. Shortly thereafter, Bruns gave a complete presentation of his new method in a lecture series in Leipzig during the winter of 1898/99.[139] In 1898, the editor of the *Kollektivmasslehre*, the philosopher and psychologist Gottlob Friedrich Lipps (1865–1931) also working in Leipzig, gave a thorough review of the *Kollektivmasslehre* in Wundt's journal *Philosophische Studien*. In later writings he tried to develop a new theory of collective objects by expanding on Fechner and

Bruns.[140] Both Bruns and Lipps explicitly formulate the view that is implicit in every page of the *Kollektivmasslehre*, that the theory of collective objects is not a science separate from probability theory. Instead, the probability calculus, as a theory of frequencies, applies to the same range of phenomena as the theory of collective objects does, namely, mass phenomena. So the theory of collective objects is nothing but a consistent consolidation of probability theory, and probability must be defined as relative frequency.[141] One of the founders of energetics, the physicist Georg Helm, proposed a similar view.[142]

Bruns and Lipps also share the opinion that chance variation is not an essential condition of a collective object and can be neglected in probability theory—thus watering down Fechner's requirements and cutting off any connection with his indeterminism. A further similarity between Lipps and Bruns was their rejection of Pearson's work. They saw a "higher flexibility" in Pearson's five types of curves but they preferred the Bruns-series (or φ-series, as it is also called) as a more general treatment.[143]

Fechner also attracted notice in France. Joseph Bertrand (1822–1900), mathematician and author of a well-known book on probability theory essentially in the classical tradition, wrote a review of Fechner's work where he expresses grave doubts concerning Fechner's program.[144] On the one hand, he does not believe that one can show the existence of pure chance variation (in the same way as it is impossible to show that the series formed by the seventh digit of the logarithms is random or not); on the other hand, it is very unlikely that one shall find one common law for all sorts of different variations.

Some psychologists also applied Fechner's theory to psychophysics, the other Fechnerian tradition. Of special note among them are Lipps and his colleague in Leipzig, Wilhelm Wirth. Work in this tradition decreased sharply after World War I and disappeared altogether with World War II.[145]

By 1908, Fechner's theory of collectives seems to have been a standard subject for anyone working in probability and statistics in Germany. Even the traditionalist Emanuel Czuber, who dominated the field at the time, included the term *Kollektivmasslehre* in the title of the reedition of one of his books on probability theory, although he is silent about the frequency view of Bruns and Lipps.[146]

By 1919, the development enters a new phase with the work of the mathematician and philosopher Richard von Mises (1883–1953), a student of Czuber in Vienna between 1901 and 1906. One can describe this new phase in three different ways: as the culmination of Fechner's reception, as the completion and final consolidation of the frequency view, and as the end of the predominance of Laplace's classical theory. Already in 1912, von Mises had given a very elegant and shortened reformulation (and thereby slight correction) of Bruns's idea by developing the distribution function as a series of Hermitean polynomials.[147] In 1919, he put forward his probability theory, which is based on the two famous postulates,[148] requiring the existence of the limiting values of the relative frequencies and the randomness (*Regellosigkeit*) by which the attributes (labels) are mapped unto the elements of the collective. Randomness is defined as the invariance of the frequencies under any place selection. These two requirements are precise mathematical formulations of

Fechner's basic intuitions. In 1928, von Mises edited his book on *Wahrscheinlichkeit, Statistik und Wahrheit*, which in 1936 received this subtitle: Introduction to the New Doctrine of Probability and Its Applications. In this book he gives a detailed and very readable exposition of his theory for the nonmathematical reader. He considers many fields of applications and draws important philosophical consequences. The book has been translated into many languages and went through many editions.

What are the main reasons for von Mises's success? Why was his idea of randomness accepted and not rejected or ignored, as Fechner's ideas of indetermination and chance were? Besides the many reasons one can give from an internal foundational standpoint, one must refer to two factors: the "crisis of mechanics" and the philosophy of Ernst Mach. Since I cannot deal here with general aspects of the crisis of classical physics in the late nineteenth century, I shall restrict myself briefly to showing how this crisis was viewed by von Mises himself, and why he thought that his new probability theory could overcome it. In a paper from 1921 entitled *Über die gegenwärtige Krise der Mechanik*[149] von Mises lists a number of problems that he thinks insoluble in the framework of classical physics. These problems include the movement of liquids, the elasticity of solids, and, most important, Brownian motion. He refers to his own theory of Brownian motion,[150] which, in contrast to the theories of von Smoluchowski and Einstein, starts from initial probabilities, does not use the "ill-famed" ergodic hypothesis, and reaches conclusions that do not possess a deterministic character. He sees the general structure of the problem exemplified in Galton's board. The distribution one obtains with such a device cannot be derived from classical mechanics; we "even lack any idea what such a derivation would look like."[151] Boltzmann's treatment would not be an alternative since it is contradictory to give a classical description of the collisions of the gas molecules and then to "thwart" these calculations by purely statistical considerations. If the kinetic theory has made us accustomed to statistical treatments anyway, why not go all the way and use a purely statistical theory if we have more success with it?[152] We are all the more justified since the new outlook does not deal with the behavior of individual entities one by one, and thus does not really come into conflict with deterministic physics, which makes specific claims about the individual cases. All these considerations aim "to replace or to complement the rigid causal structure of the classical theory."[153] Von Mises later elaborated this point of view more thoroughly in his book, in an address from 1929 and in his *Kleines Lehrbuch des Positivismus* from 1939.[154]

For all his life, von Mises was an ardent adherent to Ernst Mach's philosophy—more so perhaps than any of his friends from the Vienna Circle. All his writings that are not purely mathematical breathe a Machian spirit. In the aforementioned article on the crisis of mechanics, von Mises criticizes as unscientific the assumption that the path of a ball in a Galtonian board is determined. Since we shall never have the chance to know all the external influences acting on the balls, we cannot decide the question whether such a knowledge would enable us to determine the exact path. And assumptions that cannot be decided by experience are unscientific.[155] Besides thus strictly insisting on empirical verifiability of scientific propositions, von Mises also makes use of Mach's criticism of the a priori. Causality is not the precondition of experience but a very general result of experience itself. "The causal principle," says von Mises, "is variable and it will subordinate itself to the requirements of physics."[156] Thus, by drawing a connection between his new probability theory and Mach's philosophy, von Mises had a strategy for winning over all those scientists who saw a serious and perhaps even promising solution in Mach's thinking to the crisis of classical physics.[157]

As a last point one should note how von Mises himself saw his relation to Fechner's work. He writes that his own view was stimulated by Fechner's treatment of collectives,[158] although he thinks that there are two respects in which his theory is decisively superior to Fechner's: first, in requiring infinitely many elements of a collective; and second, in introducing randomness as a property of the collective.[159] The latter point is even up to this day taken as the "original and characteristic feature of von Mises's theory."[160] Yet, this is simply not true. The very first sentence of Fechner's book gives the definition of the collective object in terms of "chance."[161] And Fechner is also at pains to work out an estimation for how a collective object would look were it extended to infinity. For the precise validity of the "ideal laws of chance" is only given in the infinite case.[162]

Perhaps it would have been too embarrassing for a logical positivist to admit that one has learned one's decisive empiricist lesson from someone who has written "poetry by concepts" (*Dichtung in Begriffen*) or "philosophical fiction," as Wilhelm Wundt put it when he was looking for a general label for Fechner's work.[163] This would have meant that 'meaningless metaphysics' had led to one of the great triumphs of twentieth-century empiricism and scientific philosophy.

If we look from the time of von Mises back to the roots of indeterministic ideas in Neospinozism, 'half-Hegelianism', epigenesis, and doctrines of human freedom, we can agree with Stephen Brush (who alluded to the influence of ideas of irreversibility on indeterminism): "The revolution was complete ... but its history was almost obliterated."[164] And this history, I think, shows that twentieth-century probability theory is more intimately related to questions of our human existence than one would have dreamt of.

Notes

1. Richard von Mises, *Wahrscheinlichkeit, Statistik und Wahrheit* (Wien: Springer, 1951), p. 252 (also in: Richard von Mises, "Über kausale und statistische Gesetzmässigkeit in der Physik," *Die Naturwissenschaften*, 18: (1930), 145–153, on p. 146).

2. Fechner's two addresses from 1849 sometimes read as if they were a part of Charles Sanders Peirce's *Monist* series of the 1890s.

3. Recent articles on Fechner: Marilyn E. Marshall, "Physics, Metaphysics, and Fechner's Psychophysics," in William R. Woodward and Mitchell G. Ash, eds., *The Problematic Science: Psychology in Nineteenth Century Thought* (New York: Praeger, 1982), pp. 65–87; Lothar Sprung and Helga Sprung, "Gustav Theodor Fechner—Wege und Abwege in der Begründung der Psychophysik," *Zeitschrift für Psychologie*, 186 (1978), 439–454; Helmut E.

148   Michael Heidelberger

Adler, "The Vicissitudes of Fechnerian Psychophysics in America," *Annals of the New York Academy of Sciences*, 291 (1977), 21-32, also in R. W. Rieber and K. Salzinger, eds., *Psychology: Theoretical-Historical Perspectives* (New York: Academic Press, 1980). From the older literature see especially Wilhelm Wundt, "Gustav Theodor Fechner. Rede zur Feier seines hundertjährigen Geburtstages," (1901), Wilhelm Wundt, *Reden und Aufsätze* (Leipzig: Kröner, 1913), pp. 254-343; J. E. Kuntze, *Gustav Theodor Fechner (Dr. Mises). Ein deutsches Gelehrtenleben* (Leipzig: Breitkopf & Härtel, 1892).

4. William James, "The Doctrine of the Earth-Soul and of Beings Intermediate between Man and God. An Account of the Philosophy of G. T. Fechner," *The Hibbert Journal*, 7 (1909), 278-294, on p. 278. (This article became the fourth chapter of James's *A Pluralistic Universe* (New York: Longmans, Green, 1909), pp. 131-177.)

5. Friedrich Albert Lange, *Geschichte des Materialismus und Kritik seiner Bedeutung in der Gegenwart*, vol. 2 (Leipzig: Reclam, 1905), p. 102. This text follows the influential second edition from 1873/1875.

6. Matthias Jakob Schleiden, "Pflanzenchemie" [Review of Four Books on Agricultural Chemistry], *Neue Jenaische allgemeine Literatur-Zeitung*, 4(162) (1845), 645-647, on p. 646.

7. See Marilyn E. Marshall, "William James, Gustav Fechner, and the Question of Dogs and Cats in the Library," *Journal for the History of the Behavioral Sciences*, 10 (1974), 304-312.

8. Friedrich Paulsen, *Einleitung in die Philosophie*, 3rd ed. (Berlin: Wilhelm Hertz, 1895), passim.

9. This *Grundansicht* (basic view), as Fechner calls it, is treated in almost all of his works. The most important treatment is that in his *Zend-Avesta, oder über die Dinge des Himmels und des Jenseits*, 2 vols., 3rd ed. (Hamburg/Leipzig: Leopold Voss, 1906), in vol. 2, pp.129-179. (The first edition appeared in three parts at Leipzig: L. Voss 1851, the second one in 1901.) "Zend-Avesta" is supposed to mean "living word."

10. Fechner, *Zend-Avesta*, vol. 2, p. 135.

11. Fechner, *Zend-Avesta*, vol. 2, p. 135.

12. Gustav Theodor Fechner, *Elemente der Psychophysik*, 2nd unchanged ed. Wilhelm Wundt, ed., 2 vols. (Leipzig: Breitkopf & Härtel, 1889), in vol. 2, p. 135. The first edition appeared in 1860.

13. Fechner, *Zend-Avesta*, vol. 2, p. 141. Cf. also p. 152 and p. 163 in the same volume.

14. Fechner, *Psychophysik* (note 12), vol, 1, p. 8 and passim, and Fechner, *Zend-Avesta*, vol. 2, p. 133. See also vol. 2, p.153.

15. Fechner. *Psychophysik* (note 5), vol. 1, p. 8.

16. See especially Fechner, *Zend-Avesta*, vol. 1, p. 152. Note that Fechner's parallelism is different from strict Spinozism: (i) For Spinoza, mind and matter are attributes of an unknown substance. For Fechner, there is nothing unknown behind them. (ii) In Fechner's view one can construct the laws of one realm from the knowledge of the laws of the other realm. This is not true for Spinoza. Compare also note 23.

17. Fechner, *Zend-Avesta*, vol. 1, p. 153. On pp. 148-161, Fechner compares at length his theory with other alternatives.

18. Fechner's treatment of analogy seems standard for his time. A concise and clear definition is given by Fechner himself in his youthful *Katechismus der Logik oder Denklehre* (Leipzig: Baumgärtner, 1823), pp. 178ff.

19. Fechner had even written a whole book in order to defend the atomistic standpoint against many of his philosophical contemporaries. See Gustav Theodor Fechner, *Über die physikalische und philosophische Atomenlehre* (Leipzig: H. Mendelssohn, 1855). A second revised edition appeared in 1864, reprinted Frankfurt/Main: Minerva, 1982.

20. Fechner, *Zend-Avesta*, vol. 2, p. 138.

21. See Heinrich Adolph, *Die Weltanschauung Gustav Theodor Fechners* (Stuttgart: Strecker und Schröder, 1923) pp. 155–159.

22. For a quick review of those arguments see Bas C. van Fraassen, *The Scientific Image* (Oxford: Clarendon, 1980), chapter 2. For a clear example of Fechner's arguments in favor of realism concerning the psychical, see his *Zend-Avesta*, vol. 2, pp. 138ff.

23. See, e.g., Fechner, *Zend-Avesta*, vol. 1, p. 110, and vol. 2, p. 141, or his *Über die Seelenfrage* (Leipzig: C. F. Amelang, 1861) pp. 49ff. Fechner also differs from Spinoza in his teleology.

24. Fechner, *Zend-Avesta*, vol. 1, pp. 200–206, on pp. 200ff.

25. Gustav Theodor Fechner, "Die mathematische Behandlung organischer Gestalten und Processe. Öffentliche Sitzung am 18. Mai zur Feier des Geburtstags seiner Majestät des Königs," *Berichte über die Verhandlungen der Königlich sächsischen Gesellschaft der Wissenschaften zu Leipzig*, Mathematisch-physische Classe, Jahrgang 1849 (Leipzig: Weidmann, 1849), pp. 50–64, and in the same volume: Gustav Theodor Fechner, "Über das Causalgesetz. 14. November 1849. Öffentliche Sitzung zur Feier von Leibnitzens Todestage," pp. 98–120.

26. The history of the Royal Saxon Society in Leipzig is treated by Johannes Wislicenus, "Rede," *Zur fünfzigsten Jubelfeier der Königlich sächsischen Gesellschaft der Wissenschaften zu Leipzig am 1. Juli 1896*, Reden und Register (Leipzig: 1897), pp. VII–XXI, and Christa Jungnickel, "Teaching and Research in the Physical Sciences and Mathematics in Saxony, 1820-1850," *Historical Studies in the Physical Sciences*, 10 (1979), 3–47.

27. The causal law is treated in Fechner, *Atomenlehre*, 2nd ed. (note 12), pp. 125 and 170ff.; in Fechner's *Zend-Avesta*, vol. 1, pp. 207–222; vol. 2, pp. 95–129; also in Fechner's *Seelenfrage* (note 23), p. 218, and in his *Die Tagesansicht gegenüber der Nachtansicht* (Leipzig: Breitkopf & Härtel, 1879), chapter XVII.

28. Fechner, "Organische Gestalten" (note 25), p. 53.

29. Fechner, "Organische Gestalten," p. 50.

30. Fechner, "Organische Gestalten," p. 51.

31. Fechner, "Organische Gestalten," p. 53.

32. Fechner, "Organische Gestalten," p. 57.

33. Fechner, "Organische Gestalten," p. 56.

34. Hermann Helmholtz, *Über die Erhaltung der Kraft* (Leipzig: W. Engelmann, 1902), p. 4. The first edition appeared in Berlin in 1847. Kant talks of "empirical rules" in his *Prolegomena*, §29, Akademie-edition VI, p. 312.

35. Fechner, "Organische Gestalten," p. 56.

36. Fechner, "Causalgesetz" (note 25), p. 98.

37. Fechner, "Causalgesetz," p. 100.

38. These are my examples. Similar examples from Fechner himself can be found in Marilyn E. Marshall, "G. T. Fechner: Premises toward a General Theory of Organisms (1823)," *Journal for the History of the Behavioral Sciences*, 10 (1974), 438–447.

39. Fechner, "Causalgesetz," p. 102.

40. Fechner, *Zend-Avesta*, (note 9), vol. 1, pp. 207–210.

41. See, for example, William R. Woodward, "Fechner's Panpsychism: A Scientific Solution to the Mind-Body Problem," *Journal for the History of the Behavioral Sciences*, 8 (1972), 367–386, on pp. 374–376, and Marshall, "Physics" (note 3), p. 75.

42. Fechner, "Causalgesetz," pp. 110ff.

43. Fechner, "Causalgesetz," p. 111.

44. Fechner, "Causalgesetz," p. 111.

45. Fechner, "Causalgesetz," p. 112.

46. Fechner, "Causalgesetz," p. 112.

47. Fechner, "Causalgesetz," p. 113.

48. Fechner, "Causalgesetz," p. 114.

49. Fechner, "Causalgesetz," p. 114.

50. Fechner, *Zend-Avesta*, vol. 2, p. 113. Fechner had worked out the physics of Weber's and Berzelius's suggestion in his *Atomenlehre* (note 19). Weber's text as cited by Fechner can be found in Wilhelm Weber, "Elektrodynamische Maassbestimmungen. Über ein allgemeines Grundgesetz der elektrischen Wirkung" (1846), *Wilhelm Weber's Werke*, ed. by the kgl. Ges. d. Wiss. z. Göttingen, vol. 3 (Berlin: Springer, 1893), pp. 25–214, on p. 212.

51. Fechner, "Causalgesetz," p. 115.

52. Fechner, "Causalgesetz," p. 115.

53. Striking parallels can be found in the growing literature on randomness in information theory. For example: "Our results show that the ability to calculate what happened in the past is not restricted to deterministic systems. A system can evolve in a very random way but still preserve complete information about its past. It is tempting to wonder whether our own universe is like this"—J. V. Howard, "Random Sequences of Binary Digits in Which Missing Values Can Almost Certainly be Restored," *Statistics and Probability Letters*, 1 (1983), 233–238, on p.237.

54. Fechner, "Causalgesetz," pp. 117ff.

55. Fechner, "Causalgesetz," p. 118.

56. Fechner, "Causalgesetz," p. 119.

57. Fechner, "Causalgesetz," p. 119.

58. Fechner, "Causalgesetz," p. 119. Fechner stressed this idea very much in later writings and, therefore even called himself a "determinist," e.g., in his *Tagesansicht* (note 27), pp. 170ff. Many commentators of Fechner have been mislead by this.

59. Fechner, "Causalgesetz," p. 119.

60. Fechner, "Causalgesetz," p. 116.

61. See Herbert Schnädelbach, *Philosophie in Deutschland 1831–1933* (Frankfurt: Suhrkamp, 1983), p. 237. (English edition: *German Philosophy 1831–1933* (Cambridge: Cambridge University Press, 1983).)

62. See Karl Löwith, *Von Hegel zu Nietzsche*, 7th ed. (Hamburg: Meiner, 1978), pp. 125ff. and 164ff.

63. Weisse's life and work is treated by Kurt Leese, *Philosophie und Theologie im Spätidealismus* (Berlin: Juncker & Dünnhaupt, 1929), and by Albert Hartmann, *Der Spätidealismus und die Hegelsche Dialektik* (Berlin: Juncker & Dünnhaupt, 1937). The biting criticism of the orthodox Hegelian Rosenkranz is still a pleasure to read: Karl Rosenkranz, *Kritische Erläuterungen des Hegel'schen Systems* (Königsberg: Bornträger, 1840).

64. Kuntze, *Gelehrtenleben* (note 3), pp. 141ff., 168–178.

65. Max Wentscher, *Hermann Lotze*, vol. 1: *Lotzes Leben und Werke* (Heidelberg: C. Winter, 1913) pp. 26ff., Leese, *Spätidealismus* (note 63), p. 11.

66. Kuntze, *Gelehrtenleben*, p. 141, and Wentscher, *Lotze*, p. 31.

67. Christian Hermann Weisse, *Grundzüge der Metaphysik* (Hamburg: Perthes, 1835), p. IV.

68. Christian Hermann Weisse, *Über den gegenwärtigen Standpunct der philosophischen Wissenschaft. In besonderer Beziehung auf das System Hegels* (Leipzig: Barth, 1829), p. 139.

69. Weisse, *Standpunct* (previous note), p. 120.

70. Christian Hermann Weisse, *Die Idee der Gottheit. Eine philosophische Abhandlung. Als wissenschaftliche Grundlegung zur Philosophie der Religion* (Dresden: Grimmer, 1833), p. 290.

71. Johann Wolfgang von Goethe, *Dichtung und Wahrheit*, Book 11. (In the *Hamburger Ausgabe* of his *Werke*, vol. 9, 8th ed. (München: C. H. Beck 1978), pp. 490ff.)

72. Weisse, *Standpunct* (note 68), p. 149.

73. Christian Hermann Weisse, *System der Ästhetik als Wissenschaft von der Idee der Schönheit* (Leipzig: Hartmann, 1830), pp. 92ff.

74. Christian Hermann Weisse, "Ueber die metaphysische Begründung des Raumbegriffs. Antwort an Herrn Dr. Lotze," *Zeitschrift für Philosophie und spekulative Theologie*, 8 (1841), 25–70, on p. 38.

75. Weisse, *Gottheit* (note 70), p. 290.

76. Weisse, *Standpunct*, pp. 185ff.

77. Lotze, who was a trained physician, tried to make freedom and medical science compatible. His basic idea was to take the soul as a "randomizer" that sets a condition at random; the coincidence of this random condition with a material condition of the body functions as a cause for a material effect, e.g., a certain sickness.

78. See section 3 and Fechner, "Organische Gestalten," p. 52.

79. Fechner, *Zend-Avesta*, vol. 2, p. 158.

80. Owsei Temkin, "German Concepts of Ontogeny and History Around 1800," *Bulletin of the History of Medicine*, 24 (1950), 227–246 (reprinted in: Owsei Temkin, *The Double Face of Janus* (Baltimore and London: Johns Hopkins University Press, 1977), pp. 373–389).

81. Temkin, "Ontogeny," p. 387.

82. For Burdach, see Temkin, "Ontogeny," and Alan S. Kay, "Burdach, Karl Friedrich," *Dict. Sci. Biogr.*, vol. 2 (New York: 1970) pp. 594–597.

83. Fechner does not mention Burdach in his "Causalgesetz," although he was familiar with Burdach's writings.

84. Karl Friedrich Burdach, "Die persönliche Besonderheit," *Blicke ins Leben*, vol. 2 (Leipzig: Leopold Voss, 1842), pp. 232–271, on pp. 232ff.

85. Burdach, "Besonderheit," p. 235.

86. Burdach, "Besonderheit," p. 237.

87. See Ernst Mayr, "Typological versus Population Thinking," in Ernst Mayr, *Evolution and the Diversity of Life* (Cambridge, MA: Harvard University Press, 1976) pp. 26–29. Cf. also Elliott Sober, "Evolution, Population Thinking and Essentialism," *Philosophy of Science*, 47 (1980), 350–383.

88. Burdach, "Besonderheit," p. 245.

89. Burdach, "Besonderheit," p. 258.

90. Burdach, "Besonderheit," p. 257. Other references to Quetelet can be found on pp. 238, 252, and 258. On p. 249, Burdach quotes Quetelet's famous phrase about the budget of crime, "which is paid with terrifying regularity."

91. Burdach, "Besonderheit," p. 259.

92. Burdach, "Besonderheit," p. 255.

93. The only other names Fechner mentions in addition to these are Poisson (once), the Leipzig mathematician Scheibner, the statisticians E. B. Elliott, Bodio, and Gould, the anthropologist Welcker, and the meteorologists Dove and Schmid.

94. On Quetelet cf. Ted Porter's article in this volume; Peter Buck, "From Celestial Mechanics to Social Physics: Discontinuity in the Development of the Sciences in the Early Nineteenth Century," H. N. Jahnke and, M. Otte, eds., *Epistemological and Social Problems of the Sciences in the Early Nineteenth Century* (Dordrecht: Reidel, 1981), pp. 19-33; Matthias Schramm, "Some Remarks on A. Quetelet," M. Heidelberger and L. Krüger, eds., *Probability and Conceptual Change in Scientific Thought* (Report Wissenschaftsforschung, Universität Bielefeld, No. 22, Bielefeld: B. Kleine, 1982), pp. 115-127.

95. On Drobisch see Porter's chapter in this volume. Moritz Wilhelm Drobisch, *Beiträge zur Orientierung über Herbart's System der Philosophie* (Leipzig: L. Voss, 1834), p. 64. Quetelet's *mémoire* is the following: Adolphe Quetelet, "Sur la statistique morale et les principes qui doivent en former la base," *Mémoires de l'Académie royale des sciences et belles-lettres de la Belgique*, 21 (1848). Drobisch's review of it appeared in the *Leipziger Repertorium der deutschen und ausländischen Literatur*, 25 (1849), 28-39. The book mentioned is Moritz Wilhelm Drobisch, *Die moralische Statistik und die menschliche Willensfreiheit* (Leipzig: L. Voss, 1867).

96. Gustav Theodor Fechner, *Kollektivmasslehre*, ed. by Gottlob Friedrich Lipps im Auftrag der Königlich sächsischen Gesellschaft der Wissenschaften (Leipzig: W. Engelmann, 1897), p. 229.

97. Fechner, *Kollektivmasslehre*, p. 228.

98. Gustav Theodor Fechner, *Professor Schleiden und der Mond* (Leipzig: A. Gumprecht, 1856), p. V. On Fechner's relation to Schleiden see Woodward, "Panpsychism" (note 41). Poisson is cited on pp. 335ff. and 353.

99. See Bernard Singer, "Distribution-Free Methods for Non-Parametric Problems: A Classified and Selected Bibliography," *British Journal of Mathematical and Statistical Psychology*, 32 (1979), 1-60, on p. 7.

100. Fechner, *Psychophysik* (note 12), vol. 1, chapter VIII. Cf. Helen M. Walker, *Studies in the History of Statistical Method* (Baltimore: Williams & Wilkins 1929), pp. 24ff., 49ff.

101. Fechner, *Psychophysik* (note 12), vol. 2, p. 543.

102. Gustav Theodor Fechner, "Ueber die Correctionen bezüglich der Genauigkeitsbestimmung der Beobachtungen," *Berichte über die Verhandlungen der Königlich sächsischen Gesellschaft der Wissenschaften*, math.-phys. Cl. 1861; Fechner, "Ueber die Bestimmung des wahrscheinlichen Fehlers eines Beobachtungsmittels durch die Summe der einfachen Abweichungen," *Annalen der Physik und Chemie*, Jubelband (1874), 66-81. The median is developed in Fechner, "Ueber den Ausgangswerth der kleinsten Abweichungssumme," *Abhandlungen der Königlich sächsischen Gesellschaft der Wissenschaften*, 11 (1874), 1-76. (Sometimes this article is cited with the year 1878.) Walker, *Studies* (note 100) discusses Fechner's notion of the *Centralwerth* on pp. 57 and 84ff.

103. Singer, "Distribution" (note 99), p. 6.

104. Wundt, "Fechner," (note 3), p. 315.

105. Fechner, *Kollektivmasslehre* (note 96), p. VI.

106. Fechner, *Kollektivmasslehre*, p. 3. This definition already appears in the above mentioned article on the median and in another article on experimental aesthetics of 1871. Cf. Fechner, "Zur experimentalen Ästhetik," *Abhandlungen der Königlich sächsischen Gesellschaft der Wissenschaften*, 9 (1871), 555-635, on pp. 617 and 619ff. For an overview of the various

attempts in Fechner's time to use the Gaussian law as a law of variation, see F. Ludwig, "Die Variabilität der Lebewesen und das Gausssche Fehlergesetz," *Zeitschrift für Mathematik und Physik*, 43 (1898), 230-242.

107. Gustav Rümelin, "Zur Theorie der Statistik I.1863" and "Zur Theorie der Statistik II.1874," in his *Reden und Aufsätze* (Freiburg and Tübingen: Mohr, 1875), pp. 208-264, 265-284 (see esp. pp. 213-215, 241, 269-274). Rümelin stresses that "individual" does not mean "indetermined" or "outside the realm of the causal law, deprived of all explanation and reducibility to constant causes" (p. 213). Statistics as the task of characterizing human collectives by observation, and counting is interested in the properties of the collective as a whole and not of each single individual. On Rümelin cf. also the chapter by Theodore Porter in this volume.

108. Fechner, *Kollektivmasslehre* (note 96), chapter XXII.

109. Fechner, *Kollektivmasslehre*, p. 3.

110. Fechner, *Kollektivmasslehre*, p. 6.

111. Fechner, *Kollektivmasslehre*, p. 6.

112. Wundt, *Reden* (note 3), pp. 296ff.

113. Wundt, *Reden*, pp. 255 and 315.

114. Fechner, *Kollektivmasslehre*, p. VI.

115. Fechner, *Kollektivmasslehre*, p. 4.

116. Fechner, *Kollektivmasslehre*, p. 6.

117. Fechner, *Kollektivmasslehre*, p. 5.

118. Fechner, *Kollektivmasslehre*, p. 16.

119. Fechner, *Kollektivmasslehre*, p. 5.

120. Fechner, *Kollektivmasslehre*, p. 31.

121. Fechner, *Kollektivmasslehre*, p. 39.

122. Fechner, *Kollektivmasslehre*, p. 42.

123. Fechner, *Kollektivmasslehre*, p. 45.

124. Fechner, *Kollektivmasslehre*, p. 46 and again on p. 365. Fechner devised similar tests of homogeneity already in his *Professor Schleiden und der Mond* (note 98), pp. 232-235.

125. Fechner, *Kollektivmasslehre*, pp. 382-385.

126. Fechner, *Kollektivmasslehre*, pp. 372-375 and 386-398. Cf. also Singer, "Distribution-Free ... " (note 99), p. 6.

127. Fechner, *Kollektivmasslehre*, pp. 9, 20, and 95ff.

128. Fechner, *Kollektivmasslehre*, p. 85.

129. See note 102.

130. Walker, *Studies* (note 100), p. 52.

131. Bessel writes; "In the two examples ... the law of the probability of the errors is significantly different from the oft-mentioned Gaussian law. ... I only note in addition that any attempt to show that the law that underlies the method of least squares (if one wants to regard it as the most probable method) is valid *in general* [*allgemein das wirklich vorkommende*] is necessarily in vain, since the examples show that conditions that are not only mathematically possible but can also practically be fulfilled lead to completely different laws"—Friedrich Wilhelm Bessel, "Untersuchungen über die Wahrscheinlichkeit der Beobachtungs-

fehler," *Astronomische Nachrichten*, 15 (358 and 359) (1838), 369ff., repr. in: *Abhandlungen von Friedrich Bessel*, ed. by Rudolf Engelmann, vol. 2 (Leipzig: Q. Engelmann, 1876) pp. 372–391, on p. 377.

132. Adolphe Quetelet, *Lettres sur la théorie des probabilités* (Bruxelles 1846), pp. 79 and 181.

133. Fechner, *Kollektivmasslehre*, p. 64. See also chapter XII.

134. Fechner, *Kollektivmasslehre*, p. 75.

135. Fechner, *Kollektivmasslehre*, pp. 70ff. Three years before the *Kollektivmasslehre* appeared, H. de Vries had a somewhat similar idea: Hugo de Vries, "Über halbe Galton-Curven als Zeichen discontinuirlicher Variation," *Berichte der Deutschen Botanischen Gesellschaft*, 12 (1894), 197–207.

136. See Wundt, *Reden* (note 3), p. 330; Oswald Külpe, "Zu Gustav Theodor Fechners Gedächtnis," *Vierteljahrsschrift für wissenschaftliche Philosophie und Soziologie*, 25 (1901), 191–217, on p. 201; Heinrich Bruns, *Wahrscheinlichkeitsrechnung und Kollektivmasslehre* (Leipzig/Berlin: B. G. Teubner, 1906), p. 309.

137. Edmund Koenig, *Die Entwickelung des Causalproblems in der Philosophie seit Kant* (Leipzig: O. Wigand, 1890; repr. Leipzig: Zentralantiquariat, 1972), pp. 478ff.

138. This superficial reading can still be found today.

139. This lecture series was published as Bruns, *Wahrscheinlichkeitsrechnung* (note 136). Corrections to this appeared as Heinrich Bruns, "Beiträge zur Quotenrechnung," *Berichte der Königlich sächsischen Gesellschaft der Wissenschaften*, 58 (1906), 571–613, and as Heinrich Bruns, "Das Gruppenschema für zufällige Ereignisse," *Abhandlungen der Königlich sächsischen Gesellschaft der Wissenschaften*, 29 (1906), 577–628. Cf. also Heinrich Bruns, "Über die Darstellung von Fehlergesetzen," *Astronomische Nachrichten*, 143 (1897), 329–349; Heinrich Bruns, "Zur Collectiv-Masslehre," *Philosophische Studien*, 14 (1898), 339–375. Bruns's series is also called the Charlier series (Charlier's A-series) or Gram-Charlier series or Edgeworth series in the English literature, relating to the work of Edgeworth of 1905 and Charlier of 1906. A review of Charlier's and Edgeworth's methods by W. Palin Elderton appeared in *Biometrika*, 5 (1906/07), 206–210. Edgeworth seems to have been the only one from the Anglo-Saxon world who took notice of Bruns: F. Y. Edgeworth, "On the Representation of Statistical Frequency by a Series," *Journal of the Royal Statistical Society*, 70 (1907), 102–106. Cf. also Carl-Erik Särndal, "The Hypothesis of Elementary Errors and the Scandinavian School in Statistical Theory," *Biometrika* 58 (1971), 375–391, and Harald Cramér, "On the History of Certain Expansions Used in Mathematical Statistics," *Biometrika*, 59 (1972), 205–207. On Bruns' life see J. Bauschinger, "Nekrolog. Heinrich Bruns," *Vierteljahrsschrift der Astronomischen Gesellschaft*, 56 (1921), 59–69. This includes a list of Bruns's writings and a list of eight dissertations on the application of the *Kollektivmasslehre* in various fields. Bauschinger also reports that Bruns left a finished manuscript for a second edition of his *Wahrscheinlichkeitsrechnung* (note 136) and a manuscript for a textbook on error theory.

140. Gottlob Friedrich Lipps, "Über Fechner's Collectivmasslehre und die Vertheilungsgesetze der Collectivgegenstände," *Philosophische Studien*, 13 (1898), 579–612; "Die Theorie der Collectivgegenstände," *Philosophische Studien*, 17 (1901), 78–183, and 467–575; "Die Bestimmung der Abhängigkeit zwischen den Merkmalen eines Gegenstandes," *Berichte der Königlich Sächsischen Gesellschaft der Wissenschaften*, 57 (1905), 1–32.

141. See Lipps, "Collectivgegenstände," p. 107; Bruns, "Zur Collectiv-Masslehre" (note 139), pp. 341–343; Bruns, *Wahrscheinlichkeitsrechnung* (note 136) p. 95.

142 Georg Helm, "Die Wahrscheinlichkeitslehre als Theorie der Collectivbegriffe," *Annalen der Naturphilosophie*, 1 (1902), 364–381.

143. Bruns, *Wahrscheinlichkeitsrechnung* (note 136), p. 111.

144. Joseph Bertrand, [Review of] "Kollektiv-Masslehre von Gustav Theodor Fechner," *Journal des Savants* (January 1899), pp. 5–17, especially 11–17. This review includes a general survey of Fechner's writings as well as a short appreciation of Emile Dormoy's work on the divergence coefficient, which is in many ways similar to Lexis's work.

145. For example: Gottlob Friedrich Lipps, *Die psychischen Massmethoden* (Braunschweig: F. Vieweg, 1906); Wilhelm Wirth, *Psychophysik. Darstellung der Methoden der experimentellen Psychologie*, vol. III, part 5 of the *Handbuch der physiologischen Methodik*, ed. by Robert Tigerstedt (Leipzig: S. Hirzel, 1912). On the reception of Fechner's work by anthropologists, see K. E. Ranke and R. Greiner, "Das Fehlergesetz und seine Verallgemeinerungen durch Fechner und Pearson in ihrer Tragweite für die Anthropologie," *Archiv für Anthropologie*, 30(N.F. 2) (1904), 295–332. The authors came to the conclusion that Pearson's and Fechner's curves are purely empirical descriptions and therefore of no theoretical value. Pearson's response included a critical discussion of Fechner's curves: Karl Pearson, "'Das Fehlergesetz und seine Verallgemeinerungen durch Fechner und Pearson'. A Rejoinder," *Biometrika*, 4 (1905/06), 169–212. Ranke and Greiner also argued that the notion of a *Kollektivgegenstand* implies an antitypological view of species, as did Josiah Royce, "Prinzipien der Logik," *Encyclopädie der Philosophischen Wissenschaften*, in Verbindung mit Wilhelm Windelband herausgegeben von Arnold Ruge, vol. 1: *Logik* (Tübingen: Mohr, 1912), pp. 61–136, on pp. 72ff. English edition: *Encyclopedia of the Philosophical Sciences*, W. Windelband and A. Ruge, vol. I: *Logic*, Sir Henry Jones, ed. (London: Macmillan, 1913), pp. 67–135. On the antitypological view of species today see note 87. See also the chapter by John Beatty in the second volume, note 16 and the main text belonging to this note.

146. Emanuel Czuber, *Wahrscheinlichkeitsrechnung und ihre Anwendung auf Fehlerausgleich-ung, Statistik und Lebensversicherung*, vol. 1: *Wahrscheinlichkeitstheorie, Fehlerausgleichung, Kollektivmasslehre*, 2nd ed. (Leipzig: B. G. Teubner, 1908; repr. New York/London: Johnson, 1968), esp. pp. 142ff. and 344–384. From the 4th ed. of 1924 on, Czuber explicitly rejected the frequency view of von Mises (see the 5th ed., 1938, pp. 438–442). On Czuber see the chapter of Ivo Schneider in this volume.

147. Richard von Mises, "Über die Grundbegriffe der Kollektivmasslehre," *Jahresberichte der Deutschen Mathematiker-Vereinigung*, 21 (1912), 9–20, reprinted in Philipp Frank, S. Goldstein, M. Kac, et al. (eds.), *Selected Papers of Richard von Mises*, vol. 2, *Probability and Statistics, General* (Providence: American Mathematical Society, 1964), pp. 3–14. On von Mises compare also the chapter of von Plato in the second volume. After the completion of this chapter I became acquainted with Hannelore Bernhardt, "Richard von Mises und sein Beitrag zur Grundlegung der Wahrscheinlichkeitsrechnung im 20. Jahrhundert," unpublished dissertation B (Humboldt-Universität zu Berlin, GDR, 1984). Bernhardt gives a thorough account of von Mises's contributions to probability theory and of the discussions of his ideas until the present, without, however, dealing very much with the prehistory of the frequency view.

148. Richard von Mises, "Grundlagen der Wahrscheinlichkeitsrechnung," *Mathematische Zeitschrift*, 5 (1919), 52–99, reprinted in *Selected Papers* (note 147), pp. 57–105.

149. Richard von Mises, "Über die gegenwärtige Krise der Mechanik," *Zeitschrift für angewandte Mathematik und Mechanik*, 1 (1921), 425–431. This was an address to the annual meeting of the German Society of Mathematicians in Jena of September 1921. It was also reprinted in *Die Naturwissenschaften*, 10 (1922), 25–29.

150. Richard von Mises, "Ausschaltung der Ergodenhypothese in der physikalischen Statistik," *Physikalische Zeitschrift*, 21 (1920), 225–232 and 256–264.

151. Mises, "Krise" (note 149), p. 430.

152. Mises, "Krise" pp. 430ff.

153. Mises, "Krise," p. 427.

154. Richard von Mises, *Wahrscheinlichkeit, Statistik und Wahrheit*, vol. 3 of the *Schriften zur wissenschaftlichen Weltauffassung*, ed. by Ph. Frank and M. Schlick (Wien: J. Springer, 1928). The second edition appeared in 1936 and the third in 1951, also published by Springer in Vienna. The address from 1929 is printed as "Über kausale und statistische Gesetzmäßigkeit in der Physik," *Die Naturwissenschaften*, 18 (1930), 145–153. Richard von Mises, *Kleines Lehrbuch des Positivismus. Einführung in die empiristische Wissenschaftsauffassung*, vol. 1 of the Library of Unified Science, Book Series, editor-in-chief O. Neurath (The Hague: W. P. van Stockum & Zoon, 1939).

155. Mises, "Krise" (note 149), p. 430.

156. Mises, *Wahrscheinlichkeit*, 3rd ed. (note 154) p. 252.

157. The Viennese physicist and Boltzmann's successor Franz Exner wanted to synthesize Mach's and Boltzmann's physics by assuming indeterministic behavior in the microcosmic realm that levels out to an average state of the macrocosm. He says that in the case of physical mass phenomena the causes (as presupposed by the determinists) act as if there were no causes present and chance ruled instead. See Franz Exner, *Vorlesungen über die physikalischen Grundlagen der Naturwissenschaften*, 2nd augmented ed. (Leipzig/Wien: F. Deuticke, 1922), p. 680. The 1st ed. appeared in 1919. On p. 330 Exner quotes Fechner's definition of a collective. On Exner see Paul A. Hanle, "Indeterminacy before Heisenberg: The Case of Franz Exner and Erwin Schrödinger," *Historical Studies in the Physical Sciences*, 10 (1979), 225–269.

158. Mises, *Wahrscheinlichkeit* (note 154), 1st ed., p. 99. In his "Grundbegriffe" (note 147) Mises acknowledges that Fechner had coined "the concept and the term '*Kollektivmasslehre*'" (p. 9).

159. Mises, *Wahrscheinlichkeit*, 1st ed., pp. 26 and 99.

160. Hilda Geiringer, "Probability: Objective Theory," *Dictionary of the History of Ideas*, ed. Philip P. Wiener, vol. 3 (New York: Scribner's 1973), pp. 605–623, on p. 616.

161. Fechner, *Kollektivmasslehre* (note 96), p. 31; cf. also p. 466.

162. Fechner, *Kollektivmasslehre*, p. 31.

163. Wundt, *Reden* (note 3), pp. 310ff. Similar views are given on pp. 265, 294, 303, 308, 312, and in Wilhelm Wundt, "Psychologie," *Die Philosophie im Beginn des zwanzigsten Jahrhunderts. Festschrift für Kuno Fischer*, ed. Wilhelm Windelband, 2nd ed. (Heidelberg: C. Winter, 1923) pp. 1–57, on pp. 34ff. (The 1st ed. appeared in 1907.)

164. Stephen G. Brush, "Irreversibility and Indeterminism: Fourier to Heisenberg," *Journal of the History of Ideas*, 37 (1976), 603–630.

7 The Saint Petersburg Paradox 1713–1937

Gérard Jorland

The Saint Petersburg paradox is not mathematical but historical in kind. The divergence of the series representing the mathematical expectation entails that there is no expectation; and yet probabilists have all along contended that it meant an unmatchable infinite expectation, thus making it an issue of the fundamentals of probability theory. This chapter offers a comprehensive study of the long history of the paradox, focusing on how the mathematical arguments answered epistemological questions.

On the whole, three solution types have been devised, all conceived during the eighteenth century and merely refined later on. They share one feature, the contention that it is the rules of the game, by allowing for an infinite number of tosses, that make the series diverge, though in fact it is the payoff structure. Otherwise, they contradict each other on the conditions of convergence of the series: either the existence of an upper bound to the possible gains or a lower one to the chances of making them; or the substitution of another payoff or of another probability distribution. The third solution type consists in making the stakes depend on the number of games allowed.

Similarly, the Saint Petersburg paradox played a significant role in the controversies over the foundations of probability theory only in the eighteenth century: it led to the substitution of the law of large numbers for the principle of insufficient reason, as an empirical, instead of a shaky metaphysical, guarantee of the soundness of the concept of mathematical expectation. Had it not been for Laplace's endorsement of Daniel Bernoulli's concept of moral expectation, which gave him the opportunity to make use of his generating functions and to extend the application of probability calculus to the social sciences, the interest in the paradox would have faded away. Indeed, most nineteenth-century probabilists developed the fifth solution type, suggested by the law of large numbers, to answer Laplace: in that respect, Lacroix's textbook has been paradigmatic. In the 1920s and 1930s the paradox was still there in the debate over the logical and/or the ontological foundation of probability, if only as a particular example of arguments worked out elsewhere.

The paradox of the Saint Petersburg problem is that there is a paradox. The problem is to find the value of the following simple game of Heads and Tails: A tosses a coin until it falls heads up and gives B one coin if it does so at the first toss, two if at the second, four if at the third and, in general, 2^{n-1} if at the nth. The value of the game, or B's expectation, is the sum of the products of each expected gain by its probability. Since B has one chance out of two to win one coin, one out of four to win two coins and, in general, one out of 2^n to win 2^{n-1} coins, his expectation is the sum of an infinite geometric series of first term $1/2$ but of common ratio 1, hence

This chapter has been supported by a grant from the Volkswagen Foundation. I am deeply indebted to M. Bosq, Lorraine Daston, Andreas Kamlah, Lorenz Krüger, Mary Morgan, Ivo Schneider, Zeno Swijtink, and G. Th. Guilbaud for their comments at various stages of my research.

divergent. The series is not summable; there is no expectation. The real puzzle is that it took 224 years for this trivial result to be acknowledged and the nontrivial question, whether there exists a fair stake for a game without expectation, to be raised and answered, by Feller,[1] affirmatively provided variable stakes are allowed.

All along it has been contended that the expectation was infinite, as if the summation process could be actually carried through. Hence the paradox, since nobody in his right mind would stake anything but a very small amount of money on his infinite expected gain. And three solution types have been devised: the series representing expectation has been made convergent by substituting another payoff function (1) or another probability distribution (2), whether bounded (a) or unbounded (b); or by limiting the number of tosses (3). Accordingly, the stake could be infinite for an infinite wealth (2) or for an infinite number of games (3); it should never be infinite because of the ontological import of probability (1).

None of the solutions is complete. Supposing the game to be played in n tosses, the expectation is

$$\sum_{n=1}^{n} \frac{1}{2^n} 2^{n-1} = \frac{1}{2} + \frac{1}{2} + \cdots + \frac{1}{2} = n\frac{1}{2}$$

and increases without limit with n, the number of tosses required to obtain heads. It can thus be contended that what makes the series diverge is the infinite number of tosses. But this condition, if necessary, is not sufficient for divergence. The game can be looked upon as a special case of a generalized one where the payoff is x^{n-1} with probability p^n at the nth toss. The expectation is

$$\sum_{n=1}^{\infty} p^n x^{n-1} = \frac{p}{1 - px}$$

and increases without limit as px tends to 1. It can thus be contended that what makes the series diverge is the payoff function or the probability distribution, or both. Since no solution type is dealing with both the necessary and the sufficient conditions, none is complete. It will be shown moreover that none is satisfactory.

This could explain the lasting interest in the problem. However, it will be shown how it was sustained by the role the paradox played in controversies on the foundation and the nature of probability.

1   Moral Expectation and Moral Certitude

The first statement of the Saint Petersburg problem occurred in a letter written by Nicolas Bernoulli to Montmort and published in 1713.[2] It was the last of five problems and a variant of the fourth, which read as follows: A gives B one coin if he gets six at the first throw of a die, two if at the second, three if at the third, and so on; find B's mathematical expectation. The fifth problem asked the same question when the payoffs are 2^{n-1}, 3^{n-1}, n^2, or n^3 instead of the previous arithmetical progression.

N. Bernoulli showed afterward to Montmort, who missed the point, that while in the fourth problem B's mathematical expectation was finite and equal to 6 coins, in the fifth (restricted to the payoff 2^{n-1}) it was infinite and led to a paradox: For the game to be fair, B would have to stake an infinite sum, which he would be sure to lose "since it is morally impossible that he does not get six after a finite number of throws."[3]

His uncle Jacob Bernoulli had suggested that *moral certitude* was to moral impossibility as very high probability (999/1,000) to very low probability (1/1,000).[4] Accordingly, to solve the paradox, which appeared to both men as an important failure of the concept of mathematical expectation, "our lemma" as Montmort called it,[5] N. Bernoulli proposed not to take account of low probabilities even though they can yield huge gains; and he pointed out that this was relevant to the computation of the value of lottery tickets.[6]

Later, he met with Cramer, to whom he proposed his paradoxical fifth problem. Due to Cramer are the definitive formulation of the problem and both variants of solution type 1. As for the formulation,[7] he substituted a game of Heads and Tails to the game of dice, and singled out the case where the payoff function is 2^{n-1}. Then he stated the paradox: B's mathematical expectation is infinite and yet nobody would stake more than 20 coins; and the problem: Account for the discrepancy between the mathematical and the commonsense estimations.

He explained this discrepancy by a difference between the mathematical and the moral values of money: "Mathematicians value money in proportion to its quantity, commonsense men in proportion to its use."[8] That provided him with a rationale for substituting a bounded payoff function. Over a certain amount, which he put at 2^{24}, money has a constant value for the commonsense man. Thus, the series has an upper bound and B's expectation is at most equal to

$$\sum_{n=1}^{24} \frac{2^{n-1}}{2^n} + \sum_{n=25}^{\infty} \frac{2^{24}}{2^n},$$

that is, at most equal to 13. This solution (1a) was taken up later by Fontaine, then by Poisson, the maximum payoff being redefined as the whole of A's wealth.

But as a generalization of his own version of this solution, Cramer assumed that the moral value of money, its use or the pleasure it yields, is as the square root of its quantity, which enabled him to define *moral expectation* as the sum of the products of the square root of each possible gain by its probability. Thus he could substitute an unbounded payoff function, making the series representing the expectation converge:

$$\sum_{n=1}^{\infty} \frac{\sqrt{2^{n-1}}}{2^n} = \frac{1}{2} \sum_{n=1}^{\infty} \frac{1}{\sqrt{2^{n-1}}} = \frac{1}{2} \times \frac{\sqrt{2}}{\sqrt{2}-1}.$$

Now, this was not B's stake, since, by the rule of a fair game applied to the new concept, the moral expectation of loss must be equal to the moral expectation of gain. The stake is thus equal to the square of the latter:

$$\left\{\frac{\sqrt{2}}{2(\sqrt{2}-1)}\right\}^2 = \frac{2}{12-8\sqrt{2}} = 2.9\ldots < 3.$$

A more elaborated version of this solution (1b) was independently devised shortly afterward by Daniel Bernoulli and Euler, taken up later by Laplace, and became known as the theory of moral expectation, then as the Bernoullian utility theory.

Cramer's solution was at variance with N. Bernoulli's; it undertook the limiting process the other way round. So, N. Bernoulli replied[9] that since the problem was to estimate not the moral expectation, but the moral certitude, it was therefore to determine the low probability, which may be thought of as null. To do so, he argued a posteriori: The commonsense man would not stake 20 coins on account, not of his estimation of wealth, but rather of his estimation of probability. He considers it morally impossible to win 32, 64, 128, etc.; the probability to win one of these is

$$\frac{1}{64} + \frac{1}{128} + \frac{1}{256} + \cdots = \frac{1}{32}.$$

This means that he holds a probability of 1/32 as a mere impossibility and a probability of 31/32 as a certitude. N. Bernoulli could thus substitute a bounded probability distribution giving an expectation worth

$$\sum_{n=1}^{4} \frac{2^{n-1}}{2^n} = 2.$$

While in Cramer's solution (1b) the expectation was infinite for an infinite wealth, in N. Bernoulli's it could never be because of the ontological import of probability: Since a zero (one) probability means an impossibility (certitude), a very low (high) probability might well mean a quasi-impossibility (quasi-certitude) and be deemed equal to zero (one). This ontological principle provided d'Alembert with a rationale for substituting an unbounded probability distribution. Buffon and Borel[10] proposed a more general criterion for determining the lowest probability accountable in human affairs: the probability of death either overnight or by accident in the street, which they put at 1/10,000 and 1/1,000,000 respectively. They inferred that the value of a lottery ticket giving one chance out of ten thousand or one out of one million to win a single prize would be null. However, on the evidence given by Cournot[11] of an actual case of a lottery ticket withdrawn because its too low probability of winning made it too seldom purchased, the figure would be 44 times lower than Borel's. These different figures point at the arbitrariness of the procedure. This solution type (2a) might well be called the theory of moral certitude or the Bernoullian subjective probability theory.

Unsatisfied with Cramer's solution, N. Bernoulli turned to his cousin Daniel Bernoulli and put to him Cramer's formulation of the problem.[12] Daniel answered, first, that the paradox arose out of the low probability that the game will last more than 20 or 30 tosses;[13] next, that one has to estimate B's wealth.[14] Nicolas objected that the question was not whether B would be willing or able to pay an infinite stake, but whether he is compelled to by the rule of a fair game, that is, whether his expectation is actually infinite.[15] Then Daniel seemed to dismiss the case: Nobody would play with someone ready to stake an infinite sum in a game where he gets infinitely few chances to win.[16] But he researched the issue extensively, resulting in a paper sent to his cousin[17] and communicated to the Imperial Academy of Sciences at Saint Petersburg in 1731.

His paper, entitled *Specimen theoriae novae de mensura sortis*,[18] was really aiming at a new theory of expectation. He alleged against the received concept that, because of its underlying principle of insufficient reason,[19] it does not take into account people's wealth except to suppose it infinite; whereas everybody is endowed with a finite quantity of goods and contemplates any expected gain or loss relatively to it, thus differently at different levels of income. That is the basis of D. Bernoulli's hypothesis. Then he substituted this relative value, or utility (*emolumentum*), of a gain for the absolute value in the definition of expectation, which was thus no longer equal to the mean expected gain, but to the gain that yields the mean utility (*emolumentum medium*). That is his new theory of measuring chance or the theory of moral expectation.

1.   Daniel Bernoulli's Hypothesis[20]*:* "Any gain, however small (*lucrulum*), brings always a utility (*emolumentum*) inversely proportional to the whole wealth (*summa bonorum*)." It led to estimate the utility of money proportionally to the logarithm of its quantity. Analytically, the utility of a gain proportional to its relative value can be written

$$dy = k\frac{dx}{x}, \qquad \text{where} \quad \begin{cases} x = \text{absolute wealth} \\ y = \text{utility} \\ k = \text{constant.} \end{cases}$$

Integrating this differential equation gives the utility of the whole wealth:

$$y = k \log x - \log a, \tag{1}$$

where a stands for the minimum wealth of utility zero, thus giving the value of the constant of integration. Hence the logarithmic utility curve[21] (figure 1), where the utility is measured on the ordinate without specification of units. But supposing that successive gains $\alpha, \beta, \gamma, \ldots$ have probabilities $p, q, r, \ldots$, their *mean utility* is

$$y = kp \log(a + \alpha) + kq \log(a + \beta) + kr \log(a + \gamma) + \cdots - \log a, \tag{2}$$

which comes close to defining units of utility in terms of probability.

2.   The Theory of Moral Expectation[22]*:* Comparing (1) and (2), going back to exponentials, and subtracting the wealth previously owned give the moral expectation, or the expected gain, which yields the mean utility:

$$z = x - a = (a + \alpha)^p (a + \beta)^q (a + \gamma)^r \cdots - a. \tag{3}$$

In words, it reduces to taking the geometrical mean of the increased wealth rather than the arithmetical mean of the net gains.

Two corollaries have been deduced: A theory of risk-avoidance, to which we shall come back later when it will have become an issue; and the equivalence of mathe-

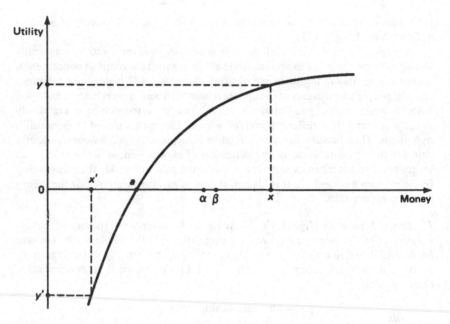

Figure 1

matical and moral expectation when the wealth a is infinite compared with the gains.[23] The latter was relevant to the Saint Petersburg problem: B's moral expectation is by (3),

$$(a + 1)^{1/2}(a + 2)^{1/4}(a + 4)^{1/8} \cdots - a.$$

D. Bernoulli computed the value for different levels of wealth: If B owns nothing, his moral expectation is worth 2 coins; approximately 3 if he owns 10; 4 if 100; 6 if 1,000; and infinity for infinite wealth.[24] But B's stake should be less, in view of the rule of a fair game restated, as Cramer did, in terms of utility: For the disutility of loss to be equal to the utility of gain, B should stake x' in figure 1, therefore less than his expected gain.[25]

Laplace computed the stake under the standard rule of a fair game.[26] Calling it z, assuming that the payoff is 2^n if heads obtains at the nth toss and nothing if it does not in n tosses, and leaving the constant of integration indeterminate, B's mean utility is

$$\frac{1}{2}k \log(a - z + 2) + \frac{1}{2^2}k \log(a - z + 2^2) + \cdots + \frac{1}{2^n}k \log(a - z + 2^n)$$

$$+ \frac{1}{2^n}k \log(a - z) + \log h.$$

By the standard rule of a fair game, B's utility of wealth must be the same after and before the game; that is, the above expression must be equal to $k \log a + \log h$. $\log h$ and k canceling out, dividing through by $a - z$ so that the penultimate term vanishes, and going back to exponentials, B's stake is given by the equation

$$1 + \alpha z = (1 + 2\alpha)^{1/2}(1 + 2^2\alpha)^{1/2^2} \cdots (1 + 2^n\alpha)^{1/2^n}, \tag{4}$$

where $\alpha = 1/a'$ and $a' = a - z$. Laplace showed that this series was decreasing and had 1 for limit when n was infinite; then he computed its value for different levels of wealth or the corresponding stakes.

The series is decreasing, that is,

$$(1 + 2^i\alpha)^{1/2^i} > (1 + 2^{i+1}\alpha)^{1/2^{i+1}},$$

since by raising both members to the power of 2^{i+1} the inequality becomes $1 + 2^{i+1}\alpha + 2^{2i}\alpha^2 > 1 + 2^{i+1}\alpha$, and thus more obvious. The series has 1 for limit since

$$\log(1 + 2^n\alpha)^{1/2^n} = \frac{n \log 2}{2^n} + \frac{1}{2^n}\log\left(\alpha + \frac{1}{2^n}\right),$$

which is null when n is infinite, which, in turn, implies that $(1 + 2^n\alpha)^{1/2^n} = 1$ when n is infinite.

To compute the value of the series, Laplace wrote equation (4),

$$\log(1 + \alpha z) = \sum_{n=1}^{n-1}\frac{1}{2^n}\log(1 + 2^n\alpha) + \sum_{n=n}^{\infty}\frac{1}{2^n}\log(1 + 2^n\alpha),$$

for, he claimed, the second partial sum, from n to ∞, was approximately equal to[27]

$$\frac{\log\alpha}{2^{n-1}} + \frac{(n + 1)\log 2}{2^{n-1}} + \frac{.4342945}{3\alpha 2^{2n-2}}.$$

So, for $n = 11$, $a' = 100$, he found that B should own $a = 107.89$ and stake $z = 7.89$; for $a' = 200$, B should own 208.78 and stake 8.78. If instead of 2^n the payoffs were 2^{n-1}, for $n = 10$ and $a' = 100$, $a = 104.38$ and $z = 4.38$, D. Bernoulli's approximate value.[28]

Thus, the answer to the problem is that B's stake is infinite if his wealth is. This is not trivial but not generally true, for it depends on the payoff function. Menger has proved the following theorem:[29] "For any evaluation of increments of wealth by an unbounded function $f(x)$ there is a Saint Petersburg game in which the mean utility of the risk-taker is infinite." The necessary and sufficient condition is the existence of a utility function $f(x)$ such that the nth increment of wealth be equal to 2^{n-1}; for then the mean utility is $\frac{1}{2}f(\alpha) + \frac{1}{4}f(\beta) + \cdots$, which is at least equal to $\frac{1}{2}1 + \frac{1}{4}2 + \cdots$, therefore infinite. As a corollary, it can be added that in such a Saint Petersburg game, the moral expectation is infinite for finite wealth as well: it is at least equal to $e \cdot e \cdot e \cdots$. Menger had given the following example: Supposing that B, owning already a, might win $ae^{2^n} - a$ at the nth toss, his mathematical expectation will be

$$\sum_{n=1}^{\infty} \frac{1}{2^n}(ae^{2^n} - a),$$

that is, infinite. Now, the utility of any gain, under D. Bernoulli's hypothesis, will be

$$k \log \frac{(a + ae^{2^n} - a)}{a} = k \log e^{2^n} = k2^n,$$

which, with probability, will be worth

$$\frac{1}{2^n} 2^n k = k,$$

and so, the mean utility will be

$$y = k + k + k + \cdots,$$

thus infinite.[30] And the moral expectation will be

$$z = (a \cdot e \cdot e \cdot e \cdots) - a,$$

therefore infinite as well even though a is finite.

However, this argument does not hold against a bounded payoff function (solution type 1a), which leads to the same answer, that the stake could be infinite if the wealth was, though A's this time. Fontaine's version of that solution type[31] rested on the idea that at any rate B cannot win more than A owns. Supposing $2^n > a + b$ (where a and b stand for A's wealth and B's stake), if B wins before or at the nth toss, his expectation will be

$$\sum_{n=1}^{n} \frac{1}{2^n}(2^{n-1} - b),$$

while if he wins after the nth, it is

$$\sum_{n}^{\infty} \frac{1}{2^{n+1}} a.$$

So, his expectation over an infinite number of tosses is

$$\frac{n}{2} - \left(1 - \frac{1}{2^n}\right)b + \frac{1}{2^{n+1}} 2a = 0,$$

and his stake

$$b = \frac{n2^{n-1} + a}{2^n - 1}.$$

This solution was not satisfactory, since the stake depended not only on wealth but also on the number of tosses.

Poisson recast it.[32] He assumed that the payoff function is 2^n and A's wealth

$$a = 2^v(1 + h),$$

where v is an integer and $0 < h < 1$. If $v = n$, A can pay whatever B wins, but if $v < n$, A can no longer pay. So, B's expectation is v for the first v tosses, and for the $n - v$ it amounts to the constant sum $a = 2^v(1 + h)$ times the sum of probabilities from $1/2^{v+1}$ to $1/2^n$; altogether it is given by the expression

$$v + 2^v(1 + h)\left(\frac{1}{2^{v+1}} + \cdots + \frac{1}{2^n}\right),$$

which can be written, after simplification and summation,

$$v + (1 + h)\left(1 - \frac{1}{2^{n-v}}\right).$$

When n tends to infinity, B cannot expect to win more than $v + 1 + h$ and should stake between $v + 1$ and $v + 2$. Poisson gave the following numerical example: if A owns 2^{26} coins (over 60 million), B should stake between 27 and 28. And so B's stake can be infinite provided A's wealth is.[33]

However, this solution (1a) is inconsistent since, under the same assumptions, B should stake all the more as the payoff function is lower, since the exponent v will be all the greater.

Hence, solution type 1a is inconsistent, 1b is not generally true, and 2a is arbitrary. Of course that was not known in the first half of the eighteenth century, and so the concepts of moral expectation and moral certitude challenged for a while the concept of mathematical expectation.

2 Equiprobability and Mean Value

In his answer to his cousin, N. Bernoulli quoted Cramer's letter at length.[34] So, when the Imperial Academy of Sciences at Saint Petersburg published his paper in 1738, D. Bernoulli reproduced Cramer's letter as an appendix. That is how his new theory became known as the theory of moral expectation and his cousin's fifth problem as the Saint Petersburg problem. D'Alembert christened it.[35]

D'Alembert's obstinate criticism of probability stemmed from his epistemological stance that physics is not reducible to mathematics.[36] In this case, his target was the concept of equiprobability: Even if all combinations are mathematically equally probable, physically they are not, and the Saint Petersburg problem is the evidence that equiprobability is valid only in the abstract, not in the actual world.[37]

He contended at first that in a simple game of chance where one bets on heads in n tosses, the number of combinations is not 2^n but $n + 1$ only, since the game ends as soon as heads obtains whatever the rank of the toss.[38] For instance, if $n = 2$, there are 3 real combinations instead of 4, H TH TT, for HH and HT cannot occur under the rule of the game. Thus, the odds should be 2 to 1 and not 3 to 1. However, applied to the Saint Petersburg game, his new rule was of no help, n remaining infinite. He stressed at that time that the paradox arose out of the infinite number of tosses and called upon Fontaine to limit this number at the toss that yields all of A's wealth. It enabled him to write B's expectation according to his new rule

$$\frac{\sum_{n=1}^{n} 2^{n-1}}{n+1},$$

A's wealth being equal to 2^{n-1}. Therefore he introduced two qualifications: Only actual cases should be retained, and the gambler's wealth should come into consideration.

But he discovered that his rule led to inconsistencies. In his simple game of Heads and Tails, the probability of H at the first toss is 1/2, of TH and of TT in two tosses 1/4, and so the odds for H in two tosses are still 3 to 1.[39] He could not change the rules of combining and leave unaffected the corresponding rules of computing probabilities. His only way out, if he were to maintain his stance, was through moral certitude, and all the more so since D. Bernoulli had ridiculed his rule. From then on he rejected all attempts at solving the Saint Petersburg paradox by taking account of the gamblers' wealth, whether A's or B's, on the grounds that the most important component of the concept of expected gain is not the gain but its probability.[40] He reformulated his criticism of the concept of equiprobability in terms of the ontological import of probability and tried all along to devise a formula for this law: "The probability of a combination where the same event obtains several times in a row is all the lower, other things equal, the greater this number of times; so that the probability is null, or as null, when this number is very large, and unaffected, or very little, when it is rather low."[41] It was given a metaphysical *ultima ratio*: Nature acts upon a principle not of sameness but of variety.[42]

He devised two different formulas. He suggested computing the probability of heads at the nth toss not by $1/2^n$, but rather by

$$\frac{1}{2^n(1+bn^2)},$$

where b is an arbitrary constant. B's expectation was thus

$$\sum_{n=1}^{\infty} \frac{2^{n-1}}{2^n(1+bn^2)} = \frac{1}{2} \sum_{n=1}^{\infty} \frac{1}{(1+bn^2)}.$$

He represented the right-hand side sum by the quarter of circle AeG in figure 2, where the radius $CA = \sqrt{1/b}$, and, on the tangent, $AE = n$ and $AF = n + 1$. Since[43]

$$\frac{1}{1+bn^2} = \frac{\sin ef(CF/CE)}{EF},$$

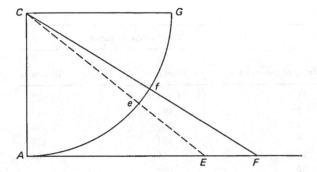

Figure 2

$EF = 1$, and CF/CE tends to 1 as n tends to infinity, the terms of the series are approximately equal to the sines of the corresponding arcs after a certain rank; these sines being replaced by the arcs, the series can be represented by the quarter of circle AeG. In other words, the expectation is finite.

Now, setting $b = 1$, so that $EF = CA$, B's stake is approximately

$$\frac{1}{2} \times \frac{AeG}{CA} = \frac{1}{2} \times \frac{90°}{57°17'44''},$$

the substitution of CA for EF enabling him to compute the value of the infinite series, as the ratio of a quarter of the circumference to the radius, by the number π. From there, he deduced the value of the stake for different values of b: since $CA = \sqrt{1/b}$ is in the denominator, if $b = 1/16$, the value of the sum that B must pay to play the game is four times greater, and if $b = 1/64$, it is eight times greater, so that the smaller b is, the greater the amount to be paid, which becomes infinite when $b = 0$. But then we are back to the standard formula of mathematical expectation. As for himself, d'Alembert assumed that $b = 1/16$, which gave a value of the game a bit over 6 coins.[44]

But this formula led to contradictory results. Since a uniform series of events is physically impossible, if heads obtained several times in a row, on the one hand one should bet on tails the following toss—that is the principle of martingale;[45] on the other hand one should rather bet on heads because there must be some hidden cause that induced the previous events and should continue to act—such was the case in the much debated issue of the distribution of planets in only one-seventeenth of the whole sphere.[46] The first case was dealt with by Béguelin,[47] the second by Laplace.[48] Thus, in the second expression of his law, d'Alembert took advantage of Laplace's formula to make Béguelin's look good.[49]

However, the solution (2b) based on this law is obviously illegitimate since it comes to substitute a biased coin for a true one. Besides, to any such probability distribution, there exists a payoff function such that the expectation increases without limit, and so the form of the probability distribution depends crucially on

Table 1

Rank of toss ending the game	Value of the game	Distribution of games	
		Empirical	Binomial
1	1	1,061	1,025
2	2	494	512
3	4	232	256
4	8	137	128
5	16	56	64
6	32	29	32
7	64	25	16
8	128	8	8
9	256	6	4
10	512		2
11	1,026		1
Total value of games		10,057	11,265

the payoff function. It would nevertheless be recast later on in terms of subjective probability.

Buffon did not take sides in the controversy between d'Alembert and D. Bernoulli;[50] on the contrary, he tried to show that both approaches lead to the same result. But more important, he was the only one in the whole history of the problem who played the game,[51] and based his solution thereupon. He played the game 2,048 times, the coin being tossed by a child (making perhaps for unbiased tosses, not for an unbiased coin), and compared the results to their binomial distribution (table 1).[52] Dividing both total values by the number of games, he arrived at an approximate value *per* game of 5 and 5.5, respectively, which matched pretty well. He could thus claim that for any given number of tosses the mean value of the game is a function of the number of games. But that was not all: he attempted the first complete solution in determining from his empirical findings the utility function of money and the lowest meaningful probability.[53]

Instead of assuming a particular law of diminishing utility to compute the moral expectation, as Cramer and D. Bernoulli had done, he proceeded the other way round and computed the utility of gains that, with probability, added up to 5 coins. He found the utility function $(9/5)^{n-1}$ for the payoff function 2^{n-1}. For instance, a twofold increase of wealth would increase the utility 1.8 times only.[54] As for moral certitude, he had found in a bill of mortality that the odds of death overnight for a 56 year-old man were 1 to 10,189, from which he inferred that the *mean man*[55] of that age does not fear death merely because of his moral certitude that a probability lower than 1/10,000 is as nought. Applied to the Saint Petersburg game, it should have led to evaluating probabilities lower than $1/2^{13}$ as zero, giving an expectation worth 6.5, which could have been reduced to 5 by computing back higher probabilities. Instead he reduced the lowest probability to 1/1,000 by ad hoc arguments, thus limiting the probabilities at $1/2^{10}$ to find his value of 5 coins.[56]

While Buffon applied his law of diminishing utility to show, oddly, that it is beneficial to divide risks,[57] he devised another formula of the relative value of gains and losses to prove that a fair game is always a losing game. Though expressed arithmetically,[58] it came to estimate the gain x proportionally to the increased wealth—$x/(a + x)$—and the loss x proportionally to the wealth previously owned—x/a. These formulas,[59] rather than his law, have been retained as his contribution to the theory of relative value. For his law was obviously inconsistent: the value of the game, from which he deduced the relative value of money, had itself been computed out of the absolute value of money. This inconsistency reveals that Buffon was halfway between the old scheme and the new: with his law of diminishing utility and his criterion for dropping low probabilities he was still arguing against the principle of insufficient reason; in his experiment he was already defining mathematical expectation as a mean value over the long run.

Condorcet took the decisive step forward. The Saint Petersburg paradox prompted him to base the concept of mathematical expectation on J. Bernoulli's theorem.[60] In all his works on probability he went through the same argument:[61] The rule of the product of the value of an uncertain event by its probability gives merely a mean value, not a real value. And he emphasized the difference by means of examples of this kind: If somebody has one chance out of three of winning 2 and two chances out of three of winning 1, his expected gain is 4/3, though he might actually win either 2 or 1, but never 4/3; similarly, if somebody has one chance out of three of winning 2 and his opponent two out of three of winning 1, the game is fair, both gamblers having an equal expected gain of 2/3, though in fact he can either win 2 or 0 and his opponent 1 or 0. The mean value is the average of all real values; it can be achieved if and only if the event is repeated and might as well never obtain in any single trial. Thus, Condorcet conceived of a game as a trade-off between certainty and uncertainty and proved by J. Bernoulli's theorem that its fair price is its mathematical expectation only over the long run.

Within that framework, the Saint Petersburg paradox was readily dismissed: B might achieve his infinite expectation if he was granted an infinite number of games, each of an infinite number of tosses. Under this form "the problem must be considered not as a real case, but as the limit of real questions of the same kind that can be asked."[62] The problem is real if the number of tosses is bounded and it becomes one of finding the number of games necessary to bring about the greatest equality between the gamblers. Condorcet assumed that the number of tosses is given and that B receives 2^n if heads does not turn up in n tosses. B's expectation is then

$$\sum_{n=1}^{n} \frac{2^{n-1}}{2^n} + 1 = \frac{n}{2} + 1.$$

B starts winning when heads obtains at a toss p such that

$$2^{p-1} > \frac{n}{2} + 1, \quad \text{or} \quad n < 2^p - 2,$$

he neither wins nor loses when

$$n = 2^p - 2,$$

and the probability that he loses is

$$1 - \frac{1}{2^{p-1}}.$$

Condorcet gave the following numerical example: for $p = 4$, $n = 14$, B's stake is 8, the probability that B loses (or that A wins) is 7/8, the probability of neither gain nor loss (or that B wins his stake) is 1/16, so that the probability that B wins is 1/16. On the other hand, A has a large chance to win but wins at most 7 if heads shows up at the first toss and a small chance to lose, but loses up to $2^{14} - 8 = 16,376$ if it does not in n tosses. Condorcet concluded that there is a great inequality between the gamblers when they play a single game, and so the problem is to find the number of games necessary to put them on a par. Hence solution type 3: B's stake is a function of the number of games and is infinite only for an infinite number of games.

Lacroix called the tune in the nineteenth century: "Considered thus as a mean value of losses and gains, the mathematical expectation formula may be used only so far as such a value can exist and replace the unknown real value; but that is not the case in the Saint Petersburg problem. A game whose outcome spans an infinite number of successive tosses, and that therefore could not be considered as having to be repeated, does not allow, for that reason, of a mean value of the various events it might bring about."[63] And he claimed that, supposing the game to be played in n tosses,[64] 2^n games should be played for the highest payoff to be won once. Fries made this point most clearly:[65] For n tosses, B should pay $n/2$ because it is his average gain over 2^n games, which yield n times 2^{n-1}.

However, that is only the first part of the theorem: The mean value is the value of highest probability. It remains to apply the second part, or to compute the number of games necessary to win the highest payoff a number of times that will not diverge from its probability by more than a given number. For Czuber, that computation evidenced the incommensurability of single and repeated events with one another.[66] He asserted that the concept of mathematical expectation owed its meaning to the law of large numbers and lost it for single events: An event might have a probability close to 1 and yet a single trial might well bring out the opposite event. And he defined mathematical expectation "as the limit that the average gain accruing to a single trial nears with an ever increasing number of trials, or as the gain accruing to a single event free from the influence of chance."[67] Accordingly, assuming a payoff function 2^n, B can expect to win 2^{30} a number of times that will not diverge from its probability by more than 1%, which will be "free from the influence of chance," if he is allowed to play 180,000 groups of 2^{30} games.[68]

Von Kries dealt with the Saint Petersburg paradox within the same framework.

He argued that mathematical expectation was a fair trade-off between certainty and uncertainty the more often the uncertain event was repeated, the repeated event being no more a mere multiple of the single event than the whole the mere sum of its parts. He gave the following example of complementary goods: Someone who wishes to buy a house with a garden will not be indifferent between a house with a garden for a certain price and either the house for two-thirds of the price or the garden for one-third. Similarly, the Saint Petersburg paradox showed that nobody would be indifferent between the low probability of a substantial gain and the certainty of a mediocre holding in a single or a small number of trials, though one could be indifferent if a large number of games was granted because one would then get the quasi-certitude of making the substantial gain sometimes.[69]

However, this solution type is inconclusive since under the first part of J. Bernoulli's theorem the game is to B's advantage,[70] while under the second part it is to A's.[71] But it led eventually to Feller's theorem.

Hence solution type 2b is illegitimate and solution type 3 is inconclusive. The challenge of moral expectation and moral certitude nevertheless prompted the substitution of the law of large numbers for the principle of insufficient reason as the foundation of mathematical expectation.

3 Subjective Probability and Objective Probability

Had it not been for Laplace's endorsement of D. Bernoulli's theory, the Saint Petersburg problem would have most likely faded away. Nineteenth-century probabilists, following Condorcet, based the concept of mathematical expectation upon the law of large numbers and retained the paradox to exemplify the idea that probability was valid only for repeatable events. But Laplace's stature was such that no textbook dispensed with a close scrutiny of the theory of moral expectation.[72]

Despite his fame, Laplace remained isolated both mathematically and philosophically. His analytical tool, generating functions, finished off classic probability calculus only to lead to a dead end. Two major late eighteenth-century achievements, which he contributed to, the probability of causes and the law of errors, shaped his negative epistemology: Knowledge was not the positive understanding of true causes but the skilled thwarting of ignorance. For chance was nothing real; it expressed merely our ignorance of efficient or final causes: "We look upon something as the effect of chance, when it exhibits nothing regular, or that reveals a purpose, and also when we ignore its causes."[73] Probability was thus a means of making up for our ignorance.

On both accounts Laplace could welcome D. Bernoulli's hypothesis and its derived theory of moral expectation: it was a case for the application of generating functions and for the subjective nature of probability. He laid the concept of moral expectation down as the tenth principle of probability,[74] and devoted a whole chapter to the theory of moral expectation.[75]

His version improved on D. Bernoulli's by showing that mathematical expectation was the limit of moral expectation not only when the wealth but also when the division of risks becomes infinite, yielding thus a complete theory of risk-avoidance.

The structure of his proofs of D. Bernoulli's three theorems is always the same: to show that moral expectation is lower than mathematical expectation.

1. A fair game is always a losing game.[76] Calling a the gambler's wealth, p his probability of winning, μ his stake and $[(1 - p)/p]\mu$ his opponent's stake, his wealth becomes $a + [(1 - p)/p]\mu$ with probability p if he wins and $a - \mu$ with probability $1 - p$ if he loses. His moral expectation is thus

$$x = \left(a + \frac{1 - p}{p}\mu\right)^{p}(a - \mu)^{1-p}. \tag{5}$$

Laplace did not subtract the original wealth a; he added it on the opposite to the mathematical expectation, written then

$$p\left(a + \frac{1 - p}{p}\mu\right) + (1 - p)(a - \mu) = a,$$

because to show that moral expectation is lower than mathematical expectation came to showing that (5) was lower than a. Or, dividing (5) through by a and taking the logarithms, it came to showing that

$$p\log\left(1 + \frac{1 - p}{p}\frac{\mu}{a}\right) + (1 - p)\log\left(1 - \frac{\mu}{a}\right) < 0.$$

Now this last inequality was readily proved since, differentiated with respect to μ, the left-side member could be written

$$\int \frac{1 - p}{a}\left[\frac{1}{1 + [(1 - p)/p](\mu/a)} - \frac{1}{1 - (\mu/a)}\right]d\mu,$$

which is obviously negative.[77]

2. There is always an advantage to dividing risks.[78] Supposing a merchant, owning already 1, to ship a sum of money ε on a single vessel whose probability of arriving safely is p: his mathematical expectation is $1 + p\varepsilon$, while his moral expectation is $(1 + \varepsilon)^{p}$. The latter is lower than the former since, taking the logarithms and differentiating with respect to ε,

$$\int \frac{p\,d\varepsilon}{1 + \varepsilon} < \int \frac{p\,d\varepsilon}{1 + p\varepsilon}.$$

Supposing now that the merchant ships equal parts of the sum ε on r vessels, by the standard theorem of Bernoulli trials his mean utility is

$$k\left\{p^{r}\log(1 + \varepsilon) + rp^{r-1}(1 - p)\log\left(1 + \frac{r - 1}{r}\varepsilon\right)\right.$$

$$\left. + \frac{r(r - 1)}{1 \cdot 2}p^{r-2}(1 - p)^{2}\log\left(1 + \frac{r - 2}{r}\varepsilon\right) + \cdots\right\} + \log h,$$

which, differentiating with respect to ε, can be written

$$kp \int \left[\frac{p^{r-1}}{1+\varepsilon} + \frac{(r-1)p^{r-2}(1-p)}{1+[(r-1)/r]\varepsilon} \right.$$

$$\left. + \frac{(r-1)(r-2)p^{r-3}(1-p)^2}{1\cdot 2\cdot\{1+[(r-2)/r]\varepsilon\}} + \cdots \right] d\varepsilon + \log h. \tag{6}$$

Had he shipped all of ε on a single vessel, his mean utility would have been, setting $r = 1$ in (6),

$$kp \int \frac{d\varepsilon}{1+\varepsilon} + \log h,$$

which can be written

$$kp \int \frac{d\varepsilon}{1+\varepsilon}\{p+(1-p)\}^{r-1} + \log h,$$

or, expanding the binomial,

$$kp \int \left[\frac{p^{r-1}}{1+\varepsilon} + \frac{(r-1)p^{r-2}(1-p)}{1+\varepsilon} \right.$$

$$\left. + \frac{(r-1)(r-2)p^{r-3}(1-p)^2}{1\cdot 2\cdot(1+\varepsilon)} + \cdots \right] d\varepsilon + \log h.$$

Subtracting this value of the mean utility from the one given in (6), the difference,

$$kp(1-p)\frac{r-1}{r} \int \frac{\varepsilon}{1+\varepsilon}\left[\frac{p^{r-2}}{1+[(r-1)/r]\varepsilon} + \frac{(r-2)p^{r-3}(1-p)}{1+[(r-2)/r]\varepsilon} + \cdots \right] d\varepsilon,$$

is positive, which means that the merchant has a moral advantage in dividing risks.

Moreover, Laplace showed that the moral advantage increases with the number of vessels and becomes approximately equal to the mathematical advantage when this number is very large. He wrote (6)

$$kp \int d\varepsilon \int_0^\infty e^{-[1+(\varepsilon/r)]x}\{pe^{-(\varepsilon/r)x} + (1-p)\}^{r-1}\, dx + \log h, \tag{6'}$$

the expression within curly brackets in (6) being equal to the integral with respect to dx in (6').[79] Since this last integral is in turn equal to [80]

$$\frac{1}{1+p\varepsilon+(1-p)(\varepsilon/r)}\left[1 + \frac{p(1-p)\varepsilon^2[1-(1/r)]}{r\{1+p\varepsilon+(1-p)(\varepsilon/r)\}} + \cdots \right],$$

the value of (6') is given by

$$k \int \frac{p\,d\varepsilon}{1+p\varepsilon+(1-p)(\varepsilon/r)}\left[1 + \frac{p(1-p)\varepsilon^2[1-(1/r)]}{r\{1+p\varepsilon+(1-p)(\varepsilon/r)\}} + \cdots \right] + \log h.$$

Now, for very large r, this value reduces approximately to

$$k \int \frac{p \, d\varepsilon}{1 + p\varepsilon} + \log h = k \log(1 + p\varepsilon) + \log h.$$

This value of the mean utility corresponds to a moral expectation equal to $1 + p\varepsilon$, that is, equal to the mathematical expectation given at the beginning.[81]

3. *There may be an advantage to insure.*[82] Supposing that the merchant insured his sum of money ε when shipped on a single vessel and paid $(1 - p)\varepsilon$ for that, Laplace was back at the previous case, for the merchant's moral expectation would be $(1 + \varepsilon)^p$ if he did not insure and $(1 + p\varepsilon)$ if he did, and it had already been shown that the former is lower than the latter. Therefore, there is a moral advantage in insuring. Instead of computing the level of wealth at which the merchant could dispense with insurance, as D. Bernoulli did,[83] Laplace determined the insurance company's benefits. Since the merchant increased his mean utility from $kp \log(1 + \varepsilon) + \log h$ to $k \log(1 + p\varepsilon) + \log h$, thanks to his insurance, he could pay more than the fair price, or $(1 - p)\varepsilon + \alpha$. But the company's benefits α should at most swallow up all of the merchant's advantage, that is, be such that $\log(1 + p\varepsilon - \alpha) = p \log(1 + \varepsilon)$, or

$$\alpha = 1 + p\varepsilon - (1 + \varepsilon)^p.$$

Laplace's theory of insurance was taken up by Fourier[84] and, most notably, by Barrois,[85] who generalized it: Since mathematical expectation is equal to moral expectation when risks are infinitely divided, he computed the insurer's expectation by the mathematical rule and the insured's by the moral rule.

Laplace had conceived of both probability and chance as subjective; Poisson distinguished between them, probability being subjective, chance objective; Cournot annulled this difference, both being objective. Having generalized J. Bernoulli's theorem in a law of large numbers that he held as a universal law of nature and society, Poisson had to make that distinction. He defined the probability of an event as "the reason we have to believe that it will or has obtained,"[86] so that it was relative to one's state of knowledge about the event. On the other hand, an event had an objective chance to happen, which he defined as the (known or unknown) "ratio of the number of favorable cases to the number of possible cases."[87]

Cournot was a staunch determinist: Everything had a cause, a sufficient reason to be such as it was and not otherwise. Randomness (*hasard*) was not an apparent lack of cause but a real combination of independent causes; chance was the set of such combinations that produce the same event; probability, the ratio of the number of cases where the event obtains to the number of all cases, when the same combination is repeated an infinite number of times.[88] Now, if the combination of independent causes cannot be repeated a large number of times, probability is subjective.[89] The distinction between objective probability and subjective probability was identical with the distinction between repeated and single events.

Lacroix's exposition of the topics had been paradigmatic for the whole century. He disconnected the Saint Petersburg problem from the theory of moral expectation, settled the former as a matter of nonrepeatable events, and discussed the latter on its own merits. Indeed, that theory seemed to provide good behavioral advice about risks and, in the last decade, D. Bernoulli's hypothesis of decreasing utility of money was thought to be at the root of the marginal utility concept in economics. As Bortkewitsch put it, "Apart from its mathematical formulation, the idea was good in itself (see the starting point of the marginal utility theory), but it was not adequate to solve the Saint Petersburg problem." [90] After Lacroix, the question of whether moral expectation might be valid for single events was tied to another question, whether D. Bernoulli's and Buffon's formulas of relative value led to the same results.

The difference between them was a matter of assumption concerning the increase of wealth, continuous *versus* discrete, and accordingly of formulation, analytical *versus* algebraical. Their compatibility was tested on whether they yielded the same utility for a loss x equivalent to a certain gain α, and that depended on how Buffon's formulas were written. Lacroix, Öttinger, and Liagre,[91] following Buffon, expressed the relative value of a loss x to someone owning a as x/a and the relative value of a gain α as $\alpha/(a + \alpha)$. Equating these expressions, the relative value of a loss equivalent to a certain gain is

$$x = \frac{\alpha a}{a + \alpha}.$$

On the other hand, the relative values of a loss and a gain were expressed, after Daniel Bernoulli, respectively, as

$$k \log(a - x) - k \log a = - k \log \frac{a - x}{a}$$

and

$$k \log(a + \alpha) - k \log a = k \log \frac{a + \alpha}{a};$$

thus, the relative value of a loss equivalent to a certain gain, given by the equation[92] $a/(a - x) = (a + \alpha)/a$, is

$$x = \frac{\alpha a}{a + \alpha}.$$

Therefore both evaluations led to the same result.

But Fries,[93] following d'Alembert, formulated the subjective value of a loss as $x/(a - x)$, and found that the subjective value of a loss equivalent to a certain gain, given by the equation $x/(a - x) = \alpha/(a + \alpha)$, was

$$x = \frac{\alpha a}{a + 2\alpha},$$

thus lower than D. Bernoulli's evaluation. Czuber[94] compared the three pairs of formulas and dismissed the original Buffonian one on the grounds that expressing the moral value of a loss as x/a entailed a definite positive utility of loss even when it exceeds the whole wealth. And he showed that the second Buffonian pair ($x/(a - x)$ for a loss and $\alpha/(a + \alpha)$ for a gain) led to the same results, that is, to the same theorem of risk-avoidance, if not to the same numerical values, as D. Bernoulli's formulas. But he remarked that the difference between them concerning the evaluation of the subjective value of a loss equivalent to a certain gain was not that significant. Then he proved the theorems of risk-avoidance by putting Buffon's formulas in Laplace's equations.[95] He concluded that since they led to the same results, Buffon's formulas were to be preferred to D. Bernoulli's, for they put more weight on losses, were easier to handle, and, most of all, they dispensed with the continuity assumption.[96] However, Timerding showed that this assumption was necessary. If the relative value of a gain α is divided in two parts, $\alpha = \alpha_1 + \alpha_2$, α_1 being made first, then α_2, the relative value of both parts will be, respectively, $\alpha_1/(a + \alpha_1)$ and $\alpha_2/(a + \alpha_1 + \alpha_2)$. So, the relative value of the sum

$$\frac{\alpha_1 + \alpha_2}{a + \alpha_1 + \alpha_2} = \frac{\alpha_1}{a + \alpha_1 + \alpha_2} + \frac{\alpha_2}{a + \alpha_1 + \alpha_2}$$

will be lower than the sum of the relative values

$$\frac{\alpha}{a + \alpha} = \frac{\alpha_1}{a + \alpha_1} + \frac{\alpha_2}{a + \alpha_1 + \alpha_2}.$$

Under D. Bernoulli's continuity assumption this contradiction does not occur:

$$\log\frac{a + \alpha}{a} = \log\frac{a + \alpha_1}{a} + \log\frac{a + \alpha_1 + \alpha_2}{a + \alpha_1} = \log\frac{a + \alpha_1 + \alpha_2}{a}$$

(neglecting the constant), since

$$\frac{a + \alpha}{a} = \frac{a + \alpha_1}{a} \cdot \frac{a + \alpha_1 + \alpha_2}{a + \alpha_1} = \frac{a + \alpha_1 + \alpha_2}{a}.$$

This time the sum of the relative values is equal to the relative value of the sum.[97]

The issue was whether moral expectation could be held as a good rule for single events. Accordingly, Lacroix, Czuber, and Timerding suggested that it could, whereas Fries denied it.[98] But even these former were ready to admit it only as a rule of the thumb, without numerical validity because of the arbitrariness of its underlying hypothesis. And that was the reason why they did not retain it as a solution of the Saint Petersburg paradox, though they explained the failure of mathematical expectation by its illegitimate application to a nonrepeatable event.

4 Subjective Probability and Subjective Value

These controversies on the nature of probability, single and repeated events, and the theory of risks raised by the Saint Petersburg paradox were pursued in the 1920s and '30s within the broader debate on the logicomathematical foundation of probability. But they focused on another issue: Whether mathematical expectation could be defined as a measuring rod of subjective probability or of subjective value.

Having defined probability as the logical theory of rational belief, rather than the mathematical theory of uncertain events, Keynes endeavored to refute "the doctrine that the mathematical expectation of alternative courses of action are the proper measures of our degrees of preference."[99] He objected that, even were it granted that quantities of goodness were measurable, (1) degrees of probability were not always measurable;[100] (2) the state of knowledge, or the evidence upon which probability is founded, was not taken into account,[101] nor (3) the risk of the venture.[102] Moreover, he claimed that mathematical expectation failed in the Saint Petersburg game because it does not take into account this element of risk. After acknowledging all previous explanations as parts of the truth, he added that "it is the great risk of the wager which deters us."[103] Calling A the possible gain, p its probability, and $E = pA$ the expectation or the stake, he formulated the risk as $R = p(A - E) = p(A - pA) = p(1 - p)A = qE$.[104] In words, it is the probability that the sacrifice made in the hope of winning has been vain. In the Saint Petersburg game, if k is the toss at which B wins $E = n/2$, the risk is given by

$$\left(1 - \frac{1}{2^k}\right)\frac{n}{2}.$$

So, Keynes could say that, for a finite number of tosses, the relative risk q might be close to 1 and should not be ignored; when there is no limit to the number of tosses, the absolute risk qE becomes infinite. However, it can be objected that mathematical expectation takes due account of the element of risk thus defined, since if the probability of winning is small, the probability of losing, or the relative risk, is large.

Borel rejoined that probability is not subjective but *intersubjective* and that there is no other way of being objective.[105] True, the probability of an event is relative to the state of knowledge concerning that event. However, on the one hand, it does not follow that probability is a measure of our ignorance: The difference between partial and complete knowledge is not akin to the difference between probability and certitude but between approximate and exact numerical values; on the other hand, when the state of knowledge is the same for everybody, relativity does not entail subjectivity but intersubjectivity. For instance, the state of knowledge concerning random devices such as coin tossing, dice throwing, or urn drawing is both complete in terms of combination rules and universal in terms of intersubjectivity. Therefore, the probability attached to any event produced by such random devices and defined by the ratio of favorable to possible cases is altogether exact and objective. That is all that is needed to compute all kinds of probability by the *wager method*.[106]

Furthermore, there is no other way of being objective, since the strong law of large numbers gives a "verification," not a "definition," of probability as the limit of frequency.[107] and is based on the ontological import of probability—his "unique law of chance"—not on statistics.[108] Hence his rejoinder to Reichenbach: "The

notion of probability of single events is the foundation of probability calculus."[109] Reichenbach had conceived of probability as a relation between classes
of events, the repeated throwing of a die and the frequency of a certain outcome, for instance, and defined it as the limit of the infinite series representing
the frequency: Probability theory was thus a theory of convergent infinite series.[110]
Accordingly, he handled the Saint Petersburg problem to show that the divergence
does not come necessarily from the infinite summation since this leads to a finite
expectation for another payoff function.[111] Supposing a payoff function n with
probability $(1 - p)^{n-1}p$, B's expectation is worth

$$\sum_{n=1}^{\infty} n(1 - p)^{n-1}p = \frac{1}{p}$$

since the sum, from 1 to ∞, of $n(1 - p)^{n-1}$ is the sum of the derivatives, within the
same interval, of $(1 - p)^n$; it is thus equal to the derivative of this latter sum, that is,
equal to $1/p^2$. It is therefore finite, though the number of tosses is infinite. If instead
the payoff function is 2^{n-1}, the series diverges for $p \leqslant 1/2$, as is the case in N.
Bernoulli's fifth problem and in the Saint Petersburg problem.

Four years later, Borel derived a Saint Petersburg game to show by the same
procedure that a game might become unfair when the number of trials increases
without limit, which reveals the ontological import of probability.[112] Suppose the
payoff function to be $(k + 1)2^{k-1}$ at the kth game, the maximum number of games
to be m, and B to withdraw as soon as he has won a game: If B wins the game of rank
k and loses the $(k - 1)$ previous games, his net gain is[113]

$$(k + 1)2^{k-1} - (k - 1)2^{k-1} + 2^k.$$

A can win only if B loses all the games; that is, A can win

$$\sum_{k=1}^{m} (k + 1)2^{k-1} = \frac{1}{2} \sum_{k=1}^{m} (k + 1)2^k = m2^m.$$

B's expectation is worth

$$\sum_{k=1}^{m} \frac{1}{2^k}2^k = m,$$

and A's

$$\frac{1}{2^m}m2^m = m;$$

therefore, the game is fair. Supposing now that the number of games is unlimited,
B's expectation becomes infinite while A's becomes null: A's is a product in which
one factor, the probability, tends towards zero as m, the number of games, increases
without limit, and a zero probability means an impossibility whatever the magnitude of the gain. Thus, the game becomes unfair.

These opposite claims about the cause of divergence of the series, whether the probability distribution or the payoff function, are the points of departure between what will be known as the French and the American Schools, and which should have rather been called the neo-Bernoullian theory of subjective probability, after Nicolas, and the neo-Bernoullian theory of subjective value, after Daniel. Both are neo-Bernoullian since the concept of mathematical expectation is no longer challenged. However, mathematical expectation could measure subjective probability through wagers if money was a standard commodity, which is not the case;[114] thus the neo-Bernoullian theory of subjective probability can only be qualitative.[115] In the same way, mathematical expectation can measure subjective value or utility if there is a standard lottery ticket.

Von Neumann and Morgenstern gave an axiomatic foundation to the neo-Bernoullian theory of subjective value.[116] They proceeded from the neoclassical analysis of utility in economics: Combinations of commodities that are not preferred one to the other are said to belong to the same indifference set, while those that are belong to different indifference sets; each indifference set is supposed to represent a definite utility level so that the order of preference can be deduced from an utility function. Neoclassical economists were content with an ordinal utility function preserving the order of preference (monotonic); von Neumann and Morgenstern went further and built a cardinal utility function, preserving the order of differences of preferences (linear), out of mathematical expectation, thus providing economics with a standard commodity, the lottery ticket.[117] However, their construction is valid only under very restrictive assumptions that mar its applicability.

The first axiom states that the set of utilities, or of indifference sets, is totally ordered by the preference relation.[118] Thus, the preference relation shares the reflexive, antisymmetric, and transitive properties of the order relation. This makes for a numerical utility function.

The second axiom states that the set of utilities is continuous in probability.[119] If there is a preference between two utilities u and v, there is the same preference between u and a lottery ticket offering either u with probability p or v with probability $(1 - p)$. That makes for a real-valued utility function.

The third axiom is the axiom of additivity.[120] If there is a preference between u and v, then for any w, there is the same preference between a lottery ticket offering either u with probability p or w with probability $(1 - p)$ and another lottery ticket offering either v with probability p or w with probability $(1 - p)$. That makes for the form of the utility function, that is, for mathematical expectation as a utility function:[121] If $u < v < w$, then

$$v = pu + (1 - p)w.$$

This axiom is secured provided there are no complementary goods.

The fourth axiom is the axiom of independence.[122] If there is a preference between v and a lottery ticket offering either u with probability p or w with probability $(1 - p)$, there is the same preference between a lottery ticket offering either v with probability q or w with probability $(1 - q)$ and another, compounded, lottery ticket offering either with probability q the first lottery ticket, which offered a chance p of winning u or a chance $(1 - p)$ of winning w, or w with probability $(1 - q)$. This axiom is required to exclude conditional probability. It is crucial since it makes for monotonicity and linearity of the utility function, hence for an inverse

$$p = \frac{v - u}{w - u},$$

that is, for probability as a cardinal measure of utility.[123] It is secured provided there is no utility to gambling,[124] so that a lottery ticket can be a fixed standard commodity. Accordingly, if a lottery ticket is valued under (over) its mathematical expectation, it does not mean preference for certitude (risk) but decreasing (increasing) marginal utility function.[125] This is too restrictive indeed to be relevant to actual behavior toward risk, as evidenced by the Allais paradox.

Allais submitted to participants at an international conference on the theory of risk the following questionnaire:[126]

Do you prefer A to B?
A {Certitude of getting 100 million
B {10 chances out of 100 to win 500 million
     {89 chances out of 100 to win 100 million
     {1 chance out of 100 to win nothing

Do you prefer C to D?
C {11 chances out of 100 to win 100 million
    {89 chances out of 100 to win nothing
D {10 chances out of 100 to win 500 million
    {90 chances out of 100 to win nothing.

Now if A is preferred to B, despite B's larger expectation, then C should be preferred to D according to the independence axiom since C is a lottery ticket offering A with probability .11 or nothing with probability .89. But it happened that participants who preferred A to B, preferred D to C in violation of the independence axiom. That is explained by a preference for certitude in the first wager that does not intervene in the second.

Savage was one of the participants who were trapped, certainly because he did not notice the link between the wagers. And to explain why, having preferred A to B he should have preferred C to D, he changed the wagers into a lottery offering either A with certitude or a chance to get B and either C with certitude or a chance to get D. Therefore the preference for certitude extends to the choice between C and D.[127] But that is begging the question of the relevance of the independence axiom. While Allais was raising the question of whether the preference relation between a certain and an uncertain event was preserved when both events become uncertain, Savage transformed it into the question of preservation of preference between two pairs of certain and uncertain events.

That Savage was trapped is not the least surprising since he interpreted von Neumann and Morgenstern's axiomatic construction in terms of subjective

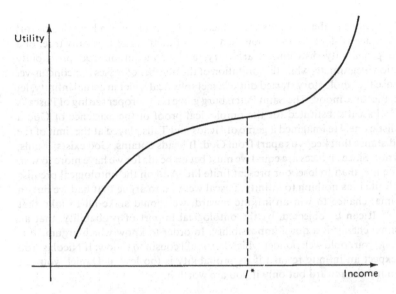

Figure 3

probability, which is inconsistent with the independence axiom.[128] And he built his neo-Bernoullian utility curve to explain the willingness to play unfair games such as lottery and insurance upon the assimilation of decreasing (increasing) marginal utility and preference for certainty (risk) (figure 3).[129]

Let I^* be current income. On the left the utility curve is concave; it rises above any of its chords representing the expected utility of a fair offer. Marginal utility is decreasing and the maximum insurance premium is determined by the point on the utility curve of equal utility as the fair offer: There will be a willingness to pay more than the fair premium up to the maximum if there is a preference for certitude, and there will be an unwillingness to do so if there is a preference for risk. On the right the utility curve is convex; it stays below any of its chords. Marginal utility is increasing, the maximum entrance fee for gambling will be determined in the same way, and there will be a willingness to pay up to this maximum price if there is a preference for risk and unwillingness to do so if there is a preference for certitude.[130] That would be perfectly legitimate if there was no attempt at measuring utility by a standard lottery ticket based upon the independence axiom for, after all, mathematical expectation, as a utility function, depends on both probability and income. But then the whole mathematical argument would be merely qualitative.

That is the last, still unsettled, development of the history of the Saint Petersburg problem, which has thus been epistemologically very fertile. It led to the substitution of the law of large numbers for the principle of insufficient reason as the foundation of mathematical expectation and it raised the question of the objective or subjective nature of probability depending on whether it applies to single events.

However, none of the solutions is satisfactory: Substitution of a bounded payoff function is inconsistent, of an unbounded payoff function not generally true, of a bounded probability distribution arbitrary, and of an unbounded probability distribution illegitimate, while the limitation of the number of tosses is inconclusive.

But since the whole story started out of a metaphysical belief in actual infinity, let us end in the same mood: The Saint Petersburg game is the proper setting of Pascal's wager.[131] Pascal substituted for the ontological proof of the existence of God a probabilistic one. He imagined a game of Heads and Tails played at the limit of the infinite distance that keeps us apart from God: If heads obtains, God exists; if tails, He does not. Since chances are equal we must bet on heads for we have more to win, an infinite life, than to lose, our present finite life. And on the ontological premise that the finite is as nothing to infinity, Pascal went on to argue that had we but an infinitesimal chance to win an infinite reward, we should stake all to take that chance.[132] It can be objected, by the ontological import of probability, that an infinitesimal chance is a quasi-impossibility. In order to know which argument is compelling, you could well ponder which Bernoulli cousin to follow: If Nicolas, you cannot expect an infinite reward if its probability is too low; if Daniel, you can expect an infinite reward but only if you are worth it.

Notes

1. William Feller, "Über das Gesetz der Grossen Zahlen," *Acta Litterarum ac Scientiarum Regiae Universitatis Hungaricae Francisco-Iosephinae*, 1936–37, VIII: 191–201; *An Introduction to Probability Theory and its Applications*, vol. I (New York: John Wiley, 1968), pp. 251–253.

2. Pierre Rémond de Montmort, *Essay d'Analyse sur les Jeux de Hazard*, (Paris: 1713), p. 402. The correspondence of N. Bernoulli with Montmort, Cramer, and D. Bernoulli on the paradox has been published in *Die Werke von Jakob Bernoulli*, vol. III (Basel: Birkhäuser, 1975), pp. 557–567.

3. Nicolas Bernoulli, letter to Montmort (2/20/1714), in *Die Werke*, p. 558 (note 2).

4. Jacob Bernoulli, *Ars Conjectandi*, in *Die Werke*, p. 240 (note 2).

5. P. de Montmort, letter to N. Bernoulli (3/24/1714), in *Die Werke*, p. 559 (note 2).

6. N. Bernoulli, letter to Montmort (2/20/1714), in *Die Werke*, pp. 558–559 (note 2).

7. Gabriel Cramer, letter to N. Bernoulli (5/21/1728), in *Die Werke*, p. 560 (note 2).

8. G. Cramer, letter to N. Bernoulli (5/21/1728), in *Die Werke*, pp. 560–561 (note 2).

9. N. Bernoulli, letter to G. Cramer (7/3/1728), in *Die Werke*, pp. 562–563 (note 2).

10. Buffon, *Essai d'Arithmétique morale* (Paris: 1777), §8; Emile Borel, *Eléments de la Théorie des Probabilités* (Paris: Albin Michel, 1950), pp. 101–102.

11. Antoine-Augustin Cournot, *Exposition de la Théorie des Chances et des Probabilités* (Paris: Hachette, 1843), pp. 105–109. In the French Lottery, five numbers were drawn out of ninety. The ninety numbers could be combined either one at a time, giving 90 *extraits*, or two at a time, giving 4,005 *ambes*, or three at a time, giving 117, 480 *ternes*, or four at a time, giving 2,555,190 *quaternes*, or five at a time, giving 43,949,268 *quines*. The five numbers drawn offered thus 5 *extraits*, 10 *ambes*, 10 *ternes*, 5 *quaternes*, and 1 *quine*. The Administration took off the *quine* because it was too seldom gambled on, the probability of winning being too low: "One imagines well that there must be a limit to the smallness of chances" (p. 106). Accordingly,

Cournot conceived of the Saint Petersburg game as a lottery comprising tickets, some of which paid 1 coin if heads obtains at the first toss, others 2 coins if at the second toss, and so on up to tickets that would be taken off because they would be too seldom bought, their chance of winning being too small. B's expectation in the Saint Petersburg game would be akin to the expectation of someone buying one ticket of every kind available in that lottery.

12. N. Bernoulli, letter to D. Bernoulli (8/27/1728), in *Die Werke*, p. 563 (note 2).

13. Daniel Bernoulli, letter to N. Bernoulli (11/5/1728), in *Die Werke*, p. 564 (note 2).

14. N. Bernoulli, letter to D. Bernoulli (2/4/1730), in *Die Werke*, p. 565 (note 2).

15. N. Bernoulli, letters to D. Bernoulli (2/5/1729 and 2/4/1730), in *Die Werke*, pp. 564–565 (note 2).

16. D. Bernoulli, letter to N. Bernoulli (1/1731), in *Die Werke*, p. 565 (note 2).

17. D. Bernoulli, letter to N. Bernoulli (7/4/1731), in *Die Werke* pp. 565–566 (note 2).

18. D. Bernoulli, "Specimen Theoriae Novae de Mensura Sortis," *Commentarii Academiae Scientiarum Imperialis Petropolitanae*, 1738, IV (1730/31), 175–192. First German translation (without the mathematics) in *Hamburgisches Magazin*, I(5) (1747), 73–90. Second German translation by Alfred Pringsheim (Leipzig: Duncker & Humblot, 1896). English translation by Louise Sommer in *Econometrica*, XXII (1954), 23–36.

19. "There is no reason why one's expectation should be worth more fulfilling than another's" (D. Bernoulli, "Specimen," §2 (note 18); see J. Bernoulli's annotation to Huygens's second proposition (*Ars Conjectandi*, in *Die Werke*, p. 111 (note 2)).

20. D. Bernoulli, "Specimen," §5 (note 18); Laplace, *Théorie Analytique des Probabilités* (Paris: Gauthier-Villars, 1886), pp. XX, 441–442.

21. D. Bernoulli, "Specimen," §7 (note 18).

22. D. Bernoulli, "Specimen," §12 (note 18); Laplace, *Théorie Analytique*, p. 442 (note 20).

23. D. Bernoulli, "Specimen," §9 (note 18); Leonard Euler, "Vera Aestimatio Sortis in Ludis," *Opera Omnia*, ser. I, vol. VII (Leipzig and Berlin: Teubner, 1862), pp. 463–464; Laplace, *Théorie Analytique*, p. XX (note 20), Sylvestre François Lacroix, *Traité élémentaire du Calcul des Probabilités* (Paris: Courcier, 1816), p. 118.

24. D. Bernoulli, "Specimen," §19 (note 18).

25. D. Bernoulli, "Specimen," §7 (note 18).

26. Laplace, *Théorie Analytique*, pp. 448–451 (note 20).

27. Indeed,

$$\sum_{n=n}^{\infty} \frac{1}{2^n} \log(1 + 2^n\alpha) = \sum_{n=n}^{\infty} \frac{\log\alpha}{2^n} + \sum_{n=n}^{\infty} \frac{n\log 2}{2^n} + \sum_{n=n}^{\infty} \frac{1}{2^n}\log\left(1 + \frac{1}{2^n\alpha}\right);$$

and

(1) $$\sum_{n=n}^{\infty} \frac{\log\alpha}{2^n} = \log\alpha \sum_{n=n}^{\infty} \frac{1}{2^n} = \frac{\log\alpha}{2^{n-1}};$$

(2) $$\sum_{n=n}^{\infty} \frac{n\log 2}{2^n} = \log 2 \sum_{n=n}^{\infty} \frac{1}{2}n\left(\frac{1}{2}\right)^{n-1}$$

$$= \log 2\left[\frac{1}{2}\sum_{n=1}^{\infty} n\left(\frac{1}{2}\right)^{n-1} - \frac{1}{2}\sum_{n=1}^{n-1} n\left(\frac{1}{2}\right)^{n-1}\right] = \frac{(n+1)\log 2}{2^{n-1}};$$

(3) $$\sum_{n=n}^{\infty} \frac{1}{2^n}\log\left(1 + \frac{1}{2^n\alpha}\right) = \sum_{n=n}^{\infty} \frac{.4342945}{2^n}\log_e\left(1 + \frac{1}{2^n\alpha}\right),$$

which is approximately equal to

$$\sum_{n=n}^{\infty} \frac{.4342945}{2^n} \left[\frac{1}{2^n\alpha} \right] = \frac{.4342945}{\alpha} \sum_{n=n}^{\infty} \frac{1}{2^{2n}}$$

$$= \frac{.4342945}{\alpha} \left[\sum_{n=1}^{\infty} \frac{1}{2^{2n}} - \sum_{n=1}^{n-1} \frac{1}{2^{2n}} \right] = \frac{.4342945}{3 \times 2^{2n-2}}.$$

28. S. F. Lacroix, *Traité*, p. 126 (note 23).

29. Karl Menger, "Das Unsicherheitsmoment in der Wertlehre," *Zeitschrift für Nationalökonomie*, IV (1934), 459–485, on p. 468.

30. K. Menger, "Das Unsicherheitsmoment," p. 467 (note 29).

31. Alexis Fontaine, "Solution d'un Problème sur les Jeux de Hasard," *Mémoires donnés à l'Académie Royale des Sciences* (Paris: 1764), pp. 429–431.

32. Simon Denis Poisson, *Recherches sur la Probabilité des Jugements en matière criminelle et en matière civile*, (Paris: Bachelier, 1837) pp. 72–75.

33. For variants of that solution, see Gauthier d'Hauteserve, *Traité élémentaire sur les Probabilités* (Paris: Bachelier, 1834), pp. 35–37; *Application de l'Algèbre élémentaire au Calcul des Probabilités* (Paris: Lambert, 1840), pp. 32–35; Eugène Catalan, "Sur le Problème de Pétersbourg," *Mémoires de la Société Royale des Sciences de Liège*, ser. 2, XV (Bruxelles: Hayez, 1888), pp. 248–249; A. Pringsheim, in a footnote to his German translation of D. Bernoulli's "Specimen," pp. 49–50 (note 18); Robert de Montessus, *Leçons élémentaires sur le Calcul des Probabilités* (Paris: Gauthier-Villars, 1908), pp. 72–73. For a variant of Laplace's and Poisson's solutions, see Henri Poincaré, *Calcul des Probabilités* (Paris: Carré & Naud, 1900), pp. 41–43.

34. N. Bernoulli, letter to D. Bernoulli (4/5/1732), in *Die Werke*, p. 566 (note 2).

35. D'Alembert, *Opuscules Mathématiques*, vol. IV (Paris: 1768), p. 78.

36. D'Alembert, *Traité de Dynamique*, (Paris: 1758), pp. VII–IX; *Traité de l'Equilibre et du Mouvement des Fluides* (Paris: 1744), pp. IV–VII; *Essai sur les Eléments de Philosophie*, in *Mélanges de Littérature, d'Histoire et de Philosophie* (Amsterdam: 1759), pp. 287–294.

37. D'Alembert, *Opuscules Mathématiques*, vol. II (Paris: 1761), pp. 10–11; vol. IV, pp. 84–85 and 287–288; vol. VII (Paris: 1780), pp. 40–41; *Mélanges*, vol. V (Amsterdam: 1767), pp. 277–278.

38. D'Alembert, "Croix ou Pile," *Encyclopédie*, vol. IV (Paris: 1754). $2^{n-1} - 1$ combinations will never occur if heads obtains at the first toss, $2^{n-2} - 1$ if at the second, ..., and 1 if at the penultimate; and so, these should be subtracted from the 2^n combinations to leave only $n + 1$ real ones.

39. D'Alembert, *Opuscules*, vol. II, p. 17 (note 37).

40. D'Alembert, *Opuscules*, vol. II, pp. 6–8 and 24–25 (note 37); vol. IV, pp. 76–78 and 82–83 (note 35).

41. D'Alembert, *Mélanges*, vol. V, p. 291 (note 37).

42. D'Alembert, *Opuscules*, vol. II, p. 15 (note 37); *Mélanges*, vol. V, pp. 283–284 (note 36).

43. Let us call u the angle ACE and v the angle ACF.
From $\tan u = n\sqrt{b}$ and $\tan v = (n + 1)\sqrt{b}$, we deduce

(1) $\tan v - \tan u = \sqrt{b}$;

and from $CE \cos u = CF \cos v = \dfrac{1}{\sqrt{b}}$

(2) $\dfrac{CF}{CE} = \dfrac{\cos u}{\cos v}$

Now we have

$\sin ef \dfrac{CF}{CE} = \sin(v - u)\dfrac{\cos u}{\cos v}$ (by 2)

$\qquad = (\sin v \cos u - \sin u \cos v)\dfrac{\cos u}{\cos v}$

$\qquad = (\tan v - \tan u)\cos^2 u = \sqrt{b}\cos^2 u$ (by 1).

Dividing through by $EF = AF - AE = \tan v - \tan u = \sqrt{b}$, we find

(3) $\dfrac{\sin ef(CF/CE)}{EF} = \cos^2 u.$

Since

(4) $\dfrac{1}{1 + bn^2} = \dfrac{1}{1 + (n\sqrt{b})^2} = \dfrac{1}{1 + \tan^2 u} = \cos^2 u,$

(3) and (4) are equal as stated in the text.

44. D'Alembert, *Opuscules*, vol. IV, pp. 74–79 (note 35). Three variants of this formula allowed for qualifications: $1/2^{n+bn}$ allowed computing the constant b out of a certain stake; $1/2^{n+b(n-1)}$ allowed giving an equal chance for heads and tails to obtain at the first toss; $1/2^n\{1 + [b/\sqrt{(k-n)^4}]\}$ allowed letting the probability become null when the number of tosses reaches a certain magnitude.

45. D'Alembert, *Mélanges*, vol. V, pp. 284–285 (note 37); *Opuscules*, vol. IV, p. 90 (note 35).

46. D'Alembert, *Mélanges*, vol. V, pp. 289, 294–300 (note 37); *Opuscules*, vol. IV, pp. 89–91 (note 35).

47. Nicolas de Béguelin, "Sur l'usage du Principe de la Raison suffisante dans le Calcul des Probabilités," *Histoire de l'Académie Royale des Sciences et Belles-Lettres de Berlin*, (Berlin: Haude & Spener, 1769), pp. 382–412.

48. Laplace, "Mémoire sur la probabilité des causes par les évènements," *Oeuvres Complètes*, vol. VIII (Paris: Gauthier-Villars, 1891), pp. 53–56.

49. D'Alembert, *Opuscules*, vol. VII, pp. 39–60 (note 37).

50. For a contemporary epistemological comment on the controversy, see Ludwig Christian Lichtenberg, "Betrachtungen über einige Methoden, eine gewisse Schwierigkeit in der Berechnung der Wahrscheinlichkeit beim Spiel zu heben," *Vermischte Schriften*, vol. IX (Göttingen: H. Dieterich, 1806), pp. 3–46.

51. Buffon, *Arithmétique morale*, §XVIII (note 10).

52. In table 1, the topmost entry under the binomial heading is, in fact, 1,024. But there remains one game where heads does not show up at all: Buffon supposes arbitrarily that this game brings only 1 coin.

53. Buffon, *Arithmétique morale*, §XVI (note 10).

54. Buffon, *Arithmétique morale*, §XIX.

55. Buffon, *Arithmétique morale* §VIII. Adolphe Quetelet, *Lettres sur la Théorie des Probabilités* (Bruxelles: Hayez, 1846), pp. 48–53; Quetelet gives only Buffon's version of the theory of moral expectation and quotes the *Arithmétique morale* extensively.

56. Buffon, *Arithmétique morale*, §XX (note 10).

57. Buffon, *Arithmétique morale*, §XXII.

58. Buffon, *Arithmétique morale*, §XIII.

59. D'Alembert wrote them more cogently as $x/(a + x)$ and $x/(a - x)$ to enhance the point, and introduced them in the formula of mathematical expectation to bring the gambler's wealth into the picture (*Opuscules*, vol. IV, pp. 79–82 and 283 (note 35)). But this kind of qualification was not characteristic of him.

60. Condorcet, "Mémoire sur le Calcul des Probabilités," *Histoire de l'Académie Royale des Sciences*, 1781 (Paris: 1784), pp. 712–713.

61. Condorcet, "Mémoire," pp. 707–708; *Essai sur l'Application de l'Analyse à la Probabilité des Décisions rendues à la Pluralité des Voix* (Paris: 1785), pp. 142–146; "Probabilité," *Encyclopédie méthodique*, vol. II (Paris: Panckoucke, 1785), pp. 652–666; *Elémens du Calcul des Probabilités* (Paris: Royez, 1805), pp. 100–120.

62. Condorcet, "Mémoire," p. 714 (note 60); "Probabilité," p. 654 (note 61).

63. S. F. Lacroix, *Traité*, pp. 129–130 (note 23).

64. Allowing for heads to obtain several times in *n* tosses does not solve the problem since the expectation remains infinite (Ludwig Öttinger, "Untersuchungen über die Wahrscheinlichkeitsrechnung," *Journal für die reine und angewandte Mathematik*, XXXVI (1848), 301–306; August Seydler, "Sur le problème de St Petersbourg," *Comptes-Rendus des Séances de l'Académie des Sciences*, vol. 110 (Paris: 1890), pp. 326–328). Moreover, in the standard problem, Öttinger and Liagre failed to apply correctly the rule of a fair game (Jean-Baptiste Joseph Liagre, *Calcul des Probabilités et Théorie des Erreurs*, (Bruxelles: Merzbach, 1852), pp. 139–140).

65. Jakob Friedrich Fries, *Versuch einer Kritik der Principien der Wahrscheinlichkeitsrechnung*, (Braunschweig: F. Vieweg, 1842), pp. 115–116.

66. Hermann Laurent had made the same point though unclearly (*Traité du Calcul des Probabilités* (Paris: Gauthier-Villars, 1873), p. 173). But he contended later that the paradox arose out of the incommensurability of the expected gains with one another, which is not very clear either (*Théorie des Jeux de Hasard* (Paris: Gauthier-Villars, 1893), p. 26).

67. Emanuel Czuber, "Das Petersburger Problem," *Archiv der Mathematik und Physik*, LXVII (1882), 1–28, on p. 16.

68. E. Czuber, "Das Petersburger Problem," p. 20.

69. Johannes von Kries, *Die Principien der Wahrscheinlichkeitsrechnung* (Freiburg: Mohr, 1886), pp. 185–191; Heinrich Bruns, *Wahrscheinlichkeitsrechnung und Kollektivmasslehre* (Leipzig and Berlin: Teubner, 1906), pp. 63–71.

70. Joseph Bertrand, *Calcul des Probabilités* (Paris: Gauthier-Villars, 1907), p. XII and 61.

71. E. Czuber, "Das Petersburger Problem," p. 18 (note 67).

72. S. F. Lacroix, *Traité*, pp. 111–122 (note 23); L. Öttinger, "Untersuchungen," pp. 260–267 and 296–300 (note 64); J.-B. J. Liagre, *Calcul*, pp. 126–137 (note 64); Antoine Meyer, *Vorlesungen über Wahrscheinlichkeitsrechnung* (Leipzig: Teubner, 1879), pp. 150–165; E. Czuber, "Vergleichung zweier Annahmen über die moralische Bedeutung von Geldsummen," *Archiv der Mathematik und Physik*, LXII (1878), 267–284; "Die Entwicklung der Wahrscheinlichkeitstheorie und ihrer Anwendungen," *Jahresbericht der Deutschen Mathematiker-Vereinigung*, VII(2) (1898), 119–122; "Wahrscheinlichkeitsrechnung," *Encyklopädie der Mathematischen Wissenschaften*, vol. I, Part 2 (Leipzig: Teubner, 1900), pp. 765–766; *Wahrscheinlichkeitsrechnung und ihre Anwendung auf Fehlerausgleichung, Statistik und Lebensversicherung* (Leipzig and Berlin: Teubner, 1903), pp. 264–273; Morgan William Crofton, "Probability," *Encyclopaedia Britannica*, vol. XIX (Edinburgh: Adam & Charles Black, 1885),

pp. 775-777; Heinrich Emil Timerding, "Die Bernoullische Wertetheorie," *Zeitschrift für Mathematik und Physik*, XLVII (1902), 321-354; II. Bruns, *Wahrscheinlichkeitsrechnung*, pp. 71-75 (note 69).

73. Laplace, "Recherches sur l'intégration des équations différentielles aux différences finies et sur leur usage dans la théorie des hasards," *Oeuvres Complètes*, vol. VIII (Paris: Gauthier-Villars, 1891), p. 145.

74. Laplace, *Théorie Analytique*, p. XX (note 20). The concept of moral expectation was no longer a substitute but a complement to the concept of mathematical expectation, their difference stemming from the distinction between the absolute and relative values of goods, the former being independent of, the latter increasing with the needs and desires for these goods ("Recherches sur l'intégration," p. 148; Leçons de Mathématiques données à l'Ecole Normale en 1795, *Oeuvres Complètes*, vol. XIV (Paris: Gauthier-Villars, 1912), p. 166; *Théorie Analytique*, pp. XIX and 189-190).

75. Laplace, *Théorie Analytique*, Chapter X of Book II (note 20).

76. Laplace, *Théorie Analytique*, p. 443 (note 20).

77. William Allen Whitworth tried to prove this theorem by the mathematical expectation rule. Considering somebody waging a fixed proportion of his wealth a with an equal chance to win or to lose, he contended that there will be exactly m successes and m failures in $2m$ wagers, so that the wagerer's wealth will be multiplied by

$$\left(1 + \frac{1}{a}\right)^m \left(1 - \frac{1}{a}\right)^m = \left(1 - \frac{1}{a^2}\right)^m < 1.$$

Of course, the mathematical expectation is worth instead

$$\sum_{k=0}^{2m} C(2m, k) \frac{1}{2^{2m}} \left(1 + \frac{1}{a}\right)^k \left(1 - \frac{1}{a}\right)^{2m-k} = 1,$$

that is, the probability times a sum expression that is the binomial expansion of

$$\left[\left(1 + \frac{1}{a}\right) + \left(1 - \frac{1}{a}\right)\right]^{2m} = 2^{2m}.$$

(*Choice and Chance* (New York and London: Hafner, 1965), p. 225). Accordingly, he tried to solve the Saint Petersburg problem (pp. 241-247) by means of the formula

$$\left(1 + \frac{P_1}{a} - \frac{X}{a}\right)^{p_1} \left(1 + \frac{P_2}{a} - \frac{X}{a}\right)^{p_2} \cdots = 1,$$

where the Ps are B's possible gains, the ps are their probability, and X is the stake. Assuming that X is small compared with a, or that B stakes but a small part of his wealth, Whitworth deduced

$$X = \frac{[1 + (P_1/a)]^{p_1} [1 + (P_2/a)]^{p_2} \cdots - 1}{[p_1/(a + P_1)] + [p_2/(a + P_2)] + \cdots};$$

but he made still another mistake, since he should have found the numerator

$$1 - \left[\left(1 + \frac{a}{P_1}\right)^{p_1} \left(1 + \frac{a}{P_2}\right)^{p_2} \cdots\right];$$

but then B's stake would have been negative.

78. Laplace, *Théorie Analytique*, pp. 443-447 (note 20).

79. By the binomial expansion of

$$e^{-[1 + (\epsilon/r)]x} \{pe^{-(\epsilon/r)x} + (1 - p)\}^{r-1}$$

and the repeated use of

$$\int_0^\infty e^{-mx}\,dx = \frac{1}{m}.$$

80. For a detailed computation, see A. Meyer, *Vorlesungen*, pp. 159–161.

81. While Lacroix (*Traité*, pp. 120-121 (note 23)) used the formula of Bernoulli trials to show that the mathematical expectation was the same whether risks were divided or not, Fries (*Versuch*, p. 114 note 65) used the formula of independent events, as N. Bernoulli did (letter to D. Bernoulli (4/5/1732), in *Die Werke*, p. 567 (note 2), to show that it was not.

82. Laplace, *Théorie Analytique*, pp. 447–448 (note 20).

83. D. Bernoulli, "Specimen," §15 (note 18).

84. Joseph Fourier, "Extrait d'un Mémoire sur la Théorie Analytique des Assurances," *Annales de Chimie et de Physique*, X (1819), 177–189.

85. Théodore Barrois, *Essai sur l'Application du Calcul des Probabilités aux Assurances contre l'Incendie*, (Lille: L. Douel, 1835).

86. S. D. Poisson, *Recherches*, p. 30 (note 32).

87. S. D. Poisson, *Recherches*, p. 80 (note 32).

88. A. A. Cournot, *Exposition*, pp. 71–83 and 437–439 (note 11).

89. A. A. Cournot, *Exposition*, pp. 84, 155, and 438–440 (note 11).

90. Ladislaus von Bortkewitsch, "Kritische Betrachtungen zür theoretischen Statistik," *Jahrbücher für Nationalökonomie und Statistik*, LXV (1895), 337. For an account of D. Bernoulli's hypothesis in economics at that time, see George J. Stigler, *Essays in the History of Economics* (Chicago: Chicago University Press, 1965), pp. 114-117, 209, and 220-221; for an account of its delayed recognition by economists, see my "Position historique de l'oeuvre économique de Cournot," *Actes Cournot* (Paris: Economica, 1978), pp. 12–22, and "Cournot et l'avènement de la théorie de la valeur-utilité," *Revue de Synthèse*, CI (1980), 221–250.

91. S. F. Lacroix, *Traité*, pp. 112-117 (note 23); L. Öttinger, "Untersuchungen," pp. 261–264 (note 64); J.-B. J. Liagre, *Calcul*, pp. 127–130 (note 64).

92. Öttinger failed to see that the expression of a loss was negative, and he was thus unable to carry out the comparison.

93. J. Fries, *Versuch*, pp. 118–120 (note 65).

94. E. Czuber, "Vergleichung," pp. 268–276 (note 72).

95. E. Czuber, "Vergleichung," pp. 276–283 (note 72).

96. E. Czuber, "Vergleichung," pp. 283–284 (note 72).

97. H. E. Timerding, "Bernoullische Wertetheorie," pp. 325–328 (note 72). While economists found a rationale for progressive taxation in D. Bernoulli's hypothesis (Emil Sax, "Die Progressivsteuer," *Zeitschrift für Volkswirtschaft, Socialpolitik und Verwaltung*, I (1892), 77–78), Timerding tried conversely to find a rationale for D. Bernoulli's hypothesis in progressive taxation (pp. 346 -351).

98. S. F. Lacroix, *Traité*, p. 128 (note 23); E. Czuber, "Vergleichung," p. 284 (note 72); H. E. Timerding, "Bernoullische Wertetheorie," p. 324 (note 72); J. Fries, *Versuch*, pp. 125–126 (note 65).

99. John M. Keynes, *A Treatise on Probability* (London: Macmillan, 1973), p. 344.

100. J. M. Keynes, *Treatise*, Part I, Chapter 3 and pp. 344–345 (note 99).

101. J. M. Keynes, *Treatise*, Part I, Chapter 6 and pp. 345–346 (note 99).

102. J. M. Keynes, *Treatise*, pp. 346–349 (note 99).

103. J. M. Keynes, *Treatise*, p. 352 (note 99).

104. Compare Hausdorff's average risk (*durchschnittliches Risiko*) $\sum_{i=1}^{n} p_i |x_i|$ and mean risk (*mittleres Risiko*) $(\sum_{i=1}^{n} p_i x_i^2)^{1/2}$ where $x_i = A_i - E$ (E. Czuber, "Wahrscheinlichkeitsrechnung," pp. 766–767 (note 72)).

105. E. Borel, *Valeur pratique et Philosophie des Probabilités* (Paris: Gauthier-Villars, 1939), pp. 136–139; *Eléménts*, pp. 264–265 (note 10).

106. To know the probability of a certain outcome of a game of strategy, Borel suggested offering somebody the following wager: would he prefer the same amount of money depending on the outcome of the throwing of n n-sided dice or on the outcome of the game of strategy? By varying the number of dice and/or of sides, his preference will determine the approximate subjective probability of the outcome of the game. By offering the same wager to several people, the mean of their subjective probability will be intersubjective and should be held as objective. Now, if the game can be repeated, by offering to let them bet on the mean value so that the wager seems favorable to everyone, those (if any) who win in the long run can be selected as a subset of subjective probabilities whose mean value is even closer to the exact probability (*Le Hasard* (Paris: Alcan, 1914), p. 36; *Valeur pratique*, pp. 84-107 and 139-145 (note 105); *Eléménts*, pp. 267–274 (note 10). For other interpretations of the wager method, see Frank Plumpton Ramsey, *The Foundations of Mathematics and other Logical Essays* (London: Routledge and Kegan Paul, 1931), pp. 166–184; K. Menger, "Das Unsicherheitsmoment," pp. 472–473 (note 29); Bruno de Finetti, "La Prévision: ses lois logiques, ses sources subjectives," *Annales de l'Institut Henri Poincaré*, VII (1937), 6-7; Hans Reichenbach, "Les fondements logiques du calcul des probabilités," *Annales de l'Institut Henri Poincaré*, VII (1937), 314-20.

107. E. Borel, *Valeur pratique*, pp. 105–106 (note 105).

108. E. Borel, *Eléménts*, pp. 100–115 and 218–219 (note 10).

109. E. Borel, *Valeur pratique*, p. 104 (note 105).

110. H. Reichenbach, *Wahrscheinlichkeitslehre* (Leiden: Sijthoff, 1935), p. 81; "Les fondements logiques," pp. 275 and 285–288, (note 106).

111. H. Reichenbach, *Wahrscheinlichkeitslehre*, pp. 183–187. This point had been made by N. Bernoulli in a letter to Montmort (*Die Werke*, p. 558 (note 2); d'Alembert in the 27th memoir of his *Opuscules Mathématiques*, vol. IV, p. 300 (note 35); E. Catalan, "Sur le problème de Pétersbourg" (note 33); Louis Bachelier, *Calcul des Probabilités* (Paris: Gauthier-Villars, 1912), p. 27.

112. E. Borel, *Valeur pratique*, pp. 65-66 (note 105); "Le paradoxe de Saint Pétersbourg," "Sur une propriété singulière de la limite d'une espérance mathématique," and "Sur une martingale mineure," *Comptes-Rendus des Séances de l'Académie des Sciences*, vol. 229 (Paris: 1949), pp. 404-405, 429-431, and 1181-1183; *Eléménts*, pp. 275–279 (note 10); *Probabilité et Certitude* (Paris: PUF, 1950), pp. 126–133. As for the original game itself, Borel contended that the number of tosses had to be limited if only because of the amount of money to be paid, and that B could pay the fair price for a single trial as in a lottery, but should not if he were to play the game repeatedly, for he would certainly lose considering the very low probability of the highest payoff or the number of games necessary for it to obtain almost certainly. Conversely, A would win if he played only with B, but could lose if he played with many Bs because the high losses would no longer be that improbable: That is the rationale for unfair State lotteries (*Valeur pratique*, pp. 60-65 (note 105); *Probabilité et Certitude*, pp. 92-96 (note 112)).

113. B's losses in the $k - 1$ first games are $\sum_{k=1}^{k-1} (k + 1)2^{k-1} = (1/2) \sum_{k=1}^{k-1} (k + 1)2^k$. The latter sum is the sum of the derivatives of 2^{k+1} and is thus equal to the derivative of $\sum_{k=1}^{k-1} 2^{k+1}$, that is,

equal to $(k - 1)2^k$. Taking half this value gives B's losses, the second term in the left-side member.

114. F. P. Ramsey, *Foundations*, p. 176 (note 74).

115. K. Menger, "Das Unsicherheitsmoment," p. 480 (note 29); see also Paul Levy's unconclusive quantitative argument in *Calcul des Probabilités* (Paris: Gauthier-Villars, 1925), pp. 122–126.

116. John von Neumann and Oskar Morgenstern, *Theory of Games and Economic Behavior* (New York: Wiley, 1953), pp. 15–31 and 617–632.

117. Milton Friedman and Leonard J. Savage, "The Utility Analysis of Choices Involving Risk", *Readings in Price Theory* (London: George Allen and Unwin, 1953), p. 69, note 20.

118. J. von Neumann and O. Morgenstern, *Theory of Games*, p. 26, axioms 3A (note 116).

119. J. von Neumann and O. Morgenstern, *Theory of Games*, p. 26, axioms 3B (note 116).

120. Jacob Marschak, "Rational Behavior, Uncertain Prospects, and Measurable Utility," *Econometrica*, XVIII (1950), 111–141, on p. 121; I. N. Herstein and John Milnor, "An Axiomatic Approach to Measurable Utility," *Econometrica*, XXI (1953), 291–297, on p. 293.

121. Edmond Malinvaud, "Note on von Neumann-Morgenstern's Strong Independence Axiom," *Econometrica*, XX (1952), 679.

122. J. von Neumann and O. Morgenstern, *Theory of Games*, p. 26, axioms 3C (note 116); Paul Anthony Samuelson, "Probability, Utility, and the Independence Axiom," *Econometrica*, XX (1952), 670–678, on p. 672.

123. J. von Neumann and O. Morgenstern, *Theory of Games*, p. 618 and p. 620 (note 116).

124. J. von Neumann and O. Morgenstern, *Theory of Games*, p. 28 and p. 632 (note 116). For an empirical evidence of the utility of gambling, see Frederick Mosteller and Philip Nogee, "An Experimental Measurement of Utility," *The Journal of Political Economy*, LIX (1951), 371–402, on pp. 386–389 and p. 402.

125. J. von Neumann and O. Morgenstern, *Theory of Games*, pp. 29–31 (note 116).

126. Maurice Allais, "Fondements d'une théorie positive des choix comportant un risque," *Econométrie* (Paris: C.N.R.S., 1953), p. 527.

127. L. J. Savage, *The Foundations of Statistics* (New York: Dover, 1972), p. 103.

128. Savage did not take it as an axiom but as a principle, which he called the "sure-thing principle" (*Foundations*, pp. 21–22 (note 127)).

129. M. Friedman and L. J. Savage, "The Utility Analysis," p. 73 (note 117).

130. In Mosteller and Nogee's empirical utility curves, the convex segment precedes the concave ("An Experimental Measurement," figures 3a–3c on pp. 387–388 (note 124)). But that corresponds to Friedman and Savage's second and third segments on their second utility curve ("The Utility Analysis," figure 3 on p. 85 (note 117)), referring both to gambling.

131. E. Borel, "Sur les probabilités dénombrables et le pari de Pascal," *Comptes-Rendus*, vol. 224 (Paris: 1947), pp. 77–78; *Probabilité et Certitude*, p. 133 (note 112).

132. Pascal, *Oeuvres Complètes* (Paris: Gallimard, 1957), p. 1214; see J. Bernoulli's note on Pascal's wager in *Die Werke*, pp. 68–69.

8 Laplace and Thereafter: The Status of Probability Calculus in the Nineteenth Century

Ivo Schneider

The influence of Laplace's Théorie analytique *persisted in France and in Europe until about the 1880s. This means that an author who died in 1827 dominated the development, or rather nondevelopment, of a subject for at least half a century following his death. The reason that Poisson, the leading candidate to succeed Laplace, failed to make an impact was his mathematical imperialism—witness his application of probability theory to the domain of human decisions. This imperialism collided with the "new" mathematics of Cauchy, which stressed the rigor and simplicity achieved by the method of restriction. The opposition to Poisson affected the whole subject, which could thereafter survive only in two forms: efforts to make Laplace's* Théorie analytique *accessible on a textbook level, and error theory. It was the latter that caught the interest of researchers.*

However, interest in Laplace's Théorie analytique *did not vanish because Laplace had presented probability theory in his* Essai philosophique *as the best means to attain the bourgeois aspirations of advancement and influence. Error theory and its most essential feature, the method of least squares, on the other hand, could claim very impressive successes, especially in astronomy.*

In the discussion between Bienaymé and Cauchy about the method of least squares, the concept of error was extended and lifted to a higher level of abstraction. Later, at the beginning of this century, this concept was subsumed under the concept of random variable. This revised error theory proved capable of extension and generalization to such an extent that the whole of the earlier probability calculus could be integrated on a more general level.

1 Laplace's *Théorie analytique* as a Paradigm for Nineteenth Century Probability Theory

The title of my paper, which involves the development of a discipline over an entire century, contains only one name, that of Laplace. But Laplace died in 1827. This date in itself should indicate that if there is any justification in mentioning his name in the title, it arises from special circumstances. If Laplace held such a dominant position in the probability calculus,[1] one should naturally ask, why and for how long did his works dominate its content and direction?

That Laplace did occupy such a central position is evident from most of the historical accounts. It has been explicitly stated in the literature in forms that range from the critically neutral to hagiographical,[2] though without offering any concrete points of departure to answer the "why" and "how long" questions.

In trying to find an answer to these questions, I want to put forth two documents that can be considered independent of any attempt at a historical evaluation of the various opinions of mathematicians, or at least those engaged in the probability calculus.

The first example is Matthieu Paul Hermann Laurent (1841–1908) (not to be confused with Pierre Adolphe Laurent, who is known for a theorem in the theory of functions named after him), a "répétiteur d'analyse" and later "examinateur" at the École Polytechnique, whose *Traité du calcul des probabilités* of 1873 initiated a series of French textbooks on the probability calculus. Among the more famous of these are the works of Bertrand and Poincaré and, in our century, those of Paul Lévy and Emile Borel. The foreword to this textbook, which was written a good fifty years after Laplace's death, gives an indication of its intended function: "Those who wish to study the probability calculus will generally encounter difficulties that are due less to the nature of the subject than to the lack of truly classical works."[3]

This is the first sentence. Laurent continues: "And, when studying the famous *Théorie analytique des probabilités* of Laplace, one should already be familiar to a certain degree with the analysis of chance, since the author, as he himself admits, deals only with the most difficult problems. As to the highly esteemed books by Lacroix and Cournot,[4] they are too elementary even to serve as an introduction to the reading of Laplace's work."

After briefly discussing the value of these elementary introductions as an approach to the probability calculus independent of Laplace, Laurent comments on the next candidate, Poisson: "Poisson's work on the probability of judgments,[5] which as the title indicates is written with a special aim in mind, nevertheless contains an introduction that can today be considered the best truly classical work. Unfortunately, it is very incomplete."

Laurent's intention is to reach all those interested in the probability calculus, especially artillery officers and candidates for the insurance professions. His aim is to provide an elementary introduction, but also one that will enable an independent study of the contents of all available literature and especially the work of Laplace. One reason why Laurent still considers Laplace's work to be so important is that its design is so comprehensive in scope that it can accommodate within itself the entire "science of chance." In view of their paradoxical or unscientific character, the concept of moral hope or expectation, which became famous through the so-called St. Petersburg problem, and the evaluation of court decisions and testimonies in the light of probability theory were explicitly excluded from Laurent's study. Considering that his aim was a relatively complete elementary introduction to the subject, there are only three chapters that stand out, following an overview of the most important mathematical methods in the first two chapters:

1. the law of large numbers and the central limit theorem;

2. the error theory, especially the method of least squares, which incorporates the mathematical discussion between Cauchy and Bienaymé, mainly in 1853, on the value and importance of this method;

3. insurance.

These five chapters are followed by a detailed bibliography of the literature on the probability calculus published up to the beginning of the 1870s. This bibliographical survey confirms that Laurent thoroughly examined the relevant works, even those of foreign origin. It also reveals that nearly all of the works listed can be placed in one of the three areas he dealt with, or, in other words, that the probability calculus, as it existed in 1873, leaving aside the applications to insurance, can be reduced to two problem areas: error theory and the so-called laws of large numbers.

Finally, this bibliography and Laurent's preface make it clear that, at least in the French literature, there was a gap of 30 years between Cournot's textbook[6] of 1843 and that of Laurent. In a paper presented in 1903, Paul Mansion[7] states that in Europe between 1837 and 1889, the years that saw the publication of Poisson's investigations and Bertrand's lectures,[8] respectively, only one reasonably comparable work came out, the lectures on the probability calculus that A. Meyer gave between 1849 and 1857 at the University of Liège. These were published in French[9] in 1874 and in German[10] in 1879. As a reason for this gap, which is particularly conspicuous in the French literature on the subject, Mansion cites the fact that the probability calculus was not firmly rooted in the French educational system. Even after the turn of the century, the probability calculus was dealt with in a very occasional fashion, as an appendage to the course on mathematical physics at the Sorbonne and in some of the lectures on analysis and astronomy at the École Polytechnique. In general, the same is true in Germany, which, in the field of mathematics, became increasingly important in the second half of nineteenth century, if one considers the position of the probability calculus the way Laplace did. This does not apply to error calculus, however; courses were available relatively soon, through the influence of Gauss, on the development of astronomy and geodesy. Of all the European countries, it was Belgium, however, where the probability calculus first became fully established. There is no doubt that the development in Belgium was due to Quetelet's influence. From 1835, the probability calculus was a constituent part of the doctoral program in physics and mathematics in Belgium; and from 1838, it was a teaching subject at the Belgian technical universities.

It seems that Belgian mathematicians like Anton Meyer and Emmanuel-Joseph Boudin, who taught at the Universtiy of Ghent and at the École du Génie for nearly half a century,[11] considered their lectures on the probability calculus, as Laurent did in Paris, to be introductions to Laplace's *Théorie analytique*. This remained the classical work until at least the 1870s. Before further interpreting the material mentioned so far in terms of the questions of "why" and "for how long" Laplace's theory determined the development, or rather nondevelopment, in the nineteenth century, I would like to introduce the second document I promised.

Its author is David Hilbert, who, in his paper at the second international congress of mathematicians in 1900, submitted 23 still unsolved mathematical problems that partly determined the development of mathematics in the present century. Convinced of the importance and solvability of these problems, Hilbert formulated them to challenge the "ignoramus et ignorabimus" attitude of Emile du Bois-Reymond, the physiologist,[12] which he thought was decadent and lamentable. Hilbert's sixth problem is as follows: "The investigations into the foundations of geometry suggest to us that, according to this model, we should treat axiomatically those physical disciplines in which mathematics already plays a prominent part: first and foremost of these are probability calculus and mechanics. As to the axioms of the probability calculus, it seems desirable to me that any logical study of them go hand in hand with a rigorous and satisfactory development of the method of mean values in mathematical physics—in particular, in kinetic gas theory."[13]

In view of the contemporary German-language literature on the probability calculus dominated by Emanuel Czuber, professor at the Technische Hochschule in Vienna, Hilbert's statement is an outrage. Just one year previous to Hilbert's paper, i.e., in 1899, Czuber had published a report entitled *Die Entwicklung der Wahrscheinlichkeitstheorie und ihrer Anwendungen* as part of the series of survey reports on the various branches of mathematics organized by the DMV (Deutsche Mathematiker Vereinigung), to which Hilbert had contributed his *Zahlbericht*.[14]

According to Czuber himself, he intended with this report to respond to the wish of the DMV not only "to present as comprehensive a survey of the literature as possible," but also "to pay more attention to the philosophical aspect of the subject."[15]

It seems that Hilbert, who understood probability calculus as a natural science, could not make much of Czuber's very thorough and careful presentation of the subject as a mathematical science. Also, in his report, Czuber had not made any mention of an application of probability theory to physics and to kinetic gas theory in particular. There are a number of reasonable explanations for this; only in his extensive bibliography does he refer to two early works by Boltzmann on the kinetic theory of gases. Apart from the fact that, at the time, a majority of physicists, lead by Ernst Mach, rejected the kinetic theory of gases, the published works of Clausius, Maxwell, and Boltzmann, who claimed the kinetic theory of gases was a new area of application of the probability calculus, had not contributed even a shred of new evidence relevant to the probability calculus.[16]

Apparently, all this did not impress Hilbert in the least. Nor did Czuber's work, which also considered historical aspects and continued in the tradition of Laplace. It seems that Hilbert had also attached great importance to the fact that physicists like Clausius based their gas models on mean values, the use of which could be justified by the probability calculus. The fact is that in the kinetic theory of gases before Boltzmann, there was no clear answer to the question of whether and to what extent the conditions of the applied theorems of probability theory also applied to the physical model under investigation. Thus Hilbert, in defiance of Czuber and all the other exponents of the probability calculus, declared the subject to be a physical discipline and implemented a program for the axiomatization of this discipline that was soon taken up by a number of German and Russian mathematicians. It was brought to a conclusion, for the time being, by the works of von Mises[17] (1918) and Kolmogorov[18] (1933).

In terms of the development of the probability calculus in the nineteenth century, this meant that by 1900, the year of Hilbert's paper, or even earlier, Laplace and the program contained in his *Théorie analytique* had been forgotten. Toward the end of the nineteenth century there was a renewed interest in probability theory in France on the part of mathematicians of the stature of Bertrand and Poincaré, even if in the latter's spectrum of mathematical activities the theory of probability plays a minor role. It would therefore have seemed inconsistent to assume that the sole purpose of their contributions to the probability calculus should have been, as had still been the case with Laurent in 1873, to prepare the way for a better understanding of Laplace, whose mathematical methods as well as his exaggerated claim for their applicability had long since become a matter for criticism, e.g., by Cauchy.

Thus, Bertrand and his *Calcul des probabilités*, which was first published in 1888/9 and was an extended version of his lectures at the Collège de France, can still be regarded as an example of the impact of Laplace; but the traditional reverence paid to the great name of Laplace no longer seems very convincing.

In his preface, Bertrand says that the wonderful work of Laplace may be one of the reasons why the probability calculus represents at one and the same time one of the most attractive and most neglected branches of mathematics. In explaining the connection between Laplace and the neglect of the probability calculus, Bertrand offers the widely held view, which Laurent made quite explicit, that any knowledge of the subject is impossible without a previous acquaintance of the *Théorie analytique*, and that, in turn, the study of Laplace requires an extensive knowledge of mathematics.

Unlike Laurent, Bertrand no longer intends merely to write an introduction to a better understanding of Laplace. Also, unlike Laplace, Bertrand has set himself two aims that to him seem desirable: first, the greatest possible simplicity in the choice of proofs, and second, a rigor that allows an exact evaluation of the reliability and domain of validity of the results obtained. In the *Calcul des probabilités* of Poincaré, which was first published in 1897, there is no trace of any statement on the model character of Laplace's book. Thus, it can be said that, at least in France, Laplace's *Théorie analytique* was the classical standard work until the 1870s.

At the same time, Laurent partly answers the question of why Laplace's work, in particular, could become a classic of the subject. Laurent's partial answer is that potential competitors like Lacroix and Cournot either remained too elementary in relation to the scientific level established by Laplace, or their presentation was incomplete. Laurent's answer concerning Poisson is not convincing because in the title of his work, Poisson had indicated the restriction of his investigation to the probability of court judgments, although in fact—and this is rightly reflected in the title of the German translation[19]—he wanted to present a fairly comprehensive survey of the probability calculus.

2   Probability theory under Attack: Poisson's Failure and the Villain Cauchy

One should therefore ask why Poisson, as Laplace's successor, remained without any appreciable success and why, after Poisson, Laplace had no serious competitors or genuine successors until well into the second half of the nineteenth century.

A partial answer to this question is provided by the relatively limited effect of Poisson's work. The reaction to Poisson's *Recherches sur la Probabilité des Jugements* of 1837 was vividly described in a history of the probability calculus[20] published by Charles Gouraud in 1848.

Apparently, Gouraud's history is a modified version of a chapter taken from a work of nearly 2,000 pages that he submitted on the subject of the "Théorie de la certitude." This topic had been chosen by the *Académie des sciences morales et politique* for the competition[21] of 1848.

This history of the probability calculus earned Gouraud the *doctorat ès lettres* in 1848. To begin with, Gouraud explains that the approach Poisson used in the *Recherches*, considering judgments and evidence in terms of probability theory, seems to have been well prepared by the ideas and theoretical approaches of Condorcet and Laplace.[22]

Unlike his predecessors, Poisson had been supplied with concrete data by the annual publication of a *Compte général de l'administration de la justice en France*, which began in 1825. The tabulated material contained in the *Comptes généraux* provided information on the number and type of offenses, the ratio of the number of charges brought to that of convictions and thus of sentences passed, the frequency of punishments imposed, and, in later editions, the age and sex of accused or convicted persons, the number of jurors fit for office, the presumed motives for capital crimes, recidivism, and finally the number of convictions obtained by the minimum majority of 7 : 5 votes.[23]

In order to understand the reaction described by Gouraud, it is enough to list the contents of the first section of the fifth chapter of Poisson's *Recherches*. This chapter, which respresents less than 20 percent of the total work, is the only one that discusses the probability of judgments and similar applications as mentioned in the title. Among the questions discussed are these:[24]

A determination of the probability that an accused person is convicted or acquitted by a fixed majority of votes of the jurors, each of whom is subject to a given probability of not being wrong, provided the probability of the guilt of the accused that occurred before the conviction is also taken into account. A determination of the probability that the accused person who is convicted or acquitted in these circumstances is guilty or innocent, according to the rule governing the probability of the causes or hypotheses.

Gouraud precedes his criticism and sometimes unusually strong rejection of this application of probability theory with a brief account of the content and importance of the law of large numbers in Poisson's work:[25]

The objects of nature as a whole, in the moral as well as in the physical world, are according to [Poisson] subject to a universal law that is approximately as follows: if one observes a very considerable number of events of the same kind that depend on constant causes, causes that, as happens in the moral world, vary in an irregular fashion, sometimes in one direction and sometimes in another, but without their variation proceeding in any particular direction, the amplitude of these irregular effects produced by the variable causes will increasingly contract proportionally as the series of experiments becomes larger; and the relationships of the observed events will noticeably approach stability and even fully attain it, if one sufficiently multiplies the experiments.

After a reference to the mathematical and philosophical significance that must be attached to the law of large numbers as so named by Poisson, Gouraud states that this law was regarded as the universal key to the application of the probability calculus on an ethical level:[26]

Armed with this principle, [Poisson] did not even shy away from the determination, however dangerous it might have been, of the mathematical probability of every human decision. Even the most complicated ones, subject to changing and capricious opinions, seemed to him to be suited, with the help of his law of large numbers, to be analyzed and judged with the utmost rigor. In the *Recherches sur la probabilité des jugements*, one finds numerical expressions that are obtained from an analysis of a large number of earlier judgments whose aim is to determine the exact future probability for each citizen of being charged, convicted, or acquitted. This is an almost incredible audacity, which exceeds that of Laplace and Condorcet.

Gouraud had thus explained that, from the point of view of Laplace's school, Poisson, with his *Recherches*, had finally extended the probability calculus to the areas of legislation, jurisprudence, and economics, something that the programs of Condorcet and Laplace had already claimed to have done. In spite of this claim, which until then had been generally felt and expressed by mathematicians, their opinion of the fifth chapter of the *Recherches* or its contents, which had already been presented to the Academy[27] in 1835, was anything but unanimous:[28]

Among the mathematicians themselves, some of them of the highest reputation dissociated themselves in a sensational manner, which was until then an unheard-of event in the history of probability theory, from the opinion generally accepted in mathematics, that there was no probabilistic decision whatsoever that could not be transformed into a calculation. They initiated a controversial debate in the Academy of Sciences, in the course of which Poisson and his colleagues were directly accused of having brought the traditional reputation of mathematics (geometry) into discredit through activities that were unworthy of the dignity of science. M. Ch. Dupin, who spoke for the united sections concerned with analysis and economics, was heard saying that the application of the calculation to an estimation of the probability of judgments was wrong and ineffectual. M. Poinsot went even further by stating that the enterprise (of Poisson) seemed to him to be a "kind of mental aberration."

In regard to these arguments at the Académie des sciences, Gouraud refers to the *Comptes rendus* of April 11 and 18, 1836. For the purposes of this paper, it suffices to summarize the objections raised by Dupin and Poinsot by saying that Poinsot categorically rejected the urn model used by Poisson because the realm of the so-called moral sciences was far too complex and unpredictable in its effects for any kind of mathematical model to be set up, while Dupin's criticism focused on the unreasonable simplification of the actual conditions by Poisson. This at least theoretically left open the possibility that useful results might also be obtained from a less simplified model.

A more interesting personality than Dupin, who did not in any way feel obliged
to provide a better model, is Poinsot. His categorical rejection of this application,
which he even supported with a warning by Laplace that applications of the
probability calculus should be undertaken with care, is a general reflection of
contemporary philosophy.[29]

Gouraud then refers not only to the reaction of the French philosophers who, in
his opinion, had accepted and even proclaimed the criticisms leveled at Poisson, but
hints at the posibility that French philosophy had prepared such a reaction a long
time before: "Even ten years ago, Royer-Collard, Laplace's successor in the Académie Française, had not hesitated, even in the *laudatio* that he had written for this
great man, to express a strong and definite reservation regarding the value of the
application of algebra to the determination of moral probabilities."[30]

Royer-Collard actually said the following: "The mathematical description of the
cosmos is different from moral science, which is concerned with men; the latter is
based on more mysterious and complex principles, which mathematical science will
not approach."[31]

Gouraud also refers to the doubts that were expressed by De Broglie and
published in the *Moniteur*, the official gazette, in 1831. De Broglie had given a
report on a new law concerning the majority at jury trials to the "Chambre de Paris"
and expressed some doubt in the premises on which Laplace had based his calculations. It is therefore not surprising that, after the publication of Poisson's *Recherches* and in the context of the controversies raging at the Académie des sciences,
the philosophers who were then led by Victor Cousin, a disciple of Royer-Collard,
unanimously protested, Gouraud reports, "in the name of the sacred rights of
liberty, against the claim put forward by mathematicians, to calculate a return of
events of a moral order."

This solemn protest by the philosophers did not find expression in a separate
publication, but is clearly reflected in the *Oeuvres* of Cousin, which were printed in
the 1840s, and particularly in his *Cours d'histoire de la philosophie morale au
dixhuitième siècle*. This was first published in 1841 from lectures delivered up to the
year 1820. In this series of lectures, Cousin undertakes a critique of Kant, e.g., the
relationship between the freedom and duty of the individual. He notes in several
places that the methods of mathematics and physics are basically irrelevant to the
social environment of man, which, as Poinsot reiterated later, is too diverse and
erratic for such an approach. More important, though, the pattern of an interplay
between chance and the regularities determined by the law of large numbers
developed by the probability calculus left no scope for the ethics-oriented approach
demanded by the philosophers. We can see from the above that the opposition to
Poisson was based on different reasons. They range from a growing resistance of
philosophy to the encroachment of mathematics to the tensions between political
liberalism and conservatism.

In very general terms, philosophers such as Victor Cousin and later, in England,
John Stuart Mill, can be considered liberal, while probabilists like Condorcet,
Laplace and Poisson, convinced as they were of an inherent calculability of all

processes of human decision making and thus of the predictability of social development, can be regarded as conservative (although this division does create problems with respect to Condorcet). This split between conservatives and liberals can be observed in the mathematical community as well, where Poisson's main opponents Poinsot and Dupin can be considered as liberals who found a good opportunity to take issue with Poisson's more conservative concept of convictability instead of guilt of the accused.

However, personal and institutional reasons were equally important for Poinsot's and Dupin's opposition to Poisson. In Poinsot's view his career had suffered from the interventions of Poisson, who had been his immediate superior, and the relations between the two men were anything but friendly. Like Dupin, Poinsot was a fervent disciple of Monge and favored geometry and geometrical methods. No wonder he found himself very often in disagreement with the French analytical school of which Poisson was one of the most prominent exponents.

Dupin was since 1832 a member of both the *Académie des sciences* and the *Académie des sciences morales et politiques*. His membership in the second Academy certainly provided him with sufficient incentive to link arms with the philosophers in their battle against the encroachments of mathematicians.

Another factor is the disintegration, beginning as early as 1820, of the scientific edifice founded by Laplace and the growing specialization in the natural sciences and the humanities.

Against the backdrop of these events, the idea of a union of the *homme des lettres* with the *homme des sciences*, which was an accepted ideal of the Enlightenment, became increasingly difficult to sustain; it tended more and more to represent a kind of dilettantism that seemed trivial in view of the professionalization that was developing in the nineteenth century.

In referring to the reaction to Poisson's *Recherches*, Gouraud thus documents the final break between the humanities and the natural sciences. His assessment of the situation in mathematics has to be judged mainly in the light of the propagandistic success of the statistics established by Quetelet; this led Gouraud to believe, wrongly, that Poisson had been successful in his attempt to enlist the support of the mathematicians. At any rate, when writing his report in 1848, Gouraud was able to see the attempt to encompass the entire social field of mathematics in full flower. At the same time, he marked the widening gap between philosophy and mathematics, as his own theses show.

In the final analysis, Gouraud is primarily a witness to the disintegration of the unity of the natural and, in a very general sense, social sciences. Condorcet was aware of the tie between these two areas in mathematics, and Laplace, Condorcet's successor, tried to deepen this unification that had been inspired by mathematics. The authority that Laplace could bring to bear, at least during his lifetime, on the implementation of his program and the claims based on it was due not least to the high social prestige that was conferred on the leading mathematicians of the Empire, when Napoleon, guided by the principle of meritocracy, established a senate and the Legion of Honor, and thus a new élite that, in the *Ancien régime*, would have corresponded to noble rank.

The high prestige of French mathematicians of the Empire, such as Lagrange, Laplace, Monge, and Fourier, which in turn enhanced the prestige of mathematicians in general, could, for a time, check the influence of those who felt the programs of Condorcet and Laplace were totally inadmissible transgressions of a mathematical science that had become too arrogant. As an example of the strong disapproval of philosophy even vis-à-vis the plans of Condorcet, Gouraud referred to La Harpe's[32] *Cours de littérature* of 1797.

The claims of mathematics, especially on behalf of the probability calculus in relation to the social sciences, which philosophers considered impermissible and exaggerated, and which may have provided Napoleon with grounds for personally patronizing individual mathematicians, were relinquished only when the philosophers were joined by some of the mathematicians.

In the debate over Poisson's *Recherches*, the division within the mathematical camp also took on an outward dimension. In a way, this reflects a break in the development of mathematics, which can be described as a break on an institutional level between the form of mathematics taught at the Académie in the eighteenth century, as represented by Lagrange and Laplace, and the form of mathematics practised at institutions of higher education like the École Polytechnique, represented by Cauchy. It reflects a break as well between the kind of commissioned mathematics of the eighteenth century, which was judged solely in terms of its potential usefulness, and the mathematics pursued at institutions of higher education, which, led by the demand for a didactics of the science, turned to such criteria as rigor and simplicity.

The *Cours d'analyse* that Cauchy gave at the École Polytechnique as an introduction to the central area of mathematics at that time, analysis, published in 1821, became the paradigm for presenting the new mathematics. In the preface to this work, which became obligatory reading for the future élite of the French administration, Cauchy emphasized that, although his work rested on the mathematical results of the eighteenth century, it was markedly different from that of his predecessors in terms of the degree of rigor applied, and that the range of applicability of mathematics is restricted. The introduction contains the following interesting passages:[33]

To begin with, Cauchy expresses gratitude for the encouragement he received in creating this course: "Some persons who had the kindness to guide me in the first steps of my scientific career, and among them I gratefully mention Laplace and Poisson, have requested the publication of the course on analysis at the *École Royale Polytechnique*, and I have therefore decided to produce a written version of this course for the benefit of the students." Having said this, Cauchy then distances himself from the view of mathematics and its applicability claimed by Laplace, although he does not mention him by name: "On the one hand, I have endeavoured to perfect mathematical analysis; but on the other hand, I am far from claiming that this analysis is sufficient for all speculative sciences." After a further paragraph explaining how the exact natural sciences rest on observation and calculation and how the existence of historical personalities is as certain as mathematical theorems, Cauchy makes a final reservation: "Let us therefore eagerly pursue the study of the mathematical sciences without letting them extend beyond their domain; and let us not imagine that we can approach history through mathematical formulas or sanction morality with algebraic theorems or integral calculus."

The argument that this verdict on the applicability of mathematics to the social world refers only to the predecessors of Laplace, especially Condorcet, seems absurd in view of the prominence of the *Théorie analytique*. A third edition had been published in the previous year and its ninth and last chapter deals with the probability of statements made by witnesses. It also seems absurd in view of the ever greater prominence of the *Essai philosophique*, which, in a section on the "Application of the probability calculus to the moral sciences," contains the following passage:[34] "Let us apply to the political and moral sciences the method founded upon observation and upon the calculus, the method that has served us so well in the natural sciences. Let us not offer in the least a useless and often dangerous resistance to the inevitable effects of the progress of knowledge."

In 1821 Cauchy considered it dangerous or perhaps just disrespectful to name Laplace in connection with a statement concerning the areas to which mathematics should definitely not be applied. To Cauchy, such a limitation in the applicability of mathematics must have seemed not just useful, but essential for the mathematics he had reformed. Cauchy had come to the conclusion that the method of limitation, e.g., designating the domain of definition and the range of a function, the domain of convergence of series, and generally the range of validity of mathematical statements, would deliver mathematics from the difficulties into which eighteenth century claims to universal validity had lead it. Mathematical imperialism ran counter to such a limitation.

It is exactly in this sense that Cauchy later applied the method of limitation to one of the centerpieces of the Laplacian calculus of probability, the method of least squares, which is the heart of error theory. As a first step Cauchy reformulated a statement by Laplace on the implicit use of the following principle: Mathematical statements have to be given in such a way that they can either be proved or disproved. As an example of his "universal" attitude, Laplace had suggested in the *Théorie analytique*, "Preference should thus be given to this method [i.e., the method of least squares] whatsoever the law of frequency of the errors may be—law on which the ratio k''/k depends."[35]

Here k'' is chiefly identifiable with the variance of the error. Of course, Laplace's formulation presupposes the existence of the variance. Moreover, Laplace's phrase cannot be disproved. In order to do this one has to transform his suggestion into a claim. This is precisely what Cauchy did.

According to Bienaymé,[36] Cauchy orally reformulated Laplace's suggestion in a session of the Academy on August 8, 1853: "Mr. Cauchy has denied the exactitude of this remarkable result, discovered and demonstrated by Laplace, which says that the method of least squares applies to the results of observations no matter what the law of probability of the errors is."

This reformulation allowed Cauchy to look for a counterexample, and he succeeded in finding the distribution later named after him, to which the method of least squares does not apply. Having thus seen the claim to the universal applicability of the method of least squares disproved that Cauchy had falsely attributed to

Laplace, Bienaymé was at pains not only to restore Laplace's reputation but also to prevent the dismantling of the theory of probability as a subject. Bienaymé saw himself pressed to defend Laplace and the theory of probability because "if, as is well known, Laplace's discovery is incorrect, a considerable part of his great work, which is the most important part in that it is the most applicable and the most practical, would be overthrown at one blow."[37]

Even if this controversy over the applicability of the method of least squares centered on a series of publications in the *Comptes rendus* of 1853, it dates back to the year 1835, the year in which mathematicians had revolted against Poisson's application of probability theory to the domain of human decisions. In this year Cauchy had published a special method of interpolation.[38] Cauchy claimed that his method was better than the interpolation methods of Lagrange and Laplace. In 1853 Cauchy again returned to his method of interpolation in order to show that it allows one to deduce in a straightforward way the same values that are given by the method of least squares.[39]

This claim was answered by Bienaymé's *Remarques sur les differences qui distinguent la méthode des moindres carrés de l'interpolation de M. Cauchy, et qui assurent la supériorité de cette methode*,[40] which in turn was followed by Cauchy's attempt to prove not only that his method of interpolation would furnish the same results as the method of least squares; it would do so with fewer calculations. Without going into the mathematical details, the modifications, and the marvelous blend of obscurity and clarity that characterizes Cauchy's procedure,[41] we can say that his aim was to deprive the method of least squares of its primary feature, namely, to deliver the most probable values. In his own words Cauchy triumphantly stated[42] that if relatively restrictive conditions are not fulfilled, "the method of least squares could furnish values for the unknowns $x, y, z, \ldots, v, w$ that differ sensibly from the most probable values."

As Bienaymé had sensed, Cauchy was eager to extend his claim to the heart of error theory, the method of least squares, in order to annex to analysis proper all the fields to which this method applies.

Cauchy's reaction should be considered against the background of a regeneration of the concepts, methods, and contents of mathematics and its resultant specialization and differentiation. Precisely because Cauchy, who had already become the doyen of French mathematics, had not named Laplace in the *Cours d'analyse* and thus left an empty space for the next person who would offend against his verdict, there was also a readiness among mathematicians to condemn Poisson at the time of the *Recherches*, which fitted in with Laplace's program of applications for the probability calculus. This situation, which Gouraud described as new and unheard-of, led to a number of repercussions: faced with the alternative of the *Théorie analytique* of Laplace and the *Recherches* of Poisson, most mathematicians sided with Laplace. In other words, Poisson, who was the only serious competitor to Laplace in determining the course of the development of the probability calculus, had more or less failed. "More or less" because even after Poisson there were mathematicians who were not influenced by the scandal of 1836/37 and examined judgments, statements made by witnesses, etc., in the light of probability theory. However, they were all authors of textbooks and not scientists engaged in research.

3   The *Essai philosophique* as a Lifeboat for Probability Theory

Leaving aside the textbook authors mentioned above, the probability calculus had forfeited its most attractive field of application, that created by Condorcet's and Laplace's expectations for political influence. Further areas of application anticipated by Laplace were likewise threatened with the same fate. Apart from the mathematical methods used by Laplace and Poisson in the probability calculus, which were all borrowed from analysis, there was hardly anything left that could have justified a young mathematician taking up probability theory after Poisson. The words "hardly anything," to which I shall return, are sufficient to explain the gap existing in the literature around the middle of the century noted by several authors, a gap of between 30 and 40 years, depending on the various opinions held regarding the question of what is an accepted contribution to the probability calculus. At the same time, these events throw light on a development that began in the second half of the nineteenth century with several retrospective assessments of Laplace, i.e., that Laplace, rather than determining the development of probability calculus in the nineteenth century, brought it to a conclusion. After Laplace, or better after the publication of Poisson's *Recherches*, the discipline of the probability calculus, as it had been created toward the end of the seventeenth century from the union of the calculus of games of chance with the philosophical concept of probability by Jakob Bernoulli, ceased to be an object of mathematical research. The fact that this discipline was able to survive up to and even beyond the turn of the twentieth century in the form given it by Laplace results from two reasons:

First, the paradise that Laplace had promised in his *Essai philosophique* tempted many scholars to become familiar with the mathematical content of the *Théorie analytique*. Second, the probability calculus was still a field of application that even to Cauchy and in spite of his reservations, seemed worthy of some scientific work, meaning the error calculus.

In regard to the first point, it seems useful to ask about the possibilities Laplace had, if any, after the developments just described, to present a make-believe paradise of the probability calculus to anyone. The answer to this question lies in the distribution of the *Essai philosophique*. The *Essai* was in fact an edited form of the last in a series of ten lectures that Laplace had given in 1795, during a period of only three months that he spent at the first École Normale. These lectures, which were presented at the large auditorium in the Jardin des Plantes before more than 1,200 persons of all ages with an educational range extending to near-illiteracy, can be considered Laplace's learning exercises. They anticipated his later mastery at popularizing his two main fields of research, i.e., astronomy and the probability calculus. The extant contemporary reports on the success of his *Exposition du système du monde*[43] of 1796 and his *Essai* of 1814, even abroad, bear witness to this. Although a major part of this evidence refers to the *Exposition du système du monde*, there are reasons to assume that the *Essai*, which, as a separate publication, ran to five editions even during Laplace's lifetime and served as an introduction to the *Théorie analytique* from its second edition, was Laplace's most successful work in terms of the impact it had on that period.[44]

The success of the *Essai*, which was addressed to educated laymen with a very limited knowledge of mathematics, is, of course, connected with the expectations entertained by the European bourgeoisie. By concentrating on the meritocratic principle in the new systems of education and training, they hoped to gain a greater influence in politics and the economy. In explaining the value of mathematics and the probability calculus in particular as a teaching subject, Laplace bolstered his arguments by asserting "that basically the probability calculus is human reason subjected to calculus." The final sentences of the *Essai* show how well the greatest political opportunist among the renowned natural scientists and mathematicians managed to exploit the probability calculus with respect to the bourgeois aspirations of advancement and influence:[45]

If we consider the analytical methods to which this theory has given birth; the truth of the principles that serve as a basis; the fine and delicate logic that their employment in the solution of problems requires; the establishments of public utility that rest upon it; the extension that it has received and which it can still receive by its application to the most important questions of natural philosophy and the moral science; if we consider again that, even in those things that cannot be submitted to calculation, it gives the surest hints to guide us in our judgments, and that it teaches us to avoid the illusions that so often confuse us, then we shall see that there is no science more worthy of our meditations, and that no more useful one could be incorporated in the system of public instruction.

Having thus explained the reason for the interest in the *Essai*, we still have to examine why this interest was hardly affected by the heated controversies and the rejection of Laplace's program by those philosophers and eminent mathematicians who were responsible for the new mathematics. The answer to this question lies in the fact that the circles addressed by Cauchy and the philosophers did not significantly overlap with the readers of the *Essai* and its translations. The majority of readers most probably belonged to those circles of the bourgeoisie who were eager to learn and who, in the first decades of the nineteenth century, acquired their knowledge, particularly in mathematics, from self-instruction. This is especially true in Germany and, in part, also in England. In Germany, until the 1850s, the gap between the level of mathematical teaching, which was rather modest at most universities, and the level of mathematical research, which was gradually being achieved, was bridged by an autodidactic or supplementary study of French textbooks, German translations of which enjoyed a wide circulation.[46]

For those European readers of Laplace who were genuinely interested in mathematics, it was clear that the *Essai*, which almost entirely avoided mathematical symbolism, could only wet their appetite for reading the *Théorie analytique*. What was to be expected by the reader who decided on taking that step is described very clearly in a paper Augustus De Morgan wrote in 1837. Although it was presented as a review of the *Théorie analytique*, it reflects the position of the probability calculus in detail and, in particular, its importance to the insurance business.[47] De Morgan calls the *Théorie analytique* the "Mont Blanc" of analysis, whose inaccessibility is all the more surprising since the fundamentals of the probability calculus can apparently be mastered very easily. However, while there were guides to assist in climbing Mont Blanc already by 1837, the student of the *Théorie analytique* was entirely on his own. According to De Morgan, the inaccessibility of the *Théorie analytique* results from the fact that it consists of a collection of Laplace's Academy publications on various problems of the probability calculus that were published between 1774 and 1812, shortened by omitting the various historical introductions, but otherwise largely unaltered.

The first book of the *Théorie analytique*, which is divided into two parts, confronts the reader with the greatest problem. This book explains in about 180 pages the theory of generating functions as it was developed by Laplace on the basis of the preceding work of de Moivre and Lagrange. It includes all the results relevant to the analysis of that time, but without any reference to the probability calculus. In comparison with other writers, Laplace thus offered a consistent theory for the mathematical treatment of problems of probability theory. By constrast, De Morgan states that generating functions are only required for the solution of some unimportant gambling problems, for which an introduction of 2 pages would have been sufficient. Even if one cannot agree with De Morgan on every point, nor with his exaggerated statement that Laplace should be considered the least sophisticated and longest-winded mathematician in comparison to Euler's incredible simplicity at presenting mathematical contexts and Lagrange's ability at generalization, he is right in one aspect that was crucial for Laplace's impact. Without an appropriate introduction, very few highly talented students—if any—were able to study the *Théorie* on their own. In view of the interest in the probability calculus, mainly created by the *Essai*, it seemed worthwhile—from a didactic and economic point of view—to produce textbooks that either obviated the difficult reading of the *Théorie* or served as a preparation for its study. De Morgan himself chose this latter way with his *Essay on Probabilities*, published in the *Cabinet Cyclopaedia*.[48] He also indicated one of the approaches to a simplified description by omitting the theory of generating functions altogether.

Up to about the end of the nineteenth century, all of the French and part of the German textbook literature dealt predominantly with the didactic adaptation of Laplace's *Théorie analytique* or made a suitable selection of applications of the probability calculus contained in it. On this level, the probability calculus continued to exist even after Poisson. In this tradition, the probability calculus ceased to be a field of research. Progress in analysis, however, was used both to replace the method of generating functions, which, according to Laurent, had almost fallen into oblivion by 1873, and to prove the results achieved by Laplace and Poisson by means of other methods in view of the increasing demands for mathematical rigor.

Laurent himself sought to introduce into Laplace's probability calculus the results of the trigonometric respresentation of functions of Fourier, Dirichlet, and Cauchy, as well as the possibilities offered by the theory of functions. These efforts, in the second half of the nineteenth century, can certainly be regarded as a first approach to new research, although it was very limited in the beginning; but it led, in our century, to the elegant and versatile method of characteristic functions. In his *Calcul des probabilités* published in 1925, Paul Lévy makes this the central instrument of the probability calculus. Thus, there was a tradition, based on the didactic adaptation of the remaining contents of Laplace's *Théorie*, which ensured the survival of Laplace's program of the probability calculus until the twentieth century, a fact that is documented by the textbook literature. But there also exists a section of the probability calculus that became an increasingly important subject of mathematical research. This is the error calculus.

4   Error Theory as the Nucleus of the New Theory of Probability

I wish to assert that error theory not only kept the probability calculus alive as a field of research, but also proved capable of extension and generalization to such a degree that the entire earlier probability calculus could be integrated on a more general level. Simultaneously, the concepts developed by the error calculus, after appropriate modifications, turned out to be of central importance to the statistics that emerged within the framework of English biometry. Since probability theory was firmly embedded in measure and integration theory as they had been developed mainly by Borel and Lebesgue, biometrical statistics became an independent field of application and separated from probability theory.

Although the error calculus had been a research topic in eighteenth century mathematics, it became prominent only with the success of the method of least squares. Gauss—probably the pioneer in the application of this method—also supplied the most widely known arguments for it in terms of probability theory. This attempt was followed by further explanations advanced by Gauss himself and by Laplace and many others. By 1877, Merriman was able to list several hundred publications dealing with the method of least squares.[49]

A discussion of this method, which is perhaps the most important and, from the point of view of historical impact, the most remarkable of its kind, can be found in the above-mentioned controversies involving Bienaymé and Cauchy. These took place mainly in the year 1853, although they had been latent for some time.

Bienaymé was born in 1796, the same year as Quetelet, and was thus seven years younger than Cauchy. He belonged to that new French élite that used the education and training received at the École Polytechnique (although in Bienaymé's case, this was admittedly very brief) not to enter an academic career, but, in accordance with the aims of the school, to take up a post in the national administration. After Bienaymé had worked as a translator and than as a lecturer in mathematics at the St. Cyr Military Academy—through circumstances forced on him by the Bourbon restoration in France of 1815–he took up a post in the financial administration in 1820 and became Inspecteur général in 1834. In 1844, he received the honorary title of an officer of the légion d'honneur and, in 1848, the year the Second Republic was proclaimed, he was retired prematurely. Shortly thereafter, Bienaymé was engaged by the Sorbonne, where, as a professor, he taught the probability calculus until 1851. He was succeeded by Gabriel Lamé (1795–1870), who was appointed to the chair of mathematical physics and the probability calculus at the Sorbonne in 1851. Upon resuming his lectures on the probability calculus on April 26, 1851, he named

Bienaymé as practically the sole exponent of probability theory in France. In the years after 1851, Bienaymé, with his sympathies for Napoleon III, to whom he occasionally referred, proved himself to be a competent consultant to the government on statistical and economic questions. In 1852, he became a member of the Académie des sciences, and it is possible that the timing of his election was connected with political changes. From a scientific point of view, Bienaymé was qualified for this election by reason of a number of publications for the Société Philomatique de Paris and, in particular, a work on the method of least squares written for the Académie des sciences.[50]

The academicians Lamé, Chasles, and Liouville, the latter in a responsible capacity, had reported about this work on January 19, 1852. One sentence is of particular interest in this connection since it both highlights the repercussions of the reaction to Poisson and the fact that Bienaymé intended to fight for the restoration of the lost areas of application of the probability calculus:[51]

Even when leaving aside any thought of an application to the important problems of natural philosophy, the mathematicians will still be reading his works with interest. But the probability calculus, which is associable with such impressive names as Pascal, Fermat, Huygens, Jakob Bernoulli, Laplace, Fourier, Poisson, etc., is not just an abstract theory. Within the appropriate limits, it can be of practical use. It has often been misused, that is true; but that is no reason to reject its application altogether.

After Bienaymé, in his new position as a member of the Académie des sciences, which also entailed considerably more respect from mathematicians, had made himself a champion of the probability calculus and of the method of least squares in particular, only a short period elapsed before Cauchy made his appearance. We have already seen that in 1853 Cauchy claimed that his method of interpolation which had been published in 1835 required less mathematical effort and produced better results.[52] With these arguments, Cauchy had aroused the curiosity of some of those concerned with applications who, because of their practical approach, did not care whether the proposed method was derived or explained by way of analytical methods or on the basis of probabilistic considerations. Cauchy's view, which was reasonable and shared by many, that the determination of a number of unknown parameters on the basis of a series of observations could be regarded as a problem of interpolations, also meant that the probability calculus would have been deprived of its last remaining research-intensive support, the area of the error calculus. It may well be that Cauchy was moved by personal reasons too, since Bienaymé was a friend of Cournot who in turn was a personal enemy of Cauchy. As we have seen in section 2, this event sparked off a whole series of exchanges and statements, particularly by Bienaymé, which became quite heated at times and are partially reflected in the *Comptes rendus* of 1853.

Cauchy's attack aimed at Bienaymé's understanding of the very heart of probability theory. This provided Bienaymé with an opportunity to reflect briefly on the position of the probability calculus. He acknowledges that, in the two centuries that have elapsed since the correspondence of Pascal and Fermat, the discipline has not progressed very far from its origins. He further concedes that this discipline, which "represents the first step of mathematics away from the realm of absolute truth and validity," has at times fallen victim to errors and misconceptions. For this very reason, he considers it his duty to put the possibilities that have occasionally been overestimated in their proper perspective, to expose the difficulties that have so far been concealed, but also to reaffirm the actual achievements, even in the face of prejudice or misjudgment, in order "to clear and secure the way so that those who follow in his steps will be able to proceed safely."

Bienaymé then refers to the "very nature of probability calculus," which enables him to renew some of the former claims regarding its applications:[53]

The very nature of the probability calculus, which deals with errors of all kinds, deviations of all sorts, inconsistencies of observations resulting from the inadequacy of the human sensory organs, the discordance of statistical data, the discordance caused by the physical variations that are usually quite strong, and the negligence or inability involved in proving these variations; the nature of this calculus, which deals with all events for which the concept of chance has been forged, the very nature of this calculus is that it is critical (. . . la nature même de ce calcul est critique).

After mentioning the fact that his own works, which he began writing 30 years before, have not been critically appraised, he discusses the dangers involved in Cauchy's attempt to limit the applicability of the method of least squares. Bienaymé names the areas of application that follow the method of least squares in the works of Laplace. They are treated in the chapters of Laplace's work "concerning research on the phenomena and their causes, the probability of causes and of future events based an observed events, the average duration of life, of marriages and of any other association, the benefactions of institutions based on the probability of events, etc."[37] Bienaymé believes that the idea of a complete work occurred to Laplace only after 1809, after he had been publishing for almost 40 years and when he finally became quite clear in his own mind about the principle underlying all "probability functions" ("fonctions des probabilités") that holds for nearly all applications.

According to Bienaymé, the reputation that Laplace acquired with the *Théorie analytique* is due mainly to the results of the analysis contained in it, but not to the principle itself. Bienaymé is convinced that otherwise, "instead of trying to doubt and upset it, one would endeavor to consolidate it and show how Laplace understood it." One would try to show "how the new results of analysis can serve to throw a light on the probability calculus."

Bienaymé reproached the successors of Laplace for having only seen the problems of the analysis and wasted their effort "in replacing the calculations of Laplace by so-called *démonstrations faciles, preuves populaires*:"[54] "Up to this point, it has not seemed possible to implement such a replacement when one wants to know the probability of a given error. The analysis used by Poisson, the analysis employed by Cauchy with a little more of the rigor that should be used in mathematics for numerical applications, which are the truly practical applications, these yield no more than the convenient formulas that Laplace had developed from his form of analysis."

If, however, the main point is not the extent of the error, but only the proof of the method of least squares, in the sense that the latter minimizes this fault, then far more modest analytical instruments will be sufficient.

Referring to the principle discovered by Laplace, Bienaymé considers it important that the distribution function of the standardized errors be approximated by the normal distribution, although in his opinion the normal distribution is only an approximation that occasionally produces results that are obviously wrong—"a pure arithmetic machine, good for numerical calculus."

As Laplace saw it, although in contrast to Gauss, who considered higher moments of an even order admissible, the method of least squares was excellent since it contained "the fundamental pre-requisite for the development of error distributions to the extent that the number of observations increases." This meant that the scene was set for the subsequent reestablishment of the method of least squares in purely mathematical terms. Naturally, in trying to save Laplace's program of applications for the probability calculus, Bienaymé had to make concessions. These were required mainly by the increased demands being made in terms of mathematical rigor.

Thus, as we have seen above, Cauchy disproved Laplace's claim that the method of least squares was applicable to all observations, regardless of the kind of error distribution, by constructing an error distribution—later named after him—that contains no moments of a higher order. In this way, it precludes any application of the method of least squares. Bienaymé thought that such a distribution was inadmissible, not only because, by definition, it could not be related to the method of least squares, but also because Laplace, against whom Cauchy had directed his attack, had at least implicitly allowed only distributions for which the variance exists. This meant that Cauchy's counterexample did not apply. In addition Bienaymé could pick at Cauchy by hinting at the fact that it was not he but Poisson who had first found this distribution function, and this finding did not shake Poisson's confidence in the method of least squares. Poisson was content to remark that one never encounters this distribution in practice.[55] This last argument had been stressed by Bienaymé in more detail before.

If we realize that Bienaymé succeeded at least in part in refuting Cauchy's claims for his method of interpolation, we should expect that error theory and with it probability theory would recover and slowly start to gain new territory. This in fact happened, but primarily outside of France. That Bienaymé did not enjoy the reaction he deserved in his own country might have to do with his position as an outsider in the French educational system. Bienaymé's strategy was more directed toward pulling political strings, that is, to influencing political decisions by launching probability-based statistical surveys. But the majority of French mathematicians were either indifferent to probability theory or sided with Cauchy and the opponents of Poisson. They had no difficulty disregarding Bienaymé, who in their eyes had become a member of the Academy for political reasons and whose scientific merit consisted of his defense of Laplace, the last hero of eighteenth-century mathematics.

However, the discussions between Cauchy and Bienaymé met on both sides the standards of mathematical rigor—which had become accepted around the middle of the century—and provided a new impetus for further research.

Cauchy's approach, as shown by his construction of the so-called Cauchy distribution, made it possible to consider error distributions and their aspects generally at the level of abstraction that had been reached in analysis around the middle of the nineteenth century. It was a manner quite detached from practical applications, which at any rate had scarcely been examined. At the same time, Bienaymé's efforts to extend the error concept to include—in colloquial speech—deviation and variation, and thus to restore the earlier claims for the applicability of the probability calculus, followed the same direction. Even though Bienaymé and Cauchy considered themselves opponents from the start and sometimes engaged in polemical exchanges, each had a limited willingness to give serious consideration to arguments put forward by the other.

Finally, one result of their different interests that Bienaymé in particular could not foresee, but that nevertheless had the same objective, was that both of them used the error concept in such a general form that Chebyshev, the father of the so-called St. Petersburg School of the probability calculus (whom Bienaymé had known since at least 1858), felt free to give a new designation of a quantity or variable, in the sense of a random variable, to this abstract error concept.[56] This partly followed up on the works of Bienaymé and paved the way for the introduction of the central concept of a random variable.[57] It thus became clear, as Bienaymé had already pointed out in 1853 in his concern about the future of the probability calculus, that the principal parts of Laplace's probability calculus could be traced back to the error calculus and the method of least squares. In the more modest form of the error clalculus, which in Germany under Gauss' influence dominated everything that was called probability theory, it continued to exist as an area of research in the nineteenth century. This concentration of the subject, coupled with new research activities, made it possible to reclaim lost territory. However, in contrast to the way of thinking still influenced by Laplace, this required a clearer separation from the classical realms of application, astronomy and geodesy. Chebyshev had already made this break. To a greater extent than Cauchy, he had taken seriously the demands for rigour and simplicity in the probability calculus, and implemented them in his school together with Markov, Lyapunov, and Bernstein.

One reason that this research tradition apparently remained unknown to Hilbert and made an impact only in this century may be that its pioneers, Bienaymé and Chebyshev, did not leave any monographs.

Largely unnoticed by the mathematicians who were engaged in adapting the probability calculus to the mainstream of mathematics in the second half of the nineteenth century, the physicists, using reasoning based on the positions of Laplace and Poisson, which had already been partly abandoned or at least criticized by mathematicians, opened up a new area of application for the probability calculus in the kinetic theory of gases. This rapidly developed into statistical mechanics.[58] This is the background explaining Hilbert's formulation of his above-mentioned sixth problem. Hilbert had seen through the vacuity and shallowness of the Laplacian tradition, the stagnation and redundancy it manifested in so many textbooks. He considered physics to be one, if not *the* new area of application for the probability calculus, to which he wanted to give new life. The linking up of his program with the scientific tradition that had continued after the reduction of the probability calculus to the error calculus finally produced this renaissance in our century.[59]

Notes

1. Laplace's main contributions to probability theory in book form are *Théorie analytique des probabilités* (Paris: 1812), 2nd ed. (Paris: 1814), 3rd ed. (Paris: 1820) with later editions, and *Essai philosophique sur les probabilités* (Paris: 1814), 5th ed. (Paris: 1825), also as introduction to 2nd and subsequent editions of the *Théorie analytique*.

2. An example of such historical accounts is Theodore Merz, *A History of European Thought in the Nineteenth Century*, 4 vols. (London: William Blackwood and Sons, 1904–1912). For Laplace, cf. vol. I, pp. 120–125.

3. Matthieu Paul Hermann Laurent, *Traité du Calcul des Probabilités* (Paris: 1873), pp. X–XII.

4. Sylvestre François Lacroix, *Traité élémentaire du calcul des probabilités* (Paris: 1816); Antoine-Augustin Cournot, *Exposition de la théorie des chances et des probabilités* (Paris: 1843).

5. Simon-Denis Poisson, *Recherches sur la probabilité des jugements en matière criminelle et en matière civile* (Paris: 1837).

6. Cournot, *Exposition*, (note 4).

7. Paul Mansion, "Sur la portée objective du calcul des probabilités," *Bulletin de l'Académie Royale de Belgique*, Classe des science, 4 (1903), 1235–1294.

8. Joseph Louis François Bertrand, *Calcul des probabilités* (Paris: 1888/9), 2nd ed. (Paris: 1897).

9. Anton Meyer and F. Folie, *Cours de calcul des probabilités fait à l'université de Liège de 1849 à 1857* (Bruxelles: 1874).

10. Anton Meyer, *Vorlesungen über Wahrscheinlichkeits-rechnung*, ed. E. Czuber (Leipzig: 1879).

11. Emmanuel-Joseph Boudin, *Leçons de calcul des probabilités faites à l'université de Gand de 1846 à 1890*, ed. Paul Mansion (Ghent: 1916).

12. Emil Heinrich du Bois-Reymond, *Über die Grenzen des Naturerkennens* (Leipzig: 1872), and *Die sieben Welträtsel* (Berlin: 1880).

13. David Hilbert, "Mathematische Probleme," *Göttinger Nachrichten* (1900), 253–257, and *Archiv für Mathematik und Physik*, 1^3 (1901), 213–237.

14. David Hilbert, "Der Zahlbericht," *Jahresbericht der DMV* (1897).

15. Emanuel Czuber, "Die Entwicklung der Wahrscheinlichkeitsrechnung und ihrer Anwendungen," *Jahresbericht der DMV*, 7.2 (1899), IV.

16. Cf. Ivo Schneider, "Rudolph Clausius' Beitrag zur Einführung wahrscheinlichkeitstheoretischer Methoden in die Physik der Gase nach 1856," *Archive for History of Exact Sciences*, 14 (1975), 237–261.

17. Richard von Mises, "Grundlagen der Wahrscheinlichkeitsrechnung," *Mathematische Zeitschrift*, 5 (1919), 52–99.

18. A. N. Kolmogorov, *Grundbegriffe der Wahrscheinlichkeitsrechnung* (Berlin: Julius Springer, 1933).

19. Simon-Denis Poisson, *Lehrbuch der Wahrscheinlichkeitsrechnung und deren wichtigsten Anwendungen*, ed. in German by C. H. Schnuse (Braunschweig: 1841).

20. Charles Gouraud, *Histoire du Calcul des Probabilités* (Paris: 1848).

21. Gérard Jorland kindly gave me this information and also read this extensive work, which was only awarded the third prize.

22. In particular, Gouraud mentions the following works: M. J. Condorcet, *Essai sur l'application de l'analyse à la probabilité des décisions rendues à la pluralité des voix* (Paris: 1785), as well as *Eléments du calcul des probabilités* (Paris: 1805); regarding the central idea of a "Mathématique sociale" of Condorcet, Gouraud refers to Condorcet's inaugural speech at the *Académie française* in 1782 on "De l'union des sciences physiques et des sciences morales"; as to Laplace: Gouraud mentions Laplace's main works (note 1), i.e., the *Théorie analytique* of 1812 and the *Essai* of 1814.

23. Cf. Lorraine Jennifer Daston, "The Reasonable Calculus: Classical Probability Theory, 1650–1840" (doctoral thesis, Harvard University, 1979), p. 415, and Poisson, *Recherches* (Paris: 1837), §§135–137, pp. 371–381.

24. Poisson, *Recherches*, p. VII (note 5).

25. Gouraud, *Histoire*, p. 132 (note 20).

26. Gouraud, *Histoire*, p. 133 (note 20).

27. S.-D. Poisson, "Recherches sur la probabilité des jugements, principalement en matière criminelle," *Comptes rendus*, 1 (1835), 473–494.

28. Gouraud, *Histoire*, pp. 134, 135 (note 20).

29. For the entire discussion of Poisson's application of probability calculus on decision-making processes including the support of Navier, see Daston, *The Reasonable Calculus*, pp. 421–429 (note 23).

30. Gouraud, *Histoire*, p. 135 (note 20).

31. Pierre Paul Royer-Collard, *Discours prononcés dans la séance publique tenue par l'Académie française pour la réception de M. Royer-Collard, le 13 november 1827* (Paris: 1827).

32. Jean Francois La Harpe, *Lycée au cours de littérature ancienne et moderne* (Paris: 1814).

33. Augustin Louis Cauchy, *Cours d'analyse de l'École Royale Polytechnique* (Paris: 1821), pp. I–V.

34. Laplace, *Théorie analytique*, 3rd ed. (Paris: 1820), p. LXXVIII (note 1); Ivor Grattan Guinness, *Development of the Foundations of Mathematical Analysis from Euler to Riemann* (Cambridge Ma: MIT Press, 1970), p. 49, finds that the last paragraph of Cauchy's introduction to the *Cours d'analyse* is understandable only in the light of Laplace's claims regarding probability theory.

35. Laplace, *Théorie analytique*, 3rd ed. (Paris: 1820), p. 326 (note 1).

36. Irenée Jules Bienaymé, "Considérations à l'appui de la découverte de Laplace sur la loi de probabilité dans la méthode des moindres carrés," *Comptes Rendus*, 37 (1853), 309–324, on p. 309.

37. Bienaymé, "Considérations," p. 311 (note 36).

38. Augustin Louis Cauchy, "Mémoire sur l'interpolation," lithographed by the *Académie des Sciences* in 1835 and published in *Journal des mathematiques pures et appliquées*, 2 (1837), 193–205, and in *Oeuvres complètes d' Augustin Cauchy*, 2nd series, vol. II (Paris: Gauthier-Villars, 1958), pp. 5–17.

39. Augustin Louis Cauchy, "Mémoire sur l'evaluation d'inconnues déterminées par un grand nombre d'équations approximatives du premier degré," *Comptes Rendus*, 36 (1853), 1114–1122.

40. *Comptes Rendus*, 37 (1853), 5–13.

41. C. C. Heyde and E. Seneta, *I. J. Bienaymé: Statistical Theory Anticipated* (New York, Heidelberg, Berlin: Springer-Verlag, 1977), pp. 71–96.

42. Augustin Louis Cauchy, "Mémoire sur les coefficients limitateurs ou restrictateurs," *Comptes Rendus*, 37 (1853), 150–162, on p. 162.

43. Pierre Simon Laplace, *Exposition du système du monde*, 2 vols. (Paris: 1796 (an IV)), 5th ed. (Paris: 1824).

44. Cf. article on Laplace, *Dictionary of Scientific Biography*, 16 vols., vol. XV (New York: Scribner, 1978), p. 374.

45. See English translation by F. W. Truscott and F. L. Emery, *A Philosophical Essay on Probability* (New York: Dover, 1951), p. 196. There are three different German editions of the *Essai*, which, at least in part, mirror the influence and impact of the *Essai* outside France. The first was done by J. W. Tönnies and appeared in Heidelberg in 1819. The second, by Norbert Schwaiger, was published in Leipzig in 1886. The latter was motivated by an increased interest on the part of German philosophers in questions concerning probability theory. A third translation was produced by Richard von Mises in 1932; it appeared as volume 233 of the series *Ostwalds Klassiker* in Leipzig.

46. See Ivo Schneider, "Christian Heinrich Schnuse als Übersetzer mathematischer, naturwissenschaftlicher und technischer Literatur," *Aus dem Antiquariat*, 6 (1982), A205–A221.

47. Augustus de Morgan, "Théorie analytique des probabilités. Par M. Le Marquis de Laplace, etc. 3ème édition, Paris, 1820," *The Dublin Review*, 2 (1837), 338–354, and *The Dublin Review*, 3 (1837), 237–248.

48. Augustus de Morgan, *An Essay on Probabilities, and on Their Application to Life Contingencies and Insurance Offices* (London: 1838).

49. M. Merriman, "A List of the Writings Relating to the Method of Least Squares, with Historical and Critical Notes," *Trans. Connecticut Acad. Arts Sci.*, 4 (1877), 151–232.

50. For all biographical details, cf. C. C. Heyde and E. Seneta, *Bienaymé*, pp. 5–18 (note 41).

51. *Journal des mathématiques pures et appliquées*, 17 (1852), 32.

52. A.L. Cauchy, "Mémoire" (note 38); Augustin Louis Cauchy, "Sur la nouvelle méthode d'interpolation comparée à la méthode des moindres carrés," *Comptes Rendus*, 37 (1853), 100–109, on p. 107 and 109.

53. Bienaymé, "Considérations," p. 310 (note 36).

54. Bienaymé "Considérations," p. 312.

55. Bienaymé "Considérations," p. 323.

56. From the 1850s on one can find in the French versions of Chebyshev's relevant publications, which were translated by different people including Bienaymé, alternately "quantité," "variable," and in one case "quantité variable." Lyapunov, Chebyshev's student, used in his demonstrations of the central limit theorem in 1900 and 1901 the term "variable" exclusively.

57. According to a remark of Eugen Slutsky, "Ueber stochastische Asymptoten und Grenzwerte" *Metron* 5 (1925), 3–89, on p. 6, the term "random variable" was introduced by P. A. Nekrasov, "New Foundations of the Doctrine of Probabilities of Sums and Mean Values" (in Russian), *Matematicesky Zbornik*, 21/22 (1901), 9, and propagated by A. A. Chuprov.

58. This development is covered in the classical article of P. and T. Ehrenfest, "Begriffliche Grundlagen der statistischen Auffassung von Mechanik," *Enzyklopädie der mathematischen Wissenschaften*, vol. IV, 32 (Leipzig: B. G. Teubner, 1911). The most extensive modern account is Stephen G. Brush, *The Kind of Motion We Call Heat*, 2 vols. (Amsterdam, New York, Oxford: Reidel, 1976). Compare also Ivo Schneider (note 16).

59. A number of other developments also contributed to this regeneration of probability calculus in the twentieth century that, in terms of this paper, which concentrates on the tradition of Laplace, are not or hardly significant, e.g., the development of measure theory and the interpretation of probability as a measure, which was a precondition of the incorporation of the so-called geometric probabilities; and, further, the research into the so-called 0–1 laws, which was carried out in Scandinavia, France (Borel), and Italy (Cantelli).

9 Emile Borel as a Probabilist

Eberhard Knobloch

In addition to many textbooks, Borel published more than fifty papers between 1905 and 1950 on the calculus of probability. They were mainly motivated or influenced by Poincaré, Bertrand, Reichenbach, and Keynes. However, he took for the most part an opposed view because of his realistic attitude toward mathematics. He stressed the important and practical value of probability theory. He emphasized the applications to the different sociological, biological, physical, and mathematical sciences. He preferred to elucidate these applications instead of looking for an axiomatization of probability theory. Its essential peculiarities were for him unpredictability, indeterminism, and discontinuity. Nevertheless, he was interested in a clarification of the probability concept.

Thus this case study is divided into three sections. The first section deals with Borel's notion of probability. He rejected Reichenbach's frequency theory. On the contrary, he concluded that the notion of the probability of a single case is the foundation of the probability calculus. He paid special attention to probabilities of judgments of value and to the foundation of a rational decision theory, two years before Ramsey.

The second section deals with Borel's attitude toward continuity and discontinuity. He wanted to deepen the study of the relations between the continuum and discontinuity. Thus he was forced to discuss the relations between measure theory and the theory of denumerable probabilities.

The third section is concerned with Borel's treatment of the determinism problem and of the irreversibility problem of statistical mechanics. He distinguished between two pairs of determinism and underlined the indeterminacy of physical data.

Preliminary Remarks on the Prehistory

Whenever we speak about Emile Borel's studies on probability, we are concerned, roughly speaking, with the period between 1905 and 1950. Borel is well known as the unchallenged founder of mathematical measure theory. But it is apparently less well known that he was a very important, prolific, and influential scholar in probability theory and probabilistic thinking. There are a few publications discussing this subject. Either the authors deal with very special questions,[1] or in my opinion their statements should be modified,[2] or they do not know the more recent research literature[3] and neglect many interesting papers of Borel. Thus we at least need a reinterpretation of Borel's contributions. It seems to me to be all the more worthwhile to prepare the following case study, because he unified the mathematical and the philosophical aspects of the underlying theory.

A few years before Borel began to write on probability, three major works had appeared, namely, Henri Poincaré's *Calcul des probabilités* in 1896, Joseph Bertrand's *Calcul des probabilités* in 1899, and Poincaré's *Science et hypothèse* in 1902.[4] They most of all stimulated Borel to deal with the subject. Only a few other authors or conceptions concerned with this theory were to preoccupy Borel to a similar degree—for example, Reichenbach's frequency interpretation of probability and Keynes's subjective probability.[5]

Around 1900 the position of probability theory, its methods, results, and foundations were still very controversial. Bertrand was so much impressed by the contradictions in geometrical probability that he wished to exclude all examples in which the number of alternatives is infinite. In particular, Bertrand argued that Maxwell had misused the principle of composed probabilities with respect to gas theory, that the intervention of chance in the formation of the universe was not acceptable, and that the application of the calculus of probability to the study of observational errors depended on a fiction, which should not be reified.[6] He declared that the question whether an event happened by chance or by causality was too inexact. He cited John Stuart Mill's view that the application of probability theory to juridical decisions is the scandal of mathematics.[7]

Poincaré said in 1896, "One can hardly give a satisfying definition of probability. One usually says: the probability of an event is the relation of the number of favorable cases for the event to the number of possible cases."[8] That means that he gave the classic definition of probability, though he was aware of the difficulties that adhere to it. He formulated in his preface to *Science et hypothèse* the need for a better foundation of probability theory and said, "In order to calculate any probability and to give a sense to this calculation we have to admit a hypothesis or a convention as the beginning that always introduces a certain arbitrariness."[9] He quoted the principle of sufficient reason as an example that is mostly manifested in the belief in continuity. It is difficult to justify this belief by an irrefutable reasoning. But without it every science is impossible.

Keynes finally remarked even in 1921, "There is still about it for scientists a smack of astrology, of alchemy."[10] He explicitly cited Bertrand's arguments.

What did Borel do? He studied these authors intensively, but he drew quite different conclusions from those of his predecessors. He underlined from the very beginning the overwhelming importance of applications, especially to physical problems. They are far more important than the principles of the theory. He objected to a normatively fixed selection of questions to which we are allowed to apply probability theory. He denied continuity and the continuum. In 1934 he said, "We can deepen and improve the calculus of probability only by simultaneously studying all of its applications, that is to say, very different sciences. Thus the unity of sciences will be reestablished in an unexpected manner. The nineteenth century looked for this unity in the field of mechanics and of differential equations. There can be no doubt that the twentieth century will find it by studying the laws of chance."[11]

Therefore I would like to discuss the following three problems: the notion of probability, continuity versus discontinuity, and determinism and physics.

1   The Notion of Probability

Hilbert's sixth problem of his celebrated list of problems dating from 1900 read as follows:[12] "We have to treat by means of axioms those physical sciences in which mathematics already plays today an important part. These are above all probability theory and mechanics."

Borel took the opposite view. The key words of his philosophy of mathematics were realities, practical value, applications. His attitude toward probability theory and especially his own contributions to this field of knowledge totally depended on his realistic conception of mathematics. This was his own expression used in 1935 when he wrote on *Documents sur la psychologie de l'invention dans le domaine de la science*.[13] He separated the mathematical objects that can be effectively defined from those the existence of which is purely hypothetical. In this respect he was an empiricist in the sense of Lebesgue, who made definability the touchstone for his empiricist philosophy of mathematics.[14]

He emphasized that the role of scholars consists in creating a science. Thus he was just as little interested in principles as Lebesgue.[15] For in 1907 he said, "The knowledge of principles is not necessary for the discovery of analytic facts and of their underlying laws. This discovery is the proper work of mathematicians."

When he discussed the paradoxes of set theory in the following years, he repeated that he wanted to avoid metaphysical or purely logical considerations, that he preferred observable realities.[16] Thus it is not astonishing that he uttered very similar opinions with regard to the foundations of probability theory. In 1924 he mentioned when reviewing Keynes's *Treatise of probability* that he was not interested in the chapter where the fundamental theorems of probability theory were reduced to logical formulas. He had never acquired a taste for such exercises and he believed that most of the mathematicians were in the same situation. It was hard enough to overcome the real difficulties of mathematical research. It was superfluous to create artificial difficulties and to introduce a hieroglyphic symbolism that had never produced a mathematical discovery proper.[17] "It would be more useful before studying its principles to elucidate the theory in its unity and to spell out all of its applications, which has never been done up to now. For the applications are the true realities. The realities are insurance premiums, samples obtained by biologists and agronomists, phenomena observed and predicted by physicists." I am inclined to call him a "panprobabilist."

Therefore he criticized Keynes, because Keynes did not say anything about the applications of the calculus of probability to the physical sciences. Keynes did not mention Maxwell, Gibbs, or Boltzmann. Borel added, "If these results can be deduced from the elementary principles of the calculus of probability ... then this calculus will be, like mechanics, founded on a solid basis, and one will be able to discuss its principles without embarrassment."

He believed that the constant contact with nature had preserved mathematical analysis from becoming a pure symbolism.[18] Already in 1925, he began to write, with the help of collaborators, the missing comprehensive presentation of probability theory and its applications. His *Traité du calcul des probabilités et de ses applications* consists of four volumes in eighteen fascicles, published between 1925 and 1939. He rightly foresaw that this publication would be a landmark in the history of science. Probability theory had ceased to be only a chapter of analysis. It had become an autonomous science and played an important part in the development of all experimental and observation sciences.[19]

In 1921, Borel said, analyzing his scientific work, that he had published the "preface" of his treatise already in 1914 when completing his book *Le hasard*.[20] There he gave Laplace's classic definition of probability and added that its critique is equally classic.[21] But he was not inclined to discuss the underlying induction problem. He simply stated that men were not able to live without relying on many syllogisms, though they had not examined all of their cases. Thus he remarked, "We do not return to the definition of probability, which is reduced to the definition of equal probability in the case of discrete probabilities. We postulate this equal probability at least as a limit case." And later on: "One may give a purely abstract definition of probability as some authors have proposed in the same way as one defines the circumference in geometry or mass in rational mechanics. One may deduce absolutely rigorous logical consequences from this definition. But whenever you want to apply these consequences to any real phenomenon you must substitute the concrete probability of a real phenomenon for the abstract probability. The incertitude that adheres to every concrete measure comes back again." It should be mentioned that Stegmüller uttered very similar ideas without naming Borel.[22]

As it is impossible to find a common measure between probability and certitude, we must try to evaluate the probability by means of the probability itself, that is to say, by means of a quantity of the same nature. Thus the theory of probability will be the set of all mathematical procedures that allow one to define by a simple manner a probability that is equivalent to a probability given in a complicate manner.[23]

Already in 1909 Borel had published his famous paper on denumerable probabilities.[24] It was not his intention to give a mathematical foundation of probability theory in this paper. Therefore von Mises's criticism is totally misleading. Von Mises said in 1919 that Borel's attempt had not passed beyond formalism.[25] Reichenbach numbered Borel among those authors who had introduced the concept of probability by a formal conception,[26] for Reichenbach distinguished between a formal and an interpreted theory of probability. The formal conception introduces the concept of probability by the method of implicit definitions and uses no properties of the concept other than those expressed in a set of formal relations placed as axioms at the beginning of the theory, leaving open various possibilities for its interpretation.

Indeed, in 1925 Borel began the first fascicle of his *Traité* in a purely abstract manner. He assigned a number p between 0 and 1 to each event that can only be favorable or not favorable. But one should take into account that this first fascicle was entitled *Principes et formules classiques du calcul des probabilités*. It does not suffice to refer only to this volume. Borel continued the discussion of axiomatization and the application of mathematics in the last fascicle, entitled *Valeur pratique et philosophie des probabilités*. Hilbert is not mentioned by name. Borel admitted this time that the axiomatization of the mathematical sciences had cleared away some philosophical difficulties and had made possible some progress in the mathematical sciences. But axiomatization presupposes a long development. If people had not studied geometry without axioms for many centuries, nobody would have had the idea to call things by arbitrary names. In mathematics there is a concrete truth as well as an abstract truth. This difference depends on the difference between the branches of mathematics, between arithmetic or algebra and geometry or mechanics. The practical certainty of physical or chemical experiences is never absolute. The calculus of probability is a science that is analogous to physics, geometry, mechanics, not to arithmetic or algebra.[27] It does not lead to any absolute certainty. Thus, in 1939, Borel replaced Hilbert's or von Mises's identification of the calculus of probability with a physical science by an analogy.[28] He mentioned that an axiomatic theory for this calculus had been proposed, as for geometry, alluding to Kolmogorov's 1933 publication.[29] He conceded the possibility of developing the calculus of probability as a purely mathematical science without any relation to reality, comparable to n-dimensional geometry. But all practical difficulties come back if this theoretical science is applied to an arbitrary real phenomenon. This was apparently the same opinion he had expressed in *Le hasard*.

Von Mises's collective seemed to be the most complete and most interesting attempt to establish a connection between purely theoretical probabilities and their applications. But there was an essential objection against the theory of collectives and every similar theory: the human mind is not able to imitate chance perfectly, that is to say, to substitute any rational mechanism for the empirical method. This method consists in effecting an indefinite series of repeated trials. We have to introduce the notion of the probability of a single case whenever we want to leave the axiomatic field in order to apply the probability to real phenomena.

Therefore Borel felt bound to comment upon Reichenbach's theses, which had appeared five years ago. Reichenbach believed that the frequency interpretation of probability could be carried through for all uses of the term "probable." He regarded the statement about the probability of the single case not as having a meaning of its own, but as representing an elliptic mode of speech: "In order to acquire meaning the statement must be translated into a statement about a frequency in a sequence of repeated occurrences."[30] Thus a fictitious meaning is given to such a statement, constructed by a transfer of meaning from the general to the particular case. He defined a posit as a statement that we treat as true, although the truth value is unknown. His method of positing makes it possible to use probability statements for decisions regarding single cases, a method that plays an important role in all practical applications.

In order to reject subjective probability of every kind, Reichenbach said in 1949, when editing the English translation of his inquiry, "There is no need for a concept of probability which is not reducible to frequency notion. Whoever wishes to reserve the right of using private meanings of a nonverifiable pattern, or of a structure useless for prediction may do so."[31]

Borel disagreed with Reichenbach in many ways, and denied that a purely empiricist solution can be given where the theory of the application of probability concepts to physical reality is concerned. At first he admitted that there are cases where Reichenbach's theory can be partly applied, but nevertheless needs to be completed. This completion consists in calculating the probability of a single case not by means of the observed frequency in a certain sequence, but rather by means of a combination of observed frequencies in many sequences.

His example was the death of an individual during a given period.[32] The mortality tables give the mean value of that probability according to the sex and age. But one obtains far more precise values if one examines successively all possible cases of death and looks for their probabilities for the given individual. Thus one regards this individual as belonging simultaneously to numerous classes that can have only this individual in common: his case would be really unique in this way.

Borel looked upon probabilities of judgments of value as the most interesting type of probabilities of a single case. It does not matter whether material objects or individuals or groups of individuals are concerned: for example, the estimate of the weight of a suitcase, or the opinion concerning which of two players will win at a sporting event. The difficulty consists in assigning numerical values to such probabilities. Borel discusses several possibilities of verifying experimentally the exactness of certain evaluations. Certainly the verification a posteriori is quite different from a frequency stated a priori.

For example, we can compare the evaluations of two persons A and B who considered isolated cases of the same nature for which they had a particular competence. But there are several practical and theoretical objections against this theory of verification. The most serious objection is a theoretical one: all judgments rely on studies of frequencies.

Borel proved that there are mental operations of special nature that are totally different from a simple observation of frequencies. His proof ran as follows: Let us assume that we are convinced by long experience that the evaluations of A are better than those of many other persons B, C, D, and so on. This confidence can be translated by a coefficient of probability, for example, by .99. This coefficient is doubtless a coefficient of frequency. But if A attributes the probability .52 to a single case, we affirm with the probability .99 that this evaluation of A is right. The utilization of the frequency concerns the coefficient .99. It does not concern the coefficient .52, which does not correspond with any observed frequency. If A makes use of his past experience, he compares the observed case with analogous, but nevertheless different, cases. He often corrects the probabilities that he had attributed to these analogous cases in a significant way, in order to take into account the differences.

Borel rejected the assertion that the a priori calculation of the probability of a single case can be justified by the possibility of repeating the same experience and of studying the statistical results. Indeed, Borel demonstrated with examples of games of chance that it is often impossible to repeat the same experiences. He gave the example of a gambler who calculates the probability that his partner can be deceived by a feint. Thus, Borel's conclusion was that the notion of the probability of a single case is the foundation of the calculus of probability.

He believed, as did Reichenbach, that a proposition is only of interest in practice for men insofar as it influences their actions. Therefore it must be possible to translate a judgment of probability into a bet. The global success of a certain number of these bets is the only criterion for the value of the judgment. Thus, the probability of a single event is subjectively defined by the conditions of a bet that one is ready to make on the occurrence of an event. These conditions must satisfy obvious conditions in the elementary case (for example, $p + q = 1$), and more complicated ones in the other cases. These conditions are now called "coherence" conditions.

Provided the probability of a single case is determined for an individual, Borel defines the objective probabilities as probabilities whose value is the same for a certain number of individuals who are equally informed about the conditions of the uncertain event. The theory of repeated trials shows that the limit of the frequency is equal to the probability. But this leads to a verification, not a definition.

Thus, Borel identified degrees of belief with a specific kind of behavior. I would like to emphasize the fact that he discussed the foundations of rational decision theory in this way not only in 1939, as was mentioned by Carnap, but already in 1924.[33] Therefore he thought of studying rational betting behavior two years before Ramsey did. Borel and Ramsey thought that the only theoretically sound way of measuring a person's degree of belief is by examining his overt behavior.[34]

Nevertheless, in 1973 Stegmüller erroneously maintained that Ramsey was the first to sketch this idea, and cited Bruno de Finetti as a second, though independent, inventor of this betting theory.[35] It is worth noting that de Finetti gave his famous lectures "Foresight: its logical laws, its subjective sources" in 1935 at the Institute Henri Poincaré, which had been founded partly at Borel's instigation in 1928.[36]

2 Continuity versus Discontinuity

In 1934 Borel remarked that during the two centuries that followed the discovery of the differential and integral calculus, mathematics was essentially the science of the continuum. The mathematicians tried to introduce the methods of the continuum into all sciences, number theory and mathematical physics included.[37] He himself, however, had tried for half a century to liberate physics from the hegemony of the continuum. He was gratified to observe[38] that the physicists replaced the theory of the continuum increasingly by discontinuous statistics, especially in molecular physics.[39] He expected that the introduction of such statistical methods would lead eventually to more delicate methods in the probability calculus.

From the mathematical viewpoint, this calculus was his main instrument. Indeed, already in 1914 he had declared that quantum theory, when studying the distribution of energy, replaces a problem concerning probabilities with a problem of discontinuous probabilities from the viewpoint of probability theory. He believed this to be the justification for rejecting the continuum. Even if the systematic introduction of discontinuity did not remove all difficulties, it at least diminished them. He even asked himself whether it might be possible to transform the problems of geometrical probabilities by leaving out any physical interpretation in such a way that the solution would be continuous, but nevertheless conserved the essential characteristics of discontinuous solutions.[40]

One has to go back to his set-theoretical papers dating from 1908 and 1909 in order to understand this research program. Every element of the continuum can be defined, but the set of all the effectively defined elements will always be denumerable. That we can never exhaust it in this way is only a negative certainty. Thus, the continuum is a purely negative notion.

It is obvious that Borel's position had changed since 1898, but only by becoming still more radical. For in 1898, he had only said that the notion of nondenumerable sets is "above all" negative.[41] Every term of a mathematical problem must be defined explicitly. If the explicit definition needs infinitely many words, it does not belong to the domain of mathematics. Borel considered, for example, the decimal numbers, which are defined by a precise method and without any possible ambiguity by means of a finite number of words. The set of these numbers is denumerable, because the definitions are denumerable.

Borel denied the existence of sets that are not denumerable. All the points that will be needed at any time in mathematical considerations will be defined by means of a finite number of words. They constitute the practical continuum that the mathematician uses.[42] The distinction between denumerable and nondenumerable sets seemed to him to be without any practical value, because all sets that can be considered will be denumerable.

But they are not all effectively enumerable. According to Borel, a set was effectively enumerable if one could state by means of a finite number of words a definite process for attributing unambiguously a unique natural number as the rank for each of its elements. We have to stress the word "effectively." For in 1926, Borel identified denumerable and enumerable sets.[43] From a practical point of view we have to distinguish between effectively enumerable sets and those that are not effectively enumerable.

This new notion is less metaphysical; that is to say, it is exclusively founded on observable realities.[44] The pretended paradoxes of set theory are a corollary of the fact that the following proposition was believed to be evident: Every denumerable set is effectively enumerable.

In 1909, Borel studied the relations between the mathematical continuum and the physical continuum.[45] He cited his paper on the paradoxes of set theory, saying that the practical mathematical continuum included in addition to the rational numbers those irrational numbers whose definition is simple. We must be satisfied with five or six decimal places for many experimental problems. Thus the arithmetical construction of the continuum is useless. The theoretical, nondenumerable continuum is a metaphysical conception we do not have to deal with.

Poincaré had defined the physical continuum by these three relations: $A = B$, $B = C, A < C$, A, B, C being weights. He had deduced from the implied contradiction the necessity of inventing the mathematical continuum.[46]

Borel rejected this statement. First, we are not allowed to regard two quantities A, B as empirically equal at a certain moment if we cannot detect any difference between them by means of any experimental investigation method that is available at that time. We can define the difference between the physical continuum and the mathematical continuum in the following way: A certain minimal difference is necessary in order to distinguish two neighboring elements, because the experiment permits always only a limited approximation. The minimal difference depends on the experimental conditions, and is not an absolute constant.

Second, Poincaré's reasoning does not need the mathematical continuum. It needs only a closed, dense subset of the continuum—for example, the set of the rational numbers.

Borel picked up the threads of these ideas in his landmark paper on denumerable probabilities.[47] It is worth quoting the opening passage at length, for it reveals his main ideas concerning the relation between probability theory and the continuity-discontinuity problem:

One generally distinguishes, in probability problems, two principal categories, according to whether the number of possible cases is finite or infinite: the first category constitutes what one calls discontinuous probabilities, or probabilities in a discontinuous domain, while the second category comprises continuous probabilities or geometric probabilities. Such a classification appears incomplete when one refers back to the results acquired in the theory of sets; between the cardinality of the continuum of finite sets and the cardinality of the continuum stands the cardinality of denumerable sets; I propose to show briefly the interest which is attached to questions of probability in whose statement such sets intervene; I will call them, for short, denumerable probabilities.

Before defining more precisely denumerable probability, I wish to indicate in a few words the reasons to study it. Principal among these reasons is the importance of the notion of denumerable sets; this importance is not contested by any mathematician; but it seems to me to be greater still than one believes.

Many analysts, indeed, put in the first rank the idea of the continuum; it is this concept which intervenes more or less explicitly in their reasoning. I have indicated recently how this notion of the continuum, considered as having a cardinality greater than that of the denumerable, seems to me to be a purely negative notion.

The cardinality of denumerable sets alone being what we may know in a positive manner, the latter alone intervenes effectively in our reasonings. It is clear, indeed, that the set of analytic elements that can be actually defined and considered can be only a denumerable set; I believe that this point of view will prevail more and more every day among mathematicians and that the continuum will prove to have been a transitory instrument, whose present-day utility is not negligible (we will supply examples at once), but it will come to be regarded only as a means of studying denumerable sets, which constitute the sole reality that we are capable of attaining.

In his notice about his own scientific works he wrote in 1912, "The distinction between continuous and discontinuous probabilities is classic. It corresponds to the distinction between finite groups (Galois groups) and continuous groups (Lie groups). But the intermediate stage, which corresponds with the discontinuous or Fuchsian groups, was not considered at all."[48] In order to fill this gap he founded the theory of denumerable probabilities.

He elaborated a research program in saying "There can be no doubt that we are thus able to deepen the study of the relations between the continuum and discontinuity, between the geometric peculiarities of space and the arithmetical peculiarities of numbers."

Roughly speaking, denumerable probabilities relate to questions of probability that are concerned with denumerable sets. Already before Borel scholars had considered infinite sequences of trials. But the classic or Bernoullian point of view discussed the asymptotic characteristics of the probability of an event that depended only on a finite number of trials.[49] Borel considered events that depended on a denumerable set of trials. By this route he found the Zero-One law. Unfortunately, the cardinality of Borel's sample spaces was that of the continuum rather than that of denumerable sets.

He regarded countable independence as the essential probabilistic ingredient of his new theory. Indeed, five years later, he explicitly said in *Le hasard*, "It is useless to continue the study of probability theory unless you have a precise understanding of the notion of independence."[50] Barone and Novikoff maintained that Borel had given a misleading emphasis to countable independence as a basic concept without relying on measure theory. According to them he had not seen the full analogy of probability with measure, except when the problem permitted a geometric interpretation. But we have seen that Borel pursued other aims. We should keep in mind that Kolmogorov said some twenty years later in his famous book dedicated to the axiomatic foundation of probability theory, "Historically, the independence of trials and random variables is the mathematical notion that gives the calculus of probability its peculiar characteristics. At least the origin of the special set of problems lies in the notion of independence."[51]

Arithmetic problems like dyadic or decimal expansions of numbers or continued fractions were and remained Borel's preferred area of application for the new theory. He came back again and again to these examples, especially in *Le hasard* and in the *Traité*. He did this all the more willingly, because he could demonstrate in this way the great range of applications of probability theory. Arithmetic was the prototype of discontinuity. Only very few scholars expected that this field of pure mathematics was appropriate to the statements of probability theory, that is to say, to approximative instead of precise statements. But such an application of probability theory would not surprise all those who had understood the essentially relative or even subjective character of probabilities.

In particular he studied normal numbers. He called a number written in the decimal system "normal" if the frequency of every decimal digit has the limit one-tenth. Borel took these numbers as an example in order to clarify his view of the relation between probability and measure theory. In 1926, he explained that the notions of measure and probability can be reduced to one another in most of the arithmetical problems that he studied. But what is more, he pointed to the equivalence of these two notions in this kind of consideration.[52] In 1939, he discussed the reasonable argument that certain theories of denumerable probabilities make double use of measure theory and are therefore superfluous.[53] That is why he demonstrated by means of the normal numbers that we come to very different results according to the point of view we have:

Let A' be the set of normal numbers written in the number system that uses only the digits 0, 1, 2, 3, 4. Let C' be the set of normal numbers written in the number system that uses only the digits 0, 1, 2, 3, 4, 5. If we sum up two such numbers $a' + c' = x$, $a' \in A'$, $c' \in C'$, the probabilities of the digits 0, 1, 2, ..., 9 of x are, respectively,

$$\frac{1}{30}, \frac{2}{30}, \frac{3}{30}, \frac{4}{30}, \frac{5}{30}, \frac{5}{30}, \frac{4}{30}, \frac{3}{30}, \frac{2}{30}, \frac{1}{30}.$$

Thus the sum x is not a normal number. Nevertheless we can deduce the following corollary:

Suppose it is possible to effect for A' and C' an infinity of arbitrary choices whose cardinality is that of the continuum. Suppose we combine these two infinities. Then we shall obtain sums $a' + c'$ whose frequencies can be very different and in particular might have the character of normal numbers.

But if we effect only a denumerable infinity of arbitrary choices, the obtained sums will not have the character of normal numbers.

These different results justify the existence of both theories. Borel even added that we can generalize the differences that hold between the standpoint of continuous probabilities, which is equivalent to measure theory, and the method of random successive selections of decimal digits, which leads to the theory of denumerable probabilities. But he remarked, "We need not be astonished at these different conclusions, though they present formal analogies in their wording. Each of these problems is interesting according to the point of view we have in mind." There is no reason to maintain that one of these problems, and therefore one of these solutions, must be preferred to the other.[54]

Unfortunately, Barone and Novikoff offered no comment on these remarks when they criticized Borel's paper, published in 1909, for leaving the readers largely unsure as to the relation between probability theory and measure theory.[54]

3 Determinism and Physics

In 1934, as president of the Institute of France, Borel delivered the year's opening address of the Academy of Sciences session. He criticized Laplace's determinism heavily without naming the French mathematician: the calculus of probability is not only a set of analytic methods of numerical results that allows us to resolve many practical problems. It is also and more particularly a scientific philosophy that at least apparently, opposes the conception of anlytical mechanism. This mechanism explains the universe by a system of differential equations whose solutions are rigorously determined in the past and in the future by the rigorous knowledge of the initial conditions. In fact, given the present state of science, these two conceptions do not absolutely oppose but rather mutually complete one another, sometimes at the price of certain contradictions. It is the task of tomorrow to eliminate these contradictions progressively, to reconcile the determinism of the laws of nature with those qualities imposed by probability theory such as unpredictability, indeterminism , and discontinuity.

Borel himself was very engaged for many years in fulfilling this task. He discussed the problem of determinism in several writings published during the period between

1914 and 1939.[55] He relied on Poincaré's statement that determinism is an indispensable postulate of scientific thinking, for the goal of science is prediction.[56] As soon as foresight is abandoned as impossible, one is beyond the limits of science. The scientist ceases to think and to act as a scientist. Therefore we must rigorously suppose the observed scientific phenomena as determined. Science is impossible without this hypothesis of determinism.

This conception of science is apparently necessary. But the question arises whether this necessity is absolute in the sense of mathematical truths or whether it contains a contingency, however small. Borel's solution to this problem consisted in distinguishing between two pairs of determinism, between the rigorous absolute or philosophical determinism,[57] and statistical, scientific determinism;[58] between global determinism and partial determinism or determinisms on different scales. The two different determinisms can be inconsistent with one another.

There is an important difference between mathematical and physical laws. Where we have to do with reality, there is no mathematical certitude. Physical laws are not absolutely rigorous, but rather statistical, approximative laws.[59] Thus the notion of scientific determinism cannot be separated from the idea of probability and chance. There is no incompatibility between the role of chance and the establishment of scientific laws. Borel very often cited the example of James Jeans: "It is only very improbable, but not impossible, that water freezes instead of evaporating when we put it on the fire."

There is doubtless a very close link between Borel's and Maxwell's statements, who made a similar distinction with regard to the second law of thermodynamics: "It is probably impossible to reduce the second law of thermodynamics to a form as axiomatic as that of the first law, for we have reason to believe, that though true, its truth is not of the same order as that of the first law .... The truth of the second law is therefore a statistical, not a mathematical, truth, for it depends on the fact that the bodies we deal with consist of millions of molecules, and that we never can get hold of single molecules."[61]

In 1914, Borel studied two consequences of the statistical explanations of physical phenomena:

1. These explanations show that the necessity of a global phenomenon is not incompatible with the freedom of partial phenomena.

2. These explanations give examples of such cases where the supposed absolute determinism of partial phenomena does not allow us to predict with absolute rigor the global phenomena.

Let us begin with the first consequence. Borel states that there is a similar tendency in different sciences: The study of single phenomena is replaced by the global statistical study of a set of numerous phenomena.[62] The detailed analysis of the determinism of the phenomena exceeds human possibilities. Some general properties can be studied, from which the statistical laws are precisely deduced. The rigorous determinism plays but a secondary role in the theoretically possible solution of a set of equations when these statistical laws are formulated. The laws remain the same even if the details of the phenomena are modified. Some of the billions of billions of molecules could satisfy laws totally different from those that are known to us, or could be free. The determinism of the observable phenomena would be not affected. The physical laws would not be modified. The hypothesis of free individuals does not influence the laws of statistics.

In 1920 Borel took radioactivity as an example in order to clarify these deliberations. Thus it is regrettable that van Brakel did not take any notice of Borel in his paper on the possible influence of the discovery of radioactive decay on the concept of physical probability.[63] According to Borel the invariance of the radioactive constant of a substance is one of the best established experimental laws. The radioactivity is an unexpected example of such natural phenomena that present an absolutely precise picture of the abstract probability, if the experimental laws are supposed to be rigorously correct. But the experimental facts do not allow us to decide between two extreme hypotheses:

a. The lifetime of every single atom is precisely determined by its own nature.

b. The lifetime is a priori the same for all atoms in every instant. Chance circumstances determine the explosion.

The experimental laws of radioactivity prove the rigor of the global determinism of radioactive phenomena. But we do not know whether this global determinism can be considered as the synthesis of a great number of partial determinisms (first hypothesis): The development of every atom is determined. Or whether the global determinism is the statistical resultant of phenomena that cannot be individually foreseen (second hypothesis): Even the notion of causality is questionable.

Let us now turn to Borel's discussion of the second consequence. The founders of the kinetic gas theory took as a starting point the determinism of molecular phenomena. Borel emphasizes that we have to take into consideration that the determinism on the molecular scale is claimed as an analogy from the human scale. Molecular determinism, however, does not lead by any means to determinism on the human scale.[64] On the contrary, it is compatible with the possibility that the phenomena on the human scale are not rigorously determined in the mathematical sense of the world.

We can study the movement of a single molecule only by using the equations that define the movements of the molecules of the whole universe. Therefore, we would have to write and integrate a real infinite number of equations. Thus the determinism of the global phenomenon can be comprehended in an abstract manner. But its detail cannot be foreseen. We can only foresee the most probable event. There can be no doubt that the degree of this probability would be universally identified with certainty, though this certainty does not have the absolute value that a deterministic philosopher must attribute to his determinism.

Certainly there is an abyss between a world where a part of this freedom exists no matter how small it is, and the world where this is not the case. We obtain in this way a determinism on our scale that is necessary in order to understand the world. Because of this determinism, our reason assumes a hidden but absolute determinism of the molecular phenomena.

Borel gives a concrete example for the probability that a noticeable variation occurs during a very short period in a space of any size, a variation from that phenomenon that is the most probable with respect to statistical mechanics. It is the often-cited miracle of the typing apes: One million apes pound at random one million typewriters ten hours daily for one year. The produced text contains an exact copy of all books of the richest libraries of the world.

Borel stated the facts in *Le hasard* in the following way:[65] We can remove the uncertainty in applying the calculus of probability, we can make use of what we may call the single law of chance: phenomena with sufficiently small probabilities do not occur at any time. It is the same law of which he spoke in great detail 36 years later.[66]

In 1950 he referred to *Le hasard*. Meanwhile he had become even more rigorous. He regretted having accepted at that time the manner of speaking of the physicists regarding those physical phenomena whose probability is very small. They say that it is very improbable that such events occur, but it is not a certainty. But such an attitude is not sufficiently realistic, it does not take into account our knowledge concerning the universe.

Therefore very small probabilities were of special interest for Borel. They must be neglected.[67] In 1930 he remarks that nobody had determined the limits beyond which probabilities can be neglected universally.[68] He deduces the limit $10^{-1,000}$. Such probabilities must be considered as rigorously equal to zero. They are not effectively equal to zero. But this fact can only be of interest to a metaphysician. They are equal to zero for a scientist. The phenomena that are related to them are absolutely impossible.[69] Later on in 1939, Borel is far more generous and declares that any probability can be neglected on the cosmic scale, that is to say, can be neglected universally, that is smaller than 10^{-50}.[70]

According to Borel, Carnot's principle of the second law of thermodynamics and the irreversibility of numerous phenomena are well-known examples in which the theoretical probability is equivalent to the practical certainty. Borel studied the mechanical explanations of the seeming irreversibility concerning Carnot's principle intensively and repeatedly. Such movements, which make Boltzmann's function H increase in the future or in the past, are infinitely less probable than those that make H decrease.[71]

In 1876, Loschmidt had criticized the method of explaining the irreversible phenomena of classic thermodynamics by means of reversible mechanical phenomena. In 1906 Borel tried to solve this paradox for the first time and in 1913 once again.[72] When he evaluated his scientific achievement in 1918, he rated his solution of this problem the highest.[73] He mentions it in another paper published in 1913. He took pride in the fact that at least Guido Castelnuovo commended his solution in 1918, speaking of an ingenious idea of Borel.[74]

In 1906 kinetic gas theory was far from being accepted by all scientists without reservation. We may think of Einstein, to mention the most prominent instance for this statement. Bertrand had remarked that the application of probability to statistical calculation concerned with molecules was similar to the problem of determining the age of the captain if we know the height of the mainmast.[75]

First of all, Borel believed that it is necessary to specify the notion of probability itself. He even said that the definition of the elementary probability is the main problem or nearly the only problem, if one takes the view of mathematical physics. Probability is a quantity of a special nature, which can only be expressed by means of quantities being of the same nature and known previously. We discussed this definition a little bit earlier. For example, if we throw dice, we assume as a hypothesis that the probability of every side is the same for all dice and that the different throws are independent of one another.

The probability of the given velocity of an arbitrary, individually chosen molecule had been calculated by Maxwell. Borel shows that in the simplest case Maxwell's law of velocity distributions is equivalent to an elementary peculiarity of n-dimensional spheres.[76] But we have to take into account the necessary indeterminism of physical data. The phenomena are never known absolutely exactly.[77] Thus statistical mechanics might be interpreted as the study of the different possibilities that can be deduced from partly undetermined data. This partial indeterminism of the data is unavoidable when we confront reality. It is a purely abstract point of view to assume a mechanical problem where the initial conditions are known with absolute precision.

The multiplicity of collisions in the kinetic gas theory leads very quickly to a constantly increasing number of small indeterminacies. Thus the representation of a gas mass by a single model built up by molecules, whose positions and velocities are rigorously determined at a given moment, is a purely abstract fiction. We can approach reality only by imagining a bundle of models, that is to say, by attributing to the initial data a certain indeterminacy. Even if this indeterminacy be ever so small, the further movement of the molecules becomes very undetermined in a few seconds: a huge number of different possibilities are a priori of equal probability.

To be sure, only a single one of these possibilities is realized at a certain moment. But the indeterminism increases to a considerable extent if we discuss the problem at a neighboring epoch, no matter how near it is. The statistical model is the only model by means of which the problem can be put forward and solved. Thus Loschmidt's argument is not applicable because of the necessarily statistical character of mechanical explanations. We do not try to determine the rigorously defined molecular mechanical phenomena, but the most probable of all reactions. Reversibility in mechanics becomes a purely abstract fiction, which vanishes at the slightest disturbance.[78]

The future is never concretely determined—that is the principle of statistical mechanics—although we cannot speak of the indeterminism of the past.

Borel's paper, dating from 1906, went largely unnoticed. He himself believed that there were two possible reasons: he had used neither the usual mathematical notation nor shown that his results were consistent with the general theorems that Gibbs had deduced from Liouville's theorem. Therefore he took up the problem again in 1913. Again, his crucial starting point was that the notion of the precise numerical value of a physical quantity is a mere mathematical abstraction that corresponds to no reality. In order to measure or to define a physical quantity it is necessary to give additional explanations. The greater the precision desired, the longer these explanations become.

He emphasized that this situation is quite different from the metaphysical issue of the relativity of our knowledge. The notion of the determinism of the phenomena is itself involved. Indeed, as we have seen, Borel rejected Laplace's determinism without citing by name his famous predecessor.[79]

Borel's statement illustrates the prevailing influence of statistical thinking on modern scientists, though Borel himself still hoped to reconcile the determinism of the laws of nature with the indeterminism imposed by probability theory.

Notes*

1. Maurice Frechet and John von Neumann, "Commentary on the Three Notes of Emile Borel," *Econometrica*, (1953), 118–127; Maurice Frechet, "Emile Borel, Initiator of the Theory of Psychological Games and Its Application," *Econometrica*, 21 (1953), 95–96.

2. Jack Barone and Albert Novikoff, "A History of the Axiomatic Formulation of Probability from Borel to Kolmogorov: Part I", *Archive for History of Exact Sciences*, 18 (1977/78), 13–190.

3. Efim Michailovich Polishchuk, *Emil Borel* (Leningrad: Isdatelstvo Nauka, 1980).

4. Henri Poincaré, *Calcul des probabilitiés, Leçons professées pendant le deuxième semestre*, ed. A. Quiquet (Paris: 1896); Joseph Bertrand, *Calcul des probabilités* (Paris: Gauthier-Villars, 1899; the 2nd ed. dating from 1907 is cited); Henri Poincaré, *Science et hypothèse* (Paris: Flammarion, 1902; the German translation, ed. F. and L. Lindemann, is cited (Leipzig: B. G. Teubner, 1914, 3rd ed.)). Poincaré's importance for Borel is especially stressed in Emile Borel, "Sur les principes de la théorie cinétique des gaz," *Annales de l'Ecole Normale*, 3rd series, 23 (1906), 9–32, reprinted in *Selecta*, pp. 243–265, and in *BO*, vol. III, 1669–1692, esp. p. 1671.

5. Hans Reichenbach, *The Theory of Probability. An Inquiry into the Logical and Mathematical Foundations of the Calculus of Probability*, English translation (of the German edition, 1934) by E. H. Hutten and M. Reichenbach (Berkeley and Los Angeles: University of California Press, 1949); John Maynard Keynes, *A Treatise on Probability* (London: Macmillan and Co., 1921).

6. Bertrand, *Calcul* (note 4), pp. 29, 166, 215.

7. Bertrand, *Calcul*, pp. XLIII, 318.

8. Poincaré, *Calcul* (note 4), p. 1.

9. Poincaré, *Science* (note 4), pp. XVII, 210–211.

10. Keynes, *Treatise* (note 5), pp. 335, 49.

11. Emile Borel, "Discours prononcé à la séance publique du 17 décembre", *C. R. Acad. Sc.*, 199 (1934), 1465–1467 = *BO*, vol. IV, 2307–2319, esp. p. 2318.

12. David Hilbert, "Mathematische Probleme," *Göttinger Nachrichten* (1900), 253–297, reprinted in *Archiv für Mathematik und Physik* 1(3) (1901), 44-63, 213–237, and in *Gesammelte Abhandlungen*, vol. III (Berlin: 1935), 290–329 (this edition is cited), esp. p. 306.

13. Emile Borel, "Documents sur la psychologie de l'invention dans le domaine de la science," *Organon*, I (1935), 33–42, reprinted in *Selecta*, pp. 388–397, and in *BO*, vol. IV, 2093–2102, esp. p. 2100; see Maurice Frechet, "La vie et l'oeuvre d'Emile Borel," *Monographies de l'Enseignement mathématique*, 14 (1965), reprinted in *BO*, vol. I, 5–98, esp. 37.

*Abbreviations: *Selecta: Selecta*, Jubilé scientifique de M. Emile Borel (Paris: Gauthier-Villars, 1940); *BO*: *Oeuvres de Emile Borel*, 4 vols. (Paris: Centre National de la Recherche Scientifique, 1972).

14. Gregory H. Moore, *Zermelo's Axiom of Choice, Its Origins, Development, and Influence* (New York, Heidelberg, Berlin: Springer-Verlag, 1982), p. 100.

15. Jean Cassinet and Michel Guillemot, *L'axiome du choix dans les mathématiques de Cauchy (1821) à Gödel (1940)* (Toulouse: Thèse présentée à l'université Paul Sabatier de Toulouse, 1983, vol. 1), p. 221; Emile Borel, "La logique et l'intuition en mathématiques," *Revue de Métaphysique et Morale*, 15 (1907), 273–283 = *BO*, vol. IV, 2081–2091, esp. p. 2082.

16. Emile Borel, "Sur les principes de la théorie des ensembles," *C. R. du Congrès de Rome*, 2 (1908), 15–17 (Rome: 1909) = *BO*, vol. III, 1267–1269, esp. pp. 1267–1268; "Les paradoxes de la théorie des ensembles," *Annales de l'Ecole Normale*, 25(3) (1908), 443–448 = *BO*, vol. III, 1271–1276, esp. pp. 1272, 1276.

17. Emile Borel, "A propos d'un traité de probabilités," *Revue philosophique*, 98 (1924), 321–326, reprinted in Emile Borel, *Traité du calcul des probabilités et de ses applications*, 4 vols. (Paris: Gauthier-Villars, 1925–1939), vol. IV, 3, *Valeur pratique et philosophie des probabilités*, note II, pp. 134–146, and in *BO*, vol. IV, 2169–2184; English translation in Henry E. Kyburg, Jr., and Howard E. Smokler (eds.), *Studies in Subjective Probability* (New York, London, Sydney: Krieger, 1964), pp. 45–60; see *BO*, 2170–2171.

18. Emile Borel, "Les théories moléculaires et les Mathématiques," *Revue générale des sciences*, 23 (1912), 842–853, reprinted in *Selecta*, pp. 316–340, and in *BO*, vol. III, 1773–1808, esp. p. 1808; English translation in *Rice Institute Pamphlet*, 1 (1915), 163–193.

19. Emile Borel, *Valeur pratique*, pp. VII, VIII.

20. Emile Borel, "Supplément (1921) à la notice (1912) sur les travaux scientifiques de M. Emile Borel" (Toulouse: 1921), reprinted in *Selecta*, pp. 381–387, and in *BO*, vol. I, 195–201, esp. p. 199; *Le Hasard* (Paris: Presses Universitaires de France, 1914; the 2nd ed., published in 1948, is cited). He reprinted in *Le hasard* largely word for word a series of papers, which he had written during the period between 1906 and 1908: for example, "La valeur pratique du calcul des probabilités," *Revue du mois*, 1 (1906), 424–437 (*BO*, vol. II, 991–1004); "Un paradoxe économique: le sophisme du tas de blé et les vérités statistiques," *Revue du mois*, 4 (1907), 688–699 (*BO*, vol. IV, 2197–2208); "Le calcul des probabilités et la méthode des majorités," *Année psychologique*, 14 (1908), 125–151 (*BO*, vol. II, 1005–1031); "Le calcul des probabilités et la mentalité individualiste," *Revue du mois*, 6 (1908), 641–650 (*BO*, vol. II, 1033–1042).

21. Emile Borel, *Le hasard* (note 20), pp. 8–9, 42, 92; see Oscar Borisovich Sheynin, "Newton and the Classical Theory of Probability," *Archive for History of Exact Sciences*, 7 (1970/1), 217–243, esp. p. 237.

22. Wolfgang Stegmüller, *Probleme und Resultate der Wissenschaftstheorie und Analytischen Philosophie* (Berlin, Heidelberg, New York: Springer Verlag, 1973), vol. IV, 1, *Personelle Wahrscheinlichkeit und rationale Entscheidung*, p. 109.

23. Emile Borel, *Le hasard* (note 20), p. 175.

24. Emile Borel, "Les probabilités dénombrables et leurs applications arithmétiques," *Rendiconti del Circolo Matematico di Palermo*, 27 (1909), 247–271, reprinted in *Selecta*, pp. 266–309, and in *BO*, vol. II, 1055–1079.

25. Richard von Mises, "Grundlagen der Wahrscheinlichkeitsrechnung," *Mathematische Zeitschrift*, 5 (1919), 52–99, esp. p. 53.

26. Reichenbach, *The Theory* (note 5), p. 121; it is the only time that he mentions Borel.

27. Emile Borel, *Valeur pratique* (note 17), p. 81.

28. von Mises, "Grundlagen," p. 52; Michel Loève, "Calcul des probabilités," in Jean Dieudonné (ed.), *Abrégé d'histoire des mathématiques 1700–1900* (Paris: Hermann, 1978), vol. II, 277–313, esp. p. 287.

29. Andrei Nikolaevich Kolmogoroff, *Grundbegriffe der Wahrscheinlichkeitsrechnung* (Berlin: Springer Verlag, 1933; reprinted in 1973).

30. Reichenbach, *The Theory* (note 5), pp. 372–376.

31. Reichenbach, *The Theory* (note 5), p. VIII.

32. Emile Borel, *Valeur pratique* (note 17), p. 91.

33. Emile Borel, "A propos" (note 17), p. 2182; Rudolf Carnap, *Induktive Logik und Wahrscheinlichkeit*, ed. W. Stegmüller (Wien: Springer-Verlag, 1959) p. 92.

34. Kyburg and Smokler, *Studies* (note 17), pp. 9–10; they should have spoken of Borel and Ramsey instead of Ramsey and Borel.

35. Stegmüller, *Personelle Wahrscheinlichkeit* (note 22), p. 288.

36. Bruno de Finetti, "La prévision: ses lois logiques, ses sources subjectives," *Annales de l'Institut Henri Poincaré*, 7 (1937), 1–68, English translation in Kyburg and Smokler, *Studies* (note 17), pp. 93–158.

37. Emile Borel, "Discours prononcé" (note 11), p. 2317.

38. Emile Borel, "Le calcul des probabilités et les sciences exactes," *Atti del Congresso*, vol. I (Bologna: 1928), 173–179, reprinted in *Journal de Mathématiques*, 1(9) (1929), 115–123, and in *BO*, vol. II, 1131–1137, esp. p. 1133.

39. Emile Borel, "Les théories moléculaires" (note 18), p. 1808.

40. Emile Borel, "Sur quelques problèmes de probabilités géométriques et les hypothèses de discontinuité," *Comptes rendus hebdomadaires des Séances de l'academie des Sciences*, 158 (1914), 27–29 = *BO*, vol. II, 1097–1099, esp. p. 1098.

41. Emile Borel, "Sur les principes" (note 16), p. 1267; Cassinet and Guillemot, *L'axiome du choix* (note 15), p. 222; Moore, *Zermelo's Axiom of Choice* (note 14), p. 103.

42. Emile Borel, "Les paradoxes" (note 16), p. 1275.

43. Emile Borel, *Traité* (note 17), vol. II, 1 *Applications à l'arithmétique et à la théorie des fonctions* (Paris: Gauthier-Villars, 1926), p. 78.

44. He continues the discussion in "Sur les ensembles effectivement énumérables et sur les définitions effectives," *Rendiconti della R. Accademia dei Lincei*, 28(15) (1919), 163–165 = *Bo*, vol. III, 1061–1063.

45. Emile Borel, "Le continu mathématique et le continu physique," *Scientia*, 6 (1909), 21–35 = *BO*, vol. IV, 2151–2165.

46. Poincaré, *Science* (note 4), p. 23; Polishchuk, *Emil Borel* (note 3), pp. 97–98.

47. Emile Borel, "Les probabilités dénombrables" (note 24), pp. 1055–1056. I use the English translation of Barone and Novikoff, "A History" (note 2), p. 133.

48. Emile Borel, *Notice sur ses travaux scientifiques* (Paris: Gauthier-Villars, 1912) = *BO*, vol. I, 119–190, esp. p. 179.

49. Paul Lévy, "Commentaire sur la théorie des probabilités dénombrables," *Selecta*, pp. 310–315 = *BO*, vol. I, 221–226, esp. p. 310.

50. Emile Borel, *Le hasard* (note 20), p. 15.

51. Kolmogoroff, *Grundbegriffe* (note 29), p. 8.

52. Emile Borel, *Applications à l'arithmétique* (note 43), pp. 79–80; I would like to mention that Popper justified his resort to the classic formalism by means of Borel's normal numbers in *Logik der Forschung*, 6th ed. (Tübingen: J. C. B. Mohr (Paul Siebeck), 1976), p. 139.

53. Emile Borel, *Valeur pratique* (note 17), pp. 110–111.

54. Barone and Novikoff, "A History" (note 2), p. 178.

55. Emile Borel, *Le hasard* (note 20), chapter 10; "Radioactivité, probabilité, déterminisme," *Revue du mois*, 21 (1920), 33–40 = *BO*, vol. IV, 2189–2196; "Les lois physiques et les

probabilités," *Revue scientifique*, 65 (1927), 225–228 = *BO*, vol. III, 1827–1837; *Valeur pratique* (note 17), pp. 40–41.

56. Emile Borel, "Radioactivité," p. 2194.

57. Emile Borel, *Valeur pratique* (note 17), p. 41.

58. Emile Borel, *Le hasard* (note 4), p. 230.

59. Emile Borel, "Les lois physiques" (note 55), p. 1831; *Valeur pratique* (note 17), p. 41.

60. For example, "Le calcul des probabilités" (note 38), pp. 1132, 1137.

61. James Clerk Maxwell, "Tait's 'Thermodynamics,'" *Nature*, 17/(1877/78), 257–259, 278–280, esp. p. 279. See Charles T. Grant, "Boltzmanns statistische Interpretation des zweiten Hauptsatzes," *Humanismus und Technik*, 20 (1976), 1–15.

62. Emile Borel, "Radioactivité" (note 55), p. 2196.

63. J. van Brakel, "The Possible Influence of the Discovery of Radio-Active Decay on the Concept of Physical Probability," *Archive for History of Exact Sciences*, 31 (1985), 369–385.

64. Emile Borel, "Les lois physiques" (note 55), p. 1835.

65. Emile Borel, *Le hasard* (note 20), p. 12.

66. Emile Borel, *Probabilité et certitude* (Paris: Presses Universitaires de France, 1950), p. 5.

67. Emile Borel, "Quelques remarques sur la théorie des résonateurs," *Bulletin de la Société française de Physique*, 29 (1912), 3–5 = *BO*, vol. III, 1694–1695, esp. p. 1693.

68. Emile Borel, "Sur les probabilités universellement négligeables." *Comptes rendus hebdomadaires des Séances de l'Académie des Sciences*, 190 (1930), 537–540, reprinted in *Valeur pratique* (note 17), note IV, and in *BO*, vol. II, 1139–1142.

69. Emile Borel, "Sur un problème de probabilités relatif aux fractions continues," *Mathematische Annalen*, 72 (1912), 578–584, reprinted in *Selecta*, pp. 302–309, and in *BO*, vol. II, 1085–1091, esp. p. 1091.

70. Emile Borel, *Valeur pratique* (note 17), pp. 6–7.

71. Emile Borel, "Modèles arithmétiques et analytiques de l'irréversibilité apparente," *Comptes rendus hebdomadaires des Séances de l'Academie des Sciences*, 154 (1912), 1148–1150 = *BO*, vol. III, 1769–1771; "Sur un problème de probabilités" (note 69), p. 1090.

72. Emile Borel, "Sur les principes de la théorie cinétique des gaz" (note 4); "La mécanique statistique et l'irréversibilité," *Journal de Physique*, 3(5) (1913), 189–196, reprinted in *Selecta*, pp. 341–349, and in *BO*, vol. III, 1697–1704; *Le hasard* (note 17), pp. 131–137; Polishchuk, *Emil Borel* (note 3), pp. 108–109.

73. Emile Borel, *Notice sur ses travaux* (note 48), p. 181; "Supplément (1921) à la notice (1912)" note 20), p. 199.

74. Guido Castelnuovo, *Calcolo delle probabilità*, 2 vols. (Bologna: Nicola Zanichelli Editore, 1918; 3rd ed. 1957), vol. II, p. 393.

75. Emile Borel, "Sur les principes de la théorie cinétique des gaz" (note 4), p. 1669.

76. Emile Borel, "Les bases géométriques de la mécanique statistique," *Comptes rendus hebdomadaires des Séances de l'Académie des Sciences*, 154 (1912), 568–570 = *BO*, vol. III, 1765–1767, esp. p. 1765.

77. Emile Borel, *Notice sur ses travaux* (note 48), p. 181.

78. Emile Borel, *Supplément sommaire à la notice (1912)* (Paris: Gauthier-Villars, 1918) = *BO*, vol. I, 191–194, esp. p. 194.

79. Sheynin, "Newton" (note 21), p. 240.

III UNCERTAINTY

10 The Domestication of Risk: Mathematical Probability and Insurance 1650–1830

Lorraine J. Daston

Despite the efforts of mathematicians to apply probability theory and mortality statistics to problems in insurance and annuities in the late seventeenth and early eighteenth centuries, the influence of this mathematical literature on the voluminous trade in annuities and insurance was negligible until the end of the eighteenth century. I argue that the combination of profitable preprobabilistic practices and a legal notion of risk as "genuine" uncertainty (as opposed to the quantified uncertainty of probabilities) was largely responsible for this neglect, and that even in the first applications of probability theory to insurance practice fiscal considerations all but overwhelmed the mathematical methods. The emphasis upon uncertainty as the defining element of an aleatory contract made for an identification of insurance, particularly life insurance, with gambling, and many insurance and annuity schemes of this period exploited the association to attract customers. Only with the advent of new middle-class attitudes that placed provision for one's family above provision for oneself; private self-sufficiency above public charity; the fear of downward above the hope of upward social mobility; and security above surprise could life insurance allegedly based upon the certainty of mathematics and the regularity of mortality statistics compete with life insurance conceived as a wager.

Introduction

Insurance and gambling are two institutionalized approaches to risk-taking. We see them as diametrically opposed approaches: gamblers pay to take unnecessary risks; buyers of insurance pay to avoid the consequences of necessary risks. Yet from the standpoint of the eighteenth and early nineteenth centuries—a period in which both gambling and insurance flourished and expanded at an unprecedented rate— the distinction was none too clear. Much ink was spilt trying to draw such a distinction, but the grounds advanced varied widely from author to author and sometimes contradicted one another. Indeed, the very scope of such efforts, which ranged from polemical pamphlets to learned treatises to legislation, suggests how blurred the line between the two must have appeared to the average reader of the period. In this essay, I shall try to explain this confusion (if indeed it was such)

I am grateful to the Zentrum für interdisziplinäre Forschung of the Universität Bielefeld under whose auspices most of the research for this chapter was carried out; also to the AMEV Insurance Company of Utrecht and the Equitable Society of London for generously making their fine collections of material on the early history of insurance available to me; and to Princeton University and to the National Endowment for Humanities for travel grants that made visits to these collections possible. My thanks to Gerd Gigerenzer, Michael Heidelberger, Mary Morgan, Joan Richards, Ivo Schneider, and Zeno Swijtink, who read and commented upon an earlier version of this chapter. Unless otherwise acknowledged, all translations are my own.

between the two approaches to risk by showing how and why legal and mathematical theory, commerical practice, and social attitudes toward risk tended to conflate them. This explanation in turn sheds light on why sellers of insurance were so slow to make use of a mathematical theory that was in many ways custom-made for them.

The relationship between mathematical probability, statistical data, and insurance was by no means a straightforward one of theory applied to practice, but rather a tangled web in which the mathematical theory, competing sets of observations, fiscal prudence and speculative savvy, changing attitudes toward charity and familial responsibility, and new values and beliefs concerning the regularity of the social and natural order were all intertwined. My focus will be the fledgling whole life insurance companies of late eighteenth-century England, for (1) they were the first branch of the then already venerable insurance industry to use mathematical probability and empirical statistics; and (2) their proponents were largely responsible for the perceived rift that opened up between gambling and insurance in the period 1760–1830. However, for purposes of background, contrast, and emphasis I shall also have occasion to refer to other forms of insurance in other periods and other countries, as well as to the early history of mathematical probability and mortality statistics.

The essay is divided into four parts: first, a brief overview of the practice of risk prior to the formulation of mathematical probability theory in the mid-seventeenth-century, and an account of the relationship between these premathematical practices and the works of the first two generations of probabilists, from Pascal through De Moivre; second, an assessment of the impact of this mathematical literature on actual eighteenth-century insurance practices; third, an examination of the first successful application of mathematical probability to insurance in the case of the Equitable Society for the Assurance of Lives; and finally, a brief epilogue on the subsequent evolution of the relationship between statistics, mathematical probability, and insurance after 1800. Throughout, I shall be concerned with changing attitudes toward and assessments of risk in both theory and practice, and the problems of rationalizing that practice.

1   Risk and the Advent of Mathematical Probability and Statistics

Certain forms of risk-taking—insurance (chiefly maritime), annuities, and gambling—were widely and successfully practiced in Europe long before the formulation of mathematical probability. Pope Gregory IX's thirteenth-century decretal *Naviganti* prohibiting the most popular form of maritime insurance as usurious stimulated jurists to distinguish between insurance (and all other forms of investment that reaped gain without labor) and usury on the basis of risk.[1] Such commercial arrangements became aleatory contracts, the legal category subsuming all agreements involving an element of chance, any trade of here-and-present certain goods for uncertain future goods: annuities, gambling, expectation of an estate, purchase of a future harvest or the next catch of a fisherman's net, insurance, and even risky business ventures all fall under this rubric. By the midsixteenth century it had become standard for lawyers to argue that those who shared risks deserved a share of the profit as much as those who had shared labor.[2] The condition of equity of such agreements was the just proportion between risk and gain,[3] but insurance manuals and legal treatises insisted that the specific amount in any given case depended on a judicious weighting of the particular circumstances: for insurance, the cargo, the season of the year, the route taken, the condition of the ship, the skill of the captain, the latest "good or bad news" concerning storms, warships, and privateers.[4]

Although experience no doubt honed the ability of the underwriter, dealer in annuities, or gambler to estimate odds, their approach to risk could hardly be described as statistical or probabilistic. A sixteenth-century insurer might have found such a statistical approach impractical, for it assumes conditions that are stable over a long period and the homogeneity of categories. Moreover, in commercial centers populous enough to support whole markets of insurers, premium prices also reacted to levels of supply and demand as well as to the latest news about the Barbary pirates. Similarly, jurists recommended that annuities be priced through a combination of rules of thumb and, above all, the consideration of a prudent judge who could weigh the specific circumstances of each case.[5] Although some of the preprobabilistic gambling puzzles from this period reveal a refined sense of very small differences in odds,[6] the combination of skill and chance in many games, the irregular casting of dice and other gambling devices, belief in streaks of good and bad luck, and sharp dealing must have all conspired to obscure the idea of equiprobable outcomes.

Although frowned upon by the church,[7] gambling was perhaps the prototype of a formalized exchange of risk in early modern Europe. Other aleatory contracts, particularly insurance on lives, were occasionally identified with gambling and therefore outlawed, as in the 1570 Code of the Low Countries, which classed "insurances of the lives of persons" together with "wagers on voyages and similar inventions" and banned the lot of them.[8] (Life insurance remained illegal in most European countries until the nineteenth century.) Even in nations that tolerated life insurance, like England, insurance on lives was closely associated with bets on the life of a third person until the latter half of the eighteenth century. The common legal framework shared by the legal risks of insurance and the illegal risks of gambling created such identifications in both regulations and practices, and the attempt to distinguish between them remained a motif of insurance law and literature through the middle of the nineteenth century.[9]

A few points concerning these preprobabilistic institutions for dealing with risk and uncertainty should be emphasized. First, largely because of the Catholic church's position on usury, risk took on a positive tinge as civil and canon lawyers made it the basis for their defense of potentially suspicious commercial practices. Of course, they trod a thin line between risk sufficient to exonerate a merchant from charges of usury, and risk sufficiently great to incur suspicion of gambling. But on the whole, the concept of risk was so important a means of harmonizing precept with practice that even gambling became more innocent by association in the works of sixteenth- and seventeenth-century casuists. Second, risk became the defining characteristic for a distinct class of legal agreements, the aleatory contract, thus combining such socially diverse practices as insurance, gambling, and annuities under a single heading, and emphasizing their similarity. Moreover, the requirement that all contracts be equitable focused legal—and later, I shall argue, mathematical—attention on the problem of determining equal expectations, i.e., some integrated weighting of the probability and outcome value of the uncertain event. Third, insurance premiums, gambling stakes, and annuity rates all represented rough quantifications of risk but not necessarily ones based on probabilistic or statistical intuitions, much less calculations and data. Annuity rates and insurance premiums certainly reflected past experience, but it was a far more nuanced experience than a simple toting up of mortality and shipwreck statistics. It was an experience sensitive to myriad individual circumstances, their weighted interrelationships, not to mention market pressures: it was not simply astatistical; it was antistatistical. Given the highly volatile conditions of both sea traffic and health in centuries notorious for warfare, pirates, plagues, and other unpredictable misfortunes, I am not persuaded that this was an unreasonable approach. In any case, it was the prevailing one. Finally, all these ways of handling and exploiting risk—insurance, annuities, gambling—evidently turned a profit.

The first two generations of mathematical probabilists—Blaise Pascal, Pierre Fermat, Christiaan Huygens, Johann De Witt, Nicholas and Jakob Bernoulli—solved problems framed within the context of aleatory contracts.[10] That is, they posed questions primarily in terms of mathematical expectation and equity rather than probability and odds. Jurists seeking the fair price of an annuity, a lottery ticket, or a partnership share thought in terms of the expectations—i.e., the product of the probability of an event and its outcome value—rather than of the risk per se, and the seventeenth-century mathematicians who attempted a more precise formulation of such questions naturally followed their lead. Moreover, the earliest versions of the mathematical theory, particularly those of Huygens and De Witt, relied heavily on an intuitive appreciation of an equitable contract to define and motivate fundamental concepts like probabilistic expectation.[11]

The doctrine of aleatory contracts provided the early probabilists with problems as well as concepts and definitions. Although games of pure chance were the most important example of such applications, mathematicians also tackled the legally related problems of annuities and reversionary payments. Huygens, De Witt, Edmund Halley, Nicholas and Jakob Bernoulli, Abraham De Moivre, and a handful of lesser figures all addressed the problem of pricing annuities in a mathematical vein, and there existed a substantial literature on the subject in Dutch, English, French, and Latin by the mideighteenth century—almost always prefaced with the proclamation that mathematical probability had outgrown its frivolous gambling phase and could now be applied to more sober, useful pursuits.[12]

However, probability theory alone was not sufficient for such laudable applications to cases where, unlike coin tossing or dice throwing, one could not plausibly assume equiprobable outcomes. Although John Graunt, the first to attempt to create a mortality table in his *Natural and Political Observations . . . upon the Bills of*

Mortality (1662), was most likely ignorant of the first treatise on mathematical probability, Huygens's *De rationciniis in ludo aleae* (1657),[13] Huygens and other mathematicians like Gottfried Wilhelm Leibniz, Jakob and Nicolas Bernoulli, and Halley were quick to see how mortality statistics could extend mathematical probability to cover applications to other sorts of aleatory contracts besides simple games of chance.[14] The almost instant alliance between the two did more than to broaden probability theory's domain of applications; it also changed what probability meant. From the outset, probabilists identified mortality statistics with probabilities without hesitation or justification. This easy substitution of number of observed instances for probabilities is both universal and also somewhat puzzling, given that the earlier interpretations of probability seem to have been derived either from the physical symmetry of gambling devices like dice and lottery urns or from the degrees of certainty of a legal proof.[15]

One can perhaps find a bridge between what appear to us as very different ways of conceiving probabilities in Jakob Bernoulli's discussion of mortality statistics in Part IV of his treatise *Ars conjectandi* (1713), in the context of the distinction between *a priori* (reasoning from causes to effects, e.g., from the physical symmetry of a coin to the equiprobability of heads or tails) and *a posteriori* (reasoning from effects to causes, e.g., from mortality statistics to probability of dying) probabilities. Bernoulli's model for the latter case is taken from the a priori case of an urn filled with different colored pebbles, drawn with replacement, which stand for, for example, diseases of the human body that bring about death.[16] Bernoulli elaborated upon this analogy in an exchange of letters (October, 1703–April 1704) with Leibniz,[17] defending it against Leibniz's objections that the cases were disanalogous in important ways: that the number of diseases, unlike the pebbles in the urn, might be indeterminate; or infinite; or variable; and that extrapolation based on a curve drawn through points representing past instances assumed one curve out of an infinite number of possibilities. Bernoulli's insistence that the human body was indeed like an urn, full of diseases as the urn is full of pebbles in some determinate and stable ratio, and that "nature follows the simplest paths" in response to Leibniz's criticisms in effect assimilated problems of statistical frequencies (drawings from the urn) to problems of physically symmetric gambling devices, with the help of certain assumptions of simplicity and regularity. It is possible that similar assumptions lay at the root of the automatic identification of mortality statistics with probabilities of mortality, Leibniz's reservations notwithstanding.

Such simplifying assumptions also entered into the construction of the statistics themselves. Graunt's mortality table was based on a combination of assumptions about what diseases killed whom at what age (a matter of guesswork, since the London bills of mortality provided only cause of death, not age), and about how regular proportions governed mortality after the age of 6, carrying off approximately 3/8 of the remaining lives each decade. Johann De Witt's pioneering attempt to apply the new theory of mathematical probability to the problem of pricing Dutch government annuities shows that the influence of mathematical probability tended to simplify the statistics of mortality even more thoroughly than Graunt's "shop arithmetique."[18] De Witt's estimation of mortality shows that Jakob Bernoulli was not alone in assimilating mortality to gambling devices with equiprobable outcomes like coins and urns. He assumed that the chances of dying are equal in any six-month period between ages 4 and 54, comparing this case to the "equality of likelihood, or chance similar to the case of a tossed penny, where there is an absolute equality of likelihood or chance that it will fall head or tail."[19] Thanks to Johannes Hudde's analysis of the ages at which Amsterdam annuitants had died,[20] De Witt was able to check his results against actual mortality data, but his calculations of the value of an annuity as a function of age are based solely upon his assumption of equal likelihood and the principle of probabilistic expectation.

Edmund Halley's 1693 memoir on mortality in Breslau provided the first complete table based on actual data for age at death. Despite his avowedly empirical approach, Halley was also convinced that "Irregularities in the Series of Ages" shown by his table "would rectify themselves, were the number of years much more considerable, as 20 instead of 5"; that is, Halley also believed, like Graunt, De Witt, and Bernoulli, more in the regularity of mortality than in his somewhat irregular data.[21] Halley used his table and probability theory to price annuities equitably; Nicholas Bernoulli used Graunt's table to the same end in his *De usu artis conjectandi in jure* (1709), at once the most thorough and most neglected (both by jurists and mathematicians) of the early mathematical treatises applying mathematical probabilities to legal problems. Thus the problem of pricing of annuities, the inverse of pricing life insurance premiums, became a part of the probabilist's repertoire of applications by the first decade of the eighteenth century.

Nicholas Bernoulli's Latin treatise was aimed at a learned audience of jurists; Abraham De Moivre's vernacular *Treatise of Annuities of Lives* (1725) was written for clerks and calculators with a little algebra, but neither seems to have much influenced the actual practice of pricing annuities. However, De Moivre's treatise became the standard reference work for eighteenth-century mathematicians, and as such can serve here as representative of the evolution of concepts and methods for applying mathematical probability to problems of mortality.[22] First, De Moivre was typical in treating only annuities and other reversionary payments (and in his *Doctrine of Chances*, gambling) among the many possible aleatory contracts that Nicholas and Jakob Bernoulli had considered grist for the mathematician's mill. No form of insurance then practiced in London (fire, life, maritime) figures in his treatises. However, De Moivre continued to regard annuities as subject "to the Rules of that Equity which ought to preside in Contracts."[23] Second, he followed Graunt, De Witt, and Halley in assuming that mortality statistics follow a simple pattern—in De Moivre's case an arithmetic progression, which he believed would be ever better approximated by more data, citing Halley's Breslau table as a good first approximation.[24] For De Moivre, this confidence was buttressed by Bernoulli's theorem, interpreted in light of natural theology: "Althou' chance produces Irregularities, still the Odds will be infinitely great, that in the process of Time, those Irregularities will bear no proportion to the recurrency of that Order which naturally results from *ORIGINAL DESIGN*."[25] In the interest of simplifying calculations, De Moivre also made assumptions that deviated more dangerously from the data, as he himself admitted.[26] Third, De Moivre made expectation of life "The Time which a person of a given Age may justly expect to continue in being,"[27] i.e., that for which there existed even odds of surviving, as opposed to the method more closely patterned on probabilistic expectation: i.e., the sum of the total number of years lived by all people divided by the total number of people. Under the assumption of an arithmetic progression in mortality, the two methods yield the same results for annuities on single lives, but never for joint lives. The confusion over how best to compute life expectancy divided writers on the subject throughout the eighteenth century.[28]

Structurally, mathematical probability retained the connections that practice and legal theory had already created between all kinds of aleatory contract, gambling and insurance in particular. In principle, both the fair price of a lottery ticket and the fair premium for insurance were reckoned by the same expectation formula: the product of the probability of the event (drawing the winning ticket, shipwreck, death, fire, etc.) and the outcome value (the payoff). Of course, probabilities for insurance were considerably harder to come by than those for games of pure chance, but Halley's 1693 mortality table made at least life insurance theoretically feasible. Any distinction between gambling and insurance would have to be made on the basis of the desirability of the event or the motives of the contracting parties, for both shared the mathematical framework of expectation.

How then did the advent of mathematical probability change the *theory* of risk-taking in the late seventeenth and early eighteenth centuries? First, it did very little to sever the connection between gambling and other forms of aleatory contracts like annuities and insurance. The probabilists took over the notions of expectation, equity, and (at least in the case of gambling and annuities) degrees of risk, and gave them a quantitative formulation that added mathematical to legal reasons for classifying them under the same heading. As the French mathematician Antoine Deparcieux argued in 1746, an annuity was simply a form of gambling more than usually advantageous to the player.[29] Second, a certain form of statistics, those concerning human mortality and later natality, provided a third interpretation of probabilities that were originally conceived of either as degrees of certainty or of equiprobability. Here the application altered the interpretation of the mathematics. Third, probabilists assumed that simple patterns underlay mortality statistics. De Moivre, Süssmilch, and others used the argument from design to justify this assumption of regularity, but the assumption seems to have been widely held decades before by writers like Graunt on other grounds.[30] It is difficult to understand why mortality should have been assumed to be regular and other phenomena of equal practical interest like the incidence of fires not, in an age where both were subject to wild fluctuations: witness the plague and Great Fire of London in 1665–1666. Yet demographers and mathematicians were not only confident enough of the regularities to collect data on mortality; they also freely supplemented that data with further simplifying assumptions. Leibniz was not the only one to balk at these impositions upon nature; the Dutchman Nicholas Struyck also complained in 1740 that "mortality doesn't listen to our suppositions" and that many of the tables allegedly based on observation were in fact "pure hypotheses."[31] Perplexity over *which* mortality table to trust grew with the data in the eighteenth century. This was the theoretical legacy of mathematical probability to institutionalized risk-taking in the eighteenth century; what was its contribution to practice?

2   The Practice of Risk-Taking in Eighteenth-Century England

Whatever we mean by modernity is in some way linked with new attitudes toward the control of the future and the possibility of a life relatively secure from the disruptions of chance. Hence, Keith Thomas sees no more revealing indicator of the decline in magical beliefs in favor of more rational ones than the rise of the insurance industry in early eighteenth-century England.[32] The argument seems on the face of it convincing. Maritime insurance expanded under the auspices of individual brokers who congregated at coffeehouses like Lloyd's; fire insurance first emerged in London in 1680; the Amicable Society for mutual insurance of lives was established in 1706; and the Royal Exchange and London assurance offices, both of which insured lives, were incorporated in 1720. Many other insurance schemes were launched and folded in short order.[33] In the feverish London insurance market of the mideighteenth century it was possible to buy insurance against cuckoldry, lying, and even losing at the lottery, the latter sold by none other than John Law himself.[34] However, *pace* Thomas, the vogue for insurance seems to have been less prudential than reckless, fueled more by the spirit of gambling than of foresight. As for the insurers, the more reputable relied on the traditional methods described in the first section of this paper; the less reputable were frankly speculators. Both insurance offices and their customers were for the most part betting on the future, not planning for it.

Indeed, gambling so riveted the imagination of Europeans after 1690 that it became a metaphor for civil society itself. A lottery craze swept England, the Netherlands, and France during the period 1690–1740, moving one chronicler of the enormously popular Dutch lotteries to exclaim that "the whole world was nothing but a lottery, that is to say a lot, a chance ...."[35] In the percolating financial center of London during the South Sea Bubble era, tea rooms were set up for ladies playing the stock market and playwrights wrote farces on the wild speculations and get-rich-quick schemes. Insurance projects were second only to the South Sea Company itself among these "bubbles" (the Royal Exchange and London Assurance offices were known as "Onslow's" and "Chetwnyd's Bubbles", respectively).[36] London underwriters issued policies on the lives of celebrities like Sir Robert Walpole, the success of battles, the succession of Louis XV's mistresses, the outcome of sensational trials, the fate of 800 German immigrants who arrived in 1765 without food and shelter, and in short served as bookmakers for all and sundry bets.[37]

Such enterprises represented the seamier side of a business that also boasted more reliable firms. But these latter depended on the application of mathematical probability no more than the "bubbles" did. The vast bulk of the practice was maritime, and although premiums responded to decreases in risk (the disappearance of marauding Turks made insuring voyages to the Levant, Spain, and Portugal considerably cheaper), statistics played no role in pricing.[38] Fire insurance was too new to be burdened with the weight of tradition, and clients were offered graduated premiums depending on the kind of building (brick *versus* wood) and trade housed therein (sugar-bakers, for example, paid especially stiff rates). Yet fire offices apparently never collected statistics on the subject.[39] In these two areas, insurers would have received little guidance from the mathematical manuals even if they had consulted them.

However, the problems of annuities and life insurance had attracted considerable mathematical attention, even if the terms upon which they were bought and sold had little, if anything, to do with mortality statistics and probability. The actuary was originally a clerical rather than a mathematical position, a combination of secretary and bookkeeper,[40] and with this audience in mind the manuals of De Moivre, Simpson, and Dodson barely required more than arithmetic, often recasting algebraic rules in verbal form and relegating demonstrations to appendices. Every such book included numerous tables of the values of annuities calculated by age, number of heads, and interest rates to ease the burden of calculation for the reader. Nonetheless, their impact upon practice appears to have been minimal prior to the establishment of the Equitable Society for the Assurance of Lives in 1762, and even then the dictates of mathematical theory were greatly tempered by other considerations.

How were the premiums of eighteenth-century life insurance policies and annuities in fact determined? The three most prominent eighteenth-century British institutions dealing in life insurance were the Amicable Society and the Royal Exchange and London Assurances. Of these, only the Amicable offered long-term insurance, and it operated more as a friendly society than as a business. Founded in 1706 by Sir Thomas Allen, Bishop of Oxford, the Amicable admitted anyone in good health between the ages of 12 and 45 years, up to 2,000 members, and charged each member an entrance fee and a fixed amount each year. The annual income was to be divided equally among the beneficiaries of all those who died in that particular year.[41] Not only did the Amicable taken no account of age, except to exclude (far too cautiously judging from the extant tables) the periods of greatest mortality; it also had something of the lottery about it, as contemporary observers noted.[42] In an age in which tontines were more popular than life insurance, it is not unusual to find annuity and insurance schemes that deliberately included an element of outright gambling.[43] Even the original plans for the Equitable included provisions for such a tontinelike arrangement to reward the survivors among the original subscribers.[44]

The Royal Exchange and London offices offered short-term (usually one-year) policies at a flat rate of 5% for every £100 insured, regardless of age. Life insurance made up only a small fraction of their trade during the eighteenth century, and those who bought policies could be divided into three categories, judging from the company's records: creditors insuring the lives of their debtors for the amount owed; gamblers betting on the life of some third person; and clergymen who insured their own lives, one year at a time. (The Amicable was also designed primarily for the benefit of the clergy.[45]) State of health is rarely mentioned (the records make occasional references as to whether the insured has had smallpox), but extraordinary risks (e.g., a voyage "beyond the Cape of Good Hope") commanded much higher premiums (here, 15%). Indeed, the life insurance registers, like the maritime ones, provide a rough mental map of risk worldwide as seen through the eyes of an eighteenth-century Englishman: for example, one might travel almost anywhere in Europe at the domestic 5% rate, but a trip to the Bahamas increased the premium to 7.1%, and to North Carolina, 9%.[46]

Most eighteenth-century annuity schemes were as innocent of probability theory and statistics as life insurance was. The only partial exceptions were the societies founded to provide annuities to widows in the 1750s and 1760s, first in the Netherlands and then in Britain.[47] Most of these societies did scale premiums to age, but they failed at such a distressing rate that they can hardly have been a good advertisement for such procedures. Richard Price ascribed their failure to an insufficient amount of mathematics, for the organizers did not understand that the claims would inevitably increase as the population of members aged, and they fixed premiums more by guesswork than by the tables.[48]

Moreover, there seem to have been other reasons for sellers of commercial annuities to ignore the mathematical manuals, for it appears from the eighteenth-century rolls that most of these annuities were essentially usurious loans disguised as aleatory contracts, particularly after the usury legislation of 1777 made any loan for interest greater than 5% illegal unless it involved some genuine risk. Borrowers circumvented this law by selling annuities on the lives of the *seller* rather than the buyer, at very low rates: for example, an annuity of £1,000 per annum might be sold at six years' purchase, £6,000, contingent on the life of the seller. Minus the cost of the life insurance, this amounted to a loan at an effective interest rate of about 12%. The overwhelming majority of annuities registered were of this form, and clearly required no recourse to the life tables for the annuity per se.[49] As late as 1793, when the Royal Exchange secured the right from Parliament to grant annuities on the grounds that the business as transacted by private individuals was rife with fraud and bankruptcy, no attempt was made to calibrate prices by mortality statistics.[50]

Why was the practice of eighteenth-century insurance and annuities so resistant to the influence of mathematical theory? It should be noted that not only businessmen but also jurists took almost no account of how the theory of aleatory contracts had been modified by mathematical probability. Robert Pothier's 1775 treatise on aleatory contracts characteristically declined to go into details about fixing premiums, aside from stipulating that they should be equitable, for "as it is not easy to determine what this just price is, one must give this just price much latitude, and hold the just price to be that which the parties have argreed to among themselves ...."[51] Nicholas Bernoulli's treatise on probability and the law was apparently unknown to him. For maritime insurance annuities, it might be argued that the inertia of an entrenched successful practice based on nonmathematical estimates worked against the application of the new mathematical and statistical methods. This is no doubt part of the answer, but it cannot explain why new forms of insurance against fire and death did not take on a more statistical cast. The case of life insurance is particularly baffling, because of the availability of mortality statistics drawn from several locales and the growing belief that they revealed, in the words of Johann Süssmilch, "a constant, general, great, complete, and beautiful order."[52] Part of the explanation perhaps lies in the lack of unanimity among the mathematicians as to the definition of life expectancy, the validity of certain simplifying assumptions, and the relative reliability of the various life tables.[53] Such controversies were far less important in England than in France, where a government-regulated economy and an official scientific body made for a smoother mesh between mathematical theory and economic practice.[54] But it was London, not Paris, that was the capital of the insurance world until the midnineteenth century, and given the elementary level of the English mathematical manuals, it is difficult to believe that London insurance merchants could fathom the mathematical issues.

I believe that the answer lies more with the theory of aleatory contracts than of mathematical probability, and more with a shift in social and familial values than with increasing rationality. The key element in an aleatory contract was risk, conceived as an exchange of certain for uncertain goods. Gambling was the paradigm aleatory contract in spirit as well as fact, as the tendency of almost every other type of aleatory contract to degenerate into a wager shows.[55] Insurers who essentially used life offices to place bets on the lives of third parties were no more interested in the probabilities than the average purchaser of a lottery ticket. Quantifying uncertainty by means of probability theory may have diluted the risk that prevented, for example, a legal annuity from becoming an illegal usurious loan: even Pascal observed that there was something paradoxical about a "géométrie de hazard."[56] This is not to say that insurers did not have a sense of the regularities governing the events upon which their trade depended—to judge from the stabilization of premiums, fire and maritime offices ironically seem to have been more sensitive to these than the early life offices were—but rather that the quantification of such risk seemed to presume too much certainty for the venture to be genuinely risky. The speculation that was rampant during the South Sea Bubble era also helped to identify insurance with gambling.[57]

In order to make quantified risk a part of the insurance trade, the link between gambling and insurance had to be broken. The literature on insurance from the latter part of the eighteenth and early nineteenth centuries is full of attempts to drive a wedge between the two, and it is no accident that the rise in long-term life insurance coincides with such distinctions. Before this period life insurance constituted only a small fraction of the insurance trade, and only the Amicable Society and the ill-fated annuity schemes for widows provided anything approximating whole life insurance. The mathematical, commercial, and legal manuals were largely about annuities and reversionary payments, and even James Dodson, who laid the mathematical foundation for the Equitable, devoted only a few pages of his three-volume *Mathematical Repository* (1755) to life insurance problems, "as it as a subject not before handled, and will show its own use."[58]

So long as life insurance—for historical, legal, and mathematical reasons—remained a gamble, whose buyers and sellers emphasized and indeed reveled in the element of risk and uncertainty; so long would the idea of a mathematical approach to either insurance or gambling seem alien. Notably, the mathematicians who attempted to find probabilistic grounds for distinguishing between gambling and insurance (1) abandoned the original definition of mathematical expectation, which was the shared method for pricing all aleatory contracts, from lottery tickets to annuities; (2) made an explicit appeal to the values of respectability and fiscal prudence; and (3) were roundly ignored. Pleas to gamblers to attend to the sobering mathematics of their passion similarly fell upon deaf ears.[59] More than mathematics was needed to create the sort of life insurance that benefited the family of the deceased. Even if mathematically minded, the investor without family sentiment could and did choose to put his money into annuities that provided for his comfort, as a rather regretful observation to this effect in the Equitable prospectus and also the preeminence of annuities over all other forms of reversionary payments in the mathematical treatises make clear. (The Amicable Society and the widows' funds are eighteenth-century examples of the sentiment without the mathematics.) At least two sets of values, not necessarily related, had to converge in order to make the new style life insurance appealing: first, a heightened sense of familial responsibility that made life insurance preferable to annuities; and second, an aversion to risk conceived along the lines of a gamble, which made the vaunted mathematical certainty of the Equitable and its imitators reassuring.

The extraordinary success of the Equitable is the result not only of its exploitation of the regularity of the mortality statistics and the mathematics of probability to fix premiums (which were in any case much padded by fiscal considerations), but also of its creation of an image of life insurance diametrically opposed to that of gambling. The prospectuses of the Equitable and the companies that imitated it made the regularity of the statistics and the certainty of the mathematics emblematic for the orderly, thrifty, prudent, farsighted *père de famille*, in contrast to the wastrel, improvident, selfish gambler. Long-term life insurance was aimed at a growing middle-class of salaried professionals—clergymen, doctors, lawyers, skilled artisans—who were respectable but not of independent means. In a world where apparently even clergymen could not count upon communal charity, the sudden death of the provider could topple the family from the middling ranks of society to the very bottom. Such reversals of fortune were the proper fate of the gambler, not the good bourgeois, and the new life insurance companies set about domesticating risk in the service of the domestic virtues: "family life and parsimony, frugality and orderliness."[60]

Late eighteenth-century authors were by no means agreed as to what exactly distinguished insurance from gambling. Nicolas Magens thought it would be sufficient for merchants to put the public good before private self-interest to prevent "so many strange inventions of unnatural and gaming insurance."[61] Johann Tetens compared life insurance to a game of chance in which one bets that the premiums and more will be paid off to one's widow, but argued that here the lucky ones were the losers.[62] In France, both the detractors and supporters of insurance thought of it in gambling terms. Brissot de Warville attacked a proposal for a Parisian fire insurance company as an entreprise where "all is left to chance";[63] Vernier defended the government lottery during the Revolution on the grounds that annuities, tontines, and fire insurance were also "games of chance based on the expectation and the probability of events."[64] In England, legislation of 1764 and 1774 (the so-called "Gambling Act") made "interest" the distinction between legitimate insurance and gambling: with respect to life insurance the holder of the policy was obliged to show a legitimate interest in the life insured that squared with the amount insured; otherwise the policy would be declared null and void.[65] Although the precise meaning of legitimate interest remained a matter of some controversy,[66] the intent of the law was clearly to distinguish between "Gaming or Wagering" and "the true Intent and Meaning" of insurance, and as such it marked a turning point in English conceptions of life insurance. Changes in practice were slower in coming, for they depended both on new values concerning familial responsibility and the stability of the social order, and on a new interpretation of the mathematical "doctrine of chances" that was consistent with those values. The early career of the Equitable Society shows that this transition occurred only gradually, and that the Equitable's phenomenal financial success owed as much to the neglect of probability and statistics as to their use.

3 The Equitable Society and the Domestication of Risk

The early records of the Society for Equitable Insurance on Lives and Survivorships (established 1762) reveal the extent to which the first full-dress attempt to apply mathematical probability and statistics to the practice of insurance both shaped and was shaped by new values that promoted family over individual welfare, by an emphasis on the predictability versus the contingency of mortality, and above all by a policy of fiscal prudence that at times threatened to make the mathematical basis of the premiums irrelevant. The history of the Equitable is a rich and intricate one; I shall here be concerned only with those aspects that relate directly to these issues.[67]

The moving spirit behind the Equitable was the mathematician James Dodson (c. 1710–1757), Fellow of the Royal Society, Master of the Royal Mathematical School, and author of several works on practical mathematics, including annuities. Denied admission to the Amicable Society on grounds of age (the Amicable admitted no one over the age 45), he formed his own project for a life insurance company in 1756 and composed *First Lecture on Insurances* (unpublished) in the same year.[68] Although Dodson's death in 1757 cut short his calculations of premiums, which were to be based on the London table of mortality, [69] other backers of the project carried forth his plan. Their petition for a Royal Charter was rejected by the Privy Council on May 1, 1760, and although the opposition of rivals like the Amicable, the London Assurance, and the Royal Exchange no doubt played a role in this decision, the reasons given by the Privy Council shed some light on extant insurance practices and the novelty of a mathematical approach. In its decision, the Privy Council worried that the Equitable's premiums were too low (in fact, they were too high, as later experience was to show) and that the starting capital was inadequate to launch such a venture, for the Council was wholly unpersuaded by the Equitable's argument that premiums alone would suffice. When it rejected the Equitable's petition for a second time on July 14, 1761, the Council expressed outright suspicion of the company's mathematical basis, "whereby the chance of mortality is attempted to be reduced to a certain standard: this is a mere speculation, never yet tried in practice, and consequently subject, like all other experiments, to various chances in the execution ...." [70]

Undeterred, the directors of the Equitable rented rooms and published a prospectus that explained the new insurance plan to the public. The Equitable was to be a mutual society, "the assured being mutually assurers one to the others," with the members entitled to dividends in case of surplus, and subject to calls for extra contributions in case of deficit. Of the eight benefits to insurers listed, only one concerned provision for widows and children; the others reflected the actual insurance market, being mostly concerned with security on loans. However, the prospectus emphasized the importance of such benefits for the "families of clergymen, counsellors, physicians, surgeons, attornies; those who have places in public and private offices; and more frequently of artificers, manufacturers, and others who support themselves by their labour"—that is, those in "a middling station of life" who "would strenuously endeavour to avert that most sensible of all distresses, which must necessarily attend their families, should they be at once reduced from a plentiful and respectable, to an indigent and deplorable situation." Moreover, life insurance could supplement or replace public charity, "and many parishes may hereby be eased of burthens, which would otherwise have fallen on them." Yet the author of the prospectus realized that such provident views were still somewhat of a rarity even amongst the salaried middle class, and hastened to add that the Equitable could serve other ends as well, since "it hath been found by experience, that a future provision for family is, in the opinion of the generality of these persons of less importance than a provision for themselves in sickness, or old age, or at a time, when they may be disabled from labour ...." [71] The prospectus was at pains to distinguish its premiums from the flat rates other companies charged, "be the life ever so young and healthy," and included sample premiums so that the reader might make the comparison himself. Above all, the prospectus stressed the certainty of the underlying principle of the new scheme, which was "grounded upon the expectancy of the continuance of life; which, although the lives of men separately taken, are uncertain, yet in an aggregate of lives is reducible to a certainty."

Provision for family versus provision for oneself; private foresight versus public charity; the uncertainty of any individual death versus the certainty of mortality en masse—the kind of life insurance offered by the Equitable threw these contrasts, real and perceived, into relief. The early prospectus could not take for granted the attitudes toward familial responsibility beyond the grave and toward the stability of the social and natural orders upon which the attractiveness of such life insurance depended. Indeed, they were as much briefs for as appeals to these attitudes. Earlier annuity and tontine schemes had beguiled subscribers with the lottery player's dream of sudden upward social mobility from a bourgeois life to a princely one,[72] while the new life insurance played upon the spectre of sudden downward social mobility. Earlier insurance schemes had deliberately emphasized the elements of uncertainty that were the essence of an aleatory contract and that had given them the allure of a gamble; the proponents of the new life insurance minimized the chance aspects. The 1788 prospectus for the French Compagnie Royale d'Assurance, which was explicitly patterned on the Equitable, is a paean to these new attitudes unadulterated by the need to conform to accepted commercial practice, for life insurance had been heretofore illegal in France. The author of the prospectus praised the moral effects of life insurance as opposed to those of annuities, for the former provides "security against misfortune without discouraging either industry or activity. On the contrary, it encourages labor and economy ... the facilities that it offers to the benefit of friendship, of filial piety, paternal tenderness, conjugal union, in a word, to generous sentiments can only tend to multiply the practice of all virtues ...." Moreover, it was preferable to charity, which tended to lead to sloth and disgrace.[73] The Compagnie Royale had taken care to employ a "profound mathematician," for it understood that such enterprises rested upon calculations: "Such indeed is the certainty of the various calculations upon which insurance is based than one can undertake it without capital, and by the simple amassing of the premiums." [74] Richard Price's evaluation of the mathematical basis of the Amicable Society makes clear the degree to which risk as uncertainty clashed with the new insurance sensibility, for he objected to the annual distribution of benefits on the basis of who had happened to die that year as "*a contingency*" that did not depend on the individual member's contribution. The regularity of mass mortality statistics had apparently made the contingency of individual deaths intolerable, even though, of course, they remained the basis for life insurance.[75]

Richard Price played an important part in the early affairs of the Equitable as a mathematical consultant, for Dodson's death had left the directors of the Equitable with a set of incomplete calculations. There was some disagreement among the directors as to which method to use for joint lives, "greatest" or "mean hazards," and judging from the early account books, the directors seemed to have adapted the premiums rather freely to what they saw as the individual exigencies of the case.[76] In 1768 the Equitable turned to Price for help in calculating reversionary payments, which may have led his interests in that direction, for his first publications on the subject followed soon thereafter. Price's treatise *Observations on Reversionary Payments* (1771) contained a full and admiring account of the Equitable's practice, which he held up as a rare example of sound planning and solvency in the dubious business of annuities and insurance. We learn from Price that the Equitable had been cautious in every respect: it had calculated interest at the lowest rate (3%); it had used the mortality table that gave the shortest lifespans (Corbyn Morris's London table); it took the further precaution of insuring only healthy lives; and finally, it added a flat percentage (6%) to all of these premiums. In words that could have been made the Equitable's motto, Price exhorted its directors to proceed "frugally, carefully, and prudently," for despite the certainty of the calculations, "at particular periods, and in particular instances, great deviations will often happen." [77] When it came to practice, even Price admitted the force of the contingent.

The directors of the Equitable preserved certain elements of the older practices oriented toward the individual case. Every candidate for a life insurance policy was interviewed in person by the directors, made to swear that he had had smallpox and was not given to intemperance, and asked to give an account on any special risks run. For these latter, added premiums ranging from 11% to 22% were summarily charged at the discretion of the directors. Some policies on these matters emerged: for example on April 21, 1779, the directors resolved to charge travelers to the West Indies an extra 5%, but exceptions and modifications were made to suit individual cases.[78] The calculations of premiums provided them with a guide, which they regularly overruled as the exigencies of practice dictated.

In 1775 Richard Price's nephew William Morgan became actuary of the Equitable after a two-year study of insurance mathematics under his uncle's tutelage. Morgan's appointment virtually transformed the position of actuary from one of secretary to one of mathematical expert, with ever increasing power within insurance companies.[79] Morgan's long tenure (until 1830) at the Equitable reinforced the sometimes exaggerated tendencies toward prudence and caution praised by Price, to the point where the company's spectacular surpluses were an endless bone of contention between the members, who insisted upon a distribution of dividends and/or decrease in premiums, and Morgan, who warned that some unforeseen disaster might flood the company with claims. In 1775, Morgan calculated the company's liabilities and discovered that 60% of its assets could be considered surplus, but he sided with his uncle against any distribution of dividends from this amount, lest "extraordinary events or a season of uncommon mortality" catch the Equitable unawares.[80] As the membership increased to over 5,000 policies in force by 1796, so did the surplus, until it had reached almost embarrasing proportions, and with it, the pressure for a distribution of dividends. By the fifth edition of *Observations on Reversionary Payments* (published posthumously by Morgan in 1803) even Price had relented, wondering whether it might not be better for the Equitable to use mortality tables "more adapted for the general state of mortality among mankind" and to calculate prices straight from the tables, thus reducing premiums by 20%, the interest being still computed at half the actual investment rate.[81] Morgan, however, held firm. Not even the salutary effects of the Law of Large Numbers on the regularity of the actual mortality experience of the Equitable would sway him, for "from the great difference in the sums assured in each life, the amount of the claims is so uncertain, that it shall often happen that events prove peculiarly unfavorable to the Society in a year which has been attended with no uncommon degree of mortality."[82]

The extent to which Morgan's legendary caution was justified by the calculations and the data was sharply challenged by mathematicians of the next generation such as Charles Babbage and Augustus De Morgan, who accused Morgan of ignoring the best available mortality data and more realistic interest rates in the name of an almost pathological prudence. De Morgan (who was Dodson's grandson) exonerated the Equitable from any intent to defraud, but still maintained that its premiums were "enormous" due to an overestimation of mortality and margin of safety: "We should write upon the door of every mutual office but one be *wary*; but upon that one should be written *be not too wary* and over it *Equitable Society*."[83] The Equitable prospered under Morgan's regime, but its prosperity seems to have been less connected with its mathematical basis, although the prospectuses made much of this aspect, than to its willingness to "always modify the exact calculations of mathematics by those of prudence," in the words of the French prospectus.[84] Although its premiums may have been inflated in light of the actual mortality statistics, the Equitable attracted members. Indeed, when competitors like the Royal Exchange threatened the Equitable's effective monopoly on whole life insurance in the 1790s, they imitated the Equitable's methods but added the stipulation that premiums be at least 20% *higher* than those of the Equitable and were nonetheless immediately successful.[85] With only slight exaggeration, one might claim that these new style life insurance companies flourished in spite of mathematical probability and mortality statistics. Not only did these concerns attempt to eliminate the element of risk that had previously been synonymous with insurance for their prudent middle-class clientele; they also attempted to eliminate any effective risk from the venture itself. In the latter case, they very nearly eliminated the mathematics and statistics that had been their claim to regularity and reliability in the former.

4 Epilogue and Conclusion

By the second decade of the nineteenth century, the divorce between insurance and gambling was almost complete. Laplace placed the full weight of his mathematical prestige behind insurance as "advantageous to morals, in favoring the gentlest tendencies of nature";[86] Quetelet echoed his master, comparing government insurance schemes favorably to lotteries;[87] and life insurance became the mathematician's favorite example for the utility of probability theory. The relationship between belief in statistical regularities and confidence in insurance was a symbiotic one: those who would persuade others of the existence of such regularities pointed to the financial success of insurance companies; insurance companies in their turn considered every new such regularity (e.g., between sunspots and epidemics) to be support for their practices.[88] Views on the applicability of statistics and probability to all forms of insurance varied, however, and despite the statistical enthusiasm of some midnineteenth-century writers, the use of such data seems to have been more or less restricted to life insurance.[89] Insurance became a pillar of the social order, guaranteeing that "a man who is rich today will not be poor tomorrow," and the equality between risk and premium was elevated to a fundamental scientific and moral principle: "every risk is represented by a number whose fairness has been checked by statistics."[90] Insurance continued to gain ground at the expense of charity as a "precise, scientific, and at the same time practical form of that unconscious solidarity that unites men," and instructed wage-earners, particularly workers, in the virtues of self-sacrifice, as opposed to the egotism fostered by saving accounts.[91] When authors of nineteenth-century insurance treatises raised the problem of the kinship between insurance and gambling, it was usually to dismiss it as a misunderstanding. Even if they agreed with the mathematician Laurent that insurance companies played "an almost equitable game with the public,"[91] they were likely to insist upon the distinction between the motives of the gambler and the policy holder, the one driven by "the gaming passion and the spirit of avarice," the other guided by "wisdom and foresight."[92]

But as the insurance market expanded and diversified in the nineteenth century, these moral advantages were placed, so to speak, at risk. New forms of accident and liability insurance introduced in the latter half of the nineteenth century created a conflict between insurance and the values of prudence, foresight, and responsibility it was supposed to cultivate. For example, the Association for Prevention of Steam Boiler Explosions, established in 1854 to inspect the notoriously dangerous high-pressure boilers, vehemently opposed boiler insurance on the grounds that it would reduce the incentive to prevent explosions.[94] Insurance for Parisian carriage drivers was opposed on the same grounds.[95] With the advent of massive workman's compensation legislation in Britain and France in the last quarter of the nineteenth century, the concepts of responsibility and culpable negligence almost disappeared in this area of the law. The idea that employers might bear a certain irreducible "professional risk" for the safety of the workplace independent of foresight and blame represented both a triumph of the faith in statistical regularities and a thorough reinterpretation of the legal notions of responsibility.[96]

One can trace the connection between the two in the works of the theory of insurance. An 1843 French treatise submitted for a prize offered by the Académie des Sciences Morales et Politiques presents the traditional division between "cas fortuit," which all of human prudence is helpless to predict; "cas imprevu," which prudent men could be reasonably expected to have foreseen; and "force majeure," which are cases of irresistible necessity. Responsibility lies between the two poles of the fortuitous and the necessary. The same author admires life insurance for its use of mortality statistics, but doubts that such statistical practices will ever invade the practice of maritime insurance, where the accidental reigns.[97] In contrast, an 1884 treatise, also submitted to a prize competition of Académie des Sciences Morales et Politiques, calls for data collection in all areas, convinced that statistical regularities determine sickness, accidents, fires, hail storms, etc. The nonstatistical practices of maritime insurers strike him as scandalous, as mere speculation. Whereas his predecessor understood insurable tisks to be those beyond the ken of human intelligence, the true believer in statistical regularities is obliged to redefine risk as the "effects of chance ... foreseen, but not yet realized."[98] Insurance without statistics promotes speculation or gambling; insurance with statistics promotes irresponsibility. The domestication of risk ultimately placed it in opposition to the domestic virtues.

The changing relationship between the theory and practice of risk in insurance might serve as an object lesson in the complexity of applying mathematics or, more generally, of rationalizing practice. The doctrine of aleatory contracts stimulated the development of a mathematics of chance and provided the fledgling theory of probability with both problems and a conceptual framework within which to solve them. The a priori conviction that mortality data reflected stable underlying probabilities arranged in a simple progression enabled mathematicians to link probability and statistics—that is, to link degrees of certainty or equiprobability with observed frequencies. One might even say that mathematical probability was tailor-made for applications to aleatory contracts like annuities, tontines, and insurance, and certainly eighteenth-century mathematicians worked hard at improving the fit. Yet the mathematical theory exerted almost no influence over eighteenth- and early nineteenth-century legal and business practices.

In order to understand why, we must clarify just what is meant by "rational" in the phrase "rationalizing practice." It is not enough simply to cite the unthinking conservatism of practice, for this period was one of enormous innovation in both law and commerce. For the lawyer, the rationality of an aleatory contract hinged upon the elements of risk and equity, interpreted as an exchange of certain for uncertain goods. That is, the legality, and therefore the legal rationality, of such a contract depended on the existence of "genuine" uncertainty about the outcome of a gamble or an annuity, an uncertainty that quantification apparently diminished for the individual purchaser of insurance and certainly diminished for the seller with a large enough clientele. For the seller of insurance and annuities, rationality meant balancing profits against market pressures, and it is certainly arguable in the case of life insurance that one might succeed better at both without statistics and probability. The nonstatistical companies in any case charged higher premiums without a drop in business, and the great success of the Equitable in attracting members seemed to owe more to its bonus system than to its mathematical methods. In fact, the Equitable itself provides an excellent example of how powerful such considerations were, for they all but swamped its mathematical procedures. For the customer, rationality initially meant simple pursuit of economic self-interest, admixed with the thrill of a gamble: hence annuities and tontines (or a combination of the two) were far more popular than mutual aid schemes like the Amicable, and life insurance was almost synonymous with gambling. Only with the emergence of a new rationality that prized security, prudence, foresight, and economic responsibility to one's family could the idea of whole life insurance purportedly grounded upon the certainties of mathematical probability and the regularity of statistics be made attractive to the practitioners of risk. In the end, it was the rationality of values rather than the rationality of ends that made it possible to "rationalize" practice, and to separate the risks of insurance from the risks of gambling.

Notes

1. Jean Halpérin, *Les Assurances Suisse et dans le monde* (Neuchâtel: Editions de la Baconnière, 1946), p. 32. For a general account of sixteenth-century maritime insurance, see Moses Amzalak, *Trois précurseurs portugais* (Paris: Librairie du Recueil Sirey, 1935). On the canonists' defense of such practices, see John T. Noonan, Jr., *The Scholastic Analysis of Usury* (Cambridge, MA: Harvard University Press, 1957), chapter 6. For the text of *Naviganti*, see *Decretales Gregorii Noni Pontificis* (Lugduni: 1558), lib. V, tit. XX, cap. XIX, p. 1023.

2. Eli F. Heckshaw, *Mercantilism*, trans. Muriel Shapiro (London: George Allen & Unwin, 1935), vol. I, p. 332; François Grimaudet, *Paraphrase des droicts des usures pignoratifs* (Paris: 1583), p. 92.

3. See, for example, Jean Domat, *Les Loix civiles dans leur ordre naturel* ed. Héricourt (Paris: 1777), p. 30.

4. Estienne Cleirac, *Les us et coutumes de la mer* (Rouen: 1671 ed.), pp. 271–272; L. A. Boiteux, *La Fortune de la mer* (Paris: École Practique des Hautes Études—VI. Section, 1968), p. 176.

5. See, for example, Charles Du Moulin, *Summaire du livre analytique des contractz usures, rentes constituées, interestez & monnoyes* (Paris: 1554), p. 187 recto. It was by no means unusual to compute annuities according to a flat rate, without any regard to age and mortality: interest rather than chance seems to have been the important temporal variable. See, for example, William Purser, *Compound Interest and Annuities* (London: 1634), which computed prices according to a flat 8% rate and number of years' purchase. The English government regularly sold annuities to raise funds in the seventeenth and eighteenth centuries; under William III they were offered at 14% for any age: John Francis, *Annals, Anecdotes, and Legends: A Chronicle of Life Assurance* (London: 1853), p. 58.

6. M. G. Kendall, "The Beginnings of a Probability Calculus," in E. S. Pearson and M. G. Kendall (eds.), *Studies in the History of Statistics and Probability* (Darien, CT: Hafner, 1970), pp. 19–34.

7. Gambling was interpreted as a profanation of God's chosen method for revealing his will on the basis of several Old Testament passages, for example, *Numbers* 33 : 54; *Proverbs* 16 : 33. For an account of the Catholic position on gambling from patristic writings through the Council of Trent (1607), see Abbé Coudrette, *Dissertation théologique sur les lotteries* (n.p.: 1742).

8. *Ordonnances, Statut et Police Novvellement Faicte par le Roy Nostre Sire, svr le faict des contractz des assevrances es Pays-Bas* (Anvers: 1571), Article 32.

9. All ordinances regulating insurance prior to 1681 that mention life insurance prohibit it and usually prohibit bets on lives in the same clause: for example, those of Amsterdam (1598), Middlebourg (1600), Rotterdam (1604), and Sweden (1666). Conversely, those who defended life insurance also defended such bets (usually placed on the lives of celebrities or on the outcome of wars, plagues, papal elections, etc.). Life insurance remained illegal in France until 1819, and some eighteenth-century writers continued to entertain the suspicion that it promoted crime. Isidore Alauzet, *Traité général des assurances* (Paris: 1843), vol. 2, pp. 466, 442–444; Nicolas Magens, *An Essay on Insurances* (London: 1755), vol. I, p. 33.

10. Ernst Coumet, "La theorie du hasard est-elle née par hasard?" *Annales: Economies, Sociétés, Civilisations*, 25 (1970), 574–598.

11. See Lorraine J. Daston, "The reasonable Calculus: Classical Probability Theory, 1650–1840," Unpublished Ph.D. dissertation (Harvard University, 1979), pp. 51–61; also Daston, "Probabilistic Expectation and Rationality in Classical Probability Theory," *Historia Mathematica* 7 (1980), 234–260.

12. See, for example, Abraham De Moivre, *The Doctrine of Chances*, 3rd ed. (London: 1756), Preface, p. 254.

13. John Graunt, *Natural and Political Observations Mentioned in a Following Index and Made upon the Bills of Mortality* (London: 1662). Although it has been suggested that Graunt thought of these proportions as probabilities, I see no evidence for this interpretation: see Daston, "Reasonable Calculus," pp. 42–44.

14. See, for example, the correspondence between Christiaan and Lodewijk Huygens, in Société Générale Néerlandaise d'Assurances sur la Vie et des Rentes Viagères, *Mémoires pour servir à l'histoire des assurances sur la vie et les rentes viagères au Pays-Bas* (Amsterdam: 1898), pp. 58ff. (Also in Société Hollandaise des Sciences, *Oeuvres complètes de Christiaan Huygens* (La Hague: 1895), vol. 6, pp. 482ff.)

15 See Daston, "Reasonable Calculus," pp. 13–37.

16. Jakob Bernoulli, *Ars conjectandi* (Basel: 1713), pp. 223-228.

17. C. I. Gerhardt, *G. W. Leibniz Mathematische Schriften* (Hildesheim: Georg Olms, 1962; reprint of 1855 ed.), vol. 3, part 1, pp. 11-89.

18. Graunt, *Natural and Political Observations* (note 13), p. 7.

19. Jean De Witt, "Treatise on Life Annuities," trans. F. Hendriks, in Robert G. Barnwell, *A Sketch on the Life and Times of Jean De Witt* (New York: 1856), p. 84.

20. Société Générale Néerlandaise, *Mémoires* (note 14), pp. 1-17, 24-33 (correspondence with Hudde).

De Witt's calculations priced an annuity on the head of someone aged 3-53 at 16 years' purchase, a substantial difference if the annuities were sold in quantity. It is likely that Hudde's data was the basis for the annuities priced by age sold by the city of Amsterdam in July 1672 and January 1673. Société Générale Néerlandaise, *Mémoires*, pp. 74-75, reproduces these tables.

21. Edmund Halley, "An Estimate of the Degrees of the Mortality of Mankind, drawn from the curious Tables of the Births and Funerals at the City of Breslau; with an Attempt to ascertain the Price of Annuities on Lives," *Philosophical Transactions of the Royal Society of London*, 17 (1693): 596-610.

22. Other important works include those of Struyck (1740), Simpson (1742), Deparcieux (1746), and Price (1771).

23. De Moivre, *Annuities for Life* (London: 1725), p. 2.

24. De Moivre, *Annuities*, p. v.

25. De Moivre, *Doctrine of Chances*, p. 251 (note 12).

26. De Moivre, *Annuities*, 3rd ed. (London: 1750), p. 326 (note 23).

27. De Moivre, *Annuities*, 3rd ed., p. 288. Christiaan Huygens had called this the gambler's, as opposed to the annuitant's, method. See note 14.

28. Thomas Simpson, for example, followed De Moivre, but Richard Price opted for the other method: Richard Price, *Observation on Reversionary Payments*, 3rd ed. (London: 1773), pp. 170ff. For an overview of these controversies, see Price, *Observations*, especially pp. 170ff., 205ff., 227ff. D'Alembert's debate with Daniel Bernoulli also turned on the problem of the true method of estimating life expectancies: see Daston, "D'Alembert's Critique of Probability Theory," *Historia Mathematica*, 6 (1979), 259-279.

29. Antoine Deparcieux, *Essai sur les probabilités de la durée de la vie humaine* (Paris: 1746), p. 123.

30. There seem to have been significant national differences in the faith in and ability to gauge the regularity of human mortality. While Dutch purchasers of annuities in the late sixteenth century cannily chose the longest lives (children age 5-12), their eighteenth-century French counterparts consistently chose elderly heads. Moreover, belief in the regularity of human mortality seems to have been fairly rare even among educated Frenchmen of the period: Société Générale Néerlandaise, *Mémoires*, p. 211 (note 14); Deparcieux, *Essai*, pp. 74-75.

31. Quoted in Société Générale Néerlandaise, *Mémoires*, p. 89 (note 14).

32. Keith Thomas, *Religion and the Decline of Magic* (New York: Scribner, 1971), pp. 651-656.

33. For an overview of the eighteenth-century British insurance scene, see H. A. L. Cockerell and Edwin Green, *The British Insurance Business, 1547-1970* (London: Heinemann, 1976).

34. Francis, *Annals*, pp. 140ff. (note 5); Magens, *Essay*, vol. I, p. 30 (note 9).

35. G. Leti, *Critique historique, politique, morale, economique, & comique sur les lotteries* (Amsterdam: 1697), vol. I, p. 1. On lotteries, see also Daniel Defoe, *The Gamester* (London:

1719); Coudrette, *Dissertation* (note 7); Claude Menestrier, *Dissertations des lotteries* (Lyon: 1700).

36. John Carswell, *The South Sea Bubble* (London: Cresset Press, 1960), pp. 155, 138.

37. Francis, *Annals*, pp. 144-145 (note 5).

38. Magens, *Essay*, vol. I, p. 84 (note 9); Samuel Marshall, *Treatise on the Law of Insurance* (Boston: 1805), Book I.

39. Cockerell and Green, *British Insurance*, p. 27 (note 33).

40. Maurice Edward Ogborn, *Equitable Assurances* (London: George Allen and Unwin, 1962), p. 48.

41. Magens, *Essay*, vol. I, p. 34 (note 9).

42. Price, *Observations*, p. 121 (note 28).

43. Ogborn, *Equitable*, p. 20 (note 40); see Deparcieux, *Essai* (note 29), pp. 50ff., on tontines; also Société Générale Néerlandaise, *Mémoires*, pp. 223ff. (note 14), for the original annuity-tontine scheme offered by the Dutch town of Kampen in 1670.

44. Ogborn, *Equitable*, p. 32 (note 40).

45. Barry Supple, *The Royal Exchange Assurance* (Cambridge: Cambridge University Press, 1970), p. 56; Royal Assurance Company, *Assurance book on Lives*, vol. I (1733–1737), MS. 8740 of the Guildhall Library, London.

46. Royal Assurance Company, *Assurance Book on Lives*, vol. 5 (1758–1771), MS. 8740 of the Guildhall Library, London.

47. Société Générale Néerlandaise, *Mémoires*, pp. 236ff. (note 14).

48. Price, *Observations*, pp. 2ff. (note 28).

49. Sybil Campbell, "Usury and Annuities of the Eighteenth Century," *Law Quarterly Review*, 44 (1928), 473–491; Campbell, "The Economic and Social Effect of the Usury Laws in the Eighteenth Century," *Transactions of the Royal Historical Society*, 16 (1933), (4th ser.) 197–210.

50. Supple, *Royal Assurance*, p. 67 (note 45).

51. Robert Pothier, *Traité des contracts aléatoires* (Paris: 1775), p. 75.

52. Johann Süssmilch, *Die göttliche Ordnung in den Veränderungen des menschlichen Geschlechts*, 3rd ed. (Berlin: 1775), vol. I, p. 49.

53. Concerning the problem of determining the true mortality curve, see Laplace and Legendre's report to the Paris Académie des Sciences on the work of Kramp, presented May 16, 1789. MS. *Procès-Verbaux*, Archives de l'Académie des Sciences, 1789, 108: pp. 137ff.

54. This meant that all insurance schemes were submitted to the Académie des Sciences for expert evaluation by mathematicians. See, for example, the reports of Laplace and Condorcet on an annuity project (March 6, 1790), of Nicole and Buffon on Deparcieux (July 21, 1745), and of Condorcet and Laplace on a proposed life insurance company (May 16, 1787), all in the MS. *Procès-Verbaux*.

55. Halpérin, *Assurances*, p. 45 (note 1); Marshall, *Treatise*, p. 672 (note 38).

56. Blaise Pascal, *Oeuvres completes*, ed. Jean Mesnard (Paris: Desclès de Brouwer, 1970), vol. I, p. 1034.

57. See note 36.

58. James Dodson, *The Mathematical Repository*, 2nd ed. (London: 1775), vol. 3, Preface.

59. See Daniel Bernoulli, "Specimen theoriae novae de mensura sortis," *Commentarii Academiae Scientiarum Imperialis Petropolitanae* 6 (1738), 175-192 (English translation by Louise

Sommer, "Exposition of a New Theory on the Measurement of Risk," *Econometrica* 22 (1954), 23–36); also George Leclerc Buffon, "Essai d'arithmétique politique," *Supplément de l'Histoire Naturelle* (Paris: 1777), vol. 4, pp. 46-148; and Defoe, "Gamester" (note 35).

60. E. A. Masius, *Lehre der Versicherung und statistische Nachweisung aller Versicherungs-Anstalten in Deutschland* (Leipzig: 1846), p. 476.

61. Magens, *Essay*, p. iv (note 9).

62. Johann Tetens, *Einleitung zur Berechnung der Leibrenten und Anwartschaften die vom Leben und Tode einer oder mehrerer Personen abhängen* (Leipzig: 1785), p. v. Tetens, however, regularly described the amount paid into a widow's fund as "gewagt, oder aufs Spiel gesetzet" (p. 160).

63. J. P. Brissot de Warville, *Seconde lettre contre la Companie d'Assurance pour les Incendies à Paris, & contre l'agiotage en général* (London: 1786), p. 30. Brissot's motives may be somewhat suspect, since he was then in the employ of the financier Étienne Clavière, who had much to lose if the company was granted a royal patent, but his Rousseauian rhetoric played upon familiar themes: see Jean Bouchary, *Les Manieurs d'argent à Paris à la fin du XVIIIᵉ siècle* (Paris: Bibliothèque d'Histoire Economique, 1939), vol. I, pp. 70–71.

64. Quoted in Jean Leonnet, *Les loteries d'état en France su XVIIIᵉ et XIXᵉ siècles* (Paris: Imprimerie Nationale, 1963), p. 44.

65. John Raithby, ed., *The Statues at Large, of England and of Great Britain* (London: 1811), vol. 13, p. 685 (Anno 14 Georg ii III.c.48).

66. The case of William Pitt's coachmaker illustrates some of the problems in interpreting the law. The coachmaker insured Pitt's life for £500 as security for Pitt's debts to him. When Pitt died in 1806 owing an amount in excess of £1,000, the premiums were paid up. However, when Parliament appropriated funds to pay off the entire debt, the insurance company refused to pay up on the grounds that it was the debt that was insured, not the life, since only the debt could be construed as legitimate insurable interest. The court decision given by Lord Ellenborough upheld the insurance company on the grounds that "this assurance . . . is in its nature a contract of indemnity, as distinguished from a contract of gaming or wagering." Quoted in Ogborn, *Equitable*, p. 148 (note 40).

67. For a complete history of the Equitable, see Ogborn, *Equitable* (note 40).

68. The Equitable Society still possesses a copy of these lectures in manuscript, along with many other documents relating to its early history.

69. Corbyn Morris, *Observations on the Past Growth and Present State of the city of London* (London: 1750).

70. Quoted in Ogborn, *Equitable*, p. 35 (note 40).

71. *A Short Account of the Society for Equitable Assurances on Lives and Survivorships* (London, August 2, 1764).

72. See, for example, the 1671 prospectus quoted in Société Générale Néerlandaise, *Mémoires*, p. 230 (note 14).

73. French opponents of insurance like Brissot de Warville worried that this might weaken the spirit of altruism and communal aid: Brissot de Warville, *Dénonciation au public d'un nouveau project d'agiotage; ou lettre a m. le Comte de S**** (London: 1786), pp. 31–34.

74. Compagnie Royale d'Assurances, *Prospectus de l'établissement des assurances sur la vie* (Paris: 1788). The mathematician was Duvillard.

75. Price, *Observations*, p. 121 (note 28).

76. Ogborn, *Equitable*, pp. 53, 81 (note 40): *Rough Minutes of the Weekly Courts*, vol. 2 (January 3, 1764—March 26, 1765), MS. volume of the Equitable Society.

77. Price, *Observations*, pp. 128–130 (note 28).

78. Equitable Society, *Orders of the Court of Directors* (1774–1848), MS. volume of the Equitable Society.

79. Francis, *Annals*, pp. 272–273 (note 5).

80. Quoted in Ogborn, *Equitable*, p. 105 (note 40).

81. Price, *Observations*, 5th ed. (London: 1803), pp. 175ff.

82. Quoted in Ogborn, *Equitable*, pp. 124–125 (note 40).

83. Quoted in Ogborn, *Equitable*, p. 206. See also Augustus De Morgan, "Reiew of Théorie Analytique des Probabilités. Par M. le Marquis de Laplace," *Dublin Review*, 2 (1837): 338–354, especially pp. 341ff.

84. Compagnie Royale, *Prospectus* (note 74), p. 53. See also Nicolas Fuss, *Éclairissements sur les éstablissemens public en faveur tant des veuves que des morts* (St. Petersburg: 1776), p. 31.

85. Supple, *Royal Exchange*, p. 66 (note 45). An 1805 overview of the London life insurance industry notes the remarkable success of the Equitable, but attributes it to the system of granting membership to all policyholders, who therefore are entitled to part of the profit. The mathematical basis of the Society is not even mentioned: Marshall, *Treatise*, p. 665 (note 38).

86. Pierre Simon Laplace, *Théorie analytique des probabilités*, in *Oeuvres* (Paris: 1847), vol. 7, p. 481; *Essai philosophique sur les probabilités* 5th ed. (Paris: 1825), pp. 192–194.

87. Adolphe Quetelet, *Instructions populaires sur le calcul des probabilités* (Brussels: 1828), pp. 195–196.

88. See, for example, Francis, *Annals*, p.282 (note 5); Underwriter's Agent of New York, *The Agent* (July 1872), p. 13.

89. I am told by currently practicing actuaries that only mortality statistics exhibit stable regularities reliable enough to fix premiums by; indeed the very regularity of mortality statistics has led some actuaries to reject probability as a basis for life insurance: see, for example, J. P. Van Rooijen, *La notion de probabilité et la science actuarielle* (Amsterdam: N.V. Noord-Hollandsche Uitgevers Maatschappi, 1935).

90. Albert Chaufton, *Les Assurances* (Paris: 1884), vol. I, p. 304.

91. Chaufton, *Assurances*, vol. I, pp. 291–297. But see Francis, *Annals*, pp. 242–247 (note 5) regarding the difficulty of persuading the workers of the virtues of life insurance over saving banks.

92. H. Laurent, *Traité du calcul des probabilités* (Paris: 1873), p. 208.

93. Chaufton, *Assurances*, vol. I, p. 208 (note 90).

94. The Association was soon rechristened the Manchester Steam User's Association. The original promoter of boiler insurance, one Thomas Forsyth, was killed in a boiler explosion while testing a locomotive shortly thereafter: Cockerell and Green, *British Insurance*, p. 51 (note 33).

95. Alauzet, *Traité*, vol. I, p. 303 (note 9).

96. D. Defert, J. Donzelot, G. Maillet, and C. Mevel, *Socialisation du risque et pouvoir dans l'entreprise*, typescript (Paris: Ministre du Travail, 1977), pp. 24–25.

97. Alauzet, *Traité*, vol. I, pp. 229, 32 (note 9).

98. Chaufton, *Assurances*, vol. I, pp. 62, 217, 4 (note 90).

11 The Objectification of Observation: Measurement and Statistical Methods in the Nineteenth Century

Zeno G. Swijtink

The application of nondeductive methods of inference, developed in the course of the nineteenth century, like the Method of Least Squares and its refinements, was made possible by an objectification of scientific measurement and has itself led to a further objectification of scientific practice. Scientific measurement has been made objective, that is, independent of the personal judgment of individual observers, through the use of precision measuring instruments that allow for unequivocal readings, or, in cases where this was not possible, by considering the human observer himself as part of the measuring apparatus. All the information obtained in an observation is thus made propositional.

It is argued that this development toward "observation without an observing subject" was a precondition for the applicability of nondeductive methods of reasoning, since these require that all the relevant information obtained in an observation can be expressed in terms that have intersubjective meaning in the relevant science. The presence of nonpropositional, ineffable information, itself incapable of entering as premises in any argument, may frustrate inference.

Introduction

There exists a legend that Gauss measured the triangle Brocken-Hoher Hagen-Inselberg, three mountaintops near his residence in Göttingen, in order to ascertain whether terrestrial geometry is Euclidean. Within the limits of experimental error, it is said, he indeed found the sum of angles equal to two right angles. Recently the story has been told by Hans Reichenbach, Carl Hempel, and others, but it must have been created during Gauss' life. It is an unlikely story, and there is no evidence for it in his published work. But it is a real legend, with some basis in fact.[1] For Gauss was opposed to the Kantian doctrine of space as only a form of our *Anschauung*, and thus a priori, but believed that general experience with geodesic measurements had shown that it is, as we would say, "locally Euclidean." In a letter of July 12, 1831, directed to Heinrich Schumacher, head of the observatory in Altona, Gauss writes[2]

Indeed, in non-Euclidean geometry the circumference of a circle with radius r is

$$\tfrac{1}{2}k(e^{r/k} - e^{-r/k}).$$

In this expression, k is a constant, which, *as we have learned from experience, has to be extremely large in comparison with everything we can measure.* In Euclid's geometry, k is infinite.

*Part of the research for this chapter was conducted while the author was a resident of the Zentrum für interdisziplinäre Forschung of the Universität Bielefeld, F.R.G., during 1982–83, supported by a stipend of the *Alexander von Humboldt-Stiftung*. Their support is gratefully acknowledged.

Gauss felt secure enough to assert that measurements had shown, within the limits of geodetic expertise of the early nineteenth century, that space is Euclidean. But this belief was a by-product of measurements made by him and others for different purposes, to determine the figure of the earth, the length of a meridian, or to obtain reliable maps.[3]

Gauss did indeed live in a period in which scientists started to impress one another, and the general public, with precise numerical determinations and precise quantitative predictions. There were determinations of the parallax of stars, and the resulting calculation of the approximate distance of those stars, by Bessel, Struve, and Henderson; the prediction of Ceres in December 1801, by Gauss himself, based on scant observations made by Piazzi earlier that year;[4] geodetic surveys of unheard-of quality, starting with Méchain and Delambre, Gauss, and Bessel and Baeyer;[5] the publication of Wollaston's "Table of Equivalents," in the *Philosophical Transactions* of 1814, which heralded an increased and systematic determination of "constants of nature": specific gravities, boiling and melting points, specific heats and coefficients for expansion by heat, and atomic weights;[6] and Wilhelm Weber's measurements of terrestrial magnetism,[7] to mention just a few.

The success of these precise numerical determinations derived from the interplay of several developments. First of all, new or improved measuring instruments, and a quantitative understanding of their systematic errors, which was then used in the reduction of the raw data. Second, an improved qualitative understanding of the causal factors that influence the outcome of experiments, which made it possible to control for these factors. Third, the abandonment of crude inductivism, implied in the recognition of measurement error, an inductivism in which experience can only be criticized and corrected by experience itself, and the adoption of a mild form of the method of hypothesis, in which no theoretical entities were introduced, but phenomenological laws of mathematical form were postulated that state exact relations between physical concepts directly based in experience.[8] This made it possible to criticize experience—for instance, by claiming that there were impurities involved, without having direct evidence for such impurities. Postulated knowledge could thus become a guide for the improvement of experiments and measuring procedures. And fourth, the development of numerical methods to process the raw data, after they had been corrected for systematic errors, in order to obtain final best estimates, and of notions concerning the reliability and accuracy of these methods, as methods for approximate calculation, rules for interpolation and extrapolation, rules for rounding off, rules for the combination of inconsistent observations, rules for the rejection of extremely discordant observations, and notions like probable error and last significant digit.

An understanding of the emergence and significance of precision measurements requires a discussion of the interplay of all four of these developments. In this paper I have to restrict myself to arguing for one thesis concerning the first and last developments: I shall try to show that the nature of the information that is obtained in the course of a measurement had to change before certain numerical methods, in particular those related to the theory of errors, could be used. Due to the nondeductive character of these rules, they require that all the information obtained in a measurement be, as I shall call it, "propositional," that is, unambiguously expressible in statements with well determined inferential relations. This change made observations and measurements more "objective"; that is, it diminished the role of personal judgments and interpretations in determining what is observed. On the other hand, the very use of rigid, numerical methods also led to an objectification of science, by eliminating personal "whim," by making observers only observe what can serve as input for these methods, by leading to a division of labor between observers and calculators, and so on. These historical changes deserve a more careful and diverse documentation than I can give here. On another occasion I intend to illustrate this development by taking two geodetic measurements, one from the late eighteenth century, the other from around 1860, and to show that in the latter much less nonpropositional, "ineffable," information was obtained, which might have been interpreted differently by different scientists. Although theory was involved in both measurements (of length of degree of arc of the meridian, and of difference in longitude), both in the correction of the raw data and the interpretation of the results, these were not determinations of theoretical parameters, or of other constants of nature. Only in this latter case would one expect the full interplay of the four factors in the emergence of precision measurements: improved instruments with propositional output; experimental control; method of hypothesis, resilience against overly quick empirical falsification; and the use of numerical methods.

1 Measuring Instruments

New or improved measuring instruments were essential to the quantitative sophistication that emerged during the last decades of the eighteenth century. Bessel's 1838 determination of a (relative) parallax, for instance, made use of a heliometer build by Fraunhofer, following Bessel's specifications.[9] As the name indicates, the heliometer was originally designed to measure the diameter of the sun, later also those of the planets or the distances between Jupiter or Saturn and their respective satellites. It is basically a telescope, with an objective that can be rotated along the optical axis of the telescope, and that has been cut along a diameter. Both halves can be shifted along the cutting line, perpendicular to the optical axis, and this displacement can be measured very accurately. Each half of the objective provides an image of an object. In the null position, these two images coincide, but when the halves are moved two noncoincident images of the same object arise. To measure, say, the diameter of the sun, one moves one half of the objective until its image lies exactly beside the other image. To measure the distance between two objects in the visual field of the telescope, one has to rotate the objective until the cutting line coincides with the imaginary line connecting the two objects, and shift one-half of the objective until the image it provides of one of the objects coincides with the image the other half provides of the second object. Parallax is the apparent displacement of an object due to a change in point of observation while the earth turns around the sun. Knowledge of a parallax makes it possible to calculate the distance of the object if one knows the diameter of the orbit of the earth around the sun, but parallaxes are hard to determine because they are so small compared with the uncertainties due to atmospheric refraction, the oscillation of the earth's axis, instrumental irregularities, and so on. But, as Galileo has pointed out, relative parallaxes avoid these problems, because both stars are similarly affected by these uncertainties. To measure relative parallax, one observes the change in apparent angular distance between a pair of stars that are very close together on the sphere, but of which one is actually much closer to the earth than the other. Before the heliometer, thread micrometers were used to measure parallaxes, for instance, by Struve, but the heliometer was more precise, as the micrometer has to be moved from pointing at the first star to pointing at the second. The interpretation of the raw data obtained with a heliometer is not straight-forward. A special dioptric theory for each individual instrument has to give the constants with which the observations have to be reduced, and in developing such a theory Bessel was greatly helped by the fundamental dioptric investigations of Gauss.[10]

Gauss' own geodetic field work, the big triangulation of Hannover, 1818–1847, was improved through his invention of the heliotrope. Its purpose was to make the points of triangulation more visible to the theodolite, a notorious problem in large-distance triangulations. It has a very simple construction, consisting of mirrors and a small telescope with which the sun's rays can be reflected exactly in the direction of the observing theodolite.

Bessel, in his survey with Lieutenant-General Baeyer of Eastern Prussia, made use of improved measuring rods that enabled him to measure the base line Trenk-Medniken, a distance of a little over $1\frac{1}{2}$ kilometer, as 1822.3447 ± 0.0027 meters, that is, with a probable error of less than 3 millimeters. The rods were made of zinc and iron, and were carried on a strecher that prohibited them from bending, while careful registration of temperature made it possible to correct the raw data for expansion.[11]

Earlier, in the triangulation by Méchain and Delambre, which led to the decimal metric system of meters and kilos, points in the net, such as cocks and crosses, were still accentuated by bunches of hay or straw, or by flags. But that investigation was also much indebted to an innovation: the repeating circle. Originally designed by Tobias Mayer around 1767, improved by Borda, the circles used in the survey were designed and built by the famous French instrument-maker Lenoir. The repeating circle is a modification of Mayer's original reflecting circle, but both are repeating instruments; that is, one and the same angle is measured repeatedly, without turning the circle to zero. By dividing the angle read off by the number of repetitions made, errors in the graduation of the circle and reading errors would be averaged out. Mayer's instrument, meant to measure the angle between two stars, uses a telescope and two mirrors. One watches through the telescope the image of one star in the mirror and turns the other mirror until the indirect image of the second star coincides with that of the first star. Lenoir's version, meant solely for geodesic measurements of angles, has two telescopes and no mirror. The two telescopes are used physically to add the angles on the graduated circle. One turns the whole instrument around its pivot and focuses with the lower telescope on the point that formerly was the point of focus of the upper telescope, unscrews the upper one, and turns it around to the other station.[12] We see here also already the beginnings of a "theory of the instrument," the calculation of correction factors due to peculiarities of the individual instruments used. For instance, the lower telescope was eccentric, not collinear with a line through the midpoint of the circle, and Delambre carefully calculated which corrections had to be made in the raw data due to this eccentricity. The raw angle between two stations with a distance to the observer of 1,000 toises, for instance, had to be corrected by $+2''15$.[13]

In chemistry, balances were constructed that would turn with a 500,000th of the greatest load that could be safely placed in either pan. Weighings were done in vacuum, and sophisticated weighing procedures were followed to eliminate errors by mechanical means, like the method of reversal. This process, also called "Double Weighing," eliminates errors arising from unequal lengths of the arms, and calls for reversing the weights, and estimating the difference by adding small weights to the deficient side.[14] Similar reversals were introduced in astronomical and geodesic measurements where a telescope would be lifted out of its support and be reversed to eliminate errors caused by imperfect adjustment of the collimation and the horizontal axis of the telescope.

Writers on scientific method exulted over the "exact measurement of phenomena" made possible by the new and more sophisticated measuring instruments.[15] The chemist Humphry Davy wrote in his *Chemical Philosophy*, "Nothing tends so much to the advancement of knowledge as the application of a new instrument. The native intellectual powers of men in different times, are not so much the causes of the different success of their labours, as the peculiar nature of the means and artificial recourses in their possession."[16] John Herschel added,[7]

But it is not merely in preserving us from exaggerated impressions that numerical precision is desirable. It is the very soul of science; and its attainment affords the only criterion, or at least the best, of the truth of theories, and the correctness of experiments. Thus, it was entirely to the omission of exact numerical determinations of quantity that the mistakes and confusion of the Stahlian chemistry were attributable,—a confusion which dissipated as morning mist as soon as precision, in this respect, came to be regarded as essential. ... we are obliged to have recourse to instrumental aids, that is, to contrivances which shall substitute for the vague impressions of sense the precise one of number, and reduce all measuring to counting.

In Germany as well, in a theoretical work on measurement that anticipates the writings of Hermann von Helmholtz on this topic, Wilhelm Weber exclaimed, "Measurement is the most essential part of the study of nature, especially of the physical sciences."[18]

"The ancient Chaldaeans recorded an eclipse to the nearest hour, and even the early Alexandrian astronomers thought it superfluous to distinguish between the edge and the center of the sun." But, William Stanley Jevons continued, thanks to the new astronomical apparatus, "we now take note of quantities, 300,000 or 400,000 times as small as in the time of the Chaldaeans."[19]

Being able to carry out precision measurements became a skill eagerly sought after by experimental scientists. Indeed, the very term "precision measurement" became a term of trade. Looking back upon his career and contrasting the two main sources of his physical thinking, Louis Paschen wrote in 1925, "As an assistant to Hittorf I got the opportunity to learn what was insufficiently emphasized in the school of Kundt, namely, to perform precision measurements as they were done by Regnault. [There I was] initiated in the technique of delicate precision measurements."[20]

The movement for precision measurements and finely tuned measuring instruments came of age with the emergence of scientific journals solely devoted to the theory of instruments irrespective of the field of science, such as the *Zeitschrift für Instrumentenkunde*, founded in 1881 with the help of the Physikalisch-Technische Reichsanstalt in Berlin.[21] The *Deutsche Gesellschaft für Mechanik und Optik* published announcements and reports of its meetings in the journal, as did the *Fachverein Berliner Mechaniker und Optiker*. In an editorial, the editor-in-chief wrote that the journal intends to "revive the fruitful contact between representatives of science and of the art of machinery, and further critical discussion of instruments and methods of measurements."[22]

Another journal that marks the professionalization of precision is the *Zeitschrift für das Vermessungswesen*, founded in 1872 as the house organ of the *Deutscher Geometerverein*. It covered all aspects of surveying, including its professional organization and the protection of its professional identity, and had, from the start via the cadaster, a strong link with society at large. Surveyors were public servants working under oath, and standards of precision for different types of triangulations were set by law. A similar connection with the society was created for the mechanical precision movement by laws stating safety levels for bridges, buildings, and equipment.

2   Objectification

Jevons remarked that his contemporaries took note of quantities 300,000 times as small as in the time of the Chaldaean era. Herschel added that precision is the only, or at least the best, criterion for truth in theory and correctness in measurement. Still, the more important epistemological point is that the data of science, thanks to instruments for observation and measurement, had become more and more impersonal. After the mechanization of our world picture we mechanized the observation of our world. In his expressive nineteenth-century voice, Charles Babbage, in that peculiar chapter "Of Observations" in his *Reflections on the Decline of Science in England*, said, "He who can see portions of matter beyond the ken of the rest of his species, confers an obligation on them, by recording what he sees; but their knowledge depends both on his testimony and on his judgment. He who contrives a method of rendering such atoms visible to ordinary observers, communicates to mankind an instrument of discovery, and stamps his own observations with a character, alike independent of testimony and of judgment."[23] Instruments of observation and measurement make scientific knowledge independent of the testimony of judicious and honest witnesses. This is not just a metaphor, but refers to a scientific practice that emerged when the investigation of nature shifted from the secret hideaway of the alchemist, with his one-on-one relationship with Nature, to the open space where the public was invited.[24] Reports of demonstrations and experiments would be signed by honorable witnesses, as we still see today with investigations into psychokinesis, clairvoyance, and so on. Boyle did it around 1660, and we still see it around 1833: "several physicists have since then seen the phenomenon in my house, and all have declared it to be decidedly electrical. Especially Captain Kater could observe it most completely, because he stayed a considerable time with me in comparison with some passing strangers who were only able to see it once."[25]

Scientific instruments create replicability. Observation is no longer the observation of the there-and-then; it is the observation of a universal, the phenomenon, the effect. The Leyden jar of 1745 demonstrated static electricity; the air pump of 1650, invented by Otto von Guericke, created the vacuum. But only around the end of the eighteenth century had technology progressed to a point that sufficient replicability was obtained. Similarly, instruments of observation brought scientific observation within the reach of every, properly trained, human being. And measuring instruments tended to diminish the role of personal judgment in recording what is observed, because it just amounts to reading a scale. At present, this development has culminated in measuring instruments that can literally be read because a digital value is presented on a screen or counter, or is directly stored in the memory of a computing machine.[26] Such direct read-out systems say, as it were, out loud what they "observe," and the observer literally reads the Book of Nature. Babbage again: "Genius marks its tract, not by the observation of quantities inappreciable to any but the acutest senses, but by placing Nature in such circumstances, that she is forced to record her minutest variations on so magnified a scale, that an observer, possessing ordinary faculties, shall find them legibly written."[27]

Although using a different technology, instruments that write down what they observe were already developed in the eighteenth century. Originally, these so-called "self-recording" instruments may just have been developed for convenience, as, for example the recording aneometer by Pajot d'Ons-en-Bray of 1734, which enabled continuous registration of wind direction and wind velocity over a period of thirty hours. Registration was on strips of paper that wound from one spool on to another, driven by clockwork.[28] This technique became standard in the nineteenth century, where we meet the self-recording water level meter of Paulsen, Prytz, and Rung, and the self-recording meter for the specific weight of sea water of the Danish Meteorological Institute.[29] Here used to avoid labor, they served elsewhere to lessen the influence of personal judgment of an observer, whose only remaining role was to determine the distance between lines on a piece of paper.[30]

In those cases where existing technology was insufficient to eliminate the human observer from scientific observation, the observer was often looked upon as himself part of the measuring apparatus. Like so much of the impetus to make precision measurements, this innovation also came out of astronomical research.[31] The discovery of the so-called "personal equation" between two observers is probably due to Bessel, who in 1823 noticed that systematic differences between observers, *in casu* the English astronomer Maskelyne and his assistant Kinnebrook, were not due to one of the two following a wrong method of observation, but to physiological differences, outside the control of the observers. Bessel reports, for instance, a difference between himself and Struve of one full second, on the average.[32] Since then it has become customary to reduce a set of data made by several observers to one observer, by correcting those data made by the other observers by an amount indicated by their personal equations, that is, the equation of the average difference between two observers. Observers and instruments were thus treated alike: just as a transit has to be corrected for collimation error, observers have to be corrected for the peculiarities arising from their physiological and psychological wiring. In a sense, these observers are the first self-recording instruments with digital output.

All these development have thus lead to, to paraphrase Karl Popper, observation without an observing subject. They have diminished and in some areas almost eliminated the use of a personal judgment in recording what is observed. It has made scientific observation objective. Scientific data do not belong to the consciousness of the perceiving subject, we believe, because different observers will obtain the same data.

3   Numerical Methods

Besides the development and improvement of measuring instruments, there is another trend, starting in about 1750, in Euler's time, that has had a decisive influence on our conception of good science. It can be broadly described as the development of numerical methods. It is an objective of this paper to argue for a thesis about the interaction between these two developments. I shall try to show that the kind of data scientific observation provides had to change before certain kinds of numerical methods could be used.

I mentioned already that Gauss had startled the scientific world in December 1801 by calculating, on the basis of a few observations made by Piazzi at the Palermo Observatory, the principle characteristics of the orbit of what turned out to be a new planet, which was named Ceres. This enabled him to write down an ephemeris or astronomical almanac for Ceres, covering a period of four months. With the almanac, respected astronomers such as Olbers were able to reidentify the planet and to make further observations. In his calculation Gauss had made use of a new numerical method, of his own invention, the so-called Method of Least Squares.[33] The Method of Least Squares defines a unique resolution of an in some sense "inconsistent" set of numerical data, and moreover (and this may be even more important) it tells you exactly how to find that resolution. In the case of Piazzi's data, there was a set of numerical observations that did not define an ellipse with the sun at one of its foci. This was partly due to measurement error, and partly due to the gravitational attraction of other planets. Least squares addresses the errors of measurement. The true errors are not known, because the true orbit of Ceres is not known. The difference between the truth-as-estimated and the observation is called the residual. The method of least squares defines that ellipse to be the resolution of this set of data that would minimize the sum of the squares of the residuals. Gauss devised a recursive or mechanical procedure to find this least squares solution. During the nineteenth century, considerable effort was devoted to the task of developing other numerical procedures that were easier to operate with in specific problems. And the problem of the perturbations that divert planets from their elliptic orbits under the influence of a third body was one to which Gauss came back several times.[34]

To illustrate the method, consider barometric pressure as a function of altitude (length of the column of air) and unit weight of air. If we disregard the latter, or correct data with an estimate based on temperature and humidity, barometric pressure should be a linear function of altitude: $B = A \cdot x + y$. We try to estimate x and y (just as Gauss calculated the principle characteristics of Ceres's ellipse) from n observations b_i, a_i), which give n linear equations

$$b_1 = a_1 \cdot x + y,$$

$$b_2 = a_2 \cdot x + y,$$

$$\vdots$$

$$b_n = a_n \cdot x + y,$$

but when there are more than two independent equations, the system has no exact solution. Plotting the observations on graph paper shows that there is no line in the plane on which all observations lie. The altitude and/or barometric pressure may have been measured with error, or linearity may only hold approximately, or on the average, especially if we have not corrected for variation in the density of air. The question is, What is the "best" linear representation of the data, the best line in the plane?

In *Gradmessung in Ostpreussen*, Bessel gives eighteen measurements of the angle Mednicken-Fuchsberg, made at the station Trenk, varying between 83°30′30″25 and 83°30′37″50. The question is, again, What is the best point on the line? "Best point" may mean best point representing the observations or, what is different, best conjecture at the underlying truth. In the latter case one wants to bring to bear all the other information one has, and the Gaussian method would only be useful in very restricted situation of "all the other information one has." Mathematically it is again a problem of a set of inconsistent linear equations

$$a_1 = x,$$

$$a_2 = x,$$

$$a_n = x.$$

To mention a very complicated example, the surface of the earth is very nearly an ellipsoid, that is, the surface of an ellipse rotating on its minor axis. To fit such an oblate spheroid to data from a geodetic survey, or from pendulum measurements via Clairaut's theorem—relating the gravity on the equator and on the poles, and the flattening of the earth—is not a linear problem, but was approximated by one, using a Taylor approximation of the first degree.[35]

Mathematically, then, the problem is one of a set of, say, n inconsistent linear equations with k unknowns (k less than n),

$$0 = a_{11} \cdot x_1 + a_{12} \cdot x_2 + \cdots + a_{1k} \cdot x_k + b_1,$$

$$0 = a_{n1} \cdot x_1 + a_{n2} \cdot x_2 + \cdots + a_{nk} \cdot x_k + b_n,$$

where the a's and b's are known, and the x's are the unknowns. No matter what resolution is obtained, that is, no matter what choices for the unknowns are made, some of the resulting e's will be unequal to zero:

$$e_1 = a_{11} \cdot \bar{x}_1 + a_{12} \cdot \bar{x}_2 + \cdots + a_{1k} \cdot \bar{x}_k + b_i,$$

$$\vdots$$

$$e_n = a_{n1} \cdot \bar{x}_1 + a_{n2} \cdot \bar{x}_2 + \cdots + a_{nk} \cdot \bar{x}_k + b_n.$$

Before least squares, other procedures to resolve the inconsistency were proposed and developed, for instance, the Method of Averages (by Euler in 1749 and, independently, by Tobias Mayer in 1750), and a more objective, that is, deterministic, method by Boscovich in 1760.

In the Method of Averages, the overdetermined set of n equations with k unknowns (k less than n) is divided into k subsets, the division being made according to one of the a's or b's, the independently known elements. The equations in each group are added together, and set equal to zero. This gives a set of k independent equations, assuming that the original equations are independent, and thus a unique solution for the unknowns. For the case of one unknown, the Method of Averages gives the arithmetical average of the observations; in general, however, the result depends on the way the equations are divided into subsets, and the method was therefore felt to be somewhat subjective and arbitrary. But the calculations involved are relatively easy, and when the partitioning is done judiciously, for instance based on a variable a that is measured without error, the method may not be too bad for real data.

Boscovich's criterion overcame the subjectivity inherent in the Euler-Mayer Method of Averages. He proposed for the first time two criteria for determining the best fitting line $b = a \cdot x + y$ through three or more points: (1) the sums of the positive and of the negative residuals e_i (in the b direction) shall be numerically equal, and (2) the sum of the absolute values of the e_i's shall be a minimum. Thus

$$e_1 = b_1 - a_{11} \cdot x - y,$$

$$\vdots$$

$$e_n = b_n - a_{1n} \cdot x - y.$$

The first criterion states that $\sum e_i = 0$, that is,

$$\sum (b_i - a_{1i} \cdot x - y) = 0,$$

or

$$\sum b_i / n - x \cdot \sum a_{1i} / n - y = 0,$$

which shows that the Boscovich line goes through the point $(\sum a_{1i}, \sum b_i)$, the centroid of the observations. The unknown y can now be expressed in terms of the observations and of x, the slope of the Boscovich line:

$$y = \sum b_i / n - x \cdot \sum a_{1i} / n.$$

The second condition states that x and y must satisfy the equation

$$\sum |b_i - a_{1i} \cdot x - y| = \text{minimum}.$$

Replacing y by its value gives a condition on the slope of the equation, x:

$$\sum |(b_i - \sum b_i / n) - x \cdot (a_{1i} - \sum a_{1i} / n)| = \text{minimum}.$$

This last equation determines x uniquely, and Boscovich gave a geometrical method for solving it in 1760. It was only in 1789 that an algebraic formulation of his algorithm was described by Laplace. Boscovich's method has the advantage that a wildly different value in a set of data that overall exhibit a linear relationship has little influence on the best fitting Boscovich line. Methods like Boscovich's, for arbitrary dimensions, have recently attracted renewed attention in statistics and operation analysis.[36]

Legendre has priority in publishing the Method of Least Squares, in 1805, but Gauss discovered it independently, and claimed to have used it already in 1801, in his work on an ephemeris for Ceres. The criterion the method stipulates is to minimize the squares of the residuals (and wildly different values have therefore a greater influence on the resolution than in Boscovich's method). The best point on the line, the resolution of n observations of one and the same quantity x, is obtained by minimizing $\sum (a_i - x)^2$. Setting the derivative equal to zero, this shows that the best point is $\bar{x} = \sum a_i / n$, that is, the arithmetic mean. Except in his astronomical work, Gauss also used his method in his triangulations, and in his work on magnetism in collaboration with Wilhelm Weber.[37] Just as the science of measuring instruments became independent of the individual sciences, there also developed a field of numerical methods, with its own textbooks, journals, and professional organizations.[38] Methods for approximate calculation were developed, rules for interpolation and extrapolation, rules for rounding off, and such notions as last significant digit, average residual, and probable error. The Method of Least Squares became the dominant one of its kind, a method for resolving inconsistent sets of observations, but it never had a monopoly, and never went unchallenged. The method was justified in several ways, but it was used by people who could not accept any of these justifications (just as induction may be used by those who think that any justification of inductive rules is less plausible than what it sets out to justify), mainly because of its simple computations, the fact that it gave unique solutions, that the computer did not have to make arbitrary choices, and that it was generally applicable.[39] But because many of the justifications were probabilistic—two had already been given by Gauss himself—the method was, in the nineteenth century, a main vehicle by which probability theory was transmitted and extended, and a main stimulus for the study of empirical frequency distributions.

A decisive step in the probabilistic justification of any method for resolving inconsistent sets of observations was taken by Thomas Simpson in 1755 when he introduced the notion of the chance to make an error of a certain size connected with a single observation independent of the true value itself.[40] This led to the idea of a probability distribution of errors connected with a certain measurement procedure, independent of what quantity is measured. Gauss, in his first justification of the Method of Least Squares, published in 1809, showed that if one accepted as a postulate that the arithmetic mean was a posteriori the most probable value in the sense of Bayes's notion of inverse probability, the true errors must have a certain probability distribution (later called the Gaussian or normal distribution). If, moreover, there is no information whatsoever about the unknowns, and they can thus be assumed to have a uniform prior distribution, then the Method of Least Squares gives the a posteriori most probable values of the unknowns.[41] This ingenious justification combines in its conclusion two probability distributions of radically different origin. The a priori uniform distribution of the unknowns, admittedly somewhat of an idealization because often something is known about the quantities that are measured, e.g., that they are positive, is an epistemological distribution, and there is no claim that it expresses the chances with which these values are realized in nature. But the Gaussian error distribution is certainly an empirical frequency distribution, again somewhat idealized. Empirical investigations were made into the frequency distributions of errors connected with actual measurement procedures, to bolster the Gaussian justification, and assumptions about the sources of measurement error were made that could explain why the distribution had the form it had.[42]

Regular, constant, or systematic error in a measurement procedure was not a problem that the Method of Least Squares was meant to address. Gauss states explicitly, in his *Theoria combinationis observationum* of 1823, that those errors should be removed from the data before the Least Squares reduction is applied.[43] This is, of course, closely connected with the Postulate of the Mean, the postulate that the arithmetic mean is the (a posteriori) most probable value. If one knowns what systematic error the measurement procedure is subject to, the arithmetic mean of the raw data is of course not the most probable value. If one suspects there to be a systematic error, but does not know of what size, or in what direction, the arithmetic mean is still the most probable value, and the Method of Least Squares still gives the general criterion and procedure for finding the set of most probable values. But another of its central notions, that of probable error of one's estimate of the true value, i.e., the probable error of the most probable value, becomes fundamentally ambiguous.

Many enthusiasts of the Method of Least Squares considered this notion of probable error of central importance. It is a measure of the reliability of the most probable value. A measurement report may, for instance, say, 3 ± 0.3. The probabilistic interpretation is that it is an even bet that the true value is in the interval [2.7, 3.3] and that the most probable value is 3. There are basically two different ways to calculate the probable error: a calculation based on the data set that led to the value 3 itself, or a calculation based on an external data set, for instance, on a broad range of measurements that were made with the same measurement procedure. The first method would be followed when it was felt that the observations under consideration were made under special circumstances, which may have affected the precision of the measurement. Gauss showed how to calculate the probable error in the first volume of the *Zeitschrift für Astronomie* of 1816, but his calculation gave only a reliable estimate of the probable error when the set of observations is large. Real small sample work had to wait until 1908.[44]

The probabilistic interpretation of the notion of probable error (that it is an even bet that the true value lies in an interval around the most probable value of twice the probable error) breaks down when one believes that there is a systematic error in the measurement without having any inkling of its size or direction. This will be the case in almost any measurement of interesting physical quantities. It has been observed by Youden that in a series of fifteen determinations of the Astronomical Unit (the average distance between the sun and the earth), made between 1895 and 1961, each determination fell outside the interval of probable error of the previous one.[45] Similar phenomena have been noticed in successive determinations of the speed of light in vacuum, and so on. The customary interpretation of the probable error of a measurement result, if calculated at all in the nineteenth century, is therefore often that of a measure of the statistical control the scientific worker was able to reach in his experiment. A large probable error means little control, and thus a bad experiment.[46]

Although the notion of probable error is thus ambiguous, the need to calculate it in complicated situations of indirect or constrained measurement stimulated research in the mathematics of probability. Similarly, controversies about the foundations of Least Squares, as between Cauchy and Bienaymé, led to further clarification of the mathematics of probability.[47]

All in all, the influence of the Method of Least Squares, and similar reduction methods, has been enormous, both on the mathematical theory of probability, and on the measuring sciences. Cleveland Abbe, the geophysicist and first official weather man to the U.S. government, called it "the most valuable arithmetical process that has been invoked to aid the progress of the exact sciences,"[48] and Czuber wrote, commenting on the strong interest the Method of Least Squares had attracted,[49]

There are two reasons the topic has received such ardent attention. In the first place its metaphysics has always been very stimulating and has led to active involvement of the most prominent mathematicians. Second, its immense, and perhaps still growing, practical significance. It happens nowadays rarely that one, in some measuring discipline, derives results from observations that have not been looked at from the point of view of the theory of errors, and accordingly corrected in order to get a sharper result with a known degree of reliability.

4 Propositional Information

What is the relation between the two historical developments that I have described? On the one hand, I have argued that measurement, the main form of observation in physical science, was mechanized to the point of "observation without an observing subject." On the other hand, there was an immense growth in the use of numerical methods that led, in Czuber's words to "sharper results with a known degree of reliability," or, we may add, a known degree of statistical control. In the next section I shall argue the now obvious point that the use of numerical methods itself has led to an objectification of science. These are standard procedures that, overall, exclude the personal judgment of the individual. Here I want to argue that the first development, the mechanization of observation and measurement, was a precondition for the use of numerical methods in a field of science. To argue this historically, much more has to be done than I have done, or than I am able to do. One has to correlate in more detail the kinds of measurement procedures used in a field, during a certain period, and the extent to which numerical methods were used and the strictness with which they were applied. One should also look into organizational features, e.g., whether observers and calculators, field workers, and desk workers are different individuals.

The argument in this section will therefore be more abstract and philosophical. But this will have at least the advantage that the conclusion will be stronger than any that can be derived from history, for the relation is stronger than correlation only: numerical methods, especially those related to the theory of error, cannot be applied deterministically without a mechanization of measurement.

I take my cue from R. A. Fisher, who, in a discussion of his notorious fiducial argument, remarks,[50]

Had knowledge a priori been available, the argument developed above would have been precluded by the consideration that some of the relevant data had been omitted. For, although in the deduction of statements of certainty it is legitimate to draw inferences from some of the axioms available while ignoring others, or, in other words to base a valid argument on a chosen subset only of the available axioms, no such liberty can be taken with statements of uncertainty, where it is essential to take the whole of the data into account, though some part of it may be shown on examination to be irrelevant, and not to affect the result.

Fisher is referring to a fundamental difference between deductively valid arguments and inductively strong arguments. An argument is a piece of discourse, consisting of a set of premises, and a single conclusion. Premises and conclusion express propositions; that is, what they say is true or false. An argument is deductively valid if its conclusion follows strictly from its premises, in the sense that its premises could not be true and its conclusion false. Logic tries to find tools with which one can establish deductive validity, and the philosophy of logic tries to explain what could possibly be meant by the "could not." Similarly, an argument is inductively strong if it is improbable, given that the premises are true, that the conclusion is false. The fundamental difference Fisher is referring to is this: if an argument is deductively valid, it remains so if more statements are added to its set of premises; but if an argument is inductively strong, it may loose its strength if more premises are added. One way to understand this is to think about the "could not" of valid deductive arguments in terms of situations or possible worlds. In a valid deductive argument, then, the conclusion is true in all situations, or possible worlds, in which the premises are true. When more premises are added, the relevant set of situations to consider will be a subset of the original set. Because the conclusion is true in all situations in that possibly larger set, it will also be true in all situations in the subset, and the extended argument thus will still be deductively valid. This is not true for inductively strong, but not deductively valid, arguments. The conclusion of such an argument may be false while all the premises are true. So if one added the negation of the conclusion as an additional premise, the extended argument would be inductively invalid. Or, given the evidence available in 1830, a strong inductive case could be made for the Newtonian theory, but with our present-day extended evidence, the theory is refuted. Inductive inference is nonmonotone.

This last example, of course, concerns the application of logic to an actual knowledge situation. Fisher did not separate these two in the passage I have quoted, but it is customary to do so now. How do we apply logic, deductive or inductive? What is its relation to our inferring and deliberating? These questions are not easy to answer, but something has to be said about them because it is central to my historical explanation. A detailed defense has to be postponed to another occasion. The explicit use of deductive logic, except in disciplines such as law or mathematics whose task it is to find closely knit arguments, is mainly critical: using logic, one tries to find contradictions, tensions, and implication in one's beliefs and those of one's fellows, in order to discuss these critically. Inductive logic may have the same use, especially in detecting the tensions. This critical use of logic can, of course, also be made in the discussion of ideas that are not held by anyone in particular. Inferring, coming to believe something, changing one's mind, are causal processes, or results of causal processes, but they are not acts (if acts are causal processes); that is, one cannot decide to believe something as one can decide to have a haircut and then go and have one. Of course, critical use of logic is an important causal factor in inferring or changing one's mind. But one should not look upon the person who, while contemplating or discovering an argument with believed premises, came to believe the conclusion as deciding to accept the conclusion. Also, contemplating an argument with believed premises may make a person change his mind about those premises, when the conclusion seems implausible.

Logic can also be used justificatorily. After having inferred something, or having come to believe something, one may try to justify the new belief by an argument with believed premises. In the case of inferring some beliefs have actually played a causal role, and one may use those beliefs in the premises to justify the inference.

When Fisher talks about it being "essential to take the whole of the data into account," he expresses what Rudolf Carnap has called "The Requirement of Total Evidence." [51] It is a requirement regarding the application of inductive logic, and states that one must use the total available evidence when constructing the relevant argument, either while criticizing an opinion or justifying one. The requirement is useful in the application of inductive logic because of the fundamental difference between deductively valid and inductively strong arguments: a conclusion may follow strongly from part of the evidence, but only weakly or not at all from all of the evidence, or vice versa.

So when Fisher, in the passage quoted, talks about "knowledge *a priori* being available," he warns us that the fiducial probability cannot be applied in a situation where other information is available that has not been used in the derivation of the fiducial probability. Or, at least, if there is other information, which, in real cases, there always will be, one has to establish that it is irrelevant.

This requirement of using all the relevant evidence available has, I think, two consequence for the use of inductive arguments. For one thing, these arguments will only be used when it can be argued that most of our knowledge is irrelevant for the conclusion in question. Otherwise, it would just become too complex from a computational point of view. This does not mean that the data sets that form the premises of the inductive argument cannot be huge, but the set of premises should be relatively homogeneous. Inductive arguments will be local.

The second implication of the requirement is that all the relevant information available is "propositional." When Fisher talks about the "available axioms" and "the whole of the data" he is obviously thinking of information that is expressed in statements that are true or false, statements that have an unambiguous meaning and can be used in communication between research workers. The premises of arguments, also, have to be statements that are true or false. The use of logic in criticism, justification, or the exploration of assumptions presupposes that arguments are entities that can be communicated. But in the individual, inference is a causal event. If there is something like nonpropositional information, this may become a causal factor when the individual makes an inference. The presence of such information may make the individual stray from the conclusion that is dictated by the standard inductive argument deemed appropriate for his data.

The information provided by an observation is partly propositional, impersonal, and objective, partly ineffable and subjective. The propositional information is extracted by any (perhaps specially) trained observer in the same way, and is expressed in vocabulary current to the field. It is a vocabulary that stands in determinate logical relation to other terms in the field and has, therefore, a well determined informational content. The ineffable information may vary from observer to observer, and often cannot be expressed in a common language, or only in terms that have no clear inferential relations to other concepts in the field. For instance, when a surveyor makes a triangulation, the propositional information will include the angular measurements, the length of the base line, and, perhaps, the angular measurements of his stations with respect to the zenith. But there will be in his observation some residual information that he cannot express in standard terms. Part of this ineffable information will influence his judgment as to the quality or reliability of his data. Likewise, experimenters may have an impression about aspects of an experiment, or about the smoothness of a run, without being able to express this otherwise than by exclaiming, "Beauty." The presence of this type of information will frustrate the use of any standarized inductive mode of arguing. The calculator who had such information would tend to select and modify his data, and would arrive at a conclusion different from that dictated by the standard argument.[52]

Before the use of inductive arguments can become general in a field of science, the methods of observation have to be reformed. This is what has happened in the objectification of observation. There is in the development of an instrument or experiment a move away from ineffability. Every aspect of an observation considered relevant is made propositional. This can be done in several ways: through the development of measuring instruments with, ideally, digital output; through the measurement of covariates that are judged to be relevant in correcting the raw data; through the development of standard vocabulary and the training of observers to use it in a standard manner; or even by considering trained observers as part of the measuring apparatus. The end result is that all the relevant information in an observation is propositional, and extracted from the observation in a manner independant of personal judgment.

The mechanization of observation made general deterministic use of the Method of Least Squares and cognate methods possible. For Least Squares is an inductive method. It is even an inductive method without the probabilistic interpretation given by Gauss, and without the calculation of the probable error of the Least Squares estimate. This may not be obvious, because it may seem that, without such an interpretation, the calculation involved in the method leads to a conclusion of the form "the Least Squares resolution of such-and-such a set of data is so-and-so," which would be an analytically true statement, a conclusion that can be validly derived from no assumptions whatsoever. But this is not correct; the conclusion is stronger. The calculation tells us what the value of the resolution is, but the method claims that this is "the best guess as to the true value," a conclusion that certainly goes beyond the data.

5   The Problem of Outliers

The use of formal rules could not have become so widespread without the objectification of observation. But on the other hand, the very introduction of formal, standarized numerical methods such as Least Squares has had itself on objectifying effect. Like a measuring instrument, a numerical method also eliminantes personal judgment. And although, as we have seen, all kinds of often abstract, probabilistic reasons have been invoked to argue against or in favor of a particular reduction method, it often seems more important that everybody follows the same method, than what method is followed. A new practice is born, and outsiders gradually conform to it by adopting the new habits.

General Schreiber, for instance, the head of the Prussian *Landesaufnahme*, writes, "The only purpose of using the Method of Least Squares in plane triangulations is to get consistent and plausible results in the least arbitrary fashion. If one does not pretend to rigorous justifications where such are impossible, one may say that the method gives results in the most simple and elegant way."[53]

Jordan believes that the introduction of Least Squares in geodetic surveys has had an enormous moral advantage: only where people can follow a strict rule, will they stop using the ambiguities in a situation to suit their own purposes:[54]

The advantages to our discipline of using the Method of Least Squares have been, to a large extent, of a *moral* nature: one has become more honest in making measurements and doing the calculations. This can be illustrated by a case from the history of geodesy: the astronomer and geodesist Méchain, it is told, had become mentally ill—an illness that led indirectly to his death—because he had suppressed some of his measurement. ... Someone has remarked that Méchain would have been saved if he could have used the Method of Least Squares. It is of course impossible to say this so categorically. But dishonest suppression of measurements and the like happens undoubtedly less often since the method was introduced.

Even for Gauss himself, to follow a deterministic method that did not permit personal caprice and arbitrariness seems to have been a source of enjoyment and a reason for pride. In a letter of 1830 he writes to Bessel, referring to his triangulation of the state of Hannover, "Today I have carefully reduced the system of main triangles: not only the sum of the angles of each triangle, but also the relations between the sides of the resulting quadrangles and pentagons are in exact agreement with each other. Without being arbitrary, without selecting or suppressing anything, I proceeded strictly by the rules of the probability calculus." [55]

But the iron law of probability theory and the strict rule of the Method of Least Squares had a hard time in getting established. For even when there is no ineffable information present, the fact that inductive arguments are not monotonic, increasing the set of premises can turn a strong inductive argument into a weak one—and can lead to further problems.

The issue is, for instance, raised by Olbers, in a letter to Gauss of April 28, 1827:[56]

I would like to get your advice, my good friend, either in a personal letter, or by way of one of your next publications. According to the regulations one should use *all* one's observations, made under similar circumstances and with the same care, when one determines the unknowns with the Method of Least Squares. However, when the unknowns have been determined and one discovers, when substituting their values in the equations, that one or a couple of the observations deviate unusually and abruptly from the others, one would discard them and make a new determination on the basis of the remaining ones. The unusual deviation indicates the presence of an error in these observations from a source of error that did not affect the other observations: incorrect reading of the scale, copying error, unnoticed shift of the instrument, and so on. But it is indefinite what should count as an unusual or too large a deviation. I would like to receive more precise directions that tell us how large the deviation, compared with the mean or the probable error, has to be before one is justified in rejecting an observation.

A simple instance of Olbers's problem is the case of one unknown x, the case of direct measurement of a quantity. We have n repetitions of the measurement with outcomes $a_1, \ldots, a_n$. The least square solution is, as we have seen, the arithmetic mean: $\bar{x} = a_i/n$. If now, Olbers argues, we notice, after calculating the mean value, that one of the outcomes appears to lie very far from the calculated mean (in comparison with the spread of the observation as measured by the mean or the probable error), we shall suspect a rogue. Something out of the ordinary must have gone wrong with this observation, and we shall be better off to throw out this deviant observation and recalculate the mean on the basis of the remaining ones. The deviant observation is not suspect on the basis of ineffable information, a feeling that something odd was happening while making the observation. There is no reason to mistrust this value other than its relation to the other values that were obtained. There is a tension between the two parts of the numerical information: the remaining ones suggest that the rejected observation was not really an observation of x. Because, for the a_i's to be observations of x, or outcomes of a measurement of x, the unknown, underlying true x has to be a major causal factor in the ith observation resulting in the value a_i. Olbers's problem illustrates the nonmonotonicity of least squares inference. It takes a slightly different form than in, say, enumerative induction. There the very same conclusion may sometimes no longer be inductively inferred when the set of premises grows. One would expect this to happen in least squares inference, because more measurements will in general correct the calculated mean and its probable error. This part of the problem of nonmonotonicity is taken care of, for the least squares calculation adapts itself to an extended set of data. But sometimes we do not feel confident in applying the Method of Least Squares at all. One could say that the set of premises is not homogeneous anymore, one part accusing the other of not being a reliable witness about the unknown x. Olbers asks Gauss to give him some modification of the Least Squares rule for those cases. He feels that he relies to much on his personal judgment.

This "problem of outliers," as it came to be called, still attracts the attention of statisticians.[57] It is a prime example of the impossibility of devising rules with an exactly described domain of application. Gauss never raised the problem of outliers in his published writings. But he answered Olbers privately, in a letter written about a week later:[58]

In applying the probability calculus to observations, to have as much knowledge of the subject matter as possible is of the utmost importance. Where this is not the case, it remains always doubtful to reject observations by reason of their large deviation, unless the number of observations is very large. It is true that all the single parts of an unavoidable observational error have bounds, even though we cannot determine them. There are many cases in which we can be certain that a large error has occurred that lies outside the range of possibility of those kinds of errors, and that an unusual mistake has been made. One should of course reject such an observation. As long as one thinks it is possible that the mistake is the result of an unlucky conspiracy of its single parts, one should not reject the observation. But sometimes one is not certain whether the deviation is of the first or of the second kind. In that case one should proceed as one thinks correct, but—and this is a law—one should not conceal anything, so that others can make their own calculation. Whether one rejects or not will often have little influence on the usefulness of the results, but when one is too quick to reject observations, one runs the risk of exaggerating the reliability of the results. Cases like these are analogous, it seems to me, to *making decisions in everyday life*. One has rarely or never mathematical rigor and certainty, and has to be satisfied to act according to one's best judgment.

But the sensible attitide expressed here by Gauss—to act according to one's best judgment when one thinks the rules to be inapplicable, but never to conceal the data—is only compatible with objectivity when the rules developed and discussed in the public domain have been more and more refined. This was what happened in the nineteenth century. Rules were proposed that gave objective criteria for the rejection of doubtful observations. The first such rule, proposed in 1852 by Benjamin Peirce, was used by B. A. Gould in reducing the data collected by the U.S. Coast Survey to determine the longitude of cities in the United States. Gould writes,[59]

Professor Peirce has given the results of the successful investigation of a singular problem, and one unquestionably among the most important of any which could be proposed in its relation to all those exact sciences to which quantitative research or measurement may be applied. This problem was nothing less than the attainment of a formula which could be legitimately derived from the fundamental principles of the Calculus of Probabilities, and furnish an exact criterion for the recognition of those observations which differ so much from the average of a series as to indicate some abnormal source of error, which would vitiate the result. The delicate task of discriminating between such observations, and those whose discordance, although great, ought not to be deemed abnormal, has hitherto been left to the arbitrary judgment of individuals; and the present introduction of a rigorous mathematical ordeal for testing the extend of tolerable discrepancy cannot fail to exercise a highly beneficial influence.

The problem of outliers is a problem at the extremity of Least Squares. Its discussion during the nineteenth century shows the need for rules of calculation that could be mechanically applied to reduce the huge sets of data the new measuring devices were spitting out, often by calculators who had no knowledge of the circumstances surrounding the observations, or even of the subject matter the numbers were related to.[60]

6 Conclusion

Two distinguishable but connected issues have been raised. The first one is the objectification of scientific observation without which a deterministic application of inductive methods is impossible. I have given an argument why this is the case. But the argument is too abstract, and has to be fine-tuned to the inferential subtleties of actual scientific practice. To do this, more information is needed about differences between disciplines—say, geodesy versus chemistry—with respect to the degree to which they have attained the state of "observation without an observing subject," and in their use of numerical methods.

The second point was that the strict use of numerical methods itself takes away a personal and thus arbitrary element from science. The existence of rules makes it easier for researchers not to be influenced by their preconceptions when they interpret their data. Here I have pointed at a problem connected with the nonmonotonicity of inductive arguments: they can only be applied to relatively poor and restricted sets of (relevant) information. Least Squares breaks down when the set of measurements is inhomogeneous. This was the problem of outliers. The Method of Least Squares can be modified, doubtful observations can be rejected, observations can be weighted differently, but the range of application of inductive rules cannot be fixed in advance, and there can always arise situations in which one cannot psychologically infer what the rule sanctions. Logic as a social, objective phenomenon breaks down.[61]

There is a third issue, connected with these two, that has only been touched upon on this paper. It is the role of probability in our story, and the influence of Least Squares on the development of probabilistic inference. The cry for *Vorschrift* (directions), to use the German expression, is not a nineteenth-century fluke, but it is still all around us. It has shaped a major school of probabilistic inference, the Neyman-Pierson school, especially via a precursor, Charles S. Peirce, and a statistician with similar ideas, Edmund Wilson.[62] And this school is still battling subjectivity.[63] That *Vorschrift* and probability got connected seems obvious. How this happened still remains to be explained.

Notes

1. See Hans Reichenbach, "The Philosophical Significance of the Theory of Relativity," in *Albert Einstein, Philosopher-Scientist*, ed. Paul A. Schilpp (La Salle: Open Court, 1949), pp. 289–311, on p. 300; Carl Hempel, "Geometry and Empirical Science," *American Mathematical Monthly*, 52 (1945), reprinted in *Readings in Philosophical Analysis*, eds. Herbert Feigl and William Sellars (New York: Appleton-Century Crofts, 1949), pp. 238–249, on p. 246. For an earlier German discussion of the legend see E. Hoppe, "C. F. Gauss und der euklidische Raum," *Die Naturwissenschaften*, 13 (1925), 743–744, and Hans Kienle, "Hugo von Seeliger," *Die Naturwissenschaften*, 13 (1925), 613–619, on p. 614.

2. "Gauss an Schumacher," letter no. 396 in *Briefwechsel zwischen C. F. Gauss und H. C. Schumacher*, 6 vols., vol. II, ed. C. A. F. Peters (Altona: Gustav Esch, 1860–65), p. 271. Emphasis added.

3. See Philipp Fischer, *Untersuchungen über die Gestalt der Erde* (Darmstadt: Diehl, 1868). The maps were often the work of the military (e.g., the Dutch lieutenant-general Krayenhoff), and the results of a survey were sometimes kept secret. See C. R. T. Krayenhoff, *Précis Historique des operations géodésiques et astronomiques faites en Hollande, pour servir de base à la topographie de cet état, exécutées par le Lieutenant-Général Krayenhoff* (The Hague: 1815).

4. See M. Brendel, "Ueber die Astronomischen Arbeiten von Gauss," *Carl Friedrich Gauss' Werke*, XI, pt. 2, sec. 3 (Göttingen: Dieterichsche Universitäts-Druckerei, 1929), pp. 3–254.

5. Méchain and Delambre, *Base du Système Mètrique Décimal, ou Mesure de l'Arc du Méridien, compris entre les Parallèles du Dunkerque et Barcelone* (Paris: Garnery, 1806); T. Gerardy, *Die Gauss'sche Triangulation des Königreichs Hannover (1821 bis 1844) und die Preussischen Grundsteuermessungen (1868 bis 1873)* (Hannover: 1952); F. W. Bessel and J. J. Bacyer, *Gradmessung in Ostpreussen und ihre Verbindung mit Preussischen und Russischen Dreieckketten* (Berlin: 1838).

6. William H. Wollaston, "A Synoptic Scale of Chemical Equivalents," *Philosophical Transactions*, 104 (1814), 1–22. The paper is agnostic on Dalton's hypothesis.

7. Wilhelm Weber, "Beschreibung eines kleinen Apparats zur Messung des Erdmagnetismus nach absolutem Maass für Reisende," *Resultate aus den Beobachtungen des magnetischen Vereins*, 4 (1836), 63–89. Reprinted in *Wilhelm Weber's Werke, herausgegeben von der Königlichen Gesellschaft der Wissenschaften zu Göttingen*, vol. 2: *Magnetismus, besorgt durch Eduart* Riecke (Berlin: Julius Springer, 1892).

8. For the transition from inductivism to the hypotheticodeductive method see Kenneth L. Caneva, "From Galvanism to Electrodynamics: The Transformation of German Physics and Its Social Context," *Historical Studies in the Physical Sciences*, 9 (1978) 63–159.

9. See H. Strassl, "Die erste Bestimmung einer Fixsternentfernung (Zum hundertsten Todestag von F. W. Bessel)," *Die Naturwissenschaften*, 33 (1946), 65–71, and L. Ambronn, "Ueber die Methoden der Distanzmessung zweier Sterne mit dem Heliometer," *Zeitschrift für Instrumentenkunde*, 13 (1893), 17–24.

10. See "Bessel an Gauss," letter no. 180. *Briefwechsel zwischen Gauss und Bessel*, ed. G. F. J. A. von Auwers (Leipzig: Engelmann, 1880), p. 532.

11. Bessel and Baeyer, *Gradmessung in Ostpreussen*, (note 5). The measuring rods are described in all details.

12. See Maurice Daumas, *Scientific Instruments of the Seventeenth and Eighteenth Centuries* (New York: Praeger, 1972), pp. 183–186.

13. Méchain and Delambre, *Base du Système Mètrique*, vol. I, p. 102 (note 5).

14. W. S. Jevons, *The Principles of Science* (London: Macmillan, 1874), 2 vols, vol. I, pp. 110–113. Jevons's book is still very readable and gives an unsurpassed discussion of basic concepts for a philosophy of science, such as measurement, error, observation, variation, experiment, and approximation.

15. "The Exact Measurement of Phenomena" is the title of a chapter in Jevons, *Principles* (note 14).

16. Humphry Davy, *Chemical Philosophy*, in *The Collected Works of Sir Humphry Davy*, ed. John Davy (London: Smith, Elder and Co., 1839–1840), 9 vols., vol. IV.

17. John F. W. Herschel, *Preliminary Discourse on the Study of Natural Philosophy* (London: Longmans, 1831; new edition 1851), section 115 and 118.

18. Wilhelm Weber, "Ueber Maassbestimmungen," *Werke*, vol. III, pp. 539–577, on p. 540 (note 7).

19. Jevons, *Principles*, vol. I, pp. 314–315 (note 14).

20. Louis Paschen, "Antrittsrede," *Sitzungsberichte der preussischen Akademie der Wissenschaften, Philosophisch-historische Klasse*, 1925, pp. cii–civ, on p. ciii.

21. "Zeitschrift für Instrumentenkunde. Organ für Mitteilungen aus dem gesammten Gebiete der Wissenschaftlichen Technik. (Berlin: Julius Springer, 1881–), herausgegeben unter Mitwirkung der zweiten (technischen) Abtheilung der Physikalisch-Technischen Reichsanstalt." The Reichsanstalt gave support beginning in 1887, when it was founded. See L. Loewenherz, "Die Aufgaben der zweiten (technischen) Abtheilung der physikalisch-technischen Reichsanstalt," *Zeitschrift für Instrumentenkunde*, 8 (1888), 153–157.

22. "An unsere Leser," *Zeitschrift für Instrumentenkunde*, 1 (1881), 1. Ernst Abbe was on the editorial board.

23. Charles Babbage, *Reflections on the Decline of Science in England, and on Some of Its Causes* (London: Fellowes, 1830), p. 169.

24. See Steven Shapin, "Pump and Circumstance: Robert Boyle's Literary Technology," *Social Studies of Science*, 14 (1984), 481–520.

25. George Wilhelm Muncke, "Bemerkungen über die Versuche des Hrn. *Lenz* in Betreff der Drehungen des Coulombschen Wagebalkens, und Nachricht von den akustischen Versuchen des Hrn. *Scheibler*," *Annalen der Physik und Chemie*, 29 (1833), 381–403, on p. 386. Part of this passage is quoted in Caneva, "Galvanism to Electrodynamics," p. 73 (note 8).

26. For example, Judith Pipher et al., "Submillimetre Observations of the Night Sky Emission above 120 Kilometres," *Nature*, 231 (1971), 375–378.

27. Babbage, *Reflections*, p. 169 (note 23). For direct read-out measuring systems, read through a recent issue of *Quality, The Magazine for Product and Service Quality* (1962–), published by the Hitchcock Publishing Co., Wheaton, Illinois.

28. See Daumas, *Scientific Instruments*, p. 216 (note 12).

29. See Sp. [full name unknown], "Neue selbstregistrierende Instrumente des Königlichen Dänischen Meteorologischen Instituts," *Zeitschrift für Instrumentenkunde*, 10 (1890), 30–32, and R. W. J. Abbe, "A Recording Compass," *Scientific American* (July 16, 1887).

30. See Benjamin A. Gould, *The Transatlantic Longitude, as Determined by the Coast Survey Expedition of 1866. Smithsonian Contributions to Knowledge*, no. 223 (Washington: Smithsonian Institution, 1869).

31. See R. Radau, "Ueber die persönlichen Gleichungen bei Beobachtungen derselben Erscheinungen durch verschiedene Beobachter," *Repertorium für Physik und Technik, für mathematische und astronomische Instrumentenkunde*, 1 (1866), 202–218.

32. F. W. Bessel, "Persönliche Gleichung bei Durchgangsbeobachtungen," *Königsberger Beobachtungen*, 8 (1823), 3–6.

33. Piazzi observed the small planet for 41 days, obtaining observations covering only 3° geocentric arc, before it moved too close to the sun to be observed. Gauss did not publish the method he had developed in 1794 until 1809 in his *Theoria motus corporum coelestium* (Hamburg: Perthes and Besser, 1809). See Galle, "Ueber die geodätischen Arbeiten von Gauss," *Gauss Werke*, XI, part 2, sect 1, pp. 3–161 (note 4).

34. See Christian Ludwig Gerling, *Die Ausgleichungs-Rechnungen der Praktische Geometrie* (Hamburg: Perthes, 1843); Gauss, "Störungen der Ceres und der Pallas," *Gauss Werke*, VII, pp. 377–610 (note 4).

35. See W. Jordan, *Handuch der Vermessungskunde* (Stuttgart: Metzler, 1895), 3 vols. vol. I, pp. 70ff., and Fischer, *Gestalt der Erde*, pp. 15–16 (note 3).

36. For a biographical overview of the pre-Least Squares period, see H. Leon Harter, "The Method of Least Squares and Some Alternatives," *International Statistical Review*, 42 (1974), 147–174, 235–264; 43 (1975), 1–44, 125–190, 269–278, on pp. 148–152 of 42 (1974).

37. C. F. Gauss, "Bericht über die in dem magnetischen Observatorium gemachten Beobachtungen," *Göttingische Gelehrte Anzeigen*, March 7, 1835; reprinted in *Gauss' Werke*, vol. V, pp. 528–536 (note 4).

38. A typical textbook from the midnineteenth century is Jacques Babinet and Charles-Pierre Housel, *Calculs Pratique Appliqués aux Sciences d'Observation* (Paris: Mallet-Bachelier, 1857).

39. See Richard Henke, *Ueber die Methode der kleinsten Quadrate* (Leipzig: Teubner, 1894, 2nd ed.; in 1868 published as Inaugural dissertation).

40. Thomas Simpson, "A Letter to the Right Honorable George Earl of Macclesfield, President of the Royal Society, on the Advantage of taking the Mean of a number of Observations, in practical Astronomy," *Philosophical Transactions, giving Some Account of the Present Undertakings, Studies and Labours, of the Ingenious in Many Considerable Parts of the World*, 49 (1755), 82–93.

41. C. F. Gauss, *Theoria motus corporum coelestium in sectionibus conicis solem ambientium* (Hamburg: Perthes and Besser, 1809).

42. F. W. Bessel, "Untersuchungen über die Wahrscheinlichkeit der Beobachtungsfehler," *Astronomische Nachrichten*, 15 (1838), 371–404; Gotthilf Hagen, *Grundzüge der Wahrscheinlichkeitsrechnung* (Berlin: Dümmler, 1837).

43. C. F. Gauss, "Theoria combinationis observationum erroribus minimis obnoxiae, pars prior," *Commentationes societatis regiae scientiarum gottingensis recentiores*, 5 (1823); reprinted in *Gauss, Werke*, vol. IV, pp. 1–26, on p. 4 (note 4). ("Errorum regularium consideratio proprie ab instituto nostro excluditur.")

44. See 'Student', "The Probable Error of a Mean," *Biometrika*, 6 (1908), 1–25.

45. W. J. Youden, "Enduring Values," *Technometrics*, 14 (1972), 1–11.

46. For a modern point of view see N. Ernest Dorsey, "The Velocity of Light," *Transactions of the American Philosophical Society*, 34 (1944), 1–110.

47. See the paper by Ivo Schneider in this volume.

48. Cleveland Abbe, "A Historical Note on the Method of Least Squares," *American Journal of Science and Arts*, 1(ser. 3) (1871), 411–415, on p. 415.

49. Emanuel Czuber, *Theorie der Beobachtungsfehler* (Leipzig: Teubner, 1891), p. v.

50. R. A. Fisher, *Statistical Methods and Scientific Inference* 3rd ed. (New York: Hafner, 1973), p. 58.

51. Rudolf Carnap, *Logical Foundations of Probability*, 2nd ed. (Chicago: University of Chicago Press, 1962), pp. 211–213.

52. For the influence ineffable information had on the way R. A. Millikan used his data on charged drops of liquid, and for "Beauty," see Gerald Holton, "Subelectrons, Presuppositions, and the Millikan-Ehrenhaft Dispute," in his *The Scientific Imagination: Case Studies* (Cambridge: Cambridge University Press, 1978), pp. 25–83, on pp. 70–71.

53. From a Report on the Surveying Conference in Nice of 1887 (Annex Xb, p. 10), quoted in Jordan, *Handbuch der Vermessungskunde*, vol. I, pp. 6–7 (note 35).

54. Jordan, *Handbuch der Vermessungskunde*, vol. I, p. 7 (note 35).

55. "Gauss an Bessel," letter no. 140, *Briefwechsel Gauss und Bessel*, p. 423 (note 10).

56. "Olbers an Gauss," letter no. 612, in C. Schilling (ed.), *Wilhelm Olbers, Sein Leben und seine Werke* (Berlin: J. Springer, 1894–1909), 2 vols., vol. II, part 2, pp. 477–478.

57. For outliers in general, including outliers in population statistics, see Vic Barnett and Toby Lewis, *Outliers in Statistical Data* (Chichester: Wiley, 1978).

58. "Gauss an Olbers," letter no. 613, *Briefwechsel Gauss-Olbers*, part 2, p. 480 (note 56).

59. B. A. Gould, "On Peirce's Criterion for the Rejection of Doubtful Observations, with Tables for Facilitating its Application," *Astronomical Journal*, 4 (1855), 81–87, on p. 81.

60. Wright, for instance, says, "Throughout this discussion it has been assumed that the observations have been reduced by the observer himself or by a computer who is at the same time a competent observer. A computer who is not an observer must of necessity employ the same criterion always." See T. W. Wright, *A Textbook on the Method of Least Squares* (New York: Wiley, 1884), p. 138.

61. The reader must not be misled by the expression 'Bayesian Inference', which seems to provide a panacea for this problem. Bayesians do not deliver rules for inference, but criteria for monitoring one's private opinions.

62. See Ian Hacking, "The Theory of Probable Inference: Neyman, Peirce and Braithwaite," in *Science, Belief and Behavior*, ed. D. H. Mellor (Cambridge: Cambridge University Press,

1980), pp. 141–160; Jerzy Neyman, "'Inductive Behavior' as a Basic Concept in the Philosophy of Science," *Revue de l'Institut Internationale de Statistique*, 25 (1957), 7–22.

63. See D. Collett and T. Lewis, "The Subjective Nature of Outlier Rejection Procedures," *Applied Statistics*, 25 (1976), 228–237, and J. Neyman, "Comments on a Paper by R. R. Braham, Titled 'Field Experimentation in Weather Modification'," *Journal of the American Statistical Association*, 74 (1979), 90–94.

12 The Measurement of Uncertainty in Nineteenth-Century Social Science

Stephen M. Stigler

The study of the history of the quantification of science can be usefully sharpened by focusing on the development of the measurement of uncertainty, on the introduction of the use of the calculus of probabilities in inference. And that development can be best understood by contrasting the path followed in astronomy and geodesy with that followed in the social sciences. By 1820, astronomy had achieved a remarkable synthesis between mathematical probability and the tradition of analyzing data through the aggregation of systems of linear equations. But while the methods of Gauss and Laplace became a commonplace in astronomy and geodesy, their use did not then extend to the social sciences. Many scientists attempted to adapt probability-based methods to social science problems, including Quetelet and Lexis, but in the end they were frustrated, Quetelet because his methods were too insensitive to segregate his data into categories amenable to statistical analysis, Lexis because his binomial models were insufficiently rich for interesting applications. Only in the last quarter of the nineteenth century were the conceptual stumbling blocks overcome, due in large part to the works of Francis Galton, Francis Edgeworth, and Karl Pearson. The arguments of this chapter are enlarged upon in the author's book on the history of statistics before 1900.

Most of the contributors to this volume come as philosophers, historians, or sociologists to discuss topics that I regard as central to statistics. I come as a statistician to discuss a topic you in turn may regard as properly belonging to philosophy, history, or sociology. It is not surprising that we bring differing perspectives to the discussion. In these brief remarks I hope to outline some of the features of my own, a statistician's perspective.

To a statistician, the whole of the history of science revolves around measurement: all sciences (and the social sciences are no exception) require at least the possibility of quantitative measurement. But measurement alone is not enough—it is essential that the measurements be susceptible to comparison, and that, in turn, requires some understanding of the accuracy of the measurements. Over the past few years I have been studying the history of several sciences from my statistician's perspective, and my goal has been an understanding of the last portion of this chain. I have focused upon the introduction of probability-based statistical methods for the quantification of uncertainty in observational data. My own view in this study[1] has been a largely internal one—I have for the most part accepted the scientific problems as defined by the scientists involved and not looked for the influence of exterior social forces upon the choice of problem or the method of attack. Rather I have sought to understand which characteristics of the objects of scientific study permit, and which inhibit, the introduction of statistical methods for the measurement of uncertainty.

In some respects my focus upon the explicit measurement of uncertainty is a narrow one. Many measurements carry with them an implicitly understood assessment of their own accuracy; indeed, it could be argued that any measurement that is accepted as worth quoting by more than one person has some commonly understood accuracy, even if the accuracy is not expressed in numerical terms. I would not dispute that argument, but I would maintain that we shall make greater progress in our analysis of the history of quantitative thought if we accept the more limited goal of explaining the explicit measurement of uncertainty via probability and statistical method. In particular, this permits us to limit our attention to a relatively narrow span of time—say, from 1700 or 1750 to 1900—and it gives us a reasonably objective criterion for progress, it being easier to recognize an explicit probability statement than to agree when a measurement is widely accepted.

Put in its plainest form, then, my goal has been to understand the how, the when, and the why or why not of the introduction of probability statements of accuracy in the sciences. "Probability statements" include, in particular, the quotation of probable errors or standard deviations, confidence or other interval statements, and tests of statistical significance or the quotation of significance probabilities as assessments of strength of evidence.

Now, we statisticians are fond of explaining to our students that, if you wish to understand the effect of various factors upon a response, the only sure course is to vary the factors purposefully and observe and measure the consequent variation in the response. This applies as well to the present problem: the factors might include the country (and individuals) involved, and the type of science and theory, and the historical and social setting. The response is the rate and manner of the development and adoption of statistical methods. Clearly the proper course is to put in for a grant to create and observe two or more new sciences, involving randomly selected scientists from randomly selected countries. Unfortunately that plan occurred to me in an unfavorable financial climate, and I have had to fall back upon the kind of analysis we warn our students is fraught with dangers and pitfalls: a retrospective observational study—in short, a more standard historical investigation, where, instead of varying the factors and observing the response, we look for differences in response and try to trace them back to differences in factor inputs.

From this point of view, we have been remarkable lucky, for the history of science from 1750 to 1900 presents a record that almost seems to signal that my currently impractical plan has been anticipated and performed by the Great Experimenter. I refer to the contrasting developments of astronomy and geodesy on the one hand, and the social sciences (particularly economics and sociology) on the other. Both might be followed from the prestatistical seventeenth century, from the times of Newton and John Graunt. Both attracted large amounts of public attention, governmental support, and some of the greatest minds of the time. In some cases they even involved the same individuals! And yet from my statistical perspective, they were a study in contrasts. From about 1750 through 1850, astronomers overcame one conceptual obstacle after another to reach a point where probability-based statistical methods were a commonplace in astronomical work. It was a heroic struggle, and Laplace was the principal hero. Jacob Bernoulli had made one early and remarkable attempt (which Ian Hacking,[2] among others, has chronicled), though Bernoulli's attempt was ultimately frustrated. De Moivre pushed further,[3] but it was only late in the eighteenth century that Laplace achieved what could be called success, and then only by building upon a large body of empirical and theoretical work in astronomy. Still, by the 1820s, the success was widespread common knowledge. The techniques were known in some form to nearly all physical scientists by midcentury, and were a standard part of what "curriculum" there was in those days.

What of the social sciences? The contrast is remarkable—I have read widely in the works of the time and can count on the fingers of one hand (indeed, almost upon the thumbs of one hand) the application of probability assessments of accuracy in the social sciences before 1880. With only a little change of the emphatic tone, I can make the same statement for the period before 1900, and it is perhaps most remarkable if we recall that those who did *not* use these techniques include such men as Quetelet, Cournot, and William Stanley Jevons, all of whom wrote books concerned with probability. We thus have a dramatic metastatistical experiment to analyze. Two different "sciences," one of the sky and one of society, were "started" in the 1600s. Astronomy developed and used statistical methods of assessing uncertainty rather early; social sciences knew well of astronomy's triumph yet lagged by nearly a century. The question is a fascinating and difficult one. I could not hope to resolve it in brief comments, but I shall try, through two illustrations, to show how a statistical perspective might help deal with it. My two illustrations involve the separate failures of Quetelet and Wilhelm Lexis to come to grips with the central question that faced them.

My description of Quetelet and Lexis as failures deserves some explanation, and some apology, for they were two of the most remarkable scientists of the time, and they are cited often as among the most influential men of the century. But if we accept the measurement of uncertainty in social data as the major methodological challenge of that century, and if we accept that both men took that challenge as a central research problem, and if we further agree that the quest was not successful, then we must indeed count them failures, even though it is evidence of the conceptual difficulty involved, rather than any sign that Quetelet and Lexis were second-rate. Let us then accept my hypotheses, and I shall try, through my two illustrations, to explain how a statistical perspective can shed light upon the nineteenth-century failure to overcome the conceptual barrier involved in the spread of statistical methods to the social sciences.

What was that barrier? One brief way of describing it is as the problem of categorizing data into homogeneous groups, that is, groups for which the major influential factors could be considered constant and residual variation was seen as due to haphazard accidental causes. Thus, counts of deaths in rural areas might not be considered homogeneous with those in cities; different observations of a transit of Venus by different astronomers might be categorized as homogeneous. Now once it was agreed that a group of measurements was homogeneous, the applicability of Laplacian theory was not in doubt; indeed what I would call mathematical statistics was called "the combination of observations" though much of that century. The difficulty lay in the classification of data for combination. And here the major distinction between astronomy and social sciences was that astronomers could often classify data a priori, disciplined by accepted mathematical theory and what were seen as uniform conditions of observation, while social scientists could not. The social scientists, unable to achieve such a categorization a priori were forced to search for it a posteriori, and that is a *much* more difficult problem. We may look at the vast empirical investigations of the nineteenth century as one way they tried to deal with that problem, but the attempted solution I want to touch upon here is different—it is the attempt to categorize data as homogeneous based upon analyses internal to the data.

The work Quetelet is best known for—the fitting of normal distributions to data—was, I would argue, an explicit attempt to deal with just this aspect of the problem. Laplace's limit theorem was the inspiration—if a constant cause is perturbed by a large number of independent accidents, the resulting data will follow what we now term a "normal curve." Quetelet's reasoning was the reverse—if a normal curve was observed, it must be due to the perturbation of a constant cause by many independent accidental causes. The observed normal curve would serve to validate both the existence of a constant cause (in some contexts this was the "average man") and the hypothesis that all other causes were accidental. In principle, had the device succeeded, a set of data exhibiting the normal distribution could be analyzed by the calculus of probabilities. But it failed. Or, as some might prefer to put it, it succeeded too well. The argument, we know now, is mathematically a non sequitur, and even Quetelet could see very soon that it was empirically a non sequitur. With scant exceptions, all data he tried passed the test. He (and we) could believe that Scotch soldiers were homogeneous, but the same could not be true for all groups of people, however large. And yet, that is what Quetelet found. We know now where the difficulty lies—the Laplacian hypothesis is not the only one leading to a normal conclusion. Far from it. Normal curves can and do arise in countless ways from the compounding of other normal curves. The mere appearance of normality is not at all sufficient to conclude homogeneity. Quetelet's test was too insensitive; it was doomed from the start (at least as far as his major purpose went), and his distributional fits were of necessity relegated to a sterile, though decorative, role until a later generation could unravel the puzzle they presented.

Somewhat after Quetelet, Wilhelm Lexis (and, independently, Emile Dormoy) made a less well known attempt at the internal characterization of homogeneity. Lexis's approach, put forth in the late 1870s, has come to be known as dispersion theory. Lexis would look at a series of counts (say, deaths in one province over many years) and ask, Was the series homogeneous (or, in this example, was it "stable")? Lexis's idea of homogeneity was binomial variation, with the chance of death remaining the same from year to year. Lexis would in effect estimate the standard deviation of his counts in two different ways: first, merely treating them as numbers and using the sum of squared deviations from the average as a measure of dispersion; second, using the binomial hypothesis and estimating dispersion via the appropriate rule for that hypothesis, using a function of $p(1 - p)$, p being the estimated chance of death. If the two estimates agreed, he would accept the binomial hypothesis and conclude that the data were homogeneous.

Lexis's test was remarkably like Quetelet's in one respect, in that it took a model for a particular form of binomial variation as representing the type of homogeneity that was amenable to analysis. But it failed for a very different reason than Quetelet's. Almost all data passed Quetelet's test; almost none passed Lexis's! Practically the only example of a stable sequence to be found was the ratio of male births to total births, hardly enough material for a serious social science! The hypothesis was too restrictive, the test too sensitive, and the result was that Lexis's approach, like Quetelet's, failed to meet its objective of providing an internal test of homogeneity that would permit the use of the probability calculus with social data.

I have spoken of the failures to measure uncertainty in nineteenth-century social data—was there no success? The problem was of such difficulty that we should not be surprised to find the solution did not come easily, and did not come through the work of one man. Rather, it came through the combined talents of three remarkable scientists, Francis Galton, Francis Edgeworth, and Karl Pearson. Together, they helped launch a statistical revolution in social science akin to that which Laplace had launched in astronomy.

Of the three, Galton was the idea man, a bold and novel thinker with energy, but he was mathematically backward and unable to extract and develop the full fruit of his own ideas. Edgeworth was the subtle theorist, and was perhaps alone among Galton's audience in appreciating the potential for the ideas.[4] Pearson was able to recognize the power in Edgeworth's formulations of Galton's ideas, and he had the zeal that in the end created a methodology from those formulations and sold it to the world.

The success of these three was not immediate, but as I see it, it involved showing how regression, using mathematical techniques borrowed from the least squares of a century earlier, could permit the use of some coordinates of multivariate data as, in effect, determining the categorization of the others. Galton, Edgeworth, and Pearson showed how multivariate data could be analyzed conditionally, given some of the coordinates as controlling or classifying variables. Much of the mathematics was identical to the least squares of an earlier century, but the conceptual standpoint was novel and acceptance did not come easily. It is a complex story, and the work of Ronald Fisher and others was needed before the end of the beginning was in sight. Neither was regression the whole story—the analysis of variance and related linear models also helped in unraveling the effects of the factors in cross-classified data. The complexity and difficulty help underscore the fact that the failures of the nineteenth century are no disgrace to the scientists involved. In summary, then, I would argue that we can learn to understand the advance of quantification best by narrowing our focus to the explicit use of probability assessments of the accuracy of measurements. With this narrowed focus we can recognize a dichotomy between astronomy, where the a priori classification of data was possible, and the social sciences, where it was not. And if we view the major nineteenth-century works, such as those of Quetelet, as attempts to grapple with the problem of providing an internal classification of data, we have both a rationale for the introduction of those methods and an explanation for why they did not work. Further, we can begin to understand the more important features of those methods that *did* work, later.

A statistical perspective such as I offer cannot answer all the more interesting external questions that other papers of this volume want to come to grips with. But if a statistical perspective helps us judge the nature and effect of what work was done, we may be better able to sharpen the external questions we can and should ask about that period.

Notes

1. See Stephen M. Stigler, *The History of Statistics: The Measurement of Uncertainty Before 1900* (Cambridge: Harvard University Press, 1986).

2. See Ian Hacking, *The Emergence of Probability* (Cambridge: Cambridge University Press, 1975).

3. Ivo Schneider, "Der Mathematiker Abraham de Moivre (1667–1754)," *Archive for History of Exact Sciences*, 5 (1968), 177–317.

4. Stephen M. Stigler, "Francis Ysidro Edgeworth, Statistician (with Discussion)," *Journal of the Royal Statistical Society* A141 (1978), 287–322.

IV SOCIETY

13 Rational Individuals versus Laws of Society: From Probability to Statistics

Lorraine J. Daston

In both the eighteenth and the nineteenth centuries, the mathematical theory of probability was intimately linked to hopes for a true science of society. Yet the "social mathematics" of Condorcet and the "social physics" of Quetelet diverge sharply with respect to their conceptions both of mathematical probability and of social science. Eighteenth-century mathematical probability took as its subject matter the judgments and decisions of an elite of reasonable men; eighteenth-century moral sciences also aimed to reveal the rational grounds for action and belief. Both were individualistic, psychological, and prescriptive in their approach. Nineteenth-century probabilists understood their theory in terms of statistical frequencies; nineteenth-century social scientists sought regularities at the macroscopic level of whole societies rather than at the microscopic level of individual action. For the eighteenth-century thinker, society was law-governed because it was an aggregate of rational individuals; for his nineteenth-century counterpart, society was law-governed in spite of its irrational individual members. I trace the reinterpretation of both mathematical probability theory and social science that made this remarriage of the two possible in the early nineteenth century.

The bond between mathematical probability and what Jakob Bernoulli called the "study of civil life" is a venerable one. Almost from its inception in the midseventeenth century down to the present, probability theory seemed to promise a calculus of the moral sciences. Yet this continuity is deceptive. The probabilist program for the eighteenth-century moral sciences diverged sharply from that for the nineteenth-century social sciences, just as a gap separates the moral from the social sciences (the shift in names marks no mere nominal distinction here), and eighteenth- and nineteenth-century versions of probability theory itself. I shall argue that whereas the model of explanation for the eighteenth-century moral sciences turned upon the psychology of rational choice and belief in jurisprudence and political economy, nineteenth-century social theorists sought macroscopic regularities of a more sociological kind. Due to its close conceptual links first with legal practices and later with associationist psychology, probability theory was admirably fitted to supply the mathematical underpinnings of the moral sciences, but required a major reinterpretation before it could perform the same role in the new social sciences of the first half of the nineteenth century. The key lay in the altered relationship between mathematical probability and statistical data, which shifted from one of opposition to one of cause and effect.

My discussion of this transition from the probability of the moral sciences to the probability of the social sciences divides into three parts: the grounds for the original alliance between mathematical probability and the Enlightenment moral sciences; the new basis for such an alliance with the social sciences of the early nineteenth century; and the changing role of statistical data vis-à-vis probability theory during this critical period. As the standards for an acceptable explanation—or even an acceptable question—in social theory shifted, so did the interpretation of mathematical probability theory that made its application plausible first to the decisions of rational individuals and later to the social laws generated by the apparent irrationality of individuals.

1

The original relationship between classical probability theory and the Enlightenment moral sciences was almost literally one of prearranged harmony. When Jakob Bernoulli turned the fledgling calculus of probabilities to the problems of civil life in Book IV of the *Ars conjectandi* (posthumous, 1713), the first such problem that he addressed was recasting the rules of evidence then current in Roman and canon law into mathematical terms. Such applications came naturally to the classical theory of probability, for its earliest formulations drew heavily upon the older legal doctrine that assigned ordered degrees of certainty to various classes of evidence and testimony. The arithmetic of proof of sixteenth- and seventeenth-century jurists assigned fractional measures to each type of evidence: for example, the corroborative testimony of how unimpeachable eyewitnesses constituted a complete proof; that of a single witness compromised by interest in the case, only a quarter. Legal treatises instructed jurists in the delicate weighting and balancing of evidence required by such addition. These fractions of legal proof, also called "probabilities," supplied the first generation of mathematical probabilists with a ready interpretation for their calculus as a quantification of degrees of certainty, which, as Jakob Bernoulli remarked, "differ from the latter (i.e., certainty) as the part from the whole."[1] Leibniz underscored this interpretation in his rebuttal to Locke's nominalism, the *Nouveaux essais sur l'entendement humain* (comp. 1703–5), by assimilating the elaborate judicial hierarchy of full and half-proofs to a new branch of logic and to the recent work of Pascal, Huygens, and DeWitt in mathematical probability.[2]

Although the probability of testimony remained a staple problem for probabilists from Bernoulli through Laplace, its real importance lay in the framework it provided for extending probability theory to other applications in the moral sciences. Bernoulli, perhaps encouraged by his correspondent Leibniz, hoped that the mathematical analysis of evidence would expand its scope beyond jurisprudence to include all decisions—the vast majority in civil life—that hinged upon partial proofs. The universal method of mathematics would translate the elements of sound conjecture, exemplified by but not restricted to legal usage, into the most general terms. In viewing their theory as the "art of conjecture," classical probabilists adopted the legal habit of thinking about probability epistemically, as a continuum of degrees of certainty. They also inherited a set of problems related to legal evidence—in particular, responsibility for establishing rational grounds for belief not only in the courtroom but in the world-at-large.

A second tie united classical probability theory to the Enlightenment moral sciences. The interpretation of probability theory derived from the legal hierarchy of proofs had been epistemic: degrees of probability or proof stemming from the evidence of things and of testimony corresponded to degrees of certainty in the mind of the judge. Probabilities and concomitant degrees of conviction had been formally correlated in the hierarchy of proofs with the type rather than the amount of evidence. The associationist psychology of Locke, Hartley, and their followers broadened this notion of evidence and probability to include objective frequencies as well as subjective beliefs. According to the associationists, the more constant and frequent the observed correlation between events, the stronger the mental association between the ideas representing these events, which in turn intensified probability and belief. Hence, the objective probabilities of experience and the subjective probabilities of belief were, in a well-ordered mind, mirror images of one another.

The classical probabilists subscribed to psychological theories that described mental operations in terms congenial to their theory. Not only did the mind naturally keep tally of frequencies and apportion belief accordingly—Hartley went so far as to propose a physiological mechanism of "cerebral vibrations" that literally etched the association more deeply with each repetition;[3] the associationists also claimed that it reasoned by the computation and comparison of probabilities. For Locke and his followers, all intellectual novelty owed to the combination and recombination of ideas, thus mimicking the kind of combinatorial enumeration of elementary probability theory. The French Lockean Condillac went so far as to describe this mental operation as a "kind of calculus,"[4] and his contemporary Condorcet echoed this view that the best intellects were those that excelled at this mental calculus and in, as Condorcet put it, "uniting more ideas in memory and in multiplying these combinations."[5] If mental reckoning was a "kind of calculus," the combinatorial calculus of probabilities could be viewed as the mathematical expression of the psychological processes that guided right reasoning. When these combinations boggled even the most agile minds, the calculus of probabilities could, as Condorcet's disciple Lacroix suggested, "correct the simple perceptions of good sense, which cannot follow the detail of combinations when they proliferate and become complicated beyond a certain point" as a natural extension of these same processes.[6] Conversely, if probabilistic computations captured the subliminal promptings of implicit reasoning too fine to be rendered explicit, it was because of an inherent similarity in structure between reasoning and the calculus.

Thus the classical interpretation of probability—that is, the classical answer to the question, "What do probabilities measure?"—made it the index of reasonable belief; initially, the reasonable belief jurists derived from legal evidence, and later reasonable belief built up from broad, constant experience via the association of ideas. For the classical probabilists from Bernoulli through Poisson, the mathematical theory was, in Laplace's words, "only good sense reduced to a calculus." This "good sense" interpretation of mathematical probability explains two striking features of the classical theory: first, its characteristic domain of applications; and second, the standards for successful application. The apparent miscellany of applications to gambling, insurance, astronomy, medicine, reliability of testimony, accuracy of tribunal judgments, economic theory of value, and reasoning from known effects to unknown causes were in fact joined by a single thread: all problems were posed in terms of reasonable belief and action based upon that belief. Is it reasonable to trust a given historical account of, say, a miracle? Is it reasonable to play a coin-toss game like the one that spawned the St. Petersburg paradox? Is it reasonable to believe that the solar system was formed by the operation of a single, uniform cause? Should we expect a phenomenon observed once to be repeated? (In this light, it is significant that expectation, rather than probability per se, was the conceptual departure point for most classical formulations of mathematical probability.)

Because its practitioners understood probability theory as a mathematical codification of good sense, the right answers to these questions were those that seconded the intuitions and practices of reasonable men. This was emphatically not a majority-rules view of rationality: if it were, there would be little need for a calculus of probabilities to recalculate the conclusions we all would have reached anyway. The probabilists believed, along with Voltaire, that "common sense is not all that common." Their calculus modeled the judgments of an elite of *hommes de lumières* who possessed "a mind clear enough ... to calculate without algebra"[7] for the benefit of those obliged to substitute mechanical reckoning for natural gifts.[8] Eighteenth-century probabilists never tired of repeating that their results "conform(ed) to that which the simplest reason would have dictated" or were "level to ... the common sense of mankind."[9]

When mathematical results clashed with the practice of reasonable men, the eighteenth-century probabilists consistently rearranged or modified the mathematics to reconcile the two. When Daniel Bernoulli argued in favor of smallpox inoculations on probabilistic grounds, d'Alembert rebuked him not because he opposed smallpox inoculations, but because Bernoulli's life expectancy calculations conflicted with the actual psychology of risk-taking.[10]

The tenacity with which eighteenth-century probabilists clung to the view that their theory should describe the way in which reasonable men conducted their affairs, even if it meant tinkering with definitions as fundamental as expectation, stemmed from related assumptions concerning the goals of "mixed mathematics." Mixed mathematics differed from latter-day applied mathematics in seldom distinguishing the formal, abstract mathematical theory from the subject matter it purportedly described. Once the calculus of probabilities adopted the "good sense" of reasonable men as its subject matter, the descriptive orientation of mixed mathematics required the probabilists to accept that good sense as the final test of its results. If a mathematical theory of lunar motion failed to predict the observed orbit, mathematicians challenged the theory; if the conventional solution of the St. Petersburg problem ran counter to the judgment of reasonable men, probabilists reexamined their definitions of expectation. In the minds of eighteenth-century mathematicians, there did not exist any theory of probabilities disembodied of subject matter.[11]

The moral sciences of the Enlightenment and the calculus of probabilities shared as their subject matter the psychology of the rational individual. Like the calculus of probabilities, the moral sciences were riven by a tension between prescriptive and descriptive aims. By describing the precepts underlying the belief and conduct of a select group of reasonable men, probability theory prescribed a code that others could follow by rote, if not by instinct. Similarly, by showing that a certain course of action maximized rational self-interest, the moral sciences sought to persuade the confused or obstinate to obey "natural law."

These "natural laws" of the moral order should not be confused with those that governed the natural order. As Montesquieu observed, "The intelligent world is far from being so well governed as the physical. For though the former has also its laws, which are of their own nature invariable, it does not conform to them so exactly as the physical world."[12] This was because human agents, unlike material objects, were by their nature liable to error and possessed free will. Even the physiocrats, who of all the schools of eighteenth-century social thought laid the heaviest emphasis upon "the machine of Nature" and its "universal law," admitted that intelligent beings *chose* to obey the laws of the moral order.[13] Quesnay and other physiocrats maintained that the natural order existed for man's benefit, and that it was therefore in man's best interests to take advantage of his privileged position in creation by obeying the laws discerned by his intelligence.[14] Failure to do so, either through ignorance or perversity, resulted in economic disaster; yet man was free to err. The natural laws that regulated the moral order exacted obedience in the sense that mathematical demonstration coerced assent, through the exercise of reason.

Thus the moral sciences, like the classical theory of probability, aimed to reveal the rational grounds for action and belief in the hopes that, once instructed in their true rational self-interest, all would bend their will to the moral order that best promoted that interest. Providentially, these natural laws coincided with natural rights, and the pursuit of self-interest also served the common weal. When Condorcet attacked the relativism of Montesquieu's *Esprit des lois* (1748) for tailoring law to the "national genius" of various peoples, the universal laws he upheld were not of the form of the universal law of gravitation, but rather "truth, reason, justice, and the rights of man, interest in property, liberty and security (which) are the same everywhere."[15]

The pervasive ambiguity in the concept of moral law, which embraced both necessity and choice, endowed the moral sciences with a double nature. On the one hand, the moral sciences, like the physical sciences, studied the immutable order of phenomena concerning human thought and action; on the other, they recommended changes in the existing social, political, and economic order. It was superfluous for physicists to urge compliance with the law of gravitation; it would have been absurd to suggest that there existed any choice in the matter. Yet eighteenth-century social thinkers complained ceaselessly that current social arrangements violated one or another law of the moral realm. These writers assumed that the natural order was best, but not necessary.

The probabilists also presented their attempts to apply mathematics to moral sciences as aids to policy making. Daniel Bernoulli claimed that his computations of moral expectation were the best guide to investment; Condorcet recommended the abolition of the death penalty on probabilistic grounds; Laplace advocated gradual social change and condemned the French judicial system in the name of probability theory; Poisson enlisted probability to argue that the true criterion of judicial success was the security of society. Because both the moral sciences and the classical theory of probability took the rational individual as their unit of analysis, their results applied to society only as an aggregate of such individuals, despite the misleading name of Condorcet's "social mathematics." Both shared an approach that was individualistic, psychological, and prescriptive and that centered upon the determination of rational criteria for action and belief.

2

The "good sense" interpretation that had characterized the classical theory of probability rested upon two assumptions: first, that associationist psychology guaranteed the direct proportion between objective experience and subjective belief; and second, that good sense was monolithic and a constant for the fortunate few who enjoyed it. Both assumptions were shattered by the end of the eighteenth century. Associationists from Condillac on devoted ever more attention to aberrations of the intellect, including the distortions in the estimation of probabilities introduced by the passions, imagination, and prejudice. In the later works of the classical probabilists, mathematical probability gradually became more of a corrective than a description of common reason. At first, probabilists considered that only "children and the people" (as Condorcet put it) were susceptible to such errors, but the events of the French Revolution apparently changed their minds. By 1793, Condorcet was urging that "since all of the truths recognized by enlightened men have been confused in the mass of uncertain changing opinions, we must fetter men to reason by the precision of ideas, by the rigor of proofs."[16] Laplace's discussion of the relationship between associationist psychology and mathematical probability some twenty years later fall under the heading of "Illusions in the estimation of probabilities." Laplace's principal motive for examining the "laws of intellectual organization" was not to demonstrate the triple parallelism between sensory evidence, mathematical probability, and rational belief, but rather to expose the psychological roots of probabilistic illusions, which included belief in astrology and divination.[17]

The French Revolution also dissolved the second assumption of a consensus among an elite of reasonable men. Dizzying shifts in political philosophy from the outbreak of the Revolution in 1789 to the restoration of the Bourbons in 1814 did what a century's worth of mathematical controversy could not: shake the confidence of *philosophes* and mathematicians in the existence of a single enlightened viewpoint among even the right-thinking. Insofar as the notion of good sense survived the Revolution and its aftermath, it had ceased to be a matter of estimating probabilities. Moral philosophers like Victor Cousin suggested that good sense was more intuitive than rational, more synthetic that analytic, more complex than abstract.[18] The classical theory of probability had lost its subject matter.

It had also lost its characteristic interpretation. Subjective belief and objective experience began as equivalents and ended as diametric opposites. Critics of the classical theory like the economist and philosopher Cournot sharply distinguished between the senses of probability as a measure of observed frequencies and as a measure of belief. Mathematicians and philosophers—notably Ellis, Boole, and Mill—advocated a strictly frequentist interpretation of mathematical probability, using "subjective" as an epithet. Once the psychological bonds between objective and subjective probabilities, and between the calculus of probabilities and good sense, were loosened, the classical theory seemed both dangerously subjective and distinctly unreasonable.

During the first half of the nineteenth-century a new program for the moral sciences also emerged. Auguste Comte's science of society epitomized the transition from the psychological framework of the eighteenth-century moral sciences to the sociological one of the nineteenth century. He pioneered the study of societies as coherent units, drawing heavily upon organismic analogies, rather than as aggregates of individuals. Indeed, Comte went so far as to deny psychology the status of an independent science. In theory, social laws exercised as ironclad a determinism as physical laws, although social theorists from Comte through Marx were still bedeviled by the role of the individual agency in advancing or hindering the course of these laws. Comte agreed with many of his contemporaries, philosophers and mathematicians alike, that the eighteenth-century applications of probability theory to the social realm were "aberrations of the intellect," and a slur upon the good name of mathematics.[19]

Thus by 1840 the Enlightenment alliance between mathematical probability and social theory seemed to lie in shambles. On the one hand, the good sense interpretation of probability had been discredited, and many of its applications along with it; on the other, the search for inexorable social laws had replaced the computation of rational self-interest in social theory. Yet a new alliance arose between mathematical probability and social theory during just this period in the form of Quetelet's "social physics." Quetelet's conceptual innovations were few; his real contributions lay in his untiring campaign for massive compilations of statistical data on all subjects and in his unshakable faith that regular patterns invisible at the individual level would emerge at the societal level. Ironically, it was probability theory that underwrote Quetelet's antipsychological, antiindividualistic creed in the social sciences: "Thus, moral phenomena observed in the mass come to resemble the order of physical phenomena, and we will be obliged to admit as the fundamental principle of these sorts of investigations that the greater the number of individuals observed, the more the individual peculiarities, be they physical or moral, disappear, leaving the series of general facts by which society exists and endures to predominate."[20]

3

A new view of the relationship between probability theory and statistical data made it possible for Quetelet to reunite the frequentist interpretation of probability and the social sciences under the arch of the normal distribution. Many of the demographic regularities in the rate of births and deaths that so impressed Quetelet had been known to eighteenth-century writers on what was called "the natural history of man." Because these regularities concerned the physical aspects of human existence, they did not involve the element of choice, and therefore of variability, that marked the moral sciences. Nor were they connected to the mathematical theory of probability, except by the actuarial techniques that estimated odds of dying at any given age from the mortality tables.

In fact, eighteenth-century collectors of these statistical regularities opposed them to the workings of probability as signal evidence of the hand of divine providence in human affairs. A typical argument of this sort, first proposed by John Arbuthnot in 1710 and repeated by De Moivre, concerned the ratio of male-to-female births. Arbuthnot noted that male births consistently exceeded female births in the ratio of 18 to 17. He argued that if this regularity were due to what he called "mere chance"—that is, assuming that the probability of either male or female equals 1/2—the probability of the observed ratio over a long period was astronomically small. Arbuthnot concluded that this was palpable evidence of design, namely, the divine provision for an equal number of men and women of marriageable age to ensure the propagation of the race via monogamy. Due to the greater "wastage" of young men, who led more hazardous lives, it was prudent to begin with a small surplus.[21] No less a mathematician that De Moivre praised this statistical version of the argument from design revealing the "wise, useful and beneficent purposes" inherent in "the stedfast [sic] Order of the Universe."[22]

Although several eighteenth-century mathematicians, including Nicholas Bernoulli and d'Alembert, attacked this argument, statistical regularities remained a favorite weapon of eighteenth-century theologians intent upon finding the signs of an "all-wise, all-powerful and good" agent in demographic data. Süssmilch's massive compilations of such data in the service of the argument from design in his Die göttliche Ordnung was simply the best known of these treatises. In order to make the argument, the stability of statistical ratios, expressing the divine order, were opposed to probabilities, representing the workings of mere chance. Specific extenuating circumstances, rather than normal dispersion, were invoked to explain the inevitable observed deviations from the purportedly stable ratio, for the model of regularity was a determinate law, rather than a distribution. Thus, the empirical study of social regularities began as an antiprobabilistic approach.

More than a familiarity with the mathematical tools of distributions was required to reconcile probability theory with the statistical study of social regularities. Laplace used the normal distribution to analyze observational error in astronomy in 1781, in the same memoir that he applied Bayes's theorem to the male-to-female birth ratio, but not until the 1830s were mathematicians and social theorists like Poisson and Quetelet able to dissociate stable statistical ratios from divine providence. By adopting a new model of both probability and the social sciences, they were able to apply these distributions to the social realm. Having abandoned the rational individuals of the eighteenth-century moral sciences in favor of macroscopic social regularities, Quetelet was able to assimilate the decision of an individual to, say, commit suicide to the haphazard circumstances that caused an astronomer to err in one or another direction in recording a celestial position. Since only the uniform causes operating in the large mattered to a "social physics," Quetelet was able to ignore the "perturbing" causes of individual decisions in a way that would have been inconceivable to a practitioner of the Enlightenment moral sciences.

Poisson severed statistical regularities from the argument from design by showing that such regularities were just what one would expect from the calculus of probabilities. Far from being a token of divine intervention, such regularities were "the natural state of things, which subsists by itself without the help of any foreign cause and which would, on the contrary, require such a cause in order to undergo a significant change."[23] The new frequentist interpretation of probability theory also brought it closer to the statistical data that had seemed a denial of chance to eighteenth probabilists.

Conclusion

Thus mathematical probability was given a second chance to become the calculus of the social sciences in the first half of the nineteenth century, but neither probability theory nor the social sciences much resembled the disciplines that had originally made common cause in the eighteenth century. Quetelet himself was a figure of transition, for he was too slavish a disciple of Laplace to abandon wholly the classical interpretation despite his preoccupation with statistical data and distributions. His famous fiction of the *l'homme moyen* epitomizes both the contrasts and continuity between the old and new alliances of mathematical probability and social theory. *L'homme moyen* was a statistical composite of the physical, moral, and intellectual traits of the entire society. All that was merely individual, particular, or specific was obliterated in *l'homme moyen*, who represented the average in all things. At first glance, nothing seems further removed from the elite of reasonable men who featured so prominently in the writings of the eighteenth-century probabilists. The reasonable man belonged to a select minority distingusihed by an unusual ability to compute probabilities intuitively; *l'homme moyen* possessed no distinction of any kind unless it was his dead-center averageness. Yet the reasonable man and *l'homme moyen* both served as standards to be described by the calculus of probabilities in order that they might be emulated by others. Quetelet held up *l'homme moyen* as a literary, moral, and intellectual ideal, literally the golden mean for a given society. The ideals represented by the reasonable man and *l'homme moyen* were indeed very different, and reflected an altered understanding of both social theory and the calculus of probabilities. Yet *l'homme moyen* and the reasonable man were still brothers under the skin: both represented social standards in mathematical dress. Neither the old nor the new alliances between mathematical probability and social theory ever quite managed to untangle the "is" and "ought" of the social realm.

Notes

1. Jakob Bernoulli, *Ars conjectandi* (1713), trans. Bing Sung in *Translations from James Bernoulli*, Harvard University Department of Statistics Technical Report, No. 2 (12 February 1966), with a Preface by A. Dempster, p. 8.

2. Gottfried Wilhelm Leibniz, *Nouveaux essais sur l'entendement humain* (comp. 1703–5), in *Sämtliche Schriften und Briefe*, Akademie der Wissenschaften. Sechste Reihe: *Philosophische Schriften* (Berlin: Akademie-Verlag, 1962), vol, 6, pp. 460–465.

3. David Hartley, *Observations on Man, His Frame, His Duty and His Expectations* (London: 1749), vol. 1, pp. 324–332.

4. Etienne Bonnot de Condillac, *Essai sur l'origine des connaissances humaines* (1746), in *Oeuvres* (Paris: An VI/1798), vol. 1, p. 109.

5. M. J. A. N. Condorcet, *Vie de Turgot* (1786), in *Oeuvres de Condorcet*, eds. F. Arago and A. Condorcet-O'Connor (Paris: 1847–49), vol. 1, p. 222.

6. Silvestre-François Lacroix, *Traité élémentaire du calcul des probabilitiés* (Paris: 1816), pp. 259–260.

7. George Leclerc Buffon, *Essais d'arithmétique morale*, in *Histoire naturelle. Supplément* (Paris: 1777), vol. 4, p. 68.

8. Condorcet, *Essai sur l'application de l'analyse à la probabilité des décisions rendues à la pluralité des voix* (Paris: 1785), p. ii.

9. Abraham De Moivre, *The Doctrine of Chances*, 3rd ed. (London: 1756), p. 253.

10. Lorraine J. Daston, "D'Alembert's Critique of Classical Probability Theory," *Historia Mathematica*, 6 (1979), 259–279.

11. Daston, "Probabilistic Expectation and Rationality in Classical Probability Theory," *Historia Mathematica*, 7 (1980), 234–260.

12. Charles de Secondat de Montesquieu, *The Spirit of the Laws* (1784), trans. Thomas Nugent (London: Colonial Press, 1900), p. 2.

13. François Quesnay and Victor de Riquetti Mirabeau, *Philosophie rurale* (Amsterdam: 1764), vol. 1, pp. 100–101.

14. Quesnay, "Despotisme de la Chine" (1767), in *Oeurvres économiques et philosophiques de Quesnay*, ed. Auguste Oncken (Paris: 1888), p. 645.

15. Condorcet, "Commentaire sur le vingt-neuvième livre de *L'Esprit des lois*," in Antoine Louis Claude Destutt de Tracy, *Commentaire sur L'Esprit des lois de Montesquieu* (Paris: 1828), pp. 380–381.

16. Condorcet, "Tableau général de la science qui a pour objet l'application du calcul aux sciences morales et politiques" (1793), *Oeuvres*, vol. 1, p. 541.

17. Pierre Simon de Laplace, *Essai philosophique sur les probabilités* (1814), in *Oeuvres complètes de Laplace*, Académie des Sciences (Paris: 1891), vol. 7, pp. cxiiff.

18. See Cousin's criticisms of Frances Hutchenson's moral calculus in his *Cours d'histoire de la philosophie moderne*, 1st series (Paris: 1846), vol. 4, pp. 169–174.

19. August Comte, *The Positive Philosophy of Auguste Comte*, trans. Harriet Martineau (London: 1875), vol. 2, p. 100.

20. Adolphe Quetelet, *Sur l'homme et le développement de ses facultés, ou Essai de physique sociale* (Paris: 1835), vol. 1, p. 12.

21. John Arbuthnot, "An Argument for Divine Providence, taken from the Constant Regularity observ'd in the Births of Both Sexes," *Philosophical Transactions of the Royal Society of London*, 27 (1710–12), 186–190.

22. De Moivre, *The Doctrine of Chances*, p. 252 (note 9).

23. Siméon-Dénis Poisson, *Recherches sur la probabilité des jugements en matière criminelle et en matière civile* (Paris: 1837), pp. 144–145.

14 Décrire, Compter, Calculer: The Debate over Statistics during the Napoleonic Period

Marie-Noëlle Bourguet

Based on a case study of the Napoleonic "Statistique générale," this chapter aims at delineating the French use of statistics and relating it to the society that produced it, as opposed to different methodological and theoretical choices made by other countries, such as German cameralism or English political arithmetic. Three topics are discussed:

1. The debate over the statistical method and matter, which led the Consulate government to reject the use of the probability calculus and, moreover, to choose an encyclopedic regional survey instead of discrete and numerical inquiries.

2. The social history of the survey and its content analysis disclose how tightly intertwined the development of knowledge about society and that society can be. By heeding Chaptal's call for help and willingly participating in the survey, the local notables, officials, and humanitarian technicians embedded the statistics in the traditional administrative framework of rural France—a very distinctive feature of French statistics, in contrast to the German academic or the English parliamentary types.

3. The progressive decline of the regional statistics as the cult of numbers developed under the later Empire indicates the extent to which the shifting of statistical methods and content was related to the changing relationship between state and society and to new conceptions of society. The surveys, then, were only intended to give the state a means of control and direction, whereas the rise of quantification, specifically in the social field, shows the birth of a new mentality that assumed social data might be referred to social causes. Social statistics became the way to analyze social problems.

It is necessary at the outset to lay out some preliminary methodological observations. When historians and social scientists talk about the history of the social sciences and statistics, they usually consider only one set of questions: how society as a subject matter is grasped by statistical investigation. But besides being a subject matter, society is also the context in which these statistics take shape: although society as a context and society as a subject overlap, they are not identical. Statisticians gather their data differently or aggregate them into different units, if they come from different social or political positions. Even those statistical theories that have universal applications bear the stamp of the particular social arrangment they were first intended to explain. Shifting from a cumulative history of cognitive steps to a properly historicist approach, we encounter this question: how can we relate society as a context and society as a subject matter? What information was regarded as relevant and obtainable in a given country, at a given period, and why?

The years 1770–1840 were a crucial period, during which intensive statistical activity developed in all European countries. As a scientific quest, the appetite for statistics was a European-wide phenomenon, which spread beyond national boundaries. Let us just mention the many foreign texts that were translated into French: W. Playfair was translated in 1789, A. Smith in 1796 by G. Garnier, and the

German statistician Hoëck in 1800 by A. Duquesnoy. Nevertheless, the development of official statistics at the level of practice remained firmly national. The reasons for statistical surveys and the methodological choices appear distinctive, arising from concerns peculiar to each country—and, one would assume, connected with the specific relationships between state and society. Such diversity (from the descriptive statistics of German cameralism to the English political arithmetic) calls for a comparative study of the national paths to statistics.[1]

The French path to statistics during the Napoleonic period offers a very intricate example, which can be roughly summarized as a contrast between the "golden age" of regional statistics of the Consulate (broad-ranging surveys conducted by both officials and experts) and the many numerical enquiries, mainly economic, of the later Empire, characterized by secrecy and specialization. But whether descriptive or numerical, the Napoleonic statistics shared an empirical orientation, which excluded the possibility of a probabilistic revolution.[2]

In order to delineate the French use of statistics and relate its specificity to the society that produced them, the grandiose attempt of the Consulate "Statistique générale" forms a basis for analysis and comparison. The vigorous debate over statistical method and subject matter among officials and theoreticians led the government to reject the use of the theory of probabilities and, moreover, to choose an encyclopedic descriptive survey instead of discrete numerical inquiries. The reactions of local officials and elites to that project shed some light on its social significance in the peculiar context of the new French nation-state. Finally, the limited role of quantification within descriptive statistics, as opposed to the cult of numbers that developed under the later Empire, reveals the extent to which the shifting of statistical methods and subject matter are related to the changing conceptions of society. Those are the three successive points I shall now consider.

1   The "Statistique Générale de la France"

In 1801, continuing the work of his predecessors, François de Neufchâteau and Lucien Bonaparte, the new minister of the Interior, the chemist J. A. Chaptal, initiated a general inquiry. Its conduct was specifically the responsability of the *préfets*, officials who had just arrived in their *départements*: in order to be trained in statistical investigation while discovering the place and people they had to administer, the new officials were asked to make a descriptive memoir covering five ordered topics—topography, population, social situation, agriculture, and industry and commerce—each of them containing a great number of items. The chapter on agriculture, for instance, asked for six tables showing the number of ploughs drawn by horses or oxen; the total area of land divided into arable, vineyards, orchards, meadows, commons, woods, roads, rivers, etc.; the area and value of each type of product (cereal, vine, meadow, etc.); the number of every animal species (cattle, sheep, goats, pigs, chickens) with the quantity and value of their produce (wool, meat, butter, cheese, honey, silk, skin) and the cost of each product; the consumption and expenses of workers and landowners, hence the net revenue; etc.[3]

In the administrative chaos of postrevolutionary France, such an encyclopedic proposal was quite quixotic: the *préfets* were overwhelmed by daily obligations; the governmental machinery was ill-equipped for collecting data; at least two-thirds of the mayors did not even know how to write. As optimistic as they were, the Parisian officials were aware of that situation. How then did they justify their attempt? Their reasoning was based on both political arguments and epistemological assumptions.

1.1 Probabilities and Politics

First of all, why not start only a limited survey by gathering a few sample data that mathematicians could use as a basis for calculation and extrapolation? Such methods had been used by English political arithmeticians since the seventeenth century. In eighteenth-century France too, demographers—Expilly, Moheau, Messance—had remedied their lack of reliable information and their inability to get complete series of data from the officials precisely by developing techniques of calculation based on birth and death rates. Social thinkers and philosophers—most prominent among them Condorcet—were so aware of the achievments of mathematics in demography, gambling, and life insurance that they drew up plans for extending its application to other areas of social life, such as the decision-making process of juries or political assemblies. Eventually, mathematics was to encompass the whole field of the political and moral sciences.[4]

In fact, there were some probabilists among Chaptal's political and scientific advisers, even within the *bureau de statistique* itself. Such was Laplace, whom the minister asked to work on a census by applying probability theory to a sample of data collected by the *préfets*. Such, too, was a minor and forgotten character, Etienne Duvillard. First an employee in the *Contrôle général* office under Turgot, then at the *Trésor* bureau, Duvillard was a member of the Society of 1789 and, in 1790, the director of a *bureau d'arithmétique politique*; he was the author of several booklets and reports—some of them discussed by the *Académie des Sciences*—on vital statistics and their use for govermental as well as private policy: life insurance, lotteries, taxes, vaccination, etc. However, when attached to the Ministry of the Interior in 1801, he occupied only a minor position during the Consulate and the first Empire, until he became temporarily the head of the *bureau*, in 1806. From the beginning Duvillard had eagerly asked for the reorganization of the *bureau*, the abandonment of extensive surveys, and the use of a mathematical method: he wanted to be in charge of a *bureau des calculs scientifiques*, where only four mathematicians ("calculateurs") would work, using their professional skills to project from the incomplete empirical reports a knowledge the government could rely on. "Ce qui ne peut être compté ou mesuré immédiatement," he argued, "le raisonnement et le calcul par l'analyse méthodique des faits le fait connaître." Therefore, the use of the theory of probability would allow governmental action to be efficient, despite the lack of factual information; the calculations on the life expectancy of the French population, for instance, would enable the state to raise money by selling life annuities.[5]

As an actuary, Duvillard was more concerned with speculative research and sophisticated mortality tables than with empirical and human realities. Still, as a government employee, he was offering the use of probabilities as a governmental technique, and that is precisely what Chaptal and his close collaborators objected to. Viewing 'arithmetic' as an instrument of executive authority, they first rejected it because of its political dangers. While that technique had permitted the *Ancien régime* monarchy to fill the gap between the ordinary tax revenue and the actual expenditure by selling lotteries or life annuities, the new political and social order should not allow such financial expedients or secret use of power: first, because the government now had to act in the open and submit all expenses to the nation's scrutiny; second, because the suppression of privileges, by spreading the tax burden among all citizens, would yield additional revenue. Hence, the government did not have to develop sophisticated skills in its bureaus: its first task was to establish a revised tax based on a general survey of the population and a new cadaster; and such was the purpose of the *Statistique générale*. Many of Chaptal's declarations make quite explicit his fear of any abusive and secret use of power by the state: "On ne doit pas perdre de vue que tout gouvernement tend à une domination arbitraire." Introducing probability theory into the central bureaus could possibly give the state too dangerous a tool.[6]

1.2   The Need for Facts

As the dismissal of Duvillard's proposal makes clear, the Consulate statisticians were to be empirically oriented: "La statistique de la France ne peut se faire à Paris. ... Mon bureau de statistique ne crée pas et ne peut pas créer. On ne peut qu'y analyser des mémoires faits sur les lieux," wrote the minister. Besides their fear of speculative research—"l'esprit de système," as Duquesnoy later called Lamarck's work on meteorology—political reasons and intellectual expectations motivated their choice of a descriptive and all-encompassing survey ("décrire la France sous tous les rapports").[7]

There was, first, a political urge: Parisian officials needed to check the impact of revolutionary reforms in the provinces and to observe the new equilibria produced by the reshaping of social organization and the remodeling of territorial divisions. At a time when France had just been born as a unified nation, they had to take a picture of its local and regional variety using data systematically gathered and classified according to new standards and uniform categories. Their request for regional statistics reveals both a practical concern for the situation of provincial France and their political worries about this very diversity. In the project of a statistical survey aiming to assess the government's task, it was regional variety that was really at stake.

Epistemological assumptions also played a part in the choice of an encyclopedic description over discrete and purely numerical tables. Coming from the natural sciences and the Linnaean tradition, the model of science as being a classificatory method was still dominant. All philosophers and experts shared the faith in the heuristic and explicative qualities inherent in the ordering of data. As Necker put it, "trouver un ordre clair et précis" was the only basis for knowledge. Even Condorcet's decimal method was mostly a formalized system of classification, a perfect language built for data not easily quantified, such as social facts. Clearly, the five chapters of Chaptal's questionnaire were so arranged as to be read as an explanatory exposition of chronological sequences: going from the ground to the population, from the natural resources to human activities, it offered a cryptophysiocratic model of the "natural order" of the economy. The juxtaposition and succession of items stood for causality and genesis.

1.3 A Liberal Use of Statistics

In its very ambition, this search for knowledge and understanding reveals the enlightened conviction on which the Consulate statistical program was based: a faith in progress as being the quasi-automatic product of science. More specifically, in the case of a society as rural and backward as France was then said to be, no improvment seemed possible without vigorous help from the top, mostly from the centralized state, whose task was to spread instruction and remove obstacles in the path of reason so that society would gradually become stronger and ultimately more autonomous. Like the creation of schools and experimental farms, the statistical survey of France, in its later publication, belonged to this enlightening function of the state. Although opposed to Duvillard's conception of statistical knowledge restricted to government use, Chaptal's ideal was a paradoxical association of liberalism and *dirigisme*: it combined a liberal conception of society (for no one contested the individual's rights to liberty, property, and privacy) and a demand for guidance from the state that, for the time being, justified its right to the close scrutiny of an encyclopedic survey. As the semiofficial head of the *bureau de statistique*, A. Duquesnoy, declared, "L'Etat sans doute ne doit pas agir souvent, mais ... doit tout savoir et tout connaître pour agir à propos."[8] The various provincial reactions to the Parisian program made that ambiguity obvious.

2 Statistics and Society

Because of the frailty of the new administrative machine, the *préfets* called upon the local elites for help. The various responses, ranging from enthusiastic participation to hostility and refusal, are most significant for judging the implications of such an inquiry. Even more than the theoretical debates and governmental declarations, the social history of the survey and the content analysis of the reports disclose the interdependence of knowledge about society and that society proper.

2.1 Collecting the Statistics

Because of his own liberalism, Chaptal expected the good will and cooperation of all manufacturers and entrepreneurs: "I do wait for you to answer my questionnaire," he wrote to some of them, "for I have no intention to fill the workshops with overseers (remplir les ateliers d'inspecteurs)."[9] But the entrepreneurs did not trust him. In any case, as the *préfets* complained, they did not bother to fill out the forms honestly, because they did not expect any good to come from the inquiry. In fact, from their point of view, anything the state would do, if given information, would only violate their privacy and interfere with the free play of a market economy.

Only the local notables, officials, and humanitarian technicians (landowners, agronomist experts, physicians, professionals) heeded Chaptal's call for help: they shared the same ideas on the use of science, the same conceptions of society. *Préfets*, local officials, elites: all of them were enlightened and responsible persons, highly aware of the importance of their task. Their participation in Chaptal's enterprise embedded the statistical survey in the "traditional" administrative and rural France Edward Fox opposes to the "other France," urban and entrepreneurial.[10] That connection between the development of statistics and administrative elites is a very distinctive feature of the history of French statistics, as roughly opposed to the German academic model or to the English parliamentary type. But how were these statistics affected and shaped by their social roots? The question might be approached by looking at the mental apparatus, the descriptive codes and categories the inquirers used for ordering, describing, and understanding the social reality they had to survey.

2.2 The Description of Society

The Revolution had substituted for the old hierarchical social system a new definition, based on equality and identity. The nation was now to be a collection of citizens, all free and equal individuals, whose positions in society were no longer determined by birth or status, but by wealth and profession. In accordance with the new social order, Chaptal gave his inquirers, in his memorandum, a tripartite classification: the landowners; the independant elite (civil servants and professionals, craftsmen and well-off peasants); and the workers and laborers. On the margin of society was a final category meant to include all sorts of poor people, beggars and vagrants. But when using this code, the local observers imposed on it many distortions and, thus, disclosed some interesting biases in their social conceptions. In fact, they hardly ordered the local society through Chaptal's strictly legal socioeconomic criteria but rather twisted them and turned back to the traditional and hierarchical categories of estate and status. For instance, the *préfet* Nogaret first arranged the inhabitants of the southern *département* of Hérault into three socioeconomic groups ("riches, médiocres, pauvres"), but then turned to a simple dichotomic division, based on cultural criteria: "The first two classes ... are the ones where instruction and politeness are to be found; the third one, having no access to educational means, is also more dependent upon the local environment (l'influence du terroir)."[11]

The implication of such a dichotomy between the elites and the people appears clearly at the very end of the questionnaire, where Chaptal asked for a detailed description of the local customs and ways of life of the various social groups: "Il est important de rechercher ici les variations survenues dans la vie des citoyens ... en distinguant ce qui se pratique dans les différentes classes de la société, dans les villes et dans les villages, etc. On ne peut entrer dans trop de détails à cet égard sur les coutumes civiles et religieuses sur les moeurs privées, etc." Despite the minister's call for a wide-ranging observation of society, the local inquirers gave quite uneven answers, depending on which part of the society they were describing. They gave almost no detailed analysis of the elites, showing a blindspot in their social insight and limiting their comment to some kind of cliché: "Les moeurs sont les mêmes dans toutes les villes ... les moeurs et le caractère des personnes éclairées ou qui ont reçu de l'éducation sont à peu près les mêmes partout." [12] Of course, an elite's way of life seemed so familiar and obvious to the official who belonged to it that there was not much to say about it. But more important perhaps for a history of social investigation is the idea that the bourgeois life, whatever it is, must not be described: that would be a violation of its privacy. By their silence, the *préfets* happened to be more "modern" than their minister: they had already accepted the new definitions and values of an individualistic society, specifically the notion of privacy—the "private sphere," as Habermas calls it. In assuming that an elite's private life need not be described and in any case ought to be beyond the gaze of the state, they anticipated the liberal definition of civil society.

Toward the other part of society, actually the most numerous one, the inquirers' attitude was quite opposite. Because these people—"ce qu'on appelle ici le peuple"—lacked all the qualities that would make them true individuals, i.e., propertied and literate citizens, their description seemed both possible (because they still belonged to the sphere of nature, their bodies and minds bearing the marks of local environmental differences) and necessary (because they had to be ruled and, it was hoped, transformed). The descriptive practices show that the peasants were perceived as a sort of natural species, a "collective individual" whose members shared all attributes, without internal differentiation or change over time: "L'habitant du Finistère est ce qu'il était il y a six siècles." In the same way as for primitive peoples, the entire group was taken as a unit of perception and description, all its members being subsumed under a single one—"l'habitant." In order to reveal the innate essence of the people, a cultural prejudice allowed the observers to bring together all sorts of moral, psychological or physical characteristics, social institutions, and cultural habits: "L'habitant de l'Ariège est naturellement bon et officieux, mais froid, grave, circonspect. ... Il est sobre, patient, peu sensible aux privations. ... Le fanatisme religieux semble faire l'essence de son caractère." [13]

2.3 A Quantified Estimate of Social Welfare

Not only did that dichotomous division of society—the elites on one hand, not subjected to a close scrutiny because of their acknowledged right to privacy and autonomy; the people on the other hand, observed and described because of their very lack of both qualities—affect the descriptive codes used by Napoleonic *préfets* in their memoirs, it also played a role in the origin of quantification applied to social facts.

Within the systematic descriptive survey requested by Chaptal, there was, in fact, some room for numerical tables, which were not limited to the strict counting of the population, the grain harvest, or the industrial production. Chaptal intended to apply the use of numbers to some social facts, too, although he gave this quantifying approach a strictly limited scope: "Vous serez bien près de la vérité si vous savez le prix des choses nécessaires à la vie, le prix des journées de travail, ... si vous avez des renseignements précis sur les hospices, si vous connaissez le degré d'instruction, si vous savez le nombre de crimes et leur nature, si vous savez combien il y a de procès ... vous pourrez facilement connaître le degré de bonheur de vos administrés." These assertions cry out for comment, for they touch on the history of social statistics: by the subject matter they assigned to quantification, by the method they indicated for reading numerical data, and by the preconceptions upon which they rested.

At first sight, this quantitative measure of social welfare—a part of the general statistics—seems paradoxically narrow in its scope and rather simple in its method, compared to the quest for a systematic description of the natural and artificial world. How could Chaptal assure his *préfets* that a few discrete variables would give them some estimate of the general well-being of society, quite sufficient for their administrative work? "You will know enough ... ," he told them. This is, first, a pragmatic injunction: because no trespassing upon individual privacy would be allowed, the government's scrutiny had to be limited to the public sphere, that is, to public institutions (hospitals, prisons, schools) and places of collective life (streets and markets), where its intervention was both necessary and legitimate. But Chaptal's assertion also had a theoretical basis, with its positivist naiveté: for the few variables chosen—the number of criminals, of poor and insane people, of literate and educated, etc.—are taken at face value. These data are not signs of some social disease that would ask for further investigation in the society itself; rather, they speak for themselves, as signs of individual deprivation—lack of education, of reason, of property: all the attributes of a free individual—and therefore indicate the solution.

Thus, the same optimistic bias, both epistemological and political, the same faith that made Chaptal call for an encyclopedic survey and believe in the autonomous functioning of an individualistic society, also drove him to reduce the administrative statistical investigation of social reality to a few simple numerical tables.

3   The Cult of Numbers

The progressive decline and abandonment of the regional and descriptive statistics began soon after Chaptal's departure, in 1804. The priority was then soon given to specialized, numerical, and national surveys.

3.1   A Utilitarian Aim

Of course, the first and immediate reason for that shift is quite clear. The Consulate had bet on time; because of their slow and unreliable data, the descriptive reports were quite inadequate for the immediate needs of the later Napoleonic Empire. The external pressures of the blockade, the continental system, the need for self-sufficiency, the series of bad harvests from 1811 on, etc., brought to the open the failure of the long-term, optimistic, and sophisticated attempt of the Consulate. A classic example of that failure was the inability of the *bureau de statistique* to provide a complete table of French manufactures within a week, when Napoleon asked for it in 1811. Hence the need for more limited and reliable statistics was brought to the fore: this urge was particularly obvious in economic matters, where the late Empire fostered intense activity.

Let us emphasize that this shift in the purpose of statistical inquiry signified a change in the relationship between the state and society: to get data "in order to improve" was no longer to be understood in the general, philosophical, and liberal sense of the enlightened Consulate statisticians; knowledge was no longer thought of as being effectual by itself. It now had a much more authoritarian and prosaic purpose: although there still was no trespassing upon the right of private property, the uses of statistics were now quite state oriented and intended to give the state its means of direction and control. Consequently, the statistics were no longer to be published.

3.2 New Methods and Contents

Some new methodological orientations of Napoleonic statistics came directly from the Consulate's failure and from the new purposes for gathering empirical data.

First, the Empire did not attempt any more multicomprehensive regional surveys, with too many items and too vaguely phrased questionnaires. Within the *bureau de statistique*, Coquebert de Montbret alone retained a curiosity for local particularities by initiating surveys of agricultural techniques and regional dialects. But the dominant trend was now toward centralization and uniformity. Most topics from the previous descriptive statistics were abandoned: thus, the whole chapter on topography disappeared from the official surveys, except the meteorological item because of its importance for agriculture. Moreover, within a given topic, the inquiries became much more specialized, focused on a particular product or activity (for instance, the 1806 survey on cotton industry, the 1811 survey on chestnut production, the 1813 survey on bees and honey, etc.).

Second, the Empire statisticians never again tried to treat change over a long period of time and laid aside the Consulate attempt to measure the historical change brought by the Revolution's reforms: the focal point became the present time, which, they expected, would be regularly updated by periodic reports. That change in the conception and practice of statistics is most important, for it implied a move from the previous search for comprehensive knowledge (in which historical analysis was essential) toward a pragmatic attitude, taking the empirical information as the only basis for action. In short, it was a move from history and politics toward policy making.

Finally, there was a rise in quantification: whereas the Consulate statisticians were anxious to arrange all the descriptive items in an ordered system and therefore did not give numerical information in tables separated from the larger framework of environmental and explicative relations, the later ones were more addicted to numbers. Of course, their greater use of quantified reports came directly from the widening range of administrative demands and the growing needs of a centralized bureaucracy, as the numerous economic surveys clearly showed.[14] But the cult of numbers extended far beyond economic affairs to developments in social surveys as well.

3.3 Toward Social Statistics

There were, then, repeated counts of the aged, the insane, the poor, the conscripts, and even the elites. Again, such a development of administrative surveys was partly the further consequence of the new social order: as already shown by Chaptal's questions on social welfare, the Revolution had brought a decline of local institutions, charitable societies, and corporate groups, so that a growing part of the society—poor and old people, abandoned infants, orphans, etc.—had become dependent on administrative care and an object of worry, hence of scrutiny, for those who feared social disorder. But as the Empire was becoming more authoritarian with time, the quantification was no longer limited to the social groups that were 'normally' the subject matter for administrative investigation: from 1807 on, a series of surveys was demanded of merchants, medical surgeons and physicians, craftsmen and farmers, members of religious sects, etc. Moreover, in 1809–1810, a special "statistique personnelle et morale" was started, focused on notables. The *préfets* were requested to make lists of all rich citizens and to detail, for each of them, the amount of the family fortune, the composition of the household, the sons' academic grades, the daughters' dowries as well as physical beauty and chances of marriage! Thus, while the 18th Brumaire had been thought of as the start of a liberal era (as Chaptal himself had at first hoped), Napoleon did not in the end trust the bourgeois good will and tried to include them in the scope of administrative surveys. By this attempt, the whole liberal conception of a statistical inquiry was defeated, although the requested inquiry was so much resented by the local officials themselves that few of these statistics were ever completed, or were later deleted after Napoleon's fall. By itself, however, that sort of inquiry reveals how the rise of quantification in social surveys was not a mere technical change or development, but also indicated a change in the representation of society and of the relationship between state and society.[15]

These changing representations of society and rising fear of social disorder had still another implication for statistical methods. For quantification in the social field not only developed because of the larger scope given to the social survey, but also because of the new types of data collected. If we consider, for instance, the series of massive statistics on crime compiled from 1809 to 1814, classified by category for each *département*, it is worth noting that they were asking for more and richer data on the criminals themselves and their social context—sex, origin, profession—than the previous tables, which only counted the various types of prisoners. Although still present, the then current assumptions about historical and environmental causes of poverty, criminality, or insanity seemed no longer to be sufficient: social data too might shed some light on those social facts. As purely deterministic and mechanicistic as it still was, the idea that some social disorders might be referred to social causes, and therefore were no longer to be read as mere individual deprivation but rather as signs of some sort of social disease, was soon to lead reformers and observers toward social statistics.[16]

Thus, from the regional survey in the 1800s to the quantified social inquiries in the 1810s, there slowly emerged a mentality that assumed quantification was the way to formulate social problems, to find their causes, and hence their solutions. There is no history of social statistics, but a social history of statistics.

Notes

1. On statistics in eighteenth-century Europe, see *Pour une histoire de la statistique*, vol. I (Paris: INSEE, ca. 1977).

2. On Napoleonic statistics, see B. Gille, *Les sources statistiques de l'histoire de France, des enquêtes du 17° siècle à 1870*, (Paris-Genève: 1964); J. Cl. Perrot, *L'âge d'or de la statistique régionale française (an IV–1804)* (Paris: 1977); *La statistique en France à l'époque napoléonienne. Journée d'études, Paris, 14 février 1980* (Bruxelles: 1981); M. N. Bourguet, "Déchiffrer la France. La statistique départementale à l'époque napoléonienne," Thèse de 3° cycle, University of Paris I, Panthéon-Sorbonne, June 1983.

3. *Recueil de lettres circulaires ... émanés du ministre de l'Intérieur*, vol. 3, pp. 464–470: Chaptal's letter and questionnaire of the 19 germinal an IX. Hereafter, unless otherwise indicated, all quotations from Chaptal refer to that letter.

4. P. Buck, "Seventeenth Century Political Arithmetic: Civil Strife and Vital Statistics," *Isis*, 68 (1977), 67–84; idem, "People Who Counted: Political Arithmetic in the 18th Century," *Isis*, 73 (1982), 28–45; E. Esmonin, *Etudes sur la France des 17° et 18° siècles* (Paris: 1964); K. M. Baker, *Condorcet: From Natural Philosophy to Social Mathematics* (Chicago: 1976). Let us note, for comparative purposes, that mathematical methods developed in absolutist France and parliamentarian England in opposite political contexts. French demographers had to turn to sampling methods and extrapolation because the tradition of monarchical secrecy deprived them of reliable series of data: Expilly was rebuked when asking the *Contrôleur général* for information on births and deaths. On the contrary, English statisticians had to elaborate "political arithmetic" because the traditional "liberties" of the English people precluded any systematic counting: liberal members of Parliament defeated a proposal for a census in 1753 for being "totally subversive of the last remains of English liberty." Meanwhile, however, the English were undertaking abroad what they could not do at home: officials of the East India Company wrote settlement reports that were descriptive surveys of regions of India (I am thankful to Meghnad Desai for this last remark).

5. Bibliothèque nationale, Paris, Fonds Duvillard, Ms. frç. nouv. acq. 20576–20591. On Duvillard, see W. G. Jonckeere, "La table de mortalité de Duvillard," *Population*, 20 (5) (1965), 865–874; K. M. Baker, *Social Mathematics*, pp. 278–280 (note 4).

6. Chaptal, "Rapport et projet de loi sur l'instruction publique," Paris, an IX; quoted by J. Pigeire, *La vie et l'oeuvre de Chaptal* (Paris: 1932), p. 187.

7. Archives nationales, Paris (hereafter, ANP), F 20 2, letter of the 10 ventôse an XI; F 20 103, Duquesnoy's report, end of an XI. Nevertheless, the Consulate never totally gave up specific and numerical statistics: regular reports on grain harvests and prices resumed as soon as 1799; the meteorological research initiated by Lamarck was welcomed by the government, which allowed him to send his form through the administrative network; the same is true of Mourgue and other scientists' research on demography and of medical inquiries on epidemics. But these data were not only to be compiled and used on a national scale by the central bureaus: taken separatly, they were to be referred to their local environmental contexts and inserted in the regional statistics.

8. ANP. F 20 103, report of the 30 frimaire an XI. The Physiocrats also combined the same two conceptions.

9. J. Pigeire, *Chaptal*, pp. 273, 348 (note 6).

10. E. Fox, *History in Geographic Perspective: The Other France* (New York: 1971).

11. ANP. F 20 196, "Mémoire statistique du département de l'Hérault," prairial an XIII, ms, p. 205.

12. ANP. F 20 242, Laboulinière, *Statistique du département des Hautes-Pyrénées*, s.d., p. 153; Borie, *Statistique du département d'Ille-et-Vilaine* (Paris: an IX), p. 8.

13. ANP. F 20 187, "Mémoire sur la statistique du département du Finistère," frimaire an XI, ms; F 20 163, "Notice sur le département de l'Ariège," 23 frimaire an IX, ms, pp. 11–12.

14. On the late Napoleonic statistics, see B. Gille, *Sources* (note 2), pp. 131–140; St. Woolf, "Contribution à l'histoire des origines de la statistique en France," in *La statistique en France* ..., pp. 81–125 (note 2).

15. M. Agulhon, "Les sources statistiques de l'histoire des notables au début du 19° siècle dans les archives d'un département: le Var," in *Actes du 84° Congrès national des sociétés savantes. Dijon, 1959* (Paris: 1960), pp. 453–469; St. Woolf, "Sur les statistiques napoléoniennes: l'Etat des pauvres et des mendiants existant dans chaque commune (département de l'Arno, 1812)," *Annales historiques de la Révolution française*, 49(230) (1977), 654-663 (first published in *Social history*, 1976).

16. On the beginnings of social statistics, see C. Duprat, "Punir et guérir. En 1819, la prison des philanthropes," *Annales historiques de la Révolution française*, 49(228) (1977), 204–246; M. Perrot, "Premières mesures des faits sociaux: les débuts de la statistique criminelle en France (1780 -1830)," in *Pour une histoire de la statistique*, pp. 125-139 (note 1); B. P. Lécuyer, "Démographie, statistique et hygiène sous la monarchie censitaire," *Annales de démographie historique* (1977), 215–245; W. Coleman, *Death is a Social Disease: Public Health and Political Economy in Early Industrial France* (Madison: 1982).

15 Probability in Vital and Social Statistics: Quetelet, Farr, and the Bertillons

Bernard-Pierre Lécuyer

The purpose of this chapter is to trace the reception of Quetelet's fundamental ideas about statistics and probability in British and French circles of practitioners of vital and social statistics, conveniently symbolized by William Farr and the two Bertillons. In the course of the analysis a further link appears between the three figures thus selected. Not only do all three refer in one way or another to Quetelet, but William Farr exerted a strong influence on Louis-Adolphe Bertillon.

As a general conclusion one can only recognize again (without any claim to originality) the tremendous impulse given to the probabilistic movement by Quetelet's personality and energetic activities. As the section of this chapter on Farr clearly shows, he played a great role in the development of the British statistical movement (to use Cullen's phrase) through the foundation of section F of the British Association for the Advancement of Science (1833), of the Statistical Society of London (1834), and the 1860 meeting of the International Statistical Congress held in London. On a more cognitive level, his staunch belief in the lawful nature of vital and social phenomena was essential to Farr as well as to Guillard and the Bertillons. This is a basically Laplacean idea. Without this powerful idea, one could venture the hypothesis that Quetelet's personal charisma, great as it was, would have been deployed in vain.

As to the specific details of Quetelet's ideas, the picture of their reception by the British and French practitioners appears quite different. Farr does not seem to have paid great attention to the "average man," to "social statistics," or to "the social system." Louis-Adolphe Bertillon, on the contrary, took them seriously enough to deny radically their validity. Thus, in France, Quetelet's basic convictions were still present, but his specific ideas were rejected.

The following paper is a provisional attempt to trace the success and/or temporary failure in the dissemination of basic and even crude concepts of probability in the field broadly defined as that of social research, then embryonic. In this respect, it may be seen as related to psychology and the biomedical sciences (see volume 2). Social research, as it is envisaged here between the 1850s and the turn of the century, was still a combination of demography, economics, and psychology: only at the end of the 1890s do Durkheim, Marshall, and Weber give sociology its specific configuration. Economics will be left aside, since it is treated elsewhere by competent authors (volume 2).

My goal is to trace the reception of Quetelet's fundamental ideas about statistics *and* probability in the circles of British and French practitioners of vital and social statistics, conveniently symbolized by the contemporary and corresponding characters of William Farr (1807–1883), Achille Guillard (1789–1876), and the two Bertillons, Louis-Adolphe (1821–1883) and Jacques (1851–1922). These four characters have for us the following advantages: (1) they present a convenient contrast between a mathematically trained and experienced theoretical statistician as well as a good practitioner like Quetelet (1796–1874), and three practitioners of "shop-statistics" like the Bertillons and Farr, both initially trained as physicians; (2) the four together present a well-balanced picture of probabilistic and statistical thinking in England and French-speaking countries; (3) they belong to successive generations that cover the entire span of the nineteenth century.

A further question to be raised may seem irrelevant or impertinent to the historian of probability in the purest sense: were not the rule-of-thumb methods and conceptions of Farr and the Bertillons eventually more productive for vital and social statistics, at least in the short and medium range, than the grandiose theories of Quetelet? Most commentators have traced the impact or rejection of Quetelet's ideas by other theoreticians and sociologists like Durkheim, Halbwachs, etc. Interesting as they may be, these discussions confine themselves to a level of theoretical and quasi-philosophical generality. This level of generality is in a sense a double injustice to Quetelet, who cannot claim to be an original contributor to probabilistic thinking. Programmatic as they are, his main writings are evidently practically oriented: thus a comparison with contemporary practitioners, who were also friends and colleagues, is of some interest. In fact, the only exception to the general tendency to discuss Quetelet's contribution on a quasi-philosophical level are the articles by Lazarsfeld and Landau and Lazarsfeld.[1] These essays focus on the pioneering aspects of Quetelet's writings, particularly as concerns the measurement of "the quality of people, which can only be assessed by their effects." As Lazarsfeld continues, "A careful reading (would) show ... that his underlying theory of measurement, partly brilliant and partly a confused foreshadowing of later developments, gives the clue to much of his work."

To this conception of the history of science, oriented toward the detection of auspicious beginnings, one may legitimately oppose the more conventional but possibly safer solution: the study of the impact of a given conception among contemporary readers and fellow-practitioners, or among those belonging to immediately subsequent generations.

We shall briefly examine Quetelet's basic ideas, at least as considered in their contemporary historical context: this is why we shall deliberately leave aside Lazarsfeld's perceptive insights into Quetelet's theory of measurement to return to the more conventional discussion of his notion of the "average man," but as we have already said, more on a practical than on a theoretical level. In addition, it should be noted that the authors and the period considered are marked not only by intensive activity in statistics at large, but also by something like a pause in probabilistic thinking as applied to the social sphere. In effect, between Laplace and Poisson at the beginning of the nineteenth century and Pareto and Walras at the turn of the twentieth, the only writer worthy of mention seems to be Cournot. The renaissance of probabilistic thinking in the social sciences, with the exception of Cournot, seems to occur only around the turn of the century (Galton, Yule, Pearson, Poincaré, Borel, etc.). Thus we also consider the relations between statistical practice and probabilistic reasoning.

1 Quetelet's Legacy: "The Average Man," "Social Physics," and "Moral Statistics"

This is not the place to relate fully Quetelet's life, career, and writings. Let us nevertheless consider him as characteristic in the striking conjunction of his humanistic, mathematical, and social interests. His scientific mission to Paris for the sake of the future observatory of Brussels arguably made him a worthy successor to Laplace, who had only hinted at (if one excepts his earlier writings in the 1780s on births and marriages in Paris) the potentially broad applications of probabilistic thinking to social matters. Needless to say, Quetelet never exhibited the creative capacities of Laplace and his direct successor Poisson in social probabilities. Nevertheless, by his writings and his actions, he was a successful propagandist for the applicability of probabilistic reasoning to social statistical data. We know that both Farr and the Bertillons actively participated in the International Congresses of Statistics created by Quetelet himself.

Interestingly enough, Quetelet's first memoranda on our subject, after four earlier publications popularizing astronomy, physics, and the calculus, are devoted simultaneously to vital and social statistics (better labeled in Quetelet's own terminology, "moral statistics"). These are "The Growth of Man" (1831), which utilizes a large number of measurements of people's sizes; a few months later "Criminal Tendencies at Different Ages"; and in 1833 a third publication giving developmental data on weight. The emphasis in these publications is deliberately on what we would today call the life cycle; still, the first two memoranda included many multivariate tabulations, such as differences in the age-specific crime rates for men and women separately, for various countries, and for different social groups. Much before the 1848 publication entitled *Du système social*, Quetelet had a tendency to mention frequently the existence of a "social system" and its "equilibrium." What he meant by these expressions is the fact that the social strata each contributed their own rates, which are constant over time but nonetheless different from one another.

By this time, the idea of social physics had slowly matured, and in 1835 he combined his earlier memoranda into a single book entitled *Sur l'homme et le développement de ses facultés* with the subtitle *Physique sociale* (translated into English by Chambers as *A Treatise on Man and the Development of his Faculties*). This publication was largely responsible for bringing Quetelet's basic ideas to the larger public.

In these early writings Quetelet evinced a primary interest in the fact that the *averages* of physical characteristics and the rates of such nonphysical characteristics as crime and marriage showed a surprisingly stable relation over time and between countries with respect to age and other demographic variables. It was these relations that he designated as the "laws" of the social world. The idea of the average man played a culminating role in his first book of 1835.[2] After 1840, however, it was not only the stability of averages and rates that aroused his interest, but more specifically the *distribution* of these characteristics. He became aware that the distribution of the heights and weights of human beings when put in graphic form looked very much like the distribution of errors of the observations that had been studied since the turn of the century. He was thus led to the firm belief that the distribution of physical characteristics could be considered as if they were normal distributions. Instead of considering stable relations between averages or rates as the "laws of the social world," he was now interested in the distributions around the average. This interest applied to intellectual or "moral" characteristics as well, if appropriate measures were used. He explicitly expressed his belief that such measurements were possible in principle, and that the lack of data was only the result of technical difficulties. A large part of his interpretation of moral statistics deals with the kind of quantitative data he uses as substitutes for those that he felt would be really desirable. He was convinced that if he could make enough observations, his distributions of physical and nonphysical characteristics as well would always have the normal form. Thus the notion of "law" was extended to the distributions themselves and their derivations, as well as to their constancy over time and place, which became "laws."

Beyond extending the concept of the "average man" to all of man's physical traits (thus forming the basis for what he called "social physics") and, thence, to all moral and intellectual qualities as well ("moral statistics"), Quetelet explicitly planned to apply it to collectivities of all sizes, ranging from the small group to the whole of mankind, and expected that it would hold equally well for any time in human history. There were suggestions of these extensions in earlier works, as we have seen. Only in 1848, however, does the grand generalization emerge with the publication of the *Social System*. The general theme is announced as follows: "There is a general law which governs our universe . . . ; it gives to everything that breathes an infinite variety . . . . That law, which science has long misunderstood and which has until now remained useless in practice, I shall call the *law of accidental causes*." [3]

This overall idea is later developed as follows: [4]

. . . among organized beings all elements vary around a mean state, and . . . variations, which arise from accidental causes, are regulated with such harmony and precision that we can classify them in advance numerically and by order of magnitude, within their limits.

One part of the present work is devoted to demonstrating the law of accidental causes, both for physical man and intellectual man, considering him individually, as well as in the aggregate . . . .

In order to understand more precisely Quetelet's idea of the average man, an explanation of what he meant by "accidental causes" and by "law" is obviously necessary. His hypothesis (as conveniently formulated in Landau and Lazarsfeld, 1968) is that every mean presented results from the operation of constant causes, while the variations about the mean were due to "perturbing" or "accidental" causes. The difference as he states it is the following: "Constant causes are those which act in a continuous manner, with the same intensity and in the same direction." Sex, age, profession, geographical latitude, as well as economic and religious institutions are some of the constant causes that he mentions. A category parallel to constant causes is sometimes called "variable causes," i.e., those that "act in a continuous manner, with energies and intensities that change." The typical case is the seasons (although Quetelet meant to include as variable causes all periodic phenomena). "Accidental causes only manifest themselves fortuitously, and act indifferently in any direction." [5] Man's free will is frequently mentioned by Quetelet as an accidental cause (although occasionally dismissed as devoid of any role), but its operation is always constrained within very narrow limits. To sum up, one could say with Landau and Lazarsfeld "that, given sufficient data over time [note: here the influence of such mathematicians as Laplace, Fourier, and Poisson can be traced] the shape and extent of variations about the mean state which result from accidental causes can be "classified in advance" with a high degree of accuracy through the application of the theory of probabilities of independent events."

As for the concept of "law" a distinction should be made according to whether it applies to man's physical attributes, to his moral features, or to all human characteristics. Thus in *Du système social*, three different meanings of the term can be distinguished. In the first pages of the book the "law of development of humanity as regards height" refers to a trend in a series of averages. A second meaning appears later with the laws of propensity to, respectively, crime and to suicide. Those laws are in fact regular patterns of correlations. They appear sharply contrasted; propensity to crime increases quite rapidly toward adulthood—it reaches a maximum and then decreases until the very end of life—whereas the law of propensity to suicide formulates a direct variation with age "until the most advanced age." [7] These two uses of the term law (a trend and a pattern of correlations) can be both considered as referring in some way to a correlation —in the first instance between height and time, and in the second between age and the occurrence of a given social act. The third type of law, the "law of accidental causes," is different: it is "the assertion that every human trait is normally distributed about the mean and that the larger the number of observations, the more closely the empirical distribution will coincide with the theoretical probability distribution."

It is this grandiose notion of the place and role of the average, as expressed in the "law of accidental causes," that set off the furious controversy over the average man. Quetelet suggested that the means of various traits could be combined to form one paradigmatic human being, who could represent the "type" for a group, a city, a nation, or even for all mankind. He claimed that such a combination would produce a "type for human beauty" or a "type for human perfection." Although he insisted that the average man be considered no more than a "fictitious being" (possibly an imaginative effect of his early interests in poetry, painting, and music), this caveat was taken as simply an evasion of the main difficulty. His attempts to reply directly to early criticism like Cournot's, who objected, on the basis of a mathematical analogy, that just as the averages of the sides of many right triangles do not form a right triangle, so the average of physical traits would not be compatible, did not succeed in convincing his opponents. [8]

What was then the attitude of the three younger practitioners of vital and social statistics (the very field in which Quetelet had proved most innovative) whom we have selected: William Farr, Louis-Adolphe Bertillon, and Jacques Bertillon? Did they react to the raging theoretical controversy over the average man and the "law of accidental causes," and if so, in what way? Did they reject the whole conceptual construction altogether? Did they operationalize it in some way, and if so, how, or were they indifferent to problems of this type due to lack of mathematical training and awareness? We shall examine first the case of Farr, and then we shall turn to Guillard and the Bertillons, father and son.

2   William Farr's Statistical Interests and Methods

Several recent books, like the general account of the British Statistical movement by Cullen[9] and more recently the comprehensive monograph by Eyler on Farr,[10] allow us to concentrate on the more fundamental aspects of Farr's career, activities, ideas, and methods. Like Quetelet, from whom he differed so much in many other respects, a striking characteristic of his early training and orientation is the influence of French scientific thinking, but this time coming from a different circle: that of the Paris medical school, which was at the time the most highly renowned in the world.

The Paris school (or, to use Ackerknecht's famous phrase, the "Paris Hospital") was not only at that time the leader in clinical medicine, but also a pioneer in the research and teaching of precisely those subjects that Farr later made his speciality: hygiene and medical statistics.

If the Revolution of 1830 brought Farr's stay in Paris to a rapid end, obviously the influence received during this brief period was a lasting one. Upon his return to Great Britain, he had a further opportunity to benefit from Quetelet's influence, although there does not seem to be any evidence of early direct relations between them.

In 1833 Quetelet attended the third meeting of the British Association for the Advancement of Science. His presence acted as a catalyst for the statistical movement already well launched in the 1830s in Britain. Quetelet persuaded a group of interested men within the British Association to act as a lobby for the creation of a statistical section, section F. This same year the Manchester Statistical Society was established, the first of its kind and for a while the most productive. The year after the same British Association group founded the Statistical Society of London, which was eventually to dominate the field. In the 1830s statistical societies also existed briefly in Bristol, Liverpool, Birmingham, Newcastle, Leeds, Glasgow, and Dublin.

Farr and his fellow members of the first generation of British career statisticians found their professional homes in the London Society and in section F. Since they were not trained as professional mathematicians of astronomers, unlike Laplace, Poisson, and Quetelet himself, they had no place in the teaching establishment. Outside government, to which we shall return later, scientific societies permitted the exchange of ideas, offered a publication outlet, and, at the beginning, some financial support for research.

The programmatic statements that the London Society issued helped to define statistics for contemporaries, as well as to explain the Society's purpose and the nature of the subject it would cultivate. Eyler cites three such statements issued in 1833, 1838, and 1840. From his analysis it appears that (1) statistics was not merely a method but a separate science; (2) the pursuit of statistics was in their view entirely a matter of the accumulation and arrangement of facts; (3) the pursuit of facts would banish speculation, opinion, and even theory.[11]

This program presents us with a most curious mixture of incompatible elements. It was dedicated to discovering solutions to social problems that were thought of as being urgent, but that abjured opinion and politics. It claimed to be a separate, autonomous, full-fledged science—a social science—but it rejected both theory and discussion of cause and effect. Concerning the relations between probabilistic theory and concrete numerical social observations, what strikes us as late twentieth-century observers is that statistics as understood in early Victorian Britain (as well, incidentally, as in many circles of social observers in France) was a science of individual facts and observations that showed remarkable indifference to mathematics.

Farr was in agreement with the founders of the statistical societies on the nature of statistics no less than on social goals. In fact, Eyler defines him as a "rather late defender of the early Victorian idea of statistics."[12] He gives as an example his presidential address to the statistical society in London in 1872, which "might have been written thirty-five years earlier." Earlier in 1855, Farr defined statistics as a bookkeeping or accounting of the state and its subject as confined of two great divisions: population and property.

Closer to our topic of the influence of probabilistic thinking on social and vital statistics is Farr's defense of the British Association's section F against the attacks of Francis Galton.[13] Galton raised criticisms against the statistical section for its lack of rigor and for its disinterest in the mathematical basis of statistics; furthermore, he showed little sympathy, to say the least, with the general reform orientation of its leading members. Naturally Farr and older statistical enthusiasts like Chadwick defended section F and the original view of statistics: apparently Farr had taken the initiative to alert Chadwick about Galton's proposed amendment and thus joined forces with his one-time adversary (see below) against the threat to their common conception of statistics. Nothing, says Eyler, better illustrates the differences between Farr and Galton. This difference between the two appears very clearly in another episode, namely, the change that occurred in the reports of the British Association's Anthropometric Committee once its chairmanship changed from Farr to Galton in 1880. Under Farr's direction, the committee had collected and printed for five years its physical measurements for various occupational and social groups and noted simple relations such as the comparison of mean values. In a sharply contrasted move, Galton published in his first report as chairman a theoretical discussion of probable error and introduced such basic statistical concepts as those of mean, quartile, and decile as applied to a normal curve: thus the work of the committee adopted a mathematically more sophisticated level typical of this new generation of statisticians.

These episodes obviously took place toward the end of Farr's career. Although he obviously lacked Galton's mathematical interests and skill, Farr never indulged in the naive overstatements typical of the early pronouncements issued by the statistical society of London. For example, in a medical editorial of 1837 he wrote, "Facts, however numerous, do not constitute a science. Like innumerable grains of sand on the sea shore, single facts appear isolated, useless, shapeless: it is only when compared, when arranged in their natural relations, when crystallized by the intellect, that they constitute the eternal truths of science." He stressed the importance for the statistician of being critical of his sources and data, investigating their accuracy, questioning the relevance and representativeness of the units that were adopted, and attempting with the help of ratios, logarithms, and the calculus of probabilities to discover relationships in order to make predictions.

This sophisticated understanding (as compared with the average level of his contemporaries) of the nature of statistics (and, moreover, a greater willingness to envisage and evaluate the practical implications of statistics) is partly explained by Farr's initial training and lasting bias as a physician. It is a well-known fact that medical men, especially physicians, had as a professional group showed a comparatively high degree of interest in the early statistical societies. The observation holds for France as well,[14] where the famous *Annales d'hygiène publique et de médecine légale* served as a model to Farr when he started editing his own journal *British Annals of Medicine, Pharmacy, Vital Statistics, and General Science*, the scope of which was intended to be even broader, extending to "all the laws of vitality capable of being observed in masses of men, expressible in numbers," with special attention to state medicine, public health, the census and vital registration, and insurance and benefit societies. In contrast to its French model, which was published well into the 1950s, the British *Annals of Medicine* appeared as a weekly journal for only nine months (January to August 1837). Still, it is worth mentioning as one of the first English medical journals devoting particular attention to statistics. Farr contributed six articles: two on medical reform and four on vital statistics.

At least three reasons may explain the presence of medical men and the discussion of health problems in statistical societies: (1) public health as a goal and hygiene as a method were of course recognized as essential elements of the well-being of the working classes; (2) the medical profession had a tradition of local investigation linked with various attempts to improve the condition of the poor: this applies more particularly to physicians associated with a dispensary, who were often the first professional men to gain extensive and direct experience with the living conditions of the poorest classes in cities; (3) contrary to Quetelet's favorite variables constituting "moral statistics," which were measurable only indirectly (all the more so when not only moral characteristics but also propensities were involved), subjects like death, sickness, and the amount spent on medical relief presented themselves more readily as useful indices.

A fourth reason may be added: doctors were more aware than their lay partners of the example of the physical sciences; they knew in particular how poorly the medicine of their day compared in its methods and predictions with those of these sciences. Eyler cites as characteristics of this greater awareness three works

published by medical men in which statistics is seen as a means of improving medical knowledge and practice.[15] In his *Elements of Vital Statistics* (London: 1829),
probably the first English textbook on vital statistics, Bisset Hawkins predicted that
the application of statistics to medicine would provide a means of assessing the
efficiency of various treatments, of compiling accurate histories of disease, of
establishing the effect of the conditions of life and labor on health, and of providing
a basis for dependable prognosis. Two other works, closer to Farr's beginnings as a
statistician (he wrote the section on vital statistics for McCulloch's 1837 statistical
digest of the British Empire) appeared about 1840. Both recommended numerical
or statistical methods to physicians on the analogy of the successful application of
these methods in the sciences. Astronomy (Quetelet's main speciality), with its
methodological rigor and extensive agreement on fundamental questions, was
contrasted by all three with the causal methods used in medicine and the disagree-
ments among physicians. All three expected from the numerical or statistical
methods benefits like those Hawkins had suggested a decade earlier.

But they were obviously interested by more than practical benefits. To Guy, it
was on the contrary the excessively practical bent of most physicians and their preoccupation with individual cases that was to be blamed for the backward state of
medical knowledge. The only way to achieve greater certainty was to gain knowledge of entire groups and of the probability of events. To him such an apparently
sophisticated practice as the calculation of chances was in fact operating, although
in a crude way, at the very center of the art of medicine: the diagnosis of an
experienced practitioner. All three authors rejected the idea that the phenomena of
life were too complex to be studied numerically, offering the success of life insurance as an example that apparently unpredictable events, such as the duration of
life, could be accurately forecast in the aggregate. More theoretically, the Griffins
referred to Laplace and Roemer to show that very complex astronomical problems
could be handled mathematically. They also referred to Louis's studies on the
effects of bloodletting, although they offered some apologies for his apparently
"absurd conclusion," which he reached by studying too few patients and by relying
on the duration of cases as a measure of the value of treatment.

More interestingly for us, Guy,[16] cited by Eyler as making "comparatively rare
use among the members of the Statistical Society of London of the example of
Quetelet's best statistical work",[17] pointed to the constancy of certain social
phenomena, such as the number of crimes committed every year or the percentage
of accused criminals found guilty, and thus showed how even complex events
involving the human will could become nearly as regular and predictable as physical
phenomena when considered in large numbers. Admittedly Guy's mathematical
vision was extremely narrow: to him the numerical method was equated with the
"method of averages." Nevertheless, he recognized Quetelet's contribution in demonstrating among social phenomena such fundamental regularities: he took the
discovery of such "laws" (see section 1 of this paper on Quetelet) as the highest aim
of statistics. One example of such a "law" was the discovery by Quetelet that what
he called the "curve of viability" (i.e., the regular change in the probability of living
at different ages) was comparable in this respect with other social phenomena, such

as the tendency to crime, which also changes with human age. As another example, he cited the recent discovery of a statistical regularity in the case of fatality from smallpox: he did not name the author of the study, but it was clear to many of his listeners at the statistical society that it was William Farr.

To Farr the lawful nature of vital phenomena was an essential principle; in this he agreed fundamentally with Quetelet. The analogy of the life table is one he constantly employed. A life table illustrated that generations of men do, in fact, succeed each other in a regular way. He used this comparison early in his career when he contributed to McCulloch's statistical abstract. He retained this conviction throughout his career, for it was essential to his entire program. As late as 1872 he praised Halley (the astronomer who had calculated as early as 1693 a life table from records of 5,869 deaths that occurred between 1687 and 1691 in Breslau) for showing that "generations of men, like the heavenly bodies, have prescribed orbits, which analysis can trace."

Between 1853 and 1876 the International Statistical Congress launched by Quetelet held nine sessions in various European capitals: Farr was usually chosen a British delegate. He was especially active at the 1860 meeting held in London. He was the one who delivered the official invitation to hold this congress in the English capital; he helped plan the program and played a part in inviting important official figures, namely, the Prince Consort and Lord Brougham, to accept honorary responsibilities in the congress. As he explained, the successive meetings of the congress gave general evidence of the twin influences of Quetelet's social physics and of the practical advances of British statistical administration.

We have seen Farr's participation in the London Statistical Society. Another development in the British statistical movement was the 1836 law, which, following the 1800 Population Act establishing the census and the establishment of the Statistical Department of the Board of Trade, established Civil Registration and entrusted it to the General Registrar Office. This series of administrative reforms reinforced the governmental statistical machinery, in connection with the Poor Law Reform and the reforms of the Factory Act. This brought a decisive turn in Farr's career: he first joined the G.R.O. unofficially to head the statistical department, and his appointment to the permanent establishment came in July 1839. We have no precise information about how he was chosen, but his articles in medical journals and his chapter on vital statistics in McCulloch's volume had presumably drawn the attention of the official establishment to him.

In his work for the Registrar General, Farr helped devise a system for collecting and abstracting the registration material. He also helped improve the census operations, notably by introducing data concerning the occupational distribution of the population. In the registration machinery, one of his main tasks was to develop successive versions of a general nosology of diseases, which he unsuccessfully proposed for adoption by the International Statistical Congress. This failure was not his sole responsibility, since until 1893 no two continental nations used precisely the same method of recording and registering the causes of deaths. In that year, however, Jacques Bertillon, the Parisian statistician, established the nomenclature that became known as the International List of the Causes of Deaths.

However, Farr's most serious statistical enterprise was his work on the rate of mortality and the life table. We have already seen the mathematical limitations of the statisticians in early Victorian Britain. Although he might at one occasion call mathematics "the soul of statistics,"[18] and despite the fact that he had taught himself the elements of calculus and was familiar with the work of Laplace and Quetelet, he made little attempt to use the most sophisticated mathematical methods and never became expert in mathematical theory. One finds only very rare examples of calculus in his published works. The life tables are the only case in which he treats phenomena as continuous functions. He never spoke explicitly of the distribution of data in the way Quetelet did, nor did he use Quetelet's law of error. For example, everything indicates that he failed to realize that his mortality figures for a smallpox epidemic reported in the *Second Annual Report* produced a normal curve. However, two authors think that this turned out for the better. Greenwood believed that Farr's amateur status as a mathematician may be considered as a blessing in disguise, since it prevented him from trying to extract at all cost good results from bad data under the cover of sophisticated manipulations.[19] Such excessive mathematical manipulation of raw data by the Galton-Pearson school aroused the suspicion of Newsholme in his article in *Economica* (1923):[20] had he lived long enough, Farr probably would have shared this suspicion.

Some salient aspects of Farr's statistical work have been conveniently singled out by Eyler. His most basic tools were simple ratios, among which the death rate (which he expresses in the now current "per thousand of living") was to him comparable to a thermometer (he even called it in 1843 a "biometer").

In 1842–1843 a controversy concerning the adequacy of the mortality rates as an index of the health or vitality of a population put him into indirect opposition to Chadwick. Farr himself allied with Guy and other statisticians and actuaries to condemn this statistical heresy and reaffirm the necessity of a life table as the unique basis for computing the average length of life

One special mortality rate that he emphasized as a standard characterizing the "healthy district" was that of 17 per 1,000, which he advanced in the mid-1850s, a proposal soon accepted by a very knowledgeable contemporary statistician, John Simon.

But, as he recognized, a more accurate picture of the relative health of a population was given by the life table. Eyler gives a very detailed account of Farr's extensive writings on this subject. As early as 1843,[21] in a very clear relation of his controversy with Chadwick, he wrote a popular account of life tables that "conveys something of Farr's enthusiasm" for the life table approach.[22] Three national life tables were issued in 1843–1844, 1859, and 1864. Special attention should be given to his health district life table. Here he assumes as a simplification that, after the first five years of life, the rate of mortality at a given age is constant during the year and that those who died during the year did so at equal intervals of time. He could thus ignore instantaneous change and was consequently freed from the necessity of using the calculus. Still, he was able to show that the error thus introduced was minor as compared with those committed in the registration and in the census.

In conclusion, let us recall that he was also active in statistical studies of diseases, particularly in describing epidemics numerically. He did this by extending the actuarial techniques of discovering the law of mortality from a life table. Thus Farr seems a good example of the sound practice but rather limited understanding and use of probability thinking in statistical computations. Still, its practice was sound enough to inspire the last terms of our comparison, namely, Guillard and the Bertillons.

3   From the Adoption of Quetelet's Legacy (Guillard) to Its Rejection (Louis-Adolphe Bertillon) and the Adoption of Farr's Notions

Three generations comprising successively an ancestor, Jean-Claude Achille Guillard (1799–1876), his son-in-law Louis-Adolphe Bertillon (1821–1883), and the latter's son Jacques Bertillon (1851–1922) constitute, in contrast to Farr and Quetelet, a dynasty of French natural scientists and physicians interested in vital and social statistics. Their published work as it appears in Moricourt is impressive by its volume;[23] their unpublished work, according to Dupâquier, also deserves consideration.[24] We shall therefore not try to give even a crude idea of these 10,000 odd pages, but shall concentrate, partly following Dupâquier, on the changing relationships of these authors to Quetelet and Farr.

A word may be said about their lives and motivations toward statistics. Very much like Quetelet in his younger years, but more consistently during their lifetimes, Guillard and Dr. Bertillon (Louis-Adolphe), the first physician in the family, were polymaths. Jacques Bertillon (also a physician) appears more specialized in vital and social statistics. More openly than Quetelet (and Farr, but the French political scene was more turbulent than the English one), all three were political activists. Their interests in vital and social statistics were rooted in philosophical, social, and political considerations that they considered central to their thinking. Thus they did not indulge in such naivetés as expelling from statistics both theory and discussion of cause and effect, as the programmatic statements of the London Society cited above had.

At the root of Guillard's decision to develop a new science designated by the word that he coined for the purpose ("démographie"), and of the efforts deployed by Louis-Adolphe Bertillon in the (at least at the beginning) same direction, lies a major political event that affected them personally: the advent of the Second Republic, which corresponded to their hopes, and its replacement by the authoritarian regime of Louis-Napoléon Bonaparte. During these events, which took place between 1848 and 1851, L. A. Bertillon was arrested three times, A. Guillard twice. On a more general level, they wanted to understand how a (so they thought) detestable regime like the one produced by the political coup of Louis-Napoléon in 1851 could overthrow the democratic Republic. These preoccupations were close to those that after 1830 inspired Quetelet's *De l'homme* and, in 1848, *Du système social*.

In his book *Eléments de statistique humaine, ou démographie comparée*, published in 1855, Achille Guillard hesitates between three definitions of the "new science":[25] Demography is in turn

1. The natural and social history of the human species. The breadth of scope reflects adequately what we know of Guillard's background, comprising the writings of Louis and the spirit of Claude Bernard. One must note that in this definition of demography, the natural sciences become the guiding model, instead of physics, as was the case in Quetelet's "physique sociale."

2. The mathematical knowledge of populations. In this second definition Guillard simply follows Quetelet. This appears in his specification of this mathematical knowledge of populations as applying to "their physical, civilian, intellectual, and moral state."

3. Demography is also, in plain Mathusian terms, defined as "the law of population." Guillard had to respond to Malthus, still the most famous author of the time, who had stated the relationship between population and environment. This last perspective cannot easily be reconciled with the preceding one.

The general trend, as it appears from Dupâquier's very minute analysis, written from a specifically demographic point of view, is a move away from the rigid generalizations formalized by Guillard either from Malthus (in which case they are deterministic) or from Quetelet to more precise computations of better defined phenomena. Most ameliorations seem due to Louis-Adolphe Bertillon's greater familiarity with Farr's techniques. This applies in turn to the state of population ("état de la population") and to all the multiple aspects describing its change ("mouvement de la population"). If one considers the advances made by Farr in the construction of mortality tables, it is hardly surprising that his writings specifically related to this point were particularly influential on Dr. Bertillon. During the 1860s, Dr. Bertillon progressively corrected Achille Guillard's confusions and erroneous computations, narrowly following Farr's conceptualizations. For example, it is under Farr's influence that Louis-Adolphe corrected Guillard's formulations of death rates. In the 1860s, he explicitly distinguished death lists from mortality tables, and again following Farr, opposed "death rates" to "death ratios" (representing the probability of death).[26] In 1865 he was able to propose to French doctors and statisticians correct methods for computing tables of survival. Finally, Louis-Adolphe Bertillon was at his best in his successive clarifications and qualifications of the concepts of birth rate, marriage rate, and fecundity. Thus he critizes the current conception of the birth rate, symbolized as N(births)$/M$(marriages), supposed to represent the mean number of children by marriage, a conception first introduced by Malthus. In his article "Mariage" in the *Dictionnaire encyclopédique des sciences médicales* (1872) he rejected N/M, the "fecundity of the marriages," in favor of the ratio of living yearly births to the number of married women aged from 15 to 40 years or from 15 to 45 years (this is termed today the global rate of legitimate fecundity). More articulated and thorough is the article "Natalité" published in the same *Dictionnaire* in 1876. He distinguished between "general natality")"natalité générale," or birth rate) and "special natality" ("natalité spéciale," termed today the global rate of general fecundity), which he based on the years 15–50 as the years of reproduction, thus following the Danish demographer Lund and departing from Farr, who preferred the years 15–55. He also defined "special legitimate natality" ("natalité spéciale légitime") as the "actual fecundity of spouses at the age of reproduction." He explains the word "actual" as designating, in opposition to the physiological and intimate capacity of spouses to conceive in the interval 15–50, that capacity which, in the social milieu to which they belong, conduces to living births. Dupâquier (1984) concludes this passage with a note of gratitude to Louis-Adolphe Bertillon for having so well clarified this part of the demographic terminology.[27]

The use of statistical methods for the study of demography also led Dr. Bertillon, in opposition to Achille Guillard, to examine and critize some of Quetelet's basic ideas. One of his most influential publications in this respect was his 1876 article "La théorie des moyennes en statistiques." In this article, after explaining in nontechnical terms how a mean is determined, he established a distinction between objective and subjective means, the first referring to specific concrete objects, the other to mere abstractions. The artificial nature of the mean appears clearly in the concept of a mean life duration: not only is this duration not characteristic of the majority of cases; it happens to be among the rarest of those registered. In the same text Adolphe Bertillon introduces a further distinction. Although he qualified the subjective mean as being artificial, since it does not relate to any specific concrete object, he maintained that it still remains close to the objective mean: this is why he also proposed the name "typical subjective mean" ("moyenne typique subjective"). On the other hand, he designated as "index mean" ("moyenne indice") that mean which does not represent a natural group in which the majority of components would be close to that mean. For example, when it is stated that the mean age of the French population is around 31, that does not mean that the greatest number of Frenchmen is grouped around the age of 30, since the largest age group is that of the children, while those who are over 30 compensate for their numerical inferiority by their greater age.

Another very different kind of mean distinguished by Dr. Bertillon, quite apart from the preceding ones, is what he called a "mean result" ("résultat moyen"). When one says, for example, that a country has 30 births per 1,000 inhabitants, this only means that it is probable that, during the 12 months of the year, there will be born in this same country as many times 30 children as there can be counted thousands of inhabitants. Density is another example: if one says that there are 50 inhabitants per square kilometer, one expresses only the mean ratio ("rapport moyen") between the size of a given territory and the sum total of its inhabitants, without supposing that this ratio can actually be observed in any given part of that same territory.

More generally, Dr. Bertillon sharply criticized not only Quetelet's notion of combining average physical characteristics, but also his ideas about moral and intellectual traits. Surely, Bertillon said, Quetelet was mistaken in believing that his average man would represent the ideal of moral virtue or intellectual perfection. Such a man would, on the contrary, be the personification of mediocrity, or, to use his apt phrase, the "type de la vulgarité." After this devastating criticism, the concept of average man was generally viewed as not worth being taken seriously. However, in 1955, Maurice Fréchet "rehabilitated" Quetelet's *homme moyen* by redefining it as the *homme typique*, this latter being defined as a *particular* individual in the group whose traits taken as a whole come closest to the average.[28] In this way be managed to avoid most of the criticism leveled against Quetelet's concept.

An interesting work by Dr. Bertillon, written shortly after Quetelet's death in 1874, is his "Considérations générales sur la démographie, appliquée tout particulièrement à la Belgique," published in the *Bulletin de l'Académie royale de médecine*, 3rd series, vol. X(8), 1876. It was presented as his reception address to the Belgian Royal Academy of Medicine and offers us theoretical and methodological considerations pertaining to Quetelet's thought, the nature of demography, and the role of statistical and quasi-probabilistic reasoning. Very representative of the spirit of the time, and more particularly of Dr. Bertillon's staunch radical position, is the undisguised way in which a scientific and technical debate on Belgian marriage and birth rates immediately turns into a direct anticlerical attack against the Catholic Church and Catholicism in general. The old spirit of the early Victorian statistical movement, in which Farr participated, although with some moderation, seems to have extended to the mid-1870s, but on specifically continental and anticlerical grounds.

Dr. Bertillon devotes the first part of his address to a thorough critique of Quetelet's phrase "physique sociale" and a justification of Guillard's concurrent invention, "démographie." To him social physics does not sufficiently delineate the specific topic, which is the study of *human collectivities* as such, much more than the study of *societies* or sociology, the only well-cultivated part of which is *Economics*. Like Quetelet, he pursued studies founded on man, on the collective man, or better, on the different peoples. This last word explains his preference for the wording "Démographie."

Quetelet's "social physics" may be criticized from still another viewpoint: "physics," according to Bertillon, is understood here in an ancient sense, one lost in the contemporary period. In its early or primitive sense it referred to natural history, to the *concrete* (or complete) study of all observable bodies; physics is today an *abstract* science (dealing only with some abstract properties of the bodies) of the best defined type. This is why diverting this word from its present meaning entails so many inconveniences. Nevertheless, the term "social physics" has at least one advantage, which is to indicate the method suitable to this science. This is the observational method—a fact that too few people suspect today.

These are some of the considerations (among many others) that prompted Bertillon to adopt the phrase proposed by Guillard: The science of *human collectivities*, or science of the peoples, insofar as they are collections of men, or *demography*. Therefore, the biological study of collectivities is the proper object of demography, not the study of society, which is the object of *sociology*. After mentioning that Germany had abandoned the word "populationistics," Bertillon pointed to statistics as the proper method for demography. He paid a renewed tribute to Quetelet despite his "mystical tendencies" for having given to these various studies the name of *natural history* or *social physics* and having thus fully indicated that social facts belong to the natural order, that their appearance and intensity are submitted to a rigorous "determinism," and that as a consequence their study requires the

rigorous methods of science. By this bold conception, despite his mental timid-
ities," Quetelet rescued social facts from the whimsical explanations of mystical
minds, and aligned them with those regulated by natural laws, which we can know,
predict, and eventually turn to our profit.

The second part of the address concerned a special point of Belgian demography
to which the author wanted to call attention. Here, statistics were openly used as a
weapon against catholicism. Dr. Bertillon wanted to analyze how in the Belgian
case the birth rate ("natalité" in French, again an expression coined by Guillard on
the model of "mortality") is related to the death rate: this is a very broad problem.

To understand this relation he introduced two further notions: what is the
fertility rate and, further, what is the proportion of women in the fertile age cohort
(15–50) that marry. To make a long story short, he noticed (1) that the global birth
rate in Belgium is among the lowest in Europe; (2) that the fertility rate is on the
contrary much higher; (3) that the cause of the low birth rate despite the high
fertility rate lies in the low marriage rate among women in the fertile age cohort; and
(4) that this low marriage rate is linked, province by province, with a high rate of
monastic life (which statistically includes mostly women): Flanders is at one extreme, Limburg at the other. In modern terms, we would say that we have a
"spurious inverse correlation" between a high fertility rate and a low birth rate, the
intervening variable being the marriage rate, itself mostly dependent upon the rate
of monastic life. It is interesting to note that the equally plausible inverse relation of
a low marriage rate causing a high rate of monastic life is not even considered by
the radical (in both French nineteenth-century and American twentieth-century
usages) Dr. Bertillon.

Of the three writers thus briefly considered, Jacques Bertillon, was the most
openly politically engaged and the most popular. In contrast to those of Guillard
and of his own father, his publications were circulated well beyond the narrow
circles of physicians and statisticians so as to reach the general public. This is
undoubtedly due to his increasing interest and involvement in two hotly debated
political issues in France after the defeat of 1870–1871: the depopulation of France,
and the morally negative effects of alcoholism. His engagement with these issues
seems at times to have led him to cross the borders of his family's political
alignments. Achille Guillard and Louis-Adolphe Bertillon were staunch repub-
licans, attached to principles of free-thinking, opposed to the church, and also
l'ordre morale. Jacques's concern with depopulation and morality undermined by
alcoholism brought him at times very close to the nationalistic themes developed by
the new right after 1900. Still, he contributed to demography proper by synthesizing
his father's contributions in his Traité de démographie(1890); he produced a comparable synthesis for statistics in his Cours élémentaire de statistique administra-
tive(1895), whose content and surprisingly limited circulation were analyzed, at
least among academic writers, by Clark.[29]

More specific contributions of Jacques Bertillon to demography include "a new
method to appreciate the frequency of mixed marriages," presented at the 4th
International Congress of Hygiene and Demography,[30] the development of observations of fecundity in France, and the sophistication of his analysis. He was well

acquainted with the best statistical work conducted abroad, notably by Kiaer in Norway, Boeckh in Germany, Korosi in Hungary, and Coghlan in Australia. This last author was among the few who had read the remarkable work by Duncan and Tait on fecundity in Scotland. It is probably through Coghlan, says Dupâquier, that Jacques Bertillon became acquainted with the calculus of probabilities of growth ("probabilités d'agrandissement"); he summarized Coghlan's discoveries in the *Journal de la société de statistique de Paris* (1904).[31] Regrettably, he was more interested in the sociological and populationistic aspects of the problem than in its fine-grained analysis. Lotka came across the problem at approximately the same time.

What interested Jacques Bertillon most was physiological sterility. He witnessed a large number of couples without children. This led him to have a question introduced in the census after 1886 on the number of legitimate children per couple, and another question on the bills of mortality of the city of Paris (in the 1880s) on the number of children procreated. Then, in order to estimate sterility, he made comparisons: thus he could establish in his *Cours élémentaire* (1895) the percentage of families without children at the death of the first spouse for Alsace-Lorraine, 1874–1875, and Paris, 1886, and then the percentage of couples without children after 15 to 25 years of marriage. He even used Boeckh's methods to build a theoretical model against which real data could be measured.[32]

Conclusion

This paper has examined Quetelet's legacy by exploring the notions of "the average man," "social physics," "moral statistics," and the "social system." These conceptions, even though they were hotly debated at the time, are really no more than ultimate and presumably excessive elaborations of basic ideas first coherently expressed by Laplace. The unquestionably deterministic bent underlying most of Quetelet's uses of probability in the study of vital and social phenomena must be traced back to Laplace.

The purpose of this paper was to trace the reception of Quetelet's fundamental ideas about statistics and probabilities in the circles of British and French practitioners of vital and social statistics, conveniently symbolized by William Farr, Guillard, and the two Bertillons. In the course of the analysis further links emerged between the three figures thus selected. Not only do all three refer in one way or another to Quetelet, but William Farr exerted a strong influence on Louis-Adolphe Bertillon.

As a general conclusion one can only reemphasize (without any claim to originality) the tremendous impulse given to the probabilistic movement by Quetelet's personality and energetic activities. As the section of this paper on Farr clearly shows, he played a great role in the development of the British statistical movement (to use Cullen's phrase) through the foundation of section F of the British Association for the Advancement of Science (1833), and of the Statistical Society of London (1834), and through the 1860 meeting of the International Statistical

Congress held in London. On a more cognitive level, his staunch belief in the lawful nature of vital and social phenomena was essential to Farr as well as to Guillard and the Bertillons. This is again a basically Laplacean idea. Without the strength of this idea, one could venture the hypothesis that Quetelet's personal charisma, great as it was, would have been deployed in vain, or, more simply, would not have existed at all.

As to the specific details of Quetelet's ideas, the picture of their reception by the British and French practitioners appears quite different. Farr does not seem to have paid great attention either to the "average man", to "social statistics," or to "social system." In contrast, Louis-Adolphe Bertillon, as we have seen, took them seriously enough to deny radically their validity. His decisive analysis of the different senses of the "mean" was a deadly blow in the French-speaking statistical circles to Quetelet's most grandiose generalizations. After the 1870s the analysis of vital statistics, or demography, as enunciated by Achille Guillard in 1855, was progressively completed by the Bertillons father and son, first under the influence of Farr, and later in connection with a new generation of statisticians and demographers. Quetelet's basic convictions and intuitions are still present, but his specific ideas have been rejected.

Notes

1. Paul F. Lazarsfeld, "Notes on the History of Quantification in Sociology: Trends, Sources and Problems," *Isis*, 52 (1961), 277–333 (translated into French by Bernard-Pierre Lécuyer, "Notes sur l'histoires de la quantification en sociologie: les sources, les tendances, les grands problèmes," in Paul Lazarsfeld, *Philosophie des sciences sociales* (Paris: Gallimard, 1970)); David Landau and Paul F. Lazarsfeld, "Quetelet, Adolphe," in *International Encyclopedia of the Social Sciences* ed. David L. Sills, 17 vols. (New York: Free Press, 1968), vol. 13, pp. 247–257.

2. Lambert-Adolphe-Jacques Quetelet, *Sur l'homme et le développement de ses facultés, ou Essai de physique sociale*, 2 vols. (Paris: Bachelier, 1835), vol. 2, part 4.

3. Lambert-Adolphe-Jacques Quetelet, *Du système social et des lois qui le régissent* (Paris: Guillaumin, 1848), p. 16, as translated in Landau and Lazarsfeld, "Quetelet," p. 252.

4. Quetelet, *Système social*, p. 17.

5. Lambert-Adolphe-Jacques Quetelet, *Letters Addressed to H.R.H. the Grand Duke of Saxe Coburg and Gotha, on the Theory of Probabilities as Applied to the Moral and Political Sciences* (London: Layton, 1846 and 1849), p. 107.

6. See Pierre-Simon de Laplace, *Théorie analytique des probabilités* (Paris: Veuve Courcier, 1820); Jean-Baptiste-Joseph Fourier, "Mémoire sur les résultats moyens déduits d'un grand nombre d'observations," *Recherches statistiques sur la ville de Paris et le département de la Seine, recueil des tableaux dressés d'après les ordres de Monsieur le comte de Chabrol, conseiller d'Etat, Préfet du Département*, vol. 3, (Paris: Imprimerie royale, 1826); Siméon Denis Poisson, *Recherches sur la probabilité des jugements en matière criminelle et en matière civile, précédées des règles générales du calcul des probabilités* (Paris: Bachelier, 1837).

7. Quetelet, *Système social*, pp. 11, 86–88 (note 3).

8. Achille-Augustin Cournot, *Exposition de la théorie des chances et des probabilités* (Paris: Hachette, 1843), pp. 213–214.

9. M. J. Cullen, *The Statistical Movement in Early Victorian Britain: The Foundations of Empirical Social Research* (New York: Barnes & Noble, 1975).

10. John M. Eyler, *Victorian Social Medicine. The Ideas and Methods of William Farr* (Baltimore and London: The Johns Hopkins University Press, 1979).

11. Eyler, *Victorian Social Medicine*, p. 15.

12. Eyler, *Victorian Social Medicine*, p. 28.

13. Francis Galton, "Considerations Adverse to the Maintenance of Section F (Economic Science and Statistics): submitted by Mr. Francis Galton to the Committee appointed by the Council to consider and report on the possibility of excluding unscientific or otherwise unsuitable papers and discussions from the Sectional Proceedings of the Association," *Journal of the Statistical Society of London*, 40 (1877), 468–473.

14. See Bernard-Pierre Lécuyer, "Médecins et observateurs sociaux: les Annales d'hygiène publique et de médecine légale (1820–1850)," in *Pour une histoire de la statistique* (Paris: INSEE, 1977), pp. 445–476; "Démographie, statistique et hygiène publique sous la monarchie censitaire," *Annales de démographie historique* (1977), 215–245.

15. Eyler, *Victorian Social Medicine*, p. 31 (note 10).

16. William A. Guy, "On the value of the Numerical Methods as applied to Science, but especially to Physiology and Medicine," *Journal of the Statistical Society of London*, 2 (1839), 25–47, on pp. 36–38.

17. Eyler, *Victorian Social Medicine*, p. 33 (note 10).

18. William Farr, *English Life Tables: Tables of Lifetimes, Annuities and Premiums* (London: 1864), pp. LXXVI-LXXXIX.

19. Major Greenwood, *The Medical Dictator and Other Biographical Studies* (London: 1936).

20. Arthur Newsholme, "The Measurement of Progress in Public Health with Special Reference to the Life and Work of William Farr," *Economica*, 3 (1923), 186–202.

21. Eyler, *Victorian Social Medicine*, pp. 572–581 (note 10).

22. Eyler, *Victorian Social Medicine*, p. 72.

23. C. Moricourt, *Bibliographie analytique des oeuvres de la Famille Bertillon (y compris Guillard), médecins et démographes, de Jean-Claude Achille Guillard (1799–1876), à Georges Bertillon (1859–1918)* (Paris: Institut National des Techniques de la Documentation, Conservatoire National des Arts et Métiers, 1962).

24. Michel Dupâquier, "La famille Bertillon et la naissance d'une nouvelle science sociale: la démographie," *Annales de démographie historique* (1983), 293–311.

25. Dupâquier, "La famille Bertillon," p. 295.

26. Dupâquier, "La famille Bertillon," p. 300.

27. Dupâquier, "La famille Bertillon," p. 302.

28. Maurice Fréchet, "Réhabilitation de la notion statistique de l'homme moyen," in Maurice Fréchet, *Les mathématiques et le concret* (Paris: Presses Universitaires de France, 1955), pp. 310–341.

29. Terry Nichols Clark, "Empirical Social Research in France, 1850–1940," Ph.D. thesis (Columbia University: 1967).

30. Dupâquier, "La famille Bertillon, p. 303 (note 24).

31. Dupâquier, "La famille Bertillon," p. 307.

32. Jacques Bertillon, "Contributions statistiques à la connaissance de la fécondité légitime," *Journal de la Société de statistique de Paris*, 7 (1905), 226–242.

16 Paupers and Numbers: The Statistical Argument for Social Reform in Britain during the Period of Industrialization

Karl H. Metz

The transformation of the basic pattern of social life, which is usually called the Industrial Revolution, produced a novel type of society in which phenomena of large numbers played the dominant role. The traditional concepts were no longer capable of describing these phenomena in a way representing social reality adequately. Statistics, however, seemed to offer a new "language," which made it possible to describe a pattern of social life that was characterized by mass phenomena and change. Industrialization and urbanization produced large numbers both by assembling many people in limited areas and by augmenting the dependences and interrelations in a social environment. Therefore, the revolution of numbers, resulting from the urban-industrial transformation of society, was the product of some decisive increase in the density of social interaction. Industrial society became dissolved into many aspects, which could best be analyzed by counting them and looking for possible correlations.

This, then, was a thoroughly practical affair that strongly appealed to the business-minded citizens of the early nineteenth century and to their interest in "useful knowledge." Moreover, the type of knowledge provided by social statistics was particularly useful since it explicitly referred to the condition of the people. Statistics on illiteracy, crime rates, figures on mortality and epidemics, and data concerning poor law expenditure pictured the actual standard of living of the greatest number, and it was difficult to maintain that these statistics had much to do with a state of great happiness. In this way social statistics became the tool of social reform, serving as a kind of social thermometer. Sanitary statistics, in particular, served this purpose, leading to public-health policy and, thus, to the hygienic rebirth of British cities.

Statistics, consequently, was more than just a methodological device. It was a viewpoint, a perspective genuinely in harmony with the statistical structure of modern society and, thus, a side effect of the greater probabilistic revolution.

1 Social Statistics: The Birth of a Perspective

The growing importance attached to statistics in the last two hundred years was bound up with the rise of a new way of looking at society, a society that was transformed by industrialization from an agricultural community based on deference and low mobility to a social organization in which people performed a great variety of roles and that was dominated by mass phenomena. Mass, however, is not simply a gathering of many people. It implies a specific state of social interaction, namely, interaction structured by flexible behavior patterns. The liberals of the era of industrialization spoke about "free individual agency" when they referred to such flexibility. Numbers mirrored diversity, and, consequently, they were conceived as a structural element of society only when diversity began to irritate people. Statistics reflected such diversity and promised to reveal uniformity in the shape of statistical regularities or "laws" governing what seemed to be chaos.

Population growth and urbanization were the two main forces behind the emergence of a world characterized by mass phenomena, and industrialization provided the economic link between the two. Statistics offered some of the concepts that made it possible to adjust to this period of rapid change.

If one was able to quantify a problem in question, one could put a handle on it. The development of unemployment statistics, for example, gave rise to the concept of unemployment in the later years of the century. That unemployment became a subject of political discussion had to do with changes in the social position of the working classes; that it could be discussed had much to do with the collection of the relevant figures and with their use as indices and analytical tools. Here, as often in the early history of statistics, figures were the by-product of some type of administration. Poor-law statistics is another example, as is mortality statistics. The statistical transformation of society brought about a new, quantitative perspective. Figures became facts. Increasingly they were assembled for their own sake because they served as mirrors of a society that was thought to be some kind of numerical entity. In this way statistics became part of the basic pattern of social thought, and, therefore, it formed a threshold between the mentalities of traditional and modern societies. In the modern statistical mentality, man is conceived as an assembly of social aspects. His so-called individuality is the fortuitous combination of a varying number of very limited social definitions or aspects. The superstructure that contains all the aspects of a given period is society. Unemployment, for example, can be such an aspect, or homosexuality, or college education, or sympathy for a political party. Because this pool of possible aspects is so huge and relatively unstable in its precise configuration, it makes statistical sense to pick out one specific aspect and to form quantitative averages of it. The old-fashioned encyclopedic type of statistics, as practiced by the German "statists," had corresponded to an essentially qualitative social structure in which everyone and everything possessed their proper place prescribed by tradition. These places were compact; i.e., the social aspects did not fluctuate but were closely interrelated according to a specific hierarchical position, a "place." German statistics or "Staaten-kunde" did not need the numerical method since the phenomena it described could still be presented in common language using a terminology that was still meaningful. With the decline of the old hierarchical order and the explosion of numbers that followed, such a method of description became obsolete.

Modern society is an urban society, and urban life, especially if combined with industrialization, is a source of diversification and quantification. This is not merely a result of markedly increased density but also of a high degree of flexibility in socioeconomic relations. The fluctuation of the aspects circumscribing a person's position in society was the product of such flexibility. It went hand in hand with the multiplication of these aspects flowing from an enlarged division of labor and exchange relations. Individualization and quantification reinforced one other. Only at first sight did it seem to be a paradox that the more man became an individual, the more he was dealt with as an average. To some extent the hero-worship of the time was a protest against that paradox. A great man demonstrated to frustrated individuals the model of a life that could not be reduced to averages and that still held up the promise of an autonomous type of social existence amidst statistical units. Nineteenth-century historical writing responded to the dilemma of individualism among averages and celebrated uniqueness in the shape of a decisive event, a great man, or the collective individuality of a nation. The struggle against the application of numerical reasoning to the study of history was an act of defense of such uniqueness. It stirred up the most passionate controversy in Germany because there the bourgeois notion of individuality was less developed than in France or Britain.

The attempts to work out a statistically based social science started from the assumption that only those things that could be counted were characteristic of modern society. The different modes of description in the prestatistical and the statistical ages refer to a new understanding of "facts," of facts representing a democratic type of social data, i.e., phenomena that could be counted and need not be weighed, to paraphrase Coleridge. Numbers, collected in this way and put in their appropriate columns, seemed to describe the world objectively. There was no longer a need to rely on opinions. Practicing statistics seemed to be a practical affair. The ups and downs of a special series of data, such as figures on mortality, the crime rate, or poor relief, could be used as indicators for trends suggesting some change of policy. One might also try to compare different sets of figures and to look for correlations. Such comparisons became the starting point of the emerging social statistics because they made it possible to shed light on relations of dependence and to reveal causes and effects. To be able to interpret causation and to express it in lawlike formulas was considered scientific work. Statistics, following Laplace's definition of probability as epistemological, moved within a deterministic universe, and the statistician might even aspire to discover Newtonian regularities. Statistics presupposed a type of quantitative individualism held together by some general laws. Liberalism had pictured a society of this type as one consisting of a great many free agents pursuing happiness but being thereby subject to some basic laws. These laws were mainly the laws formulated by political economy. Statistics moved between the levels of individual action and economic laws. Its insights could be used as inductive generalizations, lacking, however, the deductive stringency of economic laws. These economic laws formed, then, the Newtonian background to the scientific claims of statistics.

Indeterminism, on the other hand, especially as put forward by German historicism, developed into a critical argument of statistics. Indeterminism was seen here as the result of a qualitative interpretation of individualism with the uniqueness of time and people as its core. Motives were the main agents of action. External factors might modify them, but they did not produce them. Thus, two different sets of liberty were at work here. The liberty of the free agents was first and foremost freedom of competition in the market, competition for places, rewards, goods. Here, there was no need to be afraid of diversity since one knew about the existence of great regularities behind it. The notion of man's plasticity, common at the time, supplemented the concept of determinism. The impression of such plasticity reflected the novel phenomenon of a free fluctuation of social aspects. Bentham's panopticon, Owen's monitors, or the pedagogic steam engine of the educational reformers of the time all started from this assumption. Both political and economic laws or institutions could regulate the flow of these aspects, though only if the priority of economics was maintained. It was possible to measure the outcome of such interference by results in elementary education or by counting those persons who were reconvicted. Statistics measured liberal society in accordance with its doctrines, and, therefore, it did not contradict its idea of liberty. Individuals were looked upon as averages, i.e., their "private" features did not possess great significance if interpreted in a context of large numbers. The abstract notion of such an average man was the free agent in the market, as developed by the political economists, and corresponded in this respect closely to the statistical concept of averages. In both cases the stress was on a selected aspect, be it the "homo oeconomicus" or man's statistically investigated propensities, such as his propensity to crime or his average height. The romanticist's ideal of individuality as uniqueness and wholeness was incompatible with this reductionist approach. In Britain the criticism of the statistical perspective was advanced by writers such as Coleridge, Carlyle, and Dickens. They did not deny that regularities drawn from aggregate numbers existed, but they opposed the conclusion that such statistical regularities were of real interest to social philosophy. The more regularities there were, the less individualistic society was. The statisticians, on the contrary, regarded these regularities as proofs of some self-regulating social mechanism that made it possible to dispense with a state meddling in private affairs. Privacy and free agency were the elements of freedom that had nothing to do with some mysterious wholeness of the person or the community. Private actions touched only upon private interests, and only insofar as they bore upon the interests of others did they become public matters.

The two principal issues of social reform up to the late nineteenth century were public health and elementary education. Consequently social statistics was divided into vital and moral statistics. In both cases the "social" element was the result of a correlation of two sets of statistical data, each of which represented a special aspect of social life. The combination of statistics on illiteracy and crime produced moral statistics. A similar combination of figures on mortality and of data on poor law expenditure or hygienic conditions led to sanitary statistics, i.e., the interpretation of vital statistics as a variable of sanitary environment. Thus, statistics was a social science for practical men. Its figures became irresistable if one could combine them with the argument that sanitary reform was also moral reform. As E. Chadwick, the principal architect of public-health policy in Britain, put it, "The fever nests and seats of physical depravity are also the seats of moral depravity, disorder, and crime."[1] Such correlations implied a need for social reform, and only because of this did statistics become the principal means of the reformist activities during the first half of the century. This reformism reflected both the effects of urbanization and of the emergence of the condition-of-England question. Statistics served as an index of crisis, a crisis of the established pattern of social integration. It was not yet a procedure of regular tabulation as it is nowadays. As part of the crisis management of the industrializing era, its practicioners viewed statistics as a one-way street to a policy of prevention. If it was true, e.g., that there existed a close interrelationship between illiteracy and criminal behavior, then public education was proved to be a precondition of free agency. A criminal was simply someone who did not know how to live respectably: Train the child and you will get diligent and sober adults. If one knows that juvenile delinquency posed a serious threat to law and order during the Industrial Revolution, one can understand why the comparison of statistical series enabled the reformers to prove their conviction that the purpose of criminal law was prevention, not punishment. Fighting against the old-fashioned concept that only the severest punishments could protect life and property against the dangerous classes, the law reformers strengthened their case by pointing to the correlation between illiteracy and crime rates.

Statistics revealed regularities in the phenomena of large numbers, and they defined such phenomena as variables that could be manipulated by social reform. Therefore the foundation of a scientific type of domestic policy seemed to be laid by statistical inquiry, and it is no accident that it was in Britain that it first happened. Both France and Germany were still influenced by a tradition of the strong state, and here statistical information could easily be turned into a bureaucratic device strengthening government. Britain, however, had both the most advanced economic and social system of the time in Europe and a government limited in its influence by Parliament and local self-government. Although social reform had everywhere the effect of making government more powerful, this was only gradually realized by the early reformers. They thought that social reform would only remove those contingencies that hindered people in exercising their full free agency. Consequently the case for reform was a plea for more individual responsibility and in no way an argument for a paternalistic state.

2   Statistics and Social Amelioration in Britain, 1800–1850

To establish itself as a new view of society, statistics had to attract a group of practitioners and an interested public. It found them among the middle classes and their quest for practical knowledge. These men were easily bored by lengthy deductions and abstract reasoning. Statistics, however, seemed to pave the way to useful knowledge. In the true inductive spirit, it first collected facts and figures and then tried to generalize them, relying thereby on some general theory. The first influential figure in the evolution of this statistical argument was T. Malthus. His famous law of population was a statistical generalization, so at least he declared. Malthus broke with political arithmetic and its claim that population density was an index of national wealth. The relation between population and wealth was a relative one. For there existed a threshold between them, and to determine it, one had to assemble figures on the growth rates of population and food production. For Malthus statistics banished the visionary imperatives of Jacobins and traditionalists alike, and it replaced them with true knowledge. The public impact of the statisticians was already so strong that their enemies had to bolster their moral objections with some statistics supporting their case.[2] It is significant that Malthus attacked the only state institution in this field.[3] To him, as to many of his contemporaries, this law was the relic of a regressive state of society that held people in dependence. But in order to relate his argument to the liberal idea of the free agency of the poor, Malthus had to highlight his voluntary check, i.e., moral restraint. His first reading of the "law" of population was strictly Newtonian and thus pessimistic. In the second edition of his essay of 1803, a statistically more sophisticated work, moral restraint received a prominent place. Now free agency could be pointed out as a remedy. Only in such a framework was it possible to use social statistics as an index for social reform.

A. Smith's argument that society was a self-regulating system had freed society from the dominance of the state. Social sciences like economics or statistics became possible. Yet this liberation was itself the product of a profound change in the social structure, a change from weighing people according to their status to counting them according to some selected aspects, such as expectation of life or literacy. J. Bentham gave philosophical expression to this process of dissolving man into a series of aspects, a process that developed parallel to the socioeconomic explosion of numbers. He achieved this through the introduction of the "felicific" calculus, which made it possible to reduce man's impressions of pleasure and pain to six categories.[4] Each of them was to receive an index number according to the effect an external factor had on man's feelings about pain and pleasure. In this way a "moral arithmetic" was elaborated in which quantification played a crucial role. The comparison of the sums added up for pain and pleasure, respectively, decided what choice of action was to be taken. Such a calculation presupposed a mean man representing the happiness criteria of the largest number. This mean was not simply some statistical average but the expression of the existence of immutable laws of human behavior, which were as rigid as the law of gravity in physics. Statistics, then, measured the relationship between such laws and a changing social environment. It mediated between the very general and the very particular, between laws and individuals. Only insofar as their average life was concerned did they acquire social, i.e., statistical, relevance. In this way statistics became the favored instrument of observation and reasoning of the liberal social reformer, who was determined to leave alone all things individual, i.e., man's free agency, and to battle only against those collective contingencies that threatened an average man's health and vitality, contingencies that were phenomena of large numbers. In an industrializing society, large numbers implied urban working classes, and this, again, led to the condition-of-the-people question. Consequently social statistics became the principal means of investigating and discussing the social question.

In the late eighteenth century, when the explosion of numbers began, the poor law returns attracted the interest of administrators and reformers alike. In 1775, Parliament had set up a committee to investigate the rising relief expenditure, and in 1786 this national investigation was repeated. That the expenditure had increased by nearly thirty percent in these eleven years caused considerable alarm. Thus, the costs of poor relief, which were published annually after the poor law reform of 1834, became a kind of numerical thermometer of indigence through which one hoped to measure the poverty of the nation in a manner similar to using trade statistics or income tax returns to express the wealth of the nation.

The idea of employing statistics as a social indicator is, of course, much older. The French general Vauban, for example, had suggested in 1686 taking an annual count of all paupers and beggars "pour savoir l'etat où le peuple est." Ten years later, G. King wrote his "Natural and Political Observations upon the State and Condition of England," in which he expressed the social hierarchy in terms of income and number of people per social group. And a quarter of a century before King, the London merchant J. Graunt had published his celebrated analysis of the metropolitan bills of mortality. The concept of a numerical social index originated in the fear that the plague aroused in the urban communities of the late Middle Ages. It was an attempt to come to "terms" with a disease that seemed to strike blindly and against which hardly anything could be done. However, to count the number of deaths being due to the plague made it possible to construct a kind of plague thermometer, so "that the Rich might judge of the necessity of their removal, and Tradesmen might conjecture what doings they were like to have in their respective dealings."[5] Graunt also tried to use his figures to indicate the state of the poor. Comparing the number of deaths from starvation with total mortality, he concluded that only a tiny fraction of Londoners lived in extreme poverty.[6] Graunt's friend W. Petty even wanted to interpret the statistics of mortality as "Scale of Salubrity" by comparing the diverse death rates from different parts of the country. He intended to utilize the mortality returns for the introduction of a health police, which was to prevent infectious disease. The laboring poor, who were hardest hit by such disease, were the true wealth of the nation, Petty declared. Thus a wise government would protect their health in order to stimulate both the production of goods and the multiplication of the producers.

With Malthus, who subsumed all population data under his "arithmetical" notion of food resources, demography became the numerical study of poverty. But because the rise of industry replaced his direct notion of food as a limited national, agricultural resource by an indirect one, which defined food as the result of employment, Malthusianism had to give way. Its interest in numbers was to remain, however. Yet the social data available were few and unreliable. This was particularly true of vital statistics, a fact that is shown by the drawn-out debate on the question whether the population was increasing or not. An attempt to secure a national census act in 1753 came to nought.[7] Religious prejudice and the claim that such an act might threaten privacy and freedom joined together in defeating it. Not before 1801 was it possible to execute the first census. That it was made the responsibility of the officers of the poor law merely shows that they were the only existing official institution for handling large-number phenomena. Even the introduction of a civil registration of births, deaths, and marriages still had to rely on the services of the administration of the poor law (1836). Although statistics had played no role in the Poor Law reform of 1834, with the exception of the statistics on the rising relief expenditure, it became an important part of both poor law administration and social reform from the late 1830s onward. A. Quetelet's conviction that "statistical inquiry" was "the very basis on which all good legislation must be grounded"[8] was commonplace among all reformers of the following decade. The combination of statistics and public health became a powerful means of reform. And although the stress was put on sanitation as a policy for preventing destitution, it was also possible to use the mortality figures as an indicator of indigence. W. Farr, the leading statistician at the newly established Registrar General's Office, in 1839 turned Graunt's interpretation of the deaths caused by starvation upside down by viewing them as the tip of an iceberg of extreme poverty.[9] He pointed out that lack of nourishment, even if it did not lead to actual starvation, was likely to ruin a person's physical capacity to resist sickness, which, again, would help to explain why the death toll caused by disease was higher at the bottom of society than at the top. However, attempts like these did not attract much attention because they implied a critique of the poor law administration, and some of the poor law officials were among the most active public health propagandists. They supported sanitation as a measure for preventing indigence, according to their philosophy that the poor law should strive to make itself unnecessary. Therefore, the statistical correlation between rising sickness figures and rising expenditure on poor relief provided the starting point for much of the reform enthusiasm in the 1840s.

The statistical vogue of the 1830s and '40s with its unification of statistics and social reform could be called a *Baconian* reaction on the part of the professional middle classes to the emerging condition-of-the-people question. After some decades of intense ideological controversy, many people were looking for a method of conceiving reality as it "really" was. The way of reasoning in terms of number, which had been so successful in commerce and technology, seemed to provide such as method. In 1832, the Board of Trade set up its own statistical department in order to cope with the huge increase in British foreign trade.[10] The foundation of the General Registrar's Office five years later was to serve a similar purpose, though it was now the condition of the people to which the work of "collecting, arranging and publishing" statistical materials referred. Moreover, the emergence of bureaucratic structures in the fields of factory employment, public health, poor relief, and education yielded a rising tide of information, which was becoming increasingly numerical. In short, the dramatic extension of the sphere of "public" knowledge and the necessity to come to terms with the flood of empirical data favored the application of statistical reasoning. Statistics seemed to provide a method of "inductive" understanding that made it possible to eliminate "opinions" and to see things as they really were. This Baconian appeal of statistics became irresistible when the middle classes had to face the condition-of-the-people question. The formation of sanitary statistics as the major event of the statistical movement and the development of a policy of public health demonstrates the important role that ideological factors played in the transformation of statistics into a strategy of social amelioration. This is also shown by the wave of statistical societies founded in the 1830s. The older meaning of statistics, namely, to offer a survey of a country's political state to enable the statesman to act successfully, was combined with the use of figures. Statistics became a part of the debate concerning social questions that was emerging during these years.

Social statistics played a crucial role in the formation of public health policy. In this context, however, it contributed to the formulation of society as an independent conceptual field that cannot just be derived from political economy. This was the case because sanitary statistics were regarded as the statistics of the condition of the laboring poor in general. The concern about this condition, together with the wish to further "the improvement of mankind," provided an indispensable stimulus to both social statistics and early sociology. Sanitary statistics, being occupied with anonymous contingencies like infectious disease and poor housing, which could only be overcome through collective action, revealed truly societal factors that effectively limited the range of alternatives open to the "free individual agency" as recommended by the economists. The clue to the popularity enjoyed by sanitary reform among a considerable part of the middle classes can be found in its potential to improve the situation of the laboring poor without conflicting with the prescriptions of political economy. The sanitarian argument defined the social question primarily as a problem of social medicine, which made it possible to shift the focus from factory legislation and the poor law to urban health. This shift corresponded to the growing urgency posed by the mushroomlike expansion of cities during the Industrial Revolution.

Hygiene as a science is the investigation of sickness as a mass phenomenon, which has its roots in certain conditions of everyday social life, e.g., in polluted drinking water, overcrowding, lack of sanitation, and accumulation of filth. It is a type of technology rather than a type of medicine, and its stress is on prevention and not cure. Shaped by the framework of the miasma theory of epidemic disease,[12] it used statistics as its most important instrument of investigation. Epidemic disease, caused and nourished by unsanitary living conditions, could be interpreted in terms of political economy as a threat to free agency comparable to other criminal threats to the health and life of man. It reduces the free worker and his family to pauperism and is, therefore, contrary to the principles of a liberal economy. Thus, a public health policy is a means of safeguarding free agency. Reasoning of this kind provides the link between sanitary statistics, public health, social reform, and liberalism. It extends the liberal discourse to the social question, although to a question limited to the health hazards of economic inequality. W. Farr consequently defined vital statistics as "the iron index of misery," which could serve as "an arsenal for sanitary reformers to use." Petty's ideal of a "Scale of Salubrity" as a guideline for public policy was renewed and vigorously applied. The foundation of a central institution for the collection of vital statistics enabled Farr to construct such a scale by comparing mortality in different parts of the country on a uniform methodological level. Such a scale also promised the measurability of an important aspect of the happiness of the greatest number.[13] Under these circumstances it was possible for E. Chadwick, the great sanitary reformer, to replace the mercantilist poverty law of large numbers with comparative sanitary statistics as a kind of iron index of the health of large numbers.[14] The reform movements of the 1830s and 1840s derived their strength from the conviction that poverty was not the lot of the laboring classes. Statistics on pauperism and mortality revealed the poverty of the nation and whether it was rising or declining. Moreover, it allowed for social experimentation, proving or falsifying methods of reform and showing that amelioration was possible and that the large numbers were not necessarily subject to epidemics and starvation. Statistics made the social state of the nation an affair that could be measured, and in this way it contributed significantly to the bureaucratization of disease and poverty that was to pave the way for the development of the welfare state.

The discovery of an interrelationship between high mortality rates and high poor law returns[15] stimulated the sanitarian and interventionist interpretation of the poverty law of large numbers, which finally transformed it into a bureaucratic concept. The Public Health Act of 1848 prescribed a threshold of 23 deaths per 1,000 inhabitants, thus making the statistical scale of salubrity a legal one. If the death rate were higher, central government could intervene into local affairs through its novel Board of Health. Farr himself tried to show by way of statistical comparison that a rate of 17 : 1,000 could be achieved through sound sanitary management. Consequently he recommended measuring every additional death as one degree of insalubrity.[16] This would make it possible to quantify "the well-being and happiness of the masses." The publication of such tables of salubrity was supposed to provide politicians with information for intervention and to stir up public protest. If "mortality is greatly augmented wherever large masses of the people are brought together," then it was but consistent "to render the towns ... and every establishment, where large numbers are called together ... compatible with the full enjoyment of health."[17] Here, poverty, which forces people to live in unhygienic housing, became a device for sanitary legislation, which, again, implied something like a right to the protection of individual health, perhaps the first social right as such. The public effect of their statistics was very much in the mind of all social statisticians, who were often accused by their adversaries of cooking their data. Cases in which the figures collected contradicted the preconceived opinions they were meant to prove were often solved by blaming the figures. When liberals spoke up for elementary education, they always expected that more schools for more pupils would mean less crime and less intemperance. Statistics that did not support such hopes were criticized or put in their "proper" perspective.[18] The great Education Act of 1870 was welcomed by many social statisticians as a result of their work. Educational statistics was seen as a means of measuring the progress of the working classes toward higher standards of civilization, indexing how "they" were approaching "us." Although it became possible only in 1874 to set up a central agency for collecting such statistics, the working class had always been a major object of statistical work,[19] as we have already seen.

With the beginning of the Victorian boom in the early 1850s, both social reform and social statistics became less of an interest to the general public. Nearly all of the statistical societies that had sprung up in the 1830s and 1840s disappeared. Economic statistics in a more general sense became prominent and the returns from friendly societies and savings banks replaced sanitary statistics as the figures representative of working-class progress. The addition of a Labour Department to the Board of Trade in 1886 fit well into this pattern. Its task was to pool the data from trade unions, friendly societies, and cooperative associations, i.e., to mirror the advance of the self-help agencies of the working class. Until the establishment of labor exchanges in 1909, the union returns were the only index of unemployment, which began to emerge as an independent social problem in the 1880s. The friendly societies, which insured their members against sickness on a non-profit-making basis, had been a statistics-ridden agency from the very beginning. The lack of sound tables of mortality was a severe handicap for them, and, therefore, they stimulated actuarial statistics and the application of probabilistic reasoning to such material. In 1789, Dr. Price published a series of tables based on the calculus, explicitly for the use of benefit clubs. With the first Friendly Society Act of 1793, their financial soundness became a public affair. The liberals hoped that these societies could become the main instrument in making the working classes self-reliant, and they were eager to improve their actuarial basis. This led to the foundation of a central office for the collection of friendly society statistics and the provision of actuarial advice, which later became the office of the Registrar General (1825, 1829).[20] The growing interest in unemployment figures as well as the differentiation of the compact pauper statistics, providing for separate columns for the sick and, later, the aged, indicated that a change was underway. The economy had entered the phase of maturity, and the ongoing emancipation of the working classes had profoundly changed the social meaning of poverty and large numbers. The narrow bureaucratic notion of poverty was confronted with a political concept of "decent standards." The differentiation of poor law statistics and particularly the collection of old-age statistics from the 1880s on helped to clear the way for the introduction of pensions in 1908. Unemployment statistics were lagging behind for obvious reasons. Here, differentiation was needed to express the relative importance of the different causes of unemployment statistically. Such data could then supply practical guidelines rather than merely indicate the temporary state of employment.[21] With the appearance of social insurance after 1911, a new chapter in the history of social statistics began. Now, "welfare" and not poverty was to be associated with large numbers of the people. The poverty law shrank to a kind of poverty index of small numbers, an index of social assistance counting only those who lived on the fringe of society.

3 Conclusion

The obsession with the notion of social relevance provided the main impetus for the rise of social statistics up to its scientific reformulation. Because of its strong inductive appeal, it was taken to offer an empirical alternative to the clash of opinion. As a method of large numbers, it provided a conceptual framework for handling the emerging phenomena of large numbers. Its usability as an indicator appealed to the social reformer. The components of social statistics (in particular, poor law returns, mortality figures and birth rates, educational statistics, figures on unemployment, and the returns from friendly societies) added up to something like the statistics of the condition-of-the-people question, which became "the question of questions" to an increasing number of Englishmen as time went on. One might even say that social statistics as well as social reform are responses to the same basic challenge. They express the growing awareness of the new social and political significance of phenomena of large numbers. Social statistics, together with the social survey, made it possible to investigate the factual state of society and to extend the realm of sociopolitical reflection beyond the limits of political economy without being obliged to abandon it. This amalgamation of social reform, social statistics, and classical economy rendered methodological disputes unnecessary, and it may well have contributed to the slowness with which sociology developed in Britain. Social statistics was a liberal "science," and, therefore, the critique of statistical "steam-engine reasoning" went hand in hand with a rejection of liberalism as well. The attitude of counting men instead of weighing them, which was generally ascribed to the statisticians by their critics, was unbearable to a Coleridge or a Carlyle. Statistics, they claimed, reduced men to averages, to a mean man that only existed in one's fancy.[22] There was no mean happiness that could be expressed by mean wages or mean mortality. Yet, with the spread of the large number as a structural feature of modern society, accompanied as it was by the advance of the quantifying perspective, the little man, whose needs and attitudes carried importance only when expressed in large numbers, rapidly gained social and political weight.

It is no accident that, at the turn of the eighteenth century, the word "statistics" began to change its meaning. Having been an amalgamation of descriptive information on the pecularities and "curiosities" of a given country, it now became synonymous with gathering figures and arranging them to form averages. It was no longer and exclusively the statesman for whom such work was done; the public was also regarded as a legitimate audience for the statistical message. The new numerical statistics was deeply influenced by the notions of time and change. Its figures measured temporal developments. They offered no static and encyclopedic descriptions because they were gathered in order to change existing situations, not to maintain them. Statistics of the older type and political arithmetic were built upon a mercantilist understanding of the state with demography at its center. The new type of statistics, however, relied upon an idea of society and of economy as spheres that were not identical with the one occupied by the state. Society was viewed as a sphere being placed between the economy with its stringent laws and the state. It needed its own method of investigation to explore its particular problems properly, and this method was the peculiar amalgamation of statistics and social survey that lasted throughout the century. The debate on public health marks the climax of the social influence of statistical reasoning. Its scientific reputation rested upon sanitary statistics, and its interest focused explicitly on the health of the average man and not on the cure of some diseased individual, as was the case with hospital statistics, which proved to be much less influential. With the decline of the public health agitation and the decay of political economy as an explanatory model in the second half of the century, social statistics lost much of its attraction and became more of a specialist's affair. It is noteworthy that its revival as a social concern paralleled its reconstruction as a methodological concept.

The statistical method of investigation, as employed in nineteenth-century social statistics, proceeded by dividing man into the aspects of his social existence and by searching for numerical data to cover them. In this way, it reconstructed the change brought about by industrialization in the framework of society. The division of labor, the separation of production and consumption, the rise of a labor market, and the differentiation in the social roles performed by individuals are examples of this process. The concept of the mean man reflects the ruin of the traditional hierarchy with its status ascriptions. The mean man only appears as an aspect or a series of aspects that are filled with statistical data. The inside of man is excluded; viz., it is only of public or statistical interest insofar as it makes itself publicly or statistically known. The separation of private and public spheres characteristic of the liberal state supports the exclusion of individual aspects by the statistician. Gradually, the symbol of the Gaussian curve proclaiming that large numbers breed order replaced the symbol of the pyramid with its message that order was the result of ording from above. With this change, the iron nexus of poverty and large numbers was also shattered.

Notes

1. Quoted after A. S. Wohl, *Endangered Lives. Public Health in Victorian Britain* (London: J. M. Dent and Sons, 1983), p. 7.

2. Cf. H. A. Boner, *Hungry Generations. The 19th Century Case against Malthusianism* (New York: King's Crown Press, 1955), pp. 70, 116, passim.

3. I.e., the Poor Lam. Cf. J. R. Poynter, *Society and Pauperism. English Ideas on Poor Relief, 1795-1834* (London: Routledge and Kegan Paul, 1969), pp. 144-172.

4. R. D. Altick, *Victorian People and Ideas* (London: J. M. Dent and Sons, 1974), pp. 117-118.

5. John Graunt, *Natural and Political Observations ... Made upon the Bills of Mortality* (1662, 1665), in *The Economic Writings of Sir William Petty*, ed. C. H. Hull, vol. II (Cambridge: Cambridge University Press, 1899), p. 333.

6. Graunt, *Observations*, p. 352

7. David V. Glass, *Numbering the People: The Eighteentheenth Century Population Controversy and the Development of Census and Vital Statistics in Britain* (Farnborough: Gregg International Publishers, 1973), pp. 17-18.

8. Glass, *Numbering*, p. 127.

9. According to Graunt, "not above one in 4,000 deaths are starved." Farr speaks of 63 out of 143, 701 deaths (in a half-year period) as being the fatal result of undernourishment.

10. M. J. Cullen, *The Statistical Movement in Early Victorian Britain* (Hassocks: Harvester Press, 1975), pp. 19-21.

11. Cullen, *Movement*, p. II.

12. The doctrine of miasma, which was influential in the early nineteenth century, explained the origination of infectious diseases mainly through the prevalence of bad air in certain parts of a town. This "poison" was generated by the decomposition of animal and vegetable substances, including human excrement.

13. Cf. R. A. Lewis, *Edwin Chadwick and the Public Health Movement* (London: Longmans, 1952), p. 27; V. L. Hilts, Statist and Statistician: Three Studies in the History of Nineteenth Century British Statistical Thought," Ph. D. thesis Harvard University, 1967, p. 33.

14. Edwin Chadwick, *Report on the Sanitary Condition of the Labouring Population of Great Britain* (1842), ed. M. W. Flinn (Edinburgh: Edinburgh University Press, 1965), pp. 76-77, 220-266. On Chadwick's use of statistics see Lewis, *Public Health Movement*, pp. 60-65, and Cullen, *Statistical Movement* (note 10), pp. 56-63.

15. *Fourth Report of the Poor Law Commissioners, Appendix A, No. 1* (1837-1838).

16. John M. Eyler, *Victorian Social Medicine: The Ideas and Methods of William Farr* (Baltimore: Johns Hopkins University Press, 1978), pp. 71, 83.

17. William Farr, "Vital Statistics or the Statistics of Health, Sickness, Diseases and Death" (1837), in *Mortality in Mid-Nineteenth Century Britain*, ed. R. Wall (Farnborough: Gregg International Publishers, 1974), p. 601.

18. Cullen, *Statistical Movement* (note 10), pp. 139–143.

19. Cullen, *Statistical Movement*, pp. 65–69, 112–115, 139–144. An interesting piece of antireformist educational statistics is provided by the pamphlet by E. Baines on "The Social, Educational and Religious State of the Manufacturing Districts" (1843), reprinted in *The Factory Education Bill of 1843. Six Pamphlets* (New York: Arno Press, 1972).

20. P. H. J. H. Gosden, *The Friendly Societies in England, 1815–75* (Manchester: Manchester University Press, 1961), pp. 191–194.

21. Cf. Seebohm Rowntree and B. Lasker, *Unemployment: A Social Study* (London: Longmans, 1911). This is a pioneer study of the differentiating type based on regional statistics.

22. Cf. Th. Carlyle, "Chartism" (1839), in *English and Other Essays* (London: J. M. Dent and Sons, 1967), pp. 170–174.

17 Lawless Society: Social Science and the Reinterpretation of Statistics in Germany, 1850–1880

Theodore M. Porter

Statistical thinking only came to be seen as an alternative to Laplacian determinism during the 1860s. Quetelet, the leading statistical writer of the mid-nineteenth century, dismissed variability as error and stressed the possibility of establishing laws of society from knowledge of mean values alone; his science was based on the presumed analogies between social and physical phenomena. The most systematic, and perhaps also the most influential, effort to supplant that interpretation was made by German academic statisticians, many of whom were associated with the historical school of economics and the Verein für Sozialpolitik. Their statistical theorizing was intended to provide a coherent understanding and justification of their science, but also to support a program of antisocialist reform based on a view of human communities that stressed human freedom and mutual responsibility. They developed the interpretation of statistics as a method of mass observation, essential for the study of society precisely because of the diversity of individuals and the absence of determinate social laws, but applicable also to various aspects of nature.

Numerical statistics arose in Europe as the science of social facts. The science of the "statist" involved no uncertainty, for it avoided the conjectures and estimates of political arithmetic by insisting always on careful and exhaustive enumeration. The result of this activity was the discovery of what were taken to be astonishing regularities in events of all sorts, including not only births, deaths, and marriages, but even crimes and suicides. The statistician looked forward to the day when the realm of scientific law would be extended to include the whole world of society and of man. Statistical laws, it was conceded, differed in certain respects from the archetypical principles of physical or even biological science. The individual remained inaccessible to their science; statistical regularities would dissolve if subjected to a complete analysis. The causes of the underlying events were unknown, perhaps unknowable, and, in any event, too numerous and diverse to form the basis for a unified scientific theory. But the distinctive method of statistics, tallying numerical observations of a great number of individuals, and then formulating principles at the level of the collective whole, had proved so successful as to provoke widespread concern that its results were inconsistent with traditional doctrines of human freedom and divine authority.

That the accuracy of inferences from statistical data to "fixed causes" was subject to the rules of probability was clearly known to those few statisticians who cared about mathematics. This, however, involved probability only as a measure of imperfect knowledge, much like what had become standard in observational astronomy. Except for Auguste Comte and an occasional critic of medical statistics, who rejected the use of probability in science completely, nobody mentioned that statistical methods might accord poorly with the deterministic ideal of science. If some events appeared to be due to chance, this must be attributed to shifting patterns in the underlying causes. It was universally agreed that the observed regularities of the mass would be inexplicable if the constituent events were supposed subject to the whims of pure chance.

Beginning during the late 1850s, this view of statistical methods began to break down. One context in which these assumptions were challenged most pointedly was in Maxwell's interpretation of the second law of thermodynamics. The source of this "erosion of determinism," as Ian Hacking calls it, was social science. I have shown in a previous paper[1] that Maxwell was echoing, and sharpening, a widespread criticism of the concept of statistical law introduced by critics of Henry Thomas Buckle's *History of Civilization*. Such views were also developed with considerable philosophical acumen by late nineteenth-century German social scientists, who, without questioning the validity of statistics as a source of useful generalizations, denied that they could legitimately be interpreted as laws. In accordance with their view of society as a rich, heterogeneous composition, and their identification of the human domain with history and with freedom, they came to view statistics as the appropriate tool of social analysis precisely because its conclusions lacked necessity. Their conceptualization of a science that did not depend on the lawful behavior of individuals was even more important for the history of statistical thinking than for social science per se. It contributed essentially to the transition from epistemic to ontic conceptions of probability that is a central theme of these volumes.

1   The Heritage of *Statistik*

Statistics arose as a historical discipline. Its "father," by common consent, was Gottfried Achenwall, who coined the substantive *Statistik* in 1749, although the metaphor of paternity was not easily sustained given the manifest promiscuity of the disciplinary attachments that were essential to its formation. In fact, it proved impossible to agree on a definition. Nineteenth-century statisticians debated even about etymology, for while the German *Staat* is univocal, the Latin *status* can refer either to the body politic or to the present condition of something (German: *Zustand*). Accordingly, statistics might either be the descriptive science of states or a general method of characterization, applicable to any object at all.[2] In practice, statistics—Schlözer's "stationary history"[3]—was a nondescript enterprise akin to political geography, whose essence was debated with such vigor precisely because it lacked a distinctive method or well-defined subject matter.

The identification of statistics with numbers, increasingly popular in western Europe and America during the first decades of the nineteenth century, was resisted by German academic statisticians until the 1850s and 1860s. Although some, including Schlözer, thought numerical tables the most suitable means for concise statistical description, numbers were generally identified until 1860 with the official form of statistics, a "lifeless and mechanical" practice that flattened the delicate social contours and *Staatsmerkwürdigkeiten* (distinctive attributes of a given state) beneath the homogenizing force of bureaucratic centralization.[4] In the end, it was practical and bureaucratic concerns—especially the new sense of commitment to reform and the relief of rising social tensions—that converted *Statistik* into a numerical social science like the cognate disciplines in France and Great Britain.

Exemplary for the early period is Johannes Fallati, professor of statistics and contemporary history at Tübingen, and later a delegate to the Frankfurt Parliament. In 1839, two years after his habilitation, Fallati journeyed to Britain to study, as he tells us, the socialist movement there. His contacts consisted primarily of members of statistical societies in London, Manchester, Bristol, Ulster, and Liverpool (of which the latter three soon disbanded), and he returned to Württemberg greatly impressed by the facilities for the collection of useful social data in progressive England. Published numerical statistics, he reported, were characteristic products of "Self-Government," and contributed to the triumph of an integrated society over the particularizing forces of landed nobility and party prejudice. Fallati was not carried completely away by Anglophile enthusiasm. It "accords with the nature of the English spirit, to survey the surface of things more readily than to penetrate to their depths and to develop the practical rather than the theoretical side of the sciences"; "atomistic administrative statistics" were less well-suited to promote scientific knowledge than "to lay a firm foundation for reform of social relationships through the knowledge of facts." [5]

Fallati, nevertheless, devoted more effort to the low, numerical, practical aspects of statistics than to its higher, descriptive branch. The collection and study of social numbers was seen as increasingly urgent during the 1840s. In 1846, the *Deutsche Vierteljahrsschrift* lamented, "No nation has developed statistics so scientifically and systematized it so exactly while at the same time drawing so little practical benefit from it as we Germans." [6] The next year, a "Society for German Statistics" was formed by a Hanoverian bureaucrat, Freiherr von Reden, who called for the application to social problems of "indisputable numbers and facts." [7] The events of 1848 inspired a dramatic increase in official statistical activity. Before that year, only four German language states sponsored statistical agencies: Prussia (1805), Bavaria (1808/1833), Württemberg (1828), and Austria (1829). During the next two decades, statistical bureaus appeared in virtually every German state: in Hanover (1848), Saxony (1850), Bremen (1850), Mecklenburg-Schwerin (1851), Baden (1852), Brunswick (1854), Oldenburg (1855), Hessen (1861), Switzerland (1863), Thuringia (1864), Hamburg (1866), Anhalt (1866), and Lübeck (1871). [8]

The 1848 revolution provided also the context in which German thinkers "discovered" the idea of society as a distinct entity, largely autonomous of the state, which could be made the subject of its own science. Actually a French import, the concept of society had first been brought to the attention of Germans by Lorenz von Stein, whose *Socialismus und Communismus in Frankreich* was based on two years of research on labor movements as well as socialist writers in Paris, and it was for some time associated with worker radicalism. "The ill-formed socialist and barbaric communist doctrines and intrigues," wrote an anonymous author in 1854, "have at least brought one benefit to political science by obliging us to recognize and comprehend a domain that, though it underlies the whole organism of state and acts powerfully on it, has yet been almost entirely overlooked by theory—namely, society." [9] The classic statement on the science of society was an 1851 essay by the dean of the *Staatswissenschaften*, Robert von Mohl, who believed that "the false social doctrine" of the socialists "can only be overcome by a true science of society."[10] Significantly, Mohl undertook to purify the concept of society by stressing a level of organization between the individual and the state—"communities of interest," including race, inherited position, trade, property (*Besitz*), village, and religion. Thus Mohl fractured the socialists' "society" into multiple dimensions of heterogeneity, and made this very heterogeneity the key concept, just as statisticians were to do with Quetelet's homogeneous "social body" two decades later.

The discovery of social science was not immediately decisive in bringing about a transformation from descriptive to numerical statistics. Although Carl Knies argued in 1850 that numbers are essential to portray society, while verbal description is more suitable for characterizing the state,[11] both Stein and Mohl advocated continuation of a predominantly verbal approach to the science.[12] The growth of official statistical activity exerted perhaps a more specific influence, for, as Gustav Rümelin pointed out in 1863, this practical activity won the general public over to a numerical definition of statistics long before the academic community was obliged to follow.[13] Moreover, many of the leading statistical writers in Germany, as in France, Britain, and Italy, held official positions. Still, German academic statisticians were only converted to the numerical approach after they had been shown that a theory of statistics worthy of profound scholarship could be grounded in numerical tables. For this discovery, they were indebted to Quetelet, whose reputation in Germany dated from around 1860.

With this change, statistics ceased to be frozen history and became something like what it had long been in Great Britain and France, empirical political economy. The German statistical discipline was not thereby brought into perfect harmony with its neighbors, however, for German political economy had already begun to diverge from the classical tradition. In post-1848 Germany, statistics was pursued largely in the context of historical economics. It was seen as a tool by which this excessively abstract and deductive science, political economy, could be brought down to earth and grounded in empirical reality.

All three members of the so-called older historical school of economics, Wilhelm Roscher, Bruno Hildebrand, and Carl Knies, were active statisticians—although Roscher remained loyal to the older, descriptive school.[14] These men forged most of the weapons that the "younger historical school" would wield against classical economics. The approach of Adam Smith and J. B. Say was conceded to represent dialectical progress, a valid reaction against the stifling statism of the mercantilists. It had liberated the energy of individuals, and brought impressive practical results, but at a high cost, for it reduced community to self-interest and placed economic behavior outside the domain of morality. Indeed, it sanctified greed as the expression of natural law. Its irresponsibility was manifested in the growth of useless financial speculation, which led to panics and much unnecessary misery, while its faith in abstract deduction bore fruit in dogmatic socialism, the most theoretical of economic doctrines.[15]

This historical and ethical picture put forward by Hildebrand and Knies, in particular, was largely taken over by the younger historical economists. Both generations agreed also that economic principles were culture-dependent, and that the great diversity of peoples and nations demanded a similar variety of economic theories. They differed more strongly in practical politics, for whereas writers like Hildebrand stressed principally the need for personal regeneration, the younger writers were specifically concerned about the "worker question" and actively involved in promoting state-directed reform. That was the purpose of the *Verein für Sozialpolitik*, founded in 1871, whose membership included most of the prominent statisticians and historical economists of the day. To be sure, the membership was far from unanimous as to the best solutions, and there was a pronounced split between those relatively liberal members, including Wilhelm Lexis and Gustav Schmoller, who favored helping the workers help themselves by organizing them into cooperatives, and those like Adolph Wagner who preferred direct state intervention.[16] The important point, however, is that all believed in the possibility of improving economic relationships through some form of organized action. That there were still economic theorists of the classical school who denied that possibility sharpened the antipathy of the historical school to the application of "natural law" (*Naturgesetz*) to social science.

Although virtually all the prominent German interpreters of statistics were university professors, most had practical experience as well. Ernst Engel, whom Ian Hacking treats at length in a contribution to this volume, ran a statistical seminar associated with the Prussian statistical office and the University of Berlin. Among the scholars who graduated from Engel's seminar were most of the leading members of the *Verein für Sozialpolitik*, among them Gustav Schmoller, Adolf Held, Adolph Wagner, and G. F. Knapp. A similar program was instituted in Vienna later in the 1860s, and provided training for some of the most prominent Austrian statistical writers.

A significant number of the theorists discussed here also held important administrative positions in statistics at one time or another. Gustav Rümelin headed the Statistical-Topographical Office of Württemberg from 1861 to 1867, before becoming a teacher and then chancellor at Tübingen. K. T. Inama-Sternegg and F. X. von Neumann-Spallart were attached to the Austrian Central Statistical Commission, as well as the University of Vienna. F. B. W. Hermann, and then Georg Mayr, held joint positions as professors at Munich and directors of Bavarian statistics. G. F. Knapp set up the statistical bureau in Leipzig, which he continued to direct after he was offered a teaching position at the university. Bruno Hildebrand, similarly, held joint positions in Bern, Switzerland, and then at Jena. Various members of the Berlin City Statistical Bureau as well as the Prussian statistical office were simultaneously associated with the university.[17]

That all these statistical critics were professors, thus belonging, as Fritz Ringer shows for a slightly later period, to a distinct and self-conscious group, perhaps goes some way toward explaining the degree of consensus that was achieved on the interpretation of statistics. Indeed, most were associated with one or more of a small group of universities, including Berlin, Göttingen, Strassburg, Leipzig, Vienna, and Dorpat. It is also significant, though perhaps more symptomatic than explanatory, that these historicist professors entertained strikingly similar stereotypes of the history of thought as it pertained to statistics. In statistical history, as in economic history, the atomistic rationalism of Enlightenment thinking was identified as the villain. Both Adam Smith, who was placed at the head of the economic Enlightenment, and Quetelet, who carried the banner for the statistical one, were conceded to be great men. Both, however, had been followed by "schools"—David Ricardo (or *Manchestertum*) and Henry Thomas Buckle, respectively—that preached a dogmatic and exaggerated form of the doctrine of their masters.[18]

It is characteristic of the historical school viewpoint that individual authors were always interpreted as mouthpieces for whole cultures, and that the opinions of all authors understood as belonging to the same movement were assumed to be of the same fabric. Quetelet, an astronomer and a product of French traditions, was readily classified as an Enlightenment atomist.[19] In effect, authors in a given field were held responsible for the tendencies of their thoughts—for the whole system of ideas to which their own work seemed to belong. Accordingly, Quetelet was sometimes reproached for opinions that could be attributed to him only by a remarkable feat of extrapolation, perhaps in direct contradiction to ideas he had explicitly expressed. Thus it is perhaps not so strange as it seems at first that so high a level of polemical fervor could have been sustained in decades of refutations of Quetelet, when nobody who mattered was defending him. Quetelet's German critics were seeking not merely to assess the worth of his statistical contributions, but to combat the idea—with all its connotations in politics and economics—that humanity belongs to the realm of nature rather than of history, and to defend a pluralist—almost corporatist—view of human society.

2 The Light from Belgium

Quetelet was not unknown in Germany during the early nineteenth century. His principal works had been translated, and he was read by insurance writers, physicians interested in public health matters or in the already classic problem of birth ratios, and other specialists. A Prussian pastor with mathematical and astronomical interests, J. W. H. Lehmann, challenged the coherence of the average man concept in the early 1840s, and M. W. Drobisch gave his first analysis of the relation between moral statistics and free will in a review of 1849.[20] Many German statisticians became personally acquainted with Quetelet as a result of the International Statistical Congresses, which began in 1853, and German periodicals published extensive reviews of the meetings. Quetelet's notoriety in Germany, however, was not precipitated by any of his own works or activities, but rather by the publication, in English (1857) and then in German (1859), of Henry Thomas Buckle's *History of Civilization in England*.[21]

Buckle's exemplars of statistical constancy, and even his interpretation of them, were drawn from Quetelet, differing only in that Buckle issued them without Quetelet's customary reservations and qualifications. The regularity of crimes, marriages, and suicides confirmed, on Buckle's authority, that even willful human actions are subject to law. That the regularity becomes manifest only in the large militates in favor of an approach to history that views its grand sweep, without becoming lost in confusing details. Nevertheless, the existence of such laws is inconsistent with the possibility of freedom even at the level of the individual, and Buckle bravely denied that moral statistics could be reconciled either with divine authority or human free agency.

The enormous popularity of Buckle's book in Germany, as well as the uncompromising position it took on issues like these, accounts for the stimulus it gave to statistical debate. It appeared at an opportune time, and helped set the tone for the flowering of liberalism that took place in Germany during the 1860s and 1870s.[22] Mainstream German academics, however, found little to commend in this book, and tagged it with their sharpest epithet: *Dilettantismus*. Historians such as Droysen saw Buckle as an autodidact, lacking in *philosophische Bildung*, who thus was capable only of the blindest empiricism. Buckle, Droysen held, saw only the material side of things, and not the all-important spiritual and ethical. Even Buckle's translator, the aging Young-Hegelian Arnold Ruge, then in British exile and, according to Marx, committed to spend the rest of his life communicating Buckle's ideas to his fellow Germans, conceded that Buckle did not understand the concept of human freedom. But he insisted that Buckle's book expressed the *Zeitgeist* of real German philosophy, and warned his countrymen against speaking "as if Buckle were a materialist, when he is really only an Englishman."[23]

The free will debate was hardly a fair contest. On one side stood what historicist critics insisted on calling Buckle's or Quetelet's "school," whose membership consisted of Buckle himself, possibly Quetelet, and perhaps a crude German materialist like J. C. Fischer. Adolph Wagner was also sometimes included, because of a parable he related in which a meticulous ruler was imagined to declare every year how many people of every age, sex, and profession would die, marry, commit murder or suicide, with guns, knives, or poison, and so on, but though he declared this unrealizable despotism to be indistinguishable from reality, he, like Quetelet, came to no clear conclusion on the relation of all this to free will.[24] On the other side stood a great chorus of German academics, who argued in diverse and inconsistent ways that statistics could say nothing about the determination of the will, or that statistical regularity did not amount to law, or that the concept of freedom attacked by Buckle was already outdated.[25]

These arguments figured directly in the German reinterpretation of statistics. Buckle, however, was more than a foil, despite the nearly unanimous condemnation of his book by German academics. The ideas he introduced seemed to offer a promising foundation for a theoretical approach to numerical statistics, even as his presentation guaranteed that they would be marked with the stigma of materialism, fatalism, and naive empiricism. Quetelet's work, the first systematic program for converting the chaos of numbers that began appearing in profusion in western Europe during the first half of the nineteenth century into a coherent science of society, soon came to be seen in Germany as a landmark in the history of statistics. It was necessary from the beginning, however, to dissociate this science from certain doctrines that, if they were not endorsed by Quetelet himself, had arisen out of his "school" and thus indicated a likely defect in his general approach.

When German statisticians went beyond Buckle to read Quetelet, they encountered a system founded on the notion of "statistical law." This was the principle that almost any number collected systematically for a whole society—births, deaths, crimes, marriages, suicides, and so on—will display an "astonishing" regularity from year to year, and that this uniformity of effect indicates a like uniformity in the underlying causes. On the basis of this regularity, Quetelet argued for the utility of constructing an "average man," an abstract entity who possessed all the statistical properties of his society to an average degree. Social physics was to be a dynamical science of the average man, analogous to celestial physics, which analyzes the motion of imaginary masses located at centers of gravity. It was emphatically a science of the social, in which real individuals had no standing. Living persons are too unpredictable; no doubt their behavior is lawlike, but the causes that determine it are too local, too diverse, too obscure, and too numerous for the scientist to take acount of them all. Fortunately, there is no need to be troubled by all these incidental factors, for they are largely independent of one another, and thus cancel themselves out with great exactness in a whole society. As a result, the statistician is able to discover a profound social order that is concealed from observers who look only at individuals.

Although Quetelet was inordinately fond of the links between his science of society and the laws of mechanics, he was the very opposite of a reductionist. Statistical regularity, he insisted, must be attributed "not to the will of individuals, but to the customs of that concrete being that we call the people, and that we regard as endowed with its own will and customs, from which it is difficult to make it depart." [26] Far from reducing society to an aggregation of individuals, he dissolved his individuals into a monolithic, all-embracing society. Every real person was conceived to be modeled on the average man, and thus to be acted on by all the forces present in society. Each was endowed, for instance, with a "penchant for crime" proportional to the level of crime in the population. No person realized the social type in every respect, but all deviations from it could be dismissed as accidental and secondary. Quetelet's lack of interest in individuality, and his failure to recognize social patterns between the level of the individual and that of the whole society were among the features of his model that the Germans found most objectionable.

Quetelet's other principal offering to German scholars with a penchant for polemic grew out of another of his astronomical metaphors. In plain French, Quetelet's intentions were clear; he believed firmly in the progressiveness of society, was avidly committed to a positive role for government in social reform, and viewed his own demonstration of the lawlikeness of society not as proof of its fixity, but as evidence that when legislators altered the causes of social events the effects would respond predictably. The analogies of social physics, however, obscured these matters. Quetelet divided the "forces" that influence the "social body" into two categories: natural and perturbational, with the latter embracing all effects brought about by the social circumstances or conscious decisions of man, and the former comprising those produced by sex, age, climate, seasonal changes, and other natural phenomena. Following the analogy, it would seem either that "social forces" are temporary and ineffectual or that progress belongs to the realm of nature and not of society. Quetelet here ignored these implications, and made perturbational forces the source of human progress. Still, he seemed sometimes to deny that human activity could alter the course of society, for he routinely dismissed the will as an accidental cause that could have no effect on the social body.

In short, as G. F. Knapp observed, Quetelet was no philosopher. His ideas were often inconsistent, and he was wholly innocent even of those aspects of the German intellectual tradition that had become the common property of Europe. His use of mechanical and biological metaphors almost interchangeably struck German readers as pure confusion, while his attempt to apply laws from the fixed and unhistorical realm of inorganic nature to human society seemed positively dangerous. To be sure, the historicists and reformers who constituted the main part of the German statistical community found much in Quetelet to admire. When shorn of its fatalistic implications, Quetelet's idea that the assembly and arrangement of large numbers could be used to uncover the causes of social phenomena even where the behavior of individuals seemed unfathomable was regarded as highly promising. He gave them hope that statistics could attain reliable scientific knowledge; they were much taken by his belief that statistics could be a genuine science of society, and his concomitant view that society contained the causes for deviant acts such as crime. Precisely because of the importance of these contributions, it was all the more crucial to point out his errors, and those who were already inclined to attribute to French *Materialismus* every evil not directly traceable to *Manchestertum* found in his writings ample ammunition.[27]

3   Statistical Regularity and Social Law

Quetelet had been concerned to preserve some space for the exercise of human freedom by drawing a line between the lawlike regularity of society and the strong influence of accidental causes—among them the human will—at the level of individuals. For statisticians concerned about this issue, it was only a small step beyond Quetelet to assert, as Jakob Friedrich Fries had for probability, that statistical laws apply only to the mass, and are without significance for individuals. The Göttingen professor Wappäus wrote in 1859 that statistical principles "are valid only for the totality of a population, considered as a whole, or, as Quetelet puts it, for the average man of a nation."[28] This was not, however, a popular way of resolving the free will issue in Germany. German moral statisticians preferred to portray individual and community as inseparably bound together. Society was held to provide the conditions essential for the formation of the individual will, and at the same time to be made up of the totality of these wills—which do not cancel out to yield the constant effects revealed by statistics, but sum to them.[29] Moreover, there was little to be gained by adopting a sharp distinction between he levels of individual and society, for most statistical writers in Germany were at least as uncomfortable with the idea that society is governed by "natural laws" as with the denial of free will in individuals.

The idea that statistical laws apply only to collective entities was explicitly challenged in Germany as early as 1864. Adolph Wagner, who, until his "Damascus" conversion to state socialism at the end of the decade, was an unyielding classical economist and, according to Brentano, a full believer in natural laws of society, denied that lawlikeness can prevail for large numbers and yet be absent from small numbers. In a frequently quoted passage, he argued that "large numbers can only be formed from small," and that the "impulse, which is derived in the large from the lawlikeness of large numbers" must act with equal strength on individuals. If in fact the corresponding effect is not produced, this can only arise from the influence of accidental causes.[30]

Wagner's argument was turned on its head the next year by an anonymous reviewer in Hildebrand's *Jahrbücher für Nationalökonomie und Statistik*. A law, the critic argued, must be a fully determinate statement of the form, if *A* then *B*. Granting Wagner's argument that a law of the whole is only possible if it applies to all the parts, it follows that since statistical regularities do not prevail for small numbers, they are not "laws" of society at all, but only numerically expressed properties of lands or nations. It is wrong to call the individual a limb of the whole; rather, the whole is composed of individuals. For that reason, the proper strategy of a scientific statistics in search of laws is not to operate on the greatest possible numbers, but to subdivide and analyze down to the smallest.[31]

Indeed, Quetelet's idea that a statistical regularity is itself a social law never had any support in Germany, and was even challenged in the first statistical paper of his greatest German admirer, Ernst Engel. Already in the early 1840s, such authors as Fallati and J. L. Casper questioned the appropriateness of the term law for the observed constancy of statistical aggregates, on the ground that the mere existence of a past regularity furnished no guarantee that the same ratio would persist in the future.[32] Subsequent authors, perhaps a bit disingenuously, interpreted Quetelet to be implying by his use of the term law that the phenomena of the moral world are part of some natural order that cannot be altered by education or reform. This, they argued, is absurd. If the concept of law was to be used in statistics, it must involve some relation between the numbers of various events and the conditions that produced them.

Adolph Wagner published an essay as part of the introductory material to his book of 1864 on the appropriateness of the terms "law" (*Gesetz*) and "lawlikeness" (*Gesetzmässigkeit*) for statistics. Wagner felt that it was at least within the competence of statistics to demonstrate lawlikeness, and perhaps even to attain the precision necessary for a genuine law, but that mere demonstration of regularity was only the starting point. To be sure, Wagner was greatly impressed by the degree of constancy shown to prevail for willful acts such as marriage, and even irrational ones like crime and suicide. He found, however, that statistical evidence was far more compelling if it showed some social or moral quantity to vary systematically with another social or physical parameter. Here, in Wagner's opinion, was a genuine demonstration of *Gesetzmässigkeit* and, incidentally, the greatest challenge from statistics to the doctrine of free will. The procedure followed by Wagner in his own empirical work was to give tables of the phenomenon in question (here suicide) in relation to all possible extrinsic quantities—climate, weather, time of day and year, sex, age, race, profession, religion, political and economic conditions, education, and so on. A scientific explanation in statistics became a catalogue of partial causes, which would attain to the level of law when all the variation in suicide was accounted for. Wagner, however, provided no prescription by which one could tell when all causes had been discovered, and evinced no awareness that there might be interactions among them.[33]

Wagner's view, that the higher object of statistics was to seek systematic covariation, was soon held almost unanimously by German statisticians.[34] Prominent among them was Gustav Rümelin, who in 1867 argued forcefully that the term law must be reserved for expressing a relation of cause and effect that is elementary, constant, and recognizable in every case. The peculiar function of statistics is to use the so-called law of large numbers to attain knowledge in complex domains like that of human society, where a multitude of forces act simultaneously. Rümelin was skeptical not only of the appropriateness of the term law for statistical regularities, but of the coherence of such phrases as "statistical law" and "law of large numbers." Where a true law has been found, Rümelin insisted, the need for statistics vanishes, for then every case can be explained. Thus, in the domain of that great paradigm of statistical inquiry, birth ratios, a genuine law would belong not to the domain of statistics, but physiology, and would apply separately to every individual birth.[35]

While Rümelin contended that even demonstrated covariation fell short of the standard for a law, nobody seems to have explored systematically the possibility that such a statistical relationship might point to something other than linear causality. There were, however, a few cases where supposed causal relationships indicated in this way were viewed as sufficiently implausible or distasteful to be scrutinized and even doubted. One such relationship that bothered German scholars was the one between Protestantism and suicide, and we find the Göttingen economist Helferich in 1865 maintaining that it must be mediated by other factors, and hence that there was no firm basis for supposing that every increase in Protestantism must contribute to a like increase in suicide. Similarly, Gustav Schmoller adopted a position of radical skepticism on such matters as the alleged connection between suicide and the weather. Since spiritual and physical events are absolutely incommensurable, the latter can never be considered causes of the former.[36]

4 The Heterogeneity of Mass Phenomena

The belief of German statistical writers that law must be sought in variation, and not in mass regularity, reflected a commitment to the view that human diversity was genuine, and an unwillingness to follow Quetelet in dismissing variation as mere error. The critique of homogeneity was sometimes directed against the commensurability of human acts, thereby threatening the most fundamental statistical operation, counting. Thus Hermann Lotze observed in 1864 that, although true constancy—say, of theft—would imply that even the individual criminal "is not free in reference to his decision to steal, but only in whether to do so on horseback or by foot," yet the observed uniformity in the number of incidents of theft is only superficial. Crimes of the same legal category, he explained, are of very different ethical worth, and hence the regularity obtained by lumping together hundreds of such acts indicates nothing about the total quantity of evil in a society. Consequently, statistical regularity need not affect our view of free will, for it cannot be explained in terms of constancy of causes, but remains a great mystery.[37]

A more common form of attack on the related issues of philosophical necessity and natural law in human society was based on the denial that the composition of society is homogeneous, and an effort to correct Quetelet's distorted perspective by which man was seen "only as a collective concept, not in his individuality."[38] The writer who opened up this line of criticism was the Leipzig philosopher and psychologist Moritz Wilhelm Drobisch, who first suggested it in his critical review of 1849. There he objected principally to Quetelet's terminology, and specifically to such phrases as *Hang zum Verbrechen* (*penchant au crime*), because they suggested an innate tendency or leaning to crime. Quetelet's terminology, wrote Drobisch, obscured what he manifestly believed—that the level of crime results not from a fixed natural law or biological predisposition, but from an array of social circumstances that affect different people in different ways and that could certainly be changed for the better. It would be more accurate, and better accord with Quetelet's beliefs, to say *Verleitung* or *Zugänglichkeit*—that is, susceptibility—instead of *Hang*. One should also stipulate that each individual possesses the faculty to overcome the temptation to crime, though our experience of statistical regularity makes clear that as many will succumb as will resist it.[39]

The debates of the 1860s prompted—or rather tempted—Drobisch to rework his critique and to publish it as a book, which appeared in 1867. The book, which drew much more attention than the old review, was also far less generous to Quetelet. Drobisch now argued forcefully that direct inference from an average to properties of individuals was impermissible: "Only through a great failure of understanding can the mathematical fiction of an average man . . . be elaborated as if all individuals possess a real part of whatever obtains for this average person."[40] The tendency to crime, which again preoccupied Drobisch here, is not an attribute of whole age classes, for most individuals make absolutely no contribution to it. It is at most a property of those isolated persons who, due to the defects of their education or social circumstances, are tempted by crime. Individuality must not be dissolved into an all-embracing average, for the differences among members of a society are of genuine moral significance.

Drobisch's ethics were perhaps a bit more individualistic than other moral statisticians were willing to allow. The Dorpat theologian Alexander von Oettingen, whose massive and influential book on moral statistics and social ethics was intended to demonstrate social coresponsibility for crime, suicide, and the like without undermining individual freedom, thought that statistical regularity vindicated the doctrine of original sin. Drobisch, he proposed, was a Pelagian, a "proud pharisee," whose denial of the social dimension was as wrong in its way as the attempt of social physics to reduce the individual to a pure product of his environment.[41] The idea that much collective responsibility was involved in the immorality of individuals was also endorsed by G. F. Knapp. Like Oettingen, Knapp credited this doctrine to Quetelet, whose "profound thinking on the social origin of crime" Knapp regarded as "wholly within the spirit of social science." This conception, he went on, contained implicitly "the great truth that the union of beings (*Zusammensein*) in a society is more than a collection of distinct individuals (*Nebeneinandersein von Einzelnen*)."[42]

Knapp, however, like Drobisch, had no patience with the idea that real individuals differ only by accident from some general average. The proposition that humans are modeled on a type, the average man, negates the very insight that had most impressed Knapp, for it led to an interpretation of statistical regularity in terms of the similarity of the constituent individuals, rather than as an attribute of the whole community.[43] Thus Quetelet the moral statistician possessed deep insights, but lacked the philosophical sophistication to preserve the consistency of his system, which was utterly corrupted by the absurd fantasies about average men and centers of gravity of Quetelet the astronomer and anthropologist. The records of moral statistics are not like astronomical observations, for there is not and cannot be a single true value underlying the measurements of various individuals. It is presupposed by the very nature of a society that its members genuinely differ from one another, and the moral statistician disgraces his office when he rests content with an all-embracing mean.

In the concrete, as well as in the abstract, German statisticians came to distrust mean values over whole regions or societies as sufficient measures of anything. On the one hand, the same mean value can arise from utterly different distributions, and hence no single number can convey sufficient information about quantities like wealth, housing space, age of death, or grain prices.[44] More important, the use of mean values from large, miscellaneous populations was held to inhibit rather than promote the search for causes. By the 1870s virtually no statistician in Germany still thought that an average man could adequately stand for a whole society. A striking indication of the extent to which German statisticians rejected Quetelet's strategy of research may be found in Ernst Engel's eulogy for Quetelet at the 1876 Budapest Statistical Congress, which rather buried the master than praised him. Inhomogeneities, proclaimed Engel, are the most interesting objects for statistical study, and "not the largest, but relatively small averages constitute what is really worth knowing."[45]

This view of society as a union of heterogeneous individuals and groups became associated rather early with a distinct strategy of research for the social science of statistics. The foundation for this interpretation was provided by Gustav Rümelin, a Tübingen-trained pastor turned educator, Frankfurt parliamentarian, and then government administrator, who wrote his first theoretical paper on statistics in 1863, just two years after he was appointed head of the Württemberg statistical-topographical bureau and four years before he took an appointment in statistics at Tübingen. Statistics, he proposed, is properly a method for the observation of mass phenomena. For the sciences of nature, he went on—invoking a distinction made by the French statistician Dufau—statistics is scarcely necessary, for there the individual is wholly or largely typical, so that a single well observed and understood fact is sufficient to justify an induction. Society, by contrast, is the domain of individuality, where each person, though governed by law, is yet subject to so many perturbing causes that the law is completely obscured. This dichotomy is not entirely rigid; animals are less typical than plants, people than apes, moderns than ancients, adults than children, Caucasians than Negroes, men than women, and educated than uneducated. It is, however, so pronounced, that the sciences of man "could never have raised themselves above infancy, where they have been for some generations and where in part they yet remain, if there were no means of observation through which the inadequacy of individualized and idiosyncratic experience can be alleviated, and our experience grasped as a whole."[46]

That method, of course, is statistics, and it thus corresponds to the nature of statistics that it groups together heterogeneous objects. The astute statistician does not rest content with general impressions of the totality, however, but instead fractures it into tiny pieces and regroups these in various ways. The lawlikeness shown by Quetelet, then, is only a beginning. "In the end, the interest of moral statistics lies not at all in the demonstration of these regularities in willful acts of man, but rather in the perpetual movement and alteration that these numbers undergo."[47] Indeed, statistical uniformity is mostly superficial, attained only when one considers a great mass and smears together the variety of phenomena. By finding the conditions under which crime and suicide increase or decrease, and the subgroups for which they are most common, the statistician goes beyond the vague assertion of regularity and attains propositions of real worth and interest. These, Rümelin made clear, would pertain not to any particular individual, but to groups—"knowledge not of individual cases, but of the collective whole of greater or lesser groups of persons or processes." "What I say of the forest does not hold for the separate trees."[48]

Rümelin's definition of statistics as a method of mass observation won wide acceptance among German writers on the subject. German social scientists, and also some philosophers, came to regard statistics as a category of inference. Thus Etienne Laspeyres, a *Sozialpolitiker* and also an economist interested in empirical measures of changes in the value of money, argued that the statistical approach was a substitute for induction in domains like the social one, where it is impossible to set all other things equal. By letting all factors except the one in question vary freely, they cancel themselves out, permitting the effect of the cause under study to be surmised.[49] There was much excitement about the potentialities of this method among the pioneers of statistical social science in Germany. Statistical analysis, according to Georg Mayr, would permit scholars to ascertain whether effects inferred by superficial observation were genuine. Adolph Wagner anticipated that the repeated use of the statistical form of induction would lead eventually to an understanding of all sources of variation in the data, so that a complete explanation would be attained.[50] Rümelin himself was full of optimism in his early papers on the theory of statistics.

By 1878, Rümelin had become more skeptical. After two decades of statistical study, he wrote, he had yet to discover anything that could be called a social law.

That, he added, had led him to consider how different in character physical phenomena were from psychical, and how unreasonable to suppose that the same concept of law applies to each. Where the idea of freedom is involved, the method and ideal of knowledge must change.[51] Doubts like Rümelin's about the possibility of finding statistical laws, and even about their existence, had become increasingly common by the mid-1870s. In the realm of mind and of history, laws of nature had come to seem misplaced.

5 The Limitations of Statistical Knowledge

Although no prominent German statisticians were willing to accept statistical regularities as themselves laws of society, these regularities were quite commonly held up as evidence that society was lawlike. The constancy of these aggregate numbers was seen as ground for belief that statistics could become a proper science, and also as empirical demonstration of the existence of a moral community underlying human society. As Oettingen put the matter, "If all individuals were free to conduct their lives as autonomous and unfettered selves, how could this constancy in the ethical activity of the whole have arisen, and how could it be explained?"[52] That the regularity of the numbers of moral statistics was indeed remarkable was accepted virtually without question until the mid-1870s.[53]

Among the earliest and most damning criticisms of the belief in lawlike statistical regularity was a work published in 1876 by Eduard Rehnisch, a disciple of Lotze at Göttingen. Rehnisch wrote in reference to the supposed conflict between statistics and free will, an argument that, he thought, had already lost most of its credibility and deserved to be put finally to rest. Quetelet's infatuation with statistical regularity was so little based on actual facts, wrote Rehnisch, that he thoughtlessly invoked tables showing variation of 21%, 30%, 254%, 323%, and so on to support it. Indeed, his belief in undeviating uniformity blinded him to what could really be learned from the numbers. Rehnisch reprinted a table from a Quetelet paper of 1848, giving murders in France from 1826 to 1844, and observed that a new French law of 1833 had reclassified fully 45% of murders as assaults. In fact, he wrote, if one inserts the omitted crimes back into the table, the numbers become palpably more uniform. But "the penetrating eyes of the 'new Newton' who imagined himself to be on the way to laws that govern the world of the *sciences politiques et morales* in the same way as the law of gravitation governs ponderable matter—who made a reputation by his ability to prophesy how many misdeeds will be committed next year in a population—these numbers ... made no impression on him."[54] This, proposed Rehnisch, indicates how little was demanded of so-called regularities by enthusiasts in the Quetelet tradition. In fact, his tables show nothing like the degree of uniformity that is requisite for a law of nature.

Wilhelm Lexis's well-known analysis of the degree of regularity presented by statistical data, first published the same year, was set in a somewhat wider context. Lexis, trained in mathematics and natural science before he began to study economics, wrote on trade policy and labor organization as well as statistics during the 1870s and 1880s. He was closely associated with the *Verein für Sozialpolitik*, and shared the political orientation of its relatively liberal members. Like them, his concern about the relation between statistics and freedom was directed less to the issue of individual free will than to the question of the extent to which society was bound by natural laws. By 1875, a generation of German economists had renounced the deductive method of *Manchestertum* in favor of empirical and historical procedures. Deduction was not held to be useless for economics, but its applicability to the real world was regarded as problematic—requiring verification, at least. Certainly there was no ground for asserting economic propositions as "natural laws," for this implicit denial of the possibility of reform would only encourage a more radical approach to social change. Even Wagner, who defended deduction in economics, was quick to point out the limitations of economic principles if they seemed to rule out desirable reforms, while Schmoller and Knapp, like Lexis, deemed the concept of natural law utterly inapplicable to economics.[55]

The situation was similar for statistics; much of the opposition to Quetelet was inspired by distaste for his supposed application of the concept of natural law to society. G. F. Knapp wrote in 1871 that in Germany, moral statistics, like politics, had followed the path of political economy and given up the twin errors of natural law and atomism, while the French, failing to heed this lesson of "true *Quetele-tismus*," had fallen into "the frightful catastrophe of the present" (the Paris commune).[56] The argument that society was so strictly governed by natural laws as to exclude conscious economic reform had been made regularly by laissez-faire propagandists, and occasionally by serious political economists. The historical school would have none of it. Significantly, both Knapp and Lujo Brentano began moving away from deductive economics when their professor at Göttingen, Helferich, assigned them the problem of proving a particular thesis of von Thünen. That author had developed the argument that economics have a fixed wage fund, and that collective bargaining must be ineffective in raising the economic condition of workers since it can do no more than shift wages from one class to another. Both came to the opposite conclusion, as did Schmoller, who argued that the fixed wage fund had been conclusively refuted by actual historical experience. Von Thünen's principle was frequently attacked as a dangerous and misleading "natural law" in economics.[57]

Lexis, too, viewed statistics as intimately bound to political economy. Less inclined than many of his colleagues to dismiss the use of deduction in economics, he nevertheless insisted that the premises of any social science must be based on a judgment about human motives. But these are ethical, purposeful, and progressive, and hence not subject to constant natural laws. Thus it is important to have available an empirical method, statistics, to resolve whether economic deductions have any applicability to a given social situation. It follows likewise that neither statistics nor economics can attain to laws of nature.[58]

Lexis's arguments on dispersion involved first of all a criticism of the myth of "astonishing" statistical regularity. This was the idea imputed to Quetelet in 1871 by G. F. Knapp and designated the "astronomical conception of society: ... that forces act on society that, as we recognize from the regularity of their effects, seem to be independent of those acting on individual events or behaviors, and that therefore must be conceived as external forces."[59] Lexis, casting his net somewhat wider, attributed this view to Oettingen as well as to Quetelet and Buckle: "From the standpoint of mechanism, like that of mysticism, it is presupposed in the explanation of the lawlikeness of statistics that beside the conscious motives of human individuals operate unconscious ones, that are able, through their power to dominate the whole, to pursue definite numerical relationships in the total phenomena, and to compel their realization."[60]

Lexis interpreted these twin claims of *Naturgesetzlichkeit* and mystical unity as involving a concrete prediction that statistical events manifest greater regularity than can be explained by a model in which all individuals act with complete independence of one another.[61] If there is really a compensatory tendency acting in statistics, then the dispersion of sets of numbers in statistical time series ought to be lower than that of a similar series of multiple urn drawings, in which the probability for each single event equals the average for the statistical system. Both these values can be measured: the "physical" or actual dispersion using standard sums of squares of deviations from the mean, and the expected or combinatorial dispersion from a formula derived by Poisson. Quetelet's "school" is thus interpreted as predicting that the ratio of physical to combinatorial dispersion, which Lexis defined as Q, would for many time series of moral statistics be less than one. In fact, as Lexis showed for a variety of phenomena ranging from mortality records to crime and suicide statistics, it is almost always far greater. There were very few statistical series—most prominently, the male-to-female birth ratio—that displayed even enough regularity to be interpreted as consistent with a combinatorial model.[62]

The lesson to be drawn, then, was not only that no supraindividual force acts on society to compel obedience to statistical law, but that even the Manchester model of atomistic individuals could not account for the phenomena of moral statistics. For the dispersion to be as great as that observed, there must be fluctuations in the underlying probabilities (say, the probability that a given person will marry this year) that involve whole groups of people. But this, he thought, was just what we should expect, for man is a purposive, historical being, and also a highly diverse one. Lexis broke up the totality of society into a series of groups, which he called *Chancensysteme*. Within groups, individuals are considered as homogeneous, but between groups they are quite different. The *übernormale* (greater than predicted) dispersion was then explained partly as progress of the whole society and partly in terms of independent fluctuations of the various *Chancensysteme*.

Thus the mathematical investigations of statistics by Lexis led again to the pluralist model of society suggested by Mohl and by most German statisticians of the 1860s and 1870s. Statistics was not at all interested in the mere observation of regularities, and only incidentally concerned with the movement of the whole society. Instead, its task was "to form the most individualized fundamental groups possible, and, using probability relations, to characterize the *Chancensysteme* that condition their significant alterations, then to find the extent to which the differences in the *Chancensysteme* arise from a difference in the distinguishing characteristics of these fundamental groups, and finally, to resolve whether the individual *Chancensysteme* remain approximately fixed or evolve in determinate ways over time."[63] The statistician does not expect all this analysis and intercomparison to yield timeless principles, but only a sound and reliable characterization of the *Chancensysteme* that prevail in a given society.[64]

The direction in which this kind of thought was leading was first suggested by G. F. Knapp, when he argued that Quetelet's "deep insight" into the social character of crime was much to be preferred to the style of his French contemporary, Guerry, who sought "social-historical constants" such as the relation of crime to education.[65] The point was developed more fully by the Austrian statisticians F. X. Neumann-Spallart and K. T. Inama-Sternegg. These authors began with the familiar criticism of Quetelet, who had neglected the individual by thinking only of the whole society, and by a call to consider not only the freedom of the individual, but also inhomogeneities within and between states—which, not surprisingly, was an Austrian specialty.[66] Like Knapp, the Austrians went on to express skepticism about the advantages to be gained by comparisons of different states. Society is not governed by natural laws, and the changes undergone by any given community are too intricate and subtle to be subordinated to exact or cross-cultural principles.[67] These authors were enormously enthusiastic about statistics, and argued untiringly that the inchoate science of sociology could make little progress without it. They insisted, however, that statistical laws—whether of regularity or of variation—could only be empirical ones, for that is all that is accessible to social science. The demonstrations it can furnish are only probable, never certain.

It was Lexis, however, who most clearly made the general point that this uncertainty is associated not only with the social object, but also with the statistical method. That method has relatively wide applicability, but its usefulness is by much the greatest where the scholar is confronted with real, or "concrete," variability, and not just error. That is, given Rümelin's great chain of diminishing typicality, it is especially applicable to society. The statistical method, ideally suited for the study of society, thus shares the limitations of any science of society. It is not the case that social phenomena are fundamentally indeterminate; "Every single event stands in a completely closed chain of causality, which for any given case can be demonstrated." But where variation is concrete, "the system of causes is so rich and varied that the conjuncture of the single events must appear to us as chance."[68] Thus it is no accident that to study such phenomena one must have recourse to a method that gives only a numerical description of the initial and final states. The characteristic feature of statistics, as Lexis portrayed it, was precisely this—that it knows nothing of causes. Hence the only thing necessary about statistics is the uncertainty of its inferences.[69]

Conclusion

The leading historical-school economists had a greater penchant for writing abstract theory, usually in the form of a sweeping, moralistic intellectual history, than for the hard work and focused thought that would have been necessary to carry out their program for the study of social diversity. Social statistical writing at the end of the nineteenth century was, on the whole, far more sophisticated than that at midcentury, but it is not clear that the reflections of Wagner, Rümelin, Knapp, and

Lexis were instrumental to this improvement. The mathematical results of this work were perhaps more influential. Lexis's index of dispersion provided the starting point for a continental tradition of mathematical statistics that lasted well into the twentieth century, and that included such figures as Ladislaus von Bortkiewicz, Harald Westergaard, A. A. Chuprov, and even A. A. Markov. That school was subsequently influenced heavily by the British biometric tradition, to which it also contributed, though less fundamentally.

As Lorenz Krüger's paper in this volume suggests, however, it is as philosophical discussion that these German statistical writings assume their greatest importance. Like the physicists' contemporaneous analysis of the relationship between time-reversible mechanical laws and the time-directional increase in entropy, the revaluation of statistics by authors who denied the existence of social laws and who were persuaded of the importance of human diversity implied the enhancement of the status of chance in the world. This change in the interpretation was perhaps more exemplary than causal. Renouvier and Delboeuf in France and William James and C. S. Peirce in the United States were moving in a similar direction during the late nineteenth century; these German writings were not of central importance for them. Still, the historical-school interpretation of statistics was well known even outside of Germany. Within the German tradition, as Norton Wise suggests, a notion of "statistical causality" derived from these social writers may have been drawn upon by the quantum theorists of the 1920s.

That a branch of social science should have effects so far removed from the domain of its own subject matter is wholly characteristic of the development of statistical reasoning. Quetelet had little lasting influence on social science but provided much of the foundation for Maxwell's work on the kinetic gas theory and Galton's biometrics, while Galton and the biometric school contributed at least as much to mathematical statistics as to the explicit object of their study, biological inheritance.[70] Statistical reasoning evolved in the context of specific problems from the various disciplines, and grew through application to new subject matters, analysis from new viewpoints, and recognition of analogies among widely disparate fields. Social science provided an application for probability in which the reality of variation seemed most crucial, and it makes more sense than might be immediately apparent that the name for a modern branch of applied mathematics, statistics, should have been derived from the nineteenth-century numerical science of society.

Notes*

1. See T. M. Porter, "A Statistical Survey of Gases: Maxwell's Social Physics," *Historical Studies in the Physical Sciences*, 8 (1981), 77–116.

*Abbreviations: *DVS: Deutsche Vierteljahrsschrift; HdWSW2: Handwörterbuch der Staatswissenschaften* (7 vols.; Jena: Gustav Fischer, 2nd ed., 1895–1901); *HdWSW4*: ibid. (8 vols.; 4th ed., 1923–1928); *Jbb: Jahrbücher für Nationalökonomie und Statistik*; *PJ: Preussische Jahrbücher; SM: Statistische Monatschrift; ZGSW: Zeitschrift für die gesammte Staatswissenschaft; ZKPSB: Zeitschrift des Königl. Preussischen Statistischen Bureaus; ZSS: Zeitschrift für schweizerische Statistik.*

2. Some notable histories of statistics from the nineteenth century are Adolph Wagner, "Statistik," in J. C. Bluntschli et al., eds., *Deutsches Staats-Wörterbuch* (11 vols.; Stuttgart and Leipzig: Expedition des Staats-Wörterbuchs, 1867), vol. 10, pp. 400–481; Vincenz John, *Geschichte der Statistik* (Stuttgart: F. Enke, 1884); August Meitzen, *History, Theory and Technique of Statistics*, trans. Roland P. Falkner (Philadelphia: American Academy of Political and Social Science, 1891). On the name *Statistik* see Vincenz John, "Name und Wesen der Statistik," *ZSS*, 19 (1883), 97–112, and August Oncken, *Untersuchung über den Begriff der Statistik* (Leipzig: Wilhelm Engelmann, 1870).

3. August Ludwig von Schlözer, *Theorie der Statistik nebst Ideen über das Studium der Politik überhaupt. Erstes Heft. Einleitung* (Göttingen: Vandenhoek und Ruprecht, 1804), p. 86.

4. See Schlözer, *Theorie*, p. 19. For criticisms of numerical statistics see A. F. Lueder, *Kritik der Statistik und Politik nebst einer Begründung der politischen Philosophie* (Göttingen: Vandenhoek und Ruprecht, 1812), who quotes extensively from diverse sources; also C. A. Freiherr von Malchus, *Statistik und Staatenkunde: Ein Beitrag zur Staatenkunde von Europa* (Stuttgart and Tübingen: J. G. Cotta, 1826), p. 19, and (Anon.), "Die Statistik der Kultur im Geiste und nach den Forderungen des neuesten Völkerlebens," *DVS*, 1838, vol. 4, 267–308. The strongest German-language advocate of numerical statistics was Christoph Bernoulli, descended from the famous Swiss scientific family and liberal professor of political economy at Basel, whose main book was titled not *Statistik* but *Handbuch der Populationistik, oder der Völker- und Menschenkunde nach statistischen Erhebnissen* (Ulm: Stettin'sche Verlags-Buchhandlung, 1841); see also his "Vorwort," *Schweizerisches Archiv für Statistik und Nationalökonomie*, 1 (1827).

5. Johannes Fallati, *Die statistischen Vereine der Engländer* (Tübingen: Ludwig Friedrich Fues, 1840), p. 41; *Einleitung in die Wissenschaft der Statistik* (Tübingen: H. Laupp, 1843), pp. 146, 176; "Gedanken über Mittel und Wege zu Hebung der praktischen Statistik, mit besonderer Rücksicht auf Deutschland," *ZGSW*, 3 (1847), 496–557, on p. 500.

6. J. v. W., "Die Errichtung statistischer Bureaus und statistischer Privatvereine," *DVS*, 1846, vol. 3, 95–128, on p. 95.

7. See F. W. Freiherr von Reden, "Vom Nutzen der Statistik für Staat und Volk," *Zeitschrift des Vereins für deutsche Statistik*, 1 (1847), 20–23, on pp. 21–22.

8. See Friedrich Zahn, "Statistik (Amtliche)," *HdWSW4*, vol. 7, p. 895. Another testimony to the number of numbers being published in the post-1848 period is Otto Hübner's journal, *Jahrbuch für Volkswirtschaft und Statistik* (vols. 1–8, 1852–1863).

9. "Neuere deutsche Leistungen auf dem Gebiete der Staatswissenschaften," *DVS*, 1854, vol. 3, 1–78, on p. 12.

10. Robert von Mohl, "Gesellschafts-Wissenschaften und Staats-Wissenschaften," *ZGSW*, 7 (1851), on p. 28. See also Erich Angermann, *Robert von Mohl, 1799–1875: Leben und Werk eines altliberalen Staatsgelehrten* (Neuwied: Hermann Luchterhand, 1962), and Eckart Pankoke, *Sociale Bewegung—Sociale Frage—Sociale Politik, Grundfragen der deutschen "Socialwissenschaft" im 19. Jahrhundert* (Stuttgart: Ernst Klett, 1970).

11. Carl Gustav Adolph Knies, *Die Statistik als selbständige Wissenschaft: zur Lösung des Wirrsals in der Theorie und Praxis dieser Wissenschaft* (Kassel: J. Luckhardt 1850), p. 23.

12. See Robert von Mohl. *Die Geschichte und Literatur der Staatswissenschaften* (3 vols.; Erlangen: F. Enke, 1856–1858), vol. 3, pp. 637–674; Lorenz von Stein, *System der Staatswissenschaft, Erster Band: System der Statistik, der Populationistik und der Volkswirtschaftslehre* (1852; Osnabrück: Otto Zeller, 1964 reprint); also Stein's disciple Eberhard Jonak, *Theorie der Statistik in Grundzügen* (Vienna: Braumüller, 1856). That the state acts back on society was used as justification for combining the two approaches by Leopold Neumann, "Aphoristische Betrachtungen über Statistik, ihre Behandlung und ihre neueren Leistungen," *Oesterreichische Vierteljahresschrift für Rechts- und Staatswissenschaft*, 3 (1859), 87–114.

13. Gustav Rümelin, "Zur Theorie der Statistik, I" (1863), in *Reden und Aufsätze* (Freiburg: J. C. B. Mohr, 1875), pp. 208–264, on p. 225.

14. Roscher taught statistics at Leipzig from 1851 to 1869. See his review in *Göttingische gelehrte Anzeigen*, 1840, vol. 3, 1745–1760. Bruno Hildebrand organized statistical bureaus in Bern and then Jena, and founded the *Jahrbücher für Nationalökonomie und Statistik*; see his "Die wissenschaftliche Aufgabe der Statistik," *Jbb*, 6 (1886), 1–11, and Otto-Ernst Krawehl, *Die "Jahrbücher für Nationalökonomie und Statistik" unter den Herausgebern Bruno Hildebrand und Johannes Conrad (1863–1915)* (Munich: Verlag Dokumentation, 1977). Knies, *Die Statistik* (note 11) has already been mentioned.

15. This summary draws mainly from Bruno Hildebrand, *Die Nationalökonomie der Gegenwart und Zukunft* (1848), Hans Gehrig, ed. (Jena: Gustav Fischer, 1922); idem., "Vorwort" and "Die gegenwärtige Aufgabe der Wissenschaft der Nationalökonomie," *Jbb*, 1 (1863), 1–25, 137–146; Carl Knies, *Die politische Oekonomie vom geschichtlichen Standpunkte* (Braunschweig: C. A. Schwetschke, 1883; 1st ed., 1852). See also Gottfried Eisermann, *Die Grundlagen des Historismus in der Deutschen Nationalökonomie* (Stuttgart: Enke, 1956).

16. See James Sheehan, *The Career of Lujo Brentano: A Study of Liberalism and Social Reform in Imperial Germany* (Chicago: University of Chicago Press, 1966); Ulla G. Schäfer, *Historische Nationalökonomie und Sozialstatistik als Gesellschaftswissenschaften* (Cologne/Vienna: Böhlau, 1971); also Gustav Schmoller, "Die Arbeiterfrage," *PJ*, 14 (1864): 393–424, 523–547, 15 (1865); 32–63; W. Lexis, *Gewerkvereine und Unternehmerverbände in Frankreich: Ein Beitrag zur Kenntniss der socialen Bewegung* (Leipzig: Duncker und Humblot, 1879), pp. 6–10; Adolph Wagner, *Rede über die sociale Frage* (Berlin: Wiegandt & Griebner, 1872). Cooperatives on the Schulze-Delitzsch model were by much the favored alternative until the late 1870s.

17. Most of these are noted by Wilhelm Lexis, "Statistik (Allgemeine)," *HdWSW2*, vol. 8, pp. 1006–1014, p. 1011. The biographical information given here is taken largely from name entries in *HdWSW2*, *HdWSW4*, Allgemeine Deutsche Biographie (Berlin: Duncker & Humblot, 1953 et seq.); *Biographisches Lexikon des Kaiserthums Oesterreich* (60 vols.; Vienna: 1856–1891); and *Nekrologe* in *ZKPSB* and *SM*.

18. See, for example, Naum Reichesberg, *Die Statistik und die Gesellschaftswissenschaft* (Stuttgart: Ferdinand Enke, 1893); also his "Adolf Quetelet als Moralstatistiker," *ZSS*, 29 (1893), 490–498, and "Der berühmte Statistiker Adolf Quetelet, sein Leben und sein Wirken: Eine biographische Skizze," *ZSS*, 32 (1896), 418–460. The German mandarin stereotype of history is mentioned by Fritz K. Ringer, *The Decline of the German Mandarins: The German Academic Community, 1896–1933* (Cambridge: Harvard University Press, 1969), pp. 84–90.

19. In the same vein, it was explained by Adolf Held that Quetelet was incapable of seeing a role for great men or even for individual freedom because he lived in an age of leveling and of mass society. See Adolf Held, "Adam Smith und Quetelet," *Jbb*, 9 (1867), 249–279, on pp. 270–273.

20. Jacob Wilhelm Heinrich Lehmann, "Bemerkungen bei Gelegenheit der Abhandlung von Quetelet: Ueber den Menschen und die Gesetze seiner Entwicklung in diesem Jahrbuch, Jahrgang 1839," H. C. Schumacher, ed., *Jahrbuch für 1841* (Stuttgart and Tübingen: J. G. Cotta, 1841), pp. 137–219, and ibid., 1843, pp. 146–230; Moritz Wilhelm Drobisch, "Moralische Statistik," [Gersdorf's] *Leipziger Repertorium der deutschen und ausländischen Literatur*, 1849, vol. 2, pp. 28–39. See also H. Schwabe, "Ueber den Begriff der Statistik und ihr Verhalten zur politischen Arithmetik," *ZGSW*, 15 (1859), 123–142.

21. This is noted by Anthony Oberschall, *Empirical Social Research in Germany, 1848–1914* (Paris: Mouton & Co., 1965), p. 45. who cites John, *Geschichte* (note 2); see also Eduard Rehnisch, "Zur Orientierung über die Untersuchungen und Ergebnisse der Moralstatistik," *Zeitschrift für Philosophie und philosophische Kritik*, 68 (1876), 213–264; 69 (1876), 43–115, on pp. 250ff.

22. So says, for example, G. F. Knapp, *Aus der Jugend eines deutschen Gelehrten* (Stuttgart: Deutsche Verlag., 1927), p. 155. On the flowering of liberalism see James J. Sheehan, *German Liberalism in the Nineteenth Century* (Chicago: University of Chicago Press, 1978.)

23. Arnold Ruge, "Über Heinrich Thomas Buckle, und zur zweiten Auflage" (1864), in H. T. Buckle, *Geschichte der Civilisation in England* (2 vols.; Leipzig: C. F. Winter, 7th ed., 1901); Marx to Engels, 18 June 1862 in Karl Marx and Friedrich Engels, *Briefwechsel, Marx-Engels Gesamtausgabe*, vol. 3, Part III (Glashütten im Taunus: Defler Auvermann K. G., 1970), p. 78. Noteworthy Buckle criticism includes Joh. Droysen, "Die Erhebung der Geschichte zum Rang einer Wissenschaft," *Historische Zeitschrift*, 9 (1869), 1–22; R. Dieterich, "Buckle und Hegel. Ein Beitrag zur Characteristik englischer und deutscher Geschichtsphilosophie," *PJ*, 32 (1873), 257–302, 463–481; [Franz Vorländer], "Englische Geschichtsphilosophie," *PJ*, 9 (1862), 501–527.

24. Adolph Wagner, *Die Gesetzmässigkeit in den scheinbar willkührlichen menchlichen Handlungen vom Standpunkte der Statistik* (Hamburg: Boyes & Geister, 1864), pp. 44–45; J. C. Fischer, *Die Freiheit des menschlichen Willens und die Einheit der Naturgesetze* (Leipzig: Otto Wigand, 2nd ed., 1871).

25. An astonishingly copious but incomplete set of references to literature on this issue may be found in the footnotes to the introductory section of Alexander von Oettingen, *Die Moralstatistik in ihrer Bedeutung für eine Socialethik* (Erlangen: A. Deichert, 3rd ed., 1882). Among the more comprehensive and coherent are Franz Vorländer, "Die moralische Statistik und die sittliche Freiheit" *ZGSW*, 22 (1866) 477–511; L. N. [Leopold Neumann], "Zur Moralstatistik," *PJ*, 27 (1871), 223–247.

27. Adolphe Quetelet, "De l'influence de libre arbitre de l'homme sur les faits sociaux," *Bulletin de la Commission Centrale de Statistique* (de Belgique), 2 (1844), 133–155, on p. 142.

27. See T. M. Porter, "The Mathematics of Society: Variation and Error in Quetelet's Statistics," *British Journal for the History of Science*, 18 (1985), 51–69.

28. J. E. Wappäus, *Allgemeine Bevölkerungsstatistik: Vorlesungen* (2 vols.; Leipzig: J. C. Hinrichs, 1859), vol. 1, p. 17; Jakob Friedrich Fries, *Versuch einer Kritik der Principien der Wahrscheinlichkeitsrechnung* (1842), in *Sämtliche Schriften*, vol. 14 (Aalen: Scientia Verlag, 1974), p. 23. See also Franz Vorländer, "Die moralische Statistik und die sittliche Freiheit," *ZGSW*, 22 (1866), 477–511, on p. 483, and Wilhelm Windelband, *Die Lehren vom Zufall* (Berlin: A. W. Schade, 1870), p. 33.

29. Hermann Siebeck, "Das Verhältniss des Einzelwillens zur Gesammtheit im Lichte der Moralstatistik," *Jbb*, 33 (1879), 347–370.

30. Wagner, *Gesetzmässigkeit* (note 24), p. 54n22. See also his article "Statistik" (note 2), p. 460. Gustav Schmoller argued the same way: "Die neueren Ansichten über Bevölkerungs- und Moralstatistik" (1869), *Zur Litteraturgeschichte der Staats- und Sozialwissenschaften* (Leipzig: Duncker & Humblot, 1888), pp. 172–203, on p. 194.

31. Review of Wagner in *Jbb*, 4 (1865), 286–301.

32. Ernst Engel, "Mein Standpunkt der Frage gegenüber, ob die Statistik eine selbständige Wissenschaft oder nur eine Methode sei" (1851), reprinted in "Das statistische Seminar und das Studium der Statistik überhaupt," *ZKPSB*, 16 (1871), 188–194, on p. 189; Fallati, *Einleitung* (note 5), p. 54; J. L. Casper, *Über die wahrscheinliche Lebensdauer des Menschen* (Berlin: Dümmler, 1843), pp. 29–30; also Leopold Neumann, "Ueber Theorie der Statistik," *Oesterreichische Vierteljahresschrift für Rechts- und Staatswissenschaft*, 16 (1865), 40–62, p. 47.

33. See Wagner, *Gesetzmässigkeit* (note 24), especially pp. 63–80: "Ueber den Sinn und Begriff der Ausdrücke Gesetzmässigkeit und Gesetz in der Statistik;" also his review of Guerry in *ZGSW*, 21 (1865), where he argues that the variation of suicide by season is a true *Naturgesetz*. On free will and covariation see *ZGSW*, 36 (1880), 189–203, on p. 192; also "Statistik"

(note 2), p. 437, where the discovery of causes in this manner is held up as Quetelet's greatest achievement.

34. See, for example, Hildebrand, "wissenschaftliche Aufgabe" (note 14); Adolf Held's review of Quetelet in *Jbb*, 14 (1870), 81–95; Neumann, "Zur Moralstatistik" (note 25).

35. Gustav Rümelin, "Ueber den Begriff eines socialen Gesetzes," *Reden* (note 13), pp. 1–31; also "Moralstatistik und Willensfreiheit," ibid., pp. 370–377.

36. See Helferich's review of Wagner's *Gesetzmässigkeit* in *Göttingische gelehrte Anzeigen*, 1865, vol. 1, 486–506, on p. 500; Schmoller, "Ansichten" (note 30), p. 187.

37. Hermann Lotze, *Mikrokosmus: Ideen zur Naturgeschichte und Geschichte der Menschheit: Versuch einer Anthropologie* (3 vols.; Leipzig: Hirzel, 1856–58–64), vol. 3, pp. 77–80; also Lotze, *Grundzüge der praktischen Philosophie: Diktate aus den Vorlesungen* (Leipzig: Hirzel, 1884), pp. 24–28. A similar argument was given by Friedrich Albert Lange, *Geschichte des Materialismus und Kritik seiner Bedeutung in der Gegenwart* (2 vols.; Iserlohn: Baedeker, 1866) vol. 2, pp. 479–480. It was challenged on the grounds that deviations of moral worth would average out by Johannes Wahn, "Kritik der Lehre Lotzes von der menschlichen Willensfreiheit," *Zeitschrift für Philosophie und philosophische Kritik*, 94 (1888), 88–141, on pp. 111ff.

38. Paul Lippert, "Quetelet," *HdWSW2*, vol. 6, pp. 292–296, on p. 294.

39. Moritz Wilhelm Drobisch, "Moralische Statistik" (note 20). Drobisch was correct to point out that Quetelet intended his *penchant* as a purely mathematical concept. The only statistical writer I have seen who spoke of a positive drive to crime is L. Fuld, "Der Einfluss der Ehe auf die Kriminalfrequenz," *Vierteljahrsschrift für Volkswirtschaft und Kulturgeschichte*, 57 (1885), 41–45, who was writing of the urge to rape, and the possibility of curing it by marriage.

40. Drobisch, *Die moralische Statistik und die menschliche Willensfreiheit: Eine Untersuchung* (Leipzig: L. Voss, 1867), p. 18.

41. Alexander von Oettingen, *Die Moralstatistik und die Christliche Sittenlehre: Versuch einer Sozialethik auf empirischer Grundlage*, vol. 2: *Die Christliche Sittenlehre: Deductive Entwicklung der Gesetze Christlichen Heilslebens im Organismus der Menschheit* (Erlangen: Andreas Deichert, 1873), pp. 31–33, 702–706; Oettingen, *Die Moralstatistik* (note 25), p. viii, 1st ed. 1865. On Oettingen see Monika Böhme, *Die Moralstatistik: Ein Beitrag zur Geschichte der Quantifizierung der Soziologie, dargestellt an den Werken Adolphe Quetelets und Alexander von Oettingens* (Cologne and Vienna: Böhlau Verlag, 1971). Oettingen was in turn criticized by the criminologist Wilhelm Emil Wahlberg, author of *Das Princip der Individualisierung in der Strafrechtspflege* (Vienna: 1869), for weakening the concept of individual responsibility (see review in *ZGSW*, 26 (1870), 567–576).

42. G. F. Knapp, "A. Quetelet als Theoretiker," *Jbb*, 18 (1872), 89–124, on pp. 97–98.

43. Ibid., p. 105; Knapp, "Die neueren Ansichten über Moralstatistik," *Jbb*, 16 (1871), 237–250, on p. 243.

44. See Rümelin, "Zahl und Arten der Haushaltungen in Württemberg nach dem Stand der Zählung vom 3. Dec. 1864," *Württembergische Jahrbücher für Statistik und Landeskunde* (Jg. 1865), 162–217, on pp. 173, 185; idem, "Ergebnisse der Zählung der ortsanwesenden Bevölkerung nach dem Stande vom 3. December 1867," ibid. (Jg. 1867), 174–226, on p. 192; Held, "Smith und Quetelet" (note 19), p. 274; Fr. J. Neumann, "Unsere Kenntniss von den socialen Zuständen um uns," *Jbb*, 18 (1872), 279–341, on p. 288.

45. Engel, "L. A. J. Quetelet: Eine Gedächtnisrede," *ZKPSB*, 16 (1876), 207–220, on p. 217.

46. Rümelin, "Zur Theorie, I" (note 13), p. 218.

47. Rümelin, "Moralstatistik" (note 35), p. 375.

48. Rümelin, "Statistik," in Gustav Schönberg, ed., *Handbuch der politischen Oekonomie, Finanzwissenschaft und Verwaltungslehre* (3 vols.; Tübingen: H. Laupp, 3rd ed., 1891), vol. 3,

pp. 803-822, on p. 814; idem, "Zur Theorie der Statistik, II" (1874), *Reden* (note 9) pp. 265-284, on p. 270; also "Ueber den Begriff" (note 35), p. 17.

49. Etienne Laspeyres, "Die Kathedersocialisten und die statistischen Congresse," *Deutsche Zeit- und Streitfragen*, 4(51) (1875), 137-184, on p. 164. The philosophical text in which the statistical method was most fully discussed is Christoph Sigwart, *Logik* (2 vols.; Freiburg: J. C. B. Mohr, 2nd ed., 1893), vol.2.

50. Georg Mayr, *Die Gesetzmässigkeit im Gesellschaftsleben: Statistische Studien* (Munich: R. Oldenbourg, 1877), 1st section; Wagner, "Statistik" (note 2), pp. 467-472.

51. Rümelin, "Ueber Gesetze der Geschichte" (1878), *Reden und Aufsätze, Neue Folge* (Freiburg and Tübingen: J. C. B. Mohr, 1881), pp. 118-148.

52. Oettingen, *Moralstatistik* (note 41), p. 37. By 1880, however, Oettingen was working the issue from both sides. "This regularity is certainly not so undeviating that we can infer from it the existence of a necessary law of nature" *Ueber akuten und chronischen Selbstmord: Ein Zeitbild* (Dorpat: E. J. Karow, 1881), p. 12.

53. The only German-language author I have seen who pointed out before Lexis that the degree of regularity had been exaggerated is Leopold Neumann, "Zur Moralstatistik" (note 25), p. 27 (1871).

54. Rehnisch, "Orientierung" (note 21), pp. 70-71.

55. Wagner, *Rede* (note 16), p. 4; idem, *Allgemeine oder theoretische Volkswirtschaftslehre, mit Benutzung von Rau's Grundsätzen der Volkswirtschaftslehre* (Leipzig and Heidelberg: C. F. Winter, 1876), pp. 183ff.; Schmoller, "Arbeiterfrage" (note 16), who moves from discussion of the "iron law of wages" to the general issue of *Naturgesetze* in society (pp. 417ff.); idem, "Volkswirtschaft, Volkswirtschaftslehre und -methode," *HdWSW2*, vol. 7, pp. 543-580, pp. 574-575; Brentano, *Mein Leben im Kampf um die soziale Entwicklung Deutschlands* (Jena: Eugen Diederichs, 1931), pp. 73-75.

56. Knapp, "Die neueren Ansichten" (note 43), pp. 248-250.

57. See Sheehan, *Brentano* (note 16), p. 21; Schmoller, "Arbeiterfrage," (note 16), p. 413; Brentano, *Mein Leben* (note 55), p. 74.

58. See Wilhelm Lexis, "Naturwissenschaft und Sozialwissenschaft" (1874), *Abhandlungen zur Theorie der Bevölkerungs- und Moralstatistik* (Jena: Gustav Fischer, 1903), pp. 233-251, on pp. 243-248; idem, *Allgemeine Volkswirtschaftslehre* (Leipzig: B. G. Teubner, 2nd ed., 1926), pp. 15ff., 237.

59. Knapp, "Bericht über die Schriften Quetelets zur Socialstatistik und Anthropologie," *Jbb*, 17 (1871), 167-174, 342-358, 427-445, on pp. 438-439. It seems unlikely that Quetelet would have endorsed such a statement. Buckle might, and Oettingen did (see note 52, and the quote to which it refers).

60. Lexis, *Zur Theorie der Massenerscheinungen in der menschlichen Gesellschaft* (Freiburg: F. Wagner, 1877), p. 11.

61. Lexis, "Über die Wahrscheinlichkeitsrechnung und deren Anwendung auf die Statistik," *Jbb*, N.F. 13 (1886), 433-450, on p. 445.

62. See Lexis, *Einleitung in die Theorie der Bevölkerungsstatistik* (Strassburg: K. J. Trübner, 1875), pp. 93-124; idem, "Zur mathematischen Statistik," *Jbb*, 25 (1875), 158-163; idem, "Das Geschlechtsverhältnis der Geborenen und die Wahrscheinlichkeitsrechnung," (1876), *Abhandlungen* (note 58), 131-169; idem, "Ueber die Theorie der Stabilität statistischer Reihen" (1879), ibid., 170-212; idem, "Gesetz," *HdWSW2*, vol. 4, pp. 234-240; idem, "Moralstatistik," *HdWSW2*, vol. 5, pp. 865-871.

63. Lexis *Einleitung* (note 62), p. 121; also "Ueber die Ursachen der geringen Veränderlichkeit statistischer Verältniszahlen," *Abhandlungen* (note 58), pp. 85-100, on p. 92.

64. Lexis, *Einleitung* (note 62), p. 124.

65. Knapp, "Quetelet" (note 42), pp. 98–100.

66. See K. T. Inama-Sternegg, "Zur Kritik der Moralstatistik," *Jbb*, N.F.7 (1883), 505–525, on p. 512; F. X. v. Neumann-Spallart, "Sociologie und Statistik," *SM*, 4 (1878), 1–18, 57–72, on p. 16; also Ficker's *Nekrolog* for Quetelet, *SM*, 1 (1875), 6–14, on p. 10; Gustav Adolf Schimmer, "Die Statistik in ihren Beziehungen zur Anthropologie und Ethnographie," *SM*, 10 (1884), 262–267.

67. Inama-Sternegg, "Kritik" (note 66), p. 522; idem, "Vom Wesen und den Wegen der Socialwissenschaft," *SM*, 7 (1881), 481–488; idem, "Neue Beiträge zur allgemeinen Methodenlehre der Statistik" *SM*, 16 (1890), 101–110; Neumann-Spallart, "Sociologie" (note 66), p. 5.

68. Lexis, *Massenerscheinungen* (note 60), pp. 4–5.

69. Lexis, "Naturwissenschaft" (note 58), p. 242.

70. See my book, *The Rise of Statistical Thinking, 1820–1900* (Princeton: Princeton University Press, 1986).

18 Prussian Numbers 1860–1882

Ian Hacking

The chief topic of this chapter is the contrast between Western liberal and atomistic conceptions of probability and those of a German holistic and conservative tradition. They differ fundamentally in their notion of, for example, a statistical law. This is illustrated by a single but central case, that of E. Engel, who revitalized the Royal Prussian Statistical Bureau in 1860, and who directed it for 22 years. The institutional history is of more than local interest, for the organization of the Bureau became a model for most other evolving central statistical offices elsewhere. The chief concerns of Engel and his Bureau are described, with a view to grasping their underlying conception of the nature and role of statistics.

Ever since property was liberated from its feudal chains—ever since the money economy replaced the natural one—ever since machines took over the work of hands—ever since large companies came to annihilate small firms and now threaten so many more—throughout this precisely signposted stage of our cultural development we witness the inevitable consequences: the dissolution of many of the bonds that held us together, their veritable atomization. Many of the components of former social institutions floated and still float independently, no longer connected. This is the social disease of our times....

These three sentences penned in 1860 were no uncommon cry then, nor is their theme unfamiliar today. But these are not the words of a William Morris pining for an earlier world of craft guilds. They are those of Dr. Ernst Engel (1821–1896), newly appointed director of the Royal Prussian Statistical Bureau. He published them in one of the official periodicals he founded in 1860, the year he got the job. His was not an essay on moral decay, but one on savings banks.[1] Although it was a statistical essay in a government organ, its very title betrays a steady point of view. It says that the *Sparcassen* are "members in a chain" of institutions founded upon "the Principle of Self-Help."

Engel ruled the Statistical Bureau for 22 years. He possessed the numbers of the Prussian state, possessed them with a passion. The Bureau had been running continuously for half a century, but he magnified it, and created a battery of institutions—administrative, educational, communicational—that determined the history of the Bureau until it was replaced in 1934 by a new *Reichsamt* for German statistics.[2]

To study Engel's statistical work is not to study, in any direct way, the probabilistic revolution. There was no such revolution in the Bureau. There was precious little probability. The present essay supplements Norton Wise's important study of "statistical causality" in the work of Wundt and Lamprecht.[3] Wise demonstrates the essentially "conservative" and "holistic" character of much German statistical thinking, apparently typical of much work in Central Europe, and fundamentally different from the Western liberal traditions that were long dominant in France and England. I provide a parallel story of the most influential and innovative bureaucracy representing this Central European vision.

Necessarily I come to these matters with a philosopher's eye. How could a philosopher take up bureaucratic history? He must do so if he believes in conceptual analysis and believes that concepts are nothing other than the uses we make of them. Conceptual relations are formed in the material conditions of their production, and the forces that bring into being our network of usages become the momentum that maintains that structure in place, a form of thought rather than something one thinks about. If Engel's numbers are almost free of probabilistic connotations, we are not then to ignore him, turning to the probability-mongers of an earlier generation, such as Quetelet, or of a later one, such as Pearson. The deprobabilification of statistics in Central Europe is as much part of our conceptual formation as its probabilification in the West.

1   Prussian Numbers before Engel

I have elsewhere sketched the early history of Prussian statistics.[4] Around 1800 there was a remarkable transformation both in the conception of counting and in the practices of the bureaucracy. Before that time there was an epoch of secret bureaucrats and public amateurs. That is, the various ministries of the government collected vast amounts of numerical data for purposes of trade, taxation, and recruitment into the army. But although this information was intended to measure the power of the kingdom, almost all of it was highly confidential. On the other hand, there were a large number of enthusiasts who measured and counted and put it all down in local, national, and international dailies, weeklies, monthlies, and yearbooks. No travel book was complete without its little summaries of statistical information about the towns being traversed. The Prussian state neither interfered with this public curiosity, nor did it support it. The ministries and the king refused, when asked, to tell anything to the amateurs.

The advent of Napoleon changed this, like much else. A statistical bureau for the Prussian state was established in 1805. It was shortlived: the battle of Jena ended all such activity—for a moment. But the restoration of the bureau was one of the first acts of national revival in 1810. The bureau became a public institution, with a small staff, that published an incredible number of papers both in established journals and in its own pamphlets and occasional *Mittheilungen*. The director held two posts, first as head of the bureau, and second as professor of political science at the University of Berlin. In fact in the entire fifty-year period from the founding of the bureau to the appointment of Engel in 1860, there were exactly two directors. One was J. G. Hoffmann (1765–1847), who relinquished formal control only on his 80th birthday. The other was his protegé C. W. F. Dieterici (1792–1859), who had in fact assumed effective management of the bureau when his patron was 70, but was not formally appointed until Hoffmann's retirement.

The two directors had prodigious energies and published much, but their activities strike one as a trifle haphazard. Their regime of half a century was an important transitional phase in Prussian official statistics. They were head of an entirely new form of bureaucracy. What *is* a statistical bureau? That no one quite knew is indicated by the way in which it wandered through the administration. No one knew to whom it should be responsible. It started out with the police, but then was transferred to the chancellor's office. Then it went to the ministry of the interior, then to commerce, and then back to interior. There it finally stayed. When Engel assumed control in 1860 the position was stable, and a new era began. In fact one of Engel's first acts was to create a metastatistical office, which included his own office under its wing, but also supervised the statistical work of all the other ministries, coordinating their reports and publications, and overseeing their research programs. Ernst Engel heralds the beginning of modern official statistics.[5]

2 Engel's Career

Engel was not a Prussian but a Saxon, born in a village near Dresden in 1822. He trained as a mining engineer at the school of mines in Freiberg, Saxony. He spent his *Wanderjahr*, 1847, visiting the iron and steel works of France, Britain, and Belgium. Yes, he called upon Quetelet.

On returning home he first wrote a study of Saxon glass factories. But the uprisings of 1848 were upon him. Dresden was briefly held by anarchists. As soon as the old order had been restored, Engel was asked to work on a commission intended to improve relationships between Labor and Capital. He was soon to be its president. By 1850 he was organizing the Pan-German Industrial Exhibition in Leipzig, the world's first trade fair. In 1854 he was chosen to run the Saxon statistical bureau. He founded his first journals, *Zeitschrift des statistischen Bureaus des Königreichs Sachsen* and the *Jahrbuch für Statistik und Staatswirtschaft des Königreichs Sachsen*.

He stayed at the Saxon Bureau for only four years, for during his tenure there he invented a new kind of insurance: mortgage insurance. Hitherto the threat of foreclosure—always brutally enforced—had discouraged prudent small people from using mortgages. Hence property development was the preserve of exploitative monopolies. After pamphleteering for this idea from 1856, in 1858 he founded the *Dresdner Realcreditversicherungsgesellschaft*. This was, like so many of Engel's other innovations, immediately copied all over Germany. By 1860 his reputation ran far beyond Saxony, and he was called to the Prussian Statistical Bureau.

He set to work with a will. He was appointed 1 April. On 24 June he took to his minister an elaborate plan for a new Central Statistical Commission, whose members would be the Director of the Bureau, principal officers from all the ministries, and other notable ad hoc appointments. It would coordinate all statistical work of the kingdom, and determine new tasks. Thus, for example, the census was conducted by the Bureau, but its aims were set by the Commission.[6] Such a Commission had been recommended by the 1855 International Statistical Congress and was to some extent modeled on earlier Belgian practice, but Engel's tiers of bureaucracy were unparalleled.

There was some need for coordination of data. To see the scale of Prussian enterprises, one should consult a list of periodical statistical publications emanating

from Prussian government agencies (including the railways). I emphasize "periodical"—not including special reports; "publication"—not in-house documents;
and "Prussian government"—not regional or city administrations. The list of titles
takes 21 pages, and includes some 410 organs.[7]

Engel did his bit. As in Saxony, he founded a *Zeitschrift* and a *Jahrbuch*. The
Jahrbuch contained complete tables on various topics of population or trade. The
Zeitschrift, which began as a monthly in October 1860, superceded the occasional
Mittheilungen of his predecessors. Although an official publication, the *Zeitschrift*
resembles many modern scholarly journals. In the beginning Engel was writing
almost half of the signed articles. Hardly a month goes by without a major essay of
his on some topic. He also commenced the *preussische Statistik*, a public relations
arm feeding statistics to the press. Engel wanted this structure of collection and
publication to be imitated at every inferior level. Every city, each of the 25 regional
administrations (*Regierungsbezirke*), and ideally every local government (*Kreis*)
should have its own statistical office and its own vehicle of publication. Not even
Prussia could keep up with Engel's drive, but, for example, city statistical Bureaus
did spring up—Berlin, 1862; Frankfurt am Main, 1865; Hamburg, 1866; Leipzig,
1867; Lübeck, Breslau, and Chemnitz, 1871; Dresden, 1874; and so on—with 27
city offices by 1900. This is of course not a peculiarly Prussian development. For
comparison, Vienna and Rome also established offices in 1862, New York City and
Riga did so in 1866, Stockholm in 1868, Buda in 1869, and Paris in 1879.

Formally Engel's appointment was similar to that of Dieterici, his predecessor.
But Hoffman and Dieterici had been professor-directors, holding a chair of *Staatswissenschaft* in Berlin alongside the directorship. In 1860 the jobs were split, and a
man was called from Göttingen to fill the Berlin chair. Engel was not fazed by lack
of an academic lectern. One of his first acts was to establish a statistical seminar.
Engel made the proposal soon after his appointment and a syllabus was announced
in the *Zeitschrift* in 1862. It was intended to provide theoretical and practical
training in statistics both for state officials and for academics. The first meeting was
30 October 1862. Graduates of the seminar included many of the notable economists of the next generation. Vienna soon copied the idea. So did Bruno Hildebrand
in Jena, whose slightly more modest seminar opened 10 October 1865. Hildebrand
was then directing statistics for Thuringia, lecturing on the subject at the university
(so his seminar was partly preparatory for his university course), and editing his
famous *Jahrbücher für Nationalökonomie und Statistik*. There he describes Engel's
seminar as "a little statistical university"—so much for taking the chair in Berlin
away from the director of the Bureau![8]

We cannot chronicle all of Engel's industry. He could not, of course, quite
maintain the energy of his year of appointment—when in addition to creating the
institutions already mentioned, he took the time to go to London for the International Statistical Congress—and ensure that the next meeting would be in Berlin,
1863. As a passing example of his work, I may take his diligent collection of the
statistics of steam boilers and steam engines.[9] Such a piece is not as dull as one
would imagine. It begins with a "free" translation from Erasmus Darwin's poem
The Botanic Garden of 1788, and has eloquent dedications to Watt and Stephenson.

The point of the study is practical. Steam boilers explode, and different kinds explode with different frequencies and deadliness. On average an explosion kills one person and maims two more. Every year in Prussia one in every 480 cylindrical boilers blows up, although more advanced machines have better statistics; the best, the *Zweiflammrohrkessel*, has only one explosion per 3,164 boilers. Engel's statistics are international, historical, and worldwide, covering not only Europe but also America and Asia. The aim is simple; to determine the probability of an explosion (*Explosions-Wahrscheinlichkeit*), which will be used to determine insurance rates.

His 320 pages of statistics and graphs of the war with France are magnificent, if you like that sort of thing.[10] He arrived at Strasbourg immediately after it had surrendered to the long siege, and writes in his *Zeitschrift* of events. But he is cautious in dealing with the queen's personal request to him, that he will report on the support furnished by German-speaking inhabitants for the conquering troops. Earlier, after the war with Austria in 1866, he had won no popularity in high places for his studies of permanently maimed veterans, for the very large number of victims were not acknowledged even to exist.[11]

Engel did not shy away from politics. He was elected as a National-Liberal to the *Abgeordnetenhaus* 1867–1870. He was one of the founding members of the *Verein für Sozialpolitik*, whose members became known as the *Kathedersozialisten*, of whom I shall speak later. His resignation as director of the Bureau was largely caused by his pseudonymous attacks on Bismarck's protectionist grain policy. However, his real contributions to changing the state came not from overt politics but from creating new bureaucratic machineries. The work of men such as Engel made possible the "social net" of the German state, the first in the world. A political conjuncture created the acts of 1883, 1884, and 1891, providing workers with sickness insurance, workers compensation for accidents, and old-age pensions, but it was the statistical-economic proselytizing of men like Engel that was used as the groundwork for such legislation.

3 Engel's Conception of Statistics

The eighteenth-century collecting of Prussian numbers had, of course, all sorts of standard practical aims. They were relevant to the taxation base, to the opportunities for military recruitment, and to directly intelligible difficulties arising from depopulation during the Seven Years War. But these were concerns of particular ministries. The driving motive behind centralization of statistics was still this: statistics measures the might of the kingdom. With the actual creation of a central statistical office, however, the role of the statistician-in-general changes. He becomes the person who will provide information necessary for the administrators and law-givers in general. "Statistics must be obtained and presented in all particulars, so that legislators and administrators can at all times make the greatest possible use of them, both for Science and for Life."[12] Engel says, in the strongest terms, that statecraft, namely, the practical application of political science, is a mere sham without a statistical foundation: "Staatskunst, d.h. die praktische Anwendung der Staatswissenschaft ohne statistische Grundlagen, [ist] nur Staatskunstelei."[13] This passage comes in a reflective paper printed in the second year of his *Zeitschrift*. The topic is the place of the census in science, and its responsibilities to history. Here we have ample foreshadowing of Lamprecht's position, described by Wise. There is a most explicit *denial* of statistical causality. To see what was at issue, let us briefly recall some debates about statistical laws.

Many eighteenth-century thinkers, among whom the French physiocrats were most notable, firmly believed that there would arise a social science whose laws would be as firm as the laws of Newtonian mechanics. The publication of French official statistics in the 1820s and 1830s seemed the answer to this fantasy. Quetelet and others noticed that statistical proportions remain stable from year to year. The phenomenon was most noticeable where it might have been least expected, namely, in the moral statistics of deviancy: of crime, suicide, prostitution, and the like. Moreover, Quetelet urged that many social and biological phenomena are distributed exactly like the law of error. So there arose the idea that the laws of society might be statistical or even probabilistic in nature. This development seems systematically characteristic of much liberal, utilitarian, and philanthropic thinking in the West, especially in France and Britain. The conception is founded upon the liberal view of society as composed of independent atoms, which may, however, be subject to lawlike behavior, just as material atoms are subject to laws of combination, attraction, and repulsion. This had no appeal to the holistic, conservative view of society current in Germany and Central Europe after 1810.

A notorious debate provides a contrast between West and Central Europe. Quetelet was accused of "statistical determinism." The idea is that the annual proportion of suicides (for example) in a region is determined by a fixed probabilistic law. But if that is so, then it is not (nomologically) possible for the inhabitants simply and unanimously to reject suicide as an option. That is to say, they are not free, collectively at any rate, to opt out of the most final of mortal sins. This was taken to be inconsistent with free will. Statistics purveys a grim determinism.[14]

In the West this zany idea reached its climax in 1859, when Henry Buckle published his fashionable *History of England*, claiming that history was the unraveling of historical necessity by way of statistical laws. That got a rough working over from luminaries as various as Prince Albert, Victoria's consort, and Leslie Stephen, in the *Fortnightly Review*. Statistical determinism pretty well dropped out—but not in Germany. 1860 is a curious mirror-year for this debate. Before 1860, the problem is French and briefly English. After 1860 it is principally German, although in Germany it is not exactly a "debate." There is hardly a German writer on economics or statistics who is not at pains to denounce statistical determinism, often with a length and subtlety sufficient to match the apparent passions that it aroused. This is a curious symptom, which would take us far beyond the scope of Engel. Engel himself is agreeably brief on the topic, as firm as he is unsubtle.

Never loath to repeat the words of royalty, Engel starts by quoting Prince Albert on statistical determinism. Albert had rebutted it in his opening address to the 1860 International Statistical Congress—a speech that Engel heard in person. Engel's own view is even more direct than Albert's, and is founded on good primary-school

German philosophy. There is a distinction between a *rule* and a *law*. It is a statistical rule that regularly about 100 girls are born for every 105 boys. But this is not a law. A proposition can be called a law only if we know the reason for the regularity. Thus, for example, it is a law that more people die in years of famine than in years of plenty, for we know the cause. But here we have no precise statistical law, for different famines produce different effects, and there is no stable quantitative law at all.

Likewise, it may be true that "in a given population almost the same number of people commit suicide each year." [15] But that is a mere rule, for we cannot assign a cause to precisely this effect. This is not statistical *law*. But if this is not a law of nature or society, free will is not an issue. For Engel, that ends the question.

There is a more general message: there are, in effect, *no* statistical laws. It is interesting that when Lexis produced his measures of goodness of fit (or dispersion) in 1875, only one biological phenomenon was found to fit the Gaussian law of errors. It was the second oldest known statistical distribution, the distribution of births by sex. But on philosophical grounds, Engel has excluded even that as a candidate for being a statistical law.

There was room for tension in Engel's thought. Thus I mentioned his attempt to find the probability of explosion for steam boilers of various designs. He also plotted this against the age of the boiler, degrees of maintenance, regions, and the like. One might suppose that he was here investigating a lawlike probability distribution. Do we not indeed know why older boilers explode more frequently than new ones? Cannot the engineer tell us exactly why some designs are safer than others? Hence are we not in the presence of probabilistic laws? Engel does not, I think, address the question.

This has consequences for his practice of statistics. I shall show that in connection with mortality tables, he is embarrassingly more naive than his counterparts in the West. I suspect this is because he has not properly conceptualized what his tables tabulate. When he is confronted by what we would call a problem of statistical inference, he does not in general attempt to draw conclusions from a partial sample. His answer is always, Do your best to make a complete enumeration. He will, for example, produce a schedule for enumerating all pertinent facts about each and every boiler in Prussia, and urges that his opposite numbers elsewhere in Europe do the same. The idea of representative sampling requires a concept of statistical law that Engel simply lacked. When he does casually mention samples, he is, once again, deplorably confused.

Engel's conception of statistics, like that of so many of his colleagues, is roughly *historical*. It can be argued that he is a typical heir of a very old German topographical-geographical-historical tradition, in which statistics is "the history of the present"—*gegenwärtige Geschichte*: "Geschichte is fortlaufende Statistik. Statistik ist stillstehende Geschichte." [16] This historical rather than mechanistic vision of statistics fits tidily into a holistic view of society. People in Western Europe may have the liberal picture of independent atoms connected by mechanical or statistical laws. Engel's compatriots further East have that organic version of society, in which the individuals are constituted by their place within the social structure. Individualization, for the organicist, depends upon the existence of a community within which to define oneself. There may be an historical—even inevitable—development of the society as a whole. There may be regularities, "rules" about classes of individuals. But there is no place in this philosophy for such a thing as statistical law.

4 Self-Help

Engel's organic conception of the state repeatedly shows itself in his numerous proposals for reform. He believes that the "atomization" of the modern industrial nation is its chief disease. Organic unities must be created to deal with the new situation. The working classes must have institutions within which to fulfill their aspirations, and to fit into the broader social organization. Only in that way will the conflicts between capital and labor be resolved. His goal is integration directed by the educated bureaucracy and the philanthropic managing classes. Where the socialist will uproot the capitalist mode of production, Engel will instead create new institutions that fit into a harmonious whole.

A key word is "self-help." I began this essay with a quotation about the evil of our times—atomization—and noted that it came from a paper on savings banks, one of many links in the chain of institutions founded on "the Principle of Self-help." The principle is that of (among others in the preceding generation) Franz Hermann Schulze-Delitzsch (1808–1883), who did much to found the cooperative movement in Germany. He did this by creating institutions—cooperatives that he often subsidized, and cooperative banks to finance cooperatives—and by forging the law within which these institutions would operate. The philosophy was clear: leadership must come from above, but the aim must be to create self-sustaining organizations that will define the roles of workers within the state.

Thus Engel's invention of mortgage-insurance.[17] His founding of a mortgage insurance company in 1858 is a perfect example of operating on the principle of self-help. But mortgage insurance was no aid to the larger part of the working classes. Engel's *Zeitschrift* is full of denunciations of the terrible housing conditions demoralizing the populace. There are harrowing reports of the events at the beginning of each quarter of the year, when rents are raised, apartment buildings are emptied, and even the carters double their rates for moving a few sticks of furniture from bad lodgings to worse. In general Engel favored breaking up building cartels and changing the incentives for lending money to various kinds of developers.

Engel made a major statement on the housing shortage at a conference in Eisenach in October 1872.[18] This was the first meeting of the newly formed *Verein für Sozialpolitik*. Engel's presentation was one of the three main events, and he had been present at the organizational gathering, where the *Verein* was first proposed, in June in Halle.

The *Verein* was a group of up-and-coming economists who wished to break away from the laissez-faire principles that in Germany were known as Manchester economics. The *Verein* urged various kinds of state activity to promote the well-being of the working class, and to fend off the rapid progess of socialism. A wit writing in the *Nationalzeitung* dubbed them *Kathedersozialisten*—the socialists who spoke from professorial chairs. The name stuck. Engel has been described as the father of the *Kathedersozialisten*.[19] In fact he was not long active, but many of the people who passed through his Statistical Seminar showed up at the meetings. The list of founding members is a list of economist-statisticians nearly all of whom occur in our ZiF studies of the probabilistic revolution: Engel, W. Roscher, B. Hildebrand, A. Wagner, J. Conrad, G. F. Knapp, L. Brentano, J. V. Eckhardt, G. Schmoller, and E. Meir.

The work of the *Kathedersozialisten* had much to do with the insurance reforms of 1883–1891, in that they laid the groundwork for these radical changes intended to preserve German society. The professors did not, however, have very good relations with the regime. In the 1890s there was a great outcry: how dare a government allow these men to teach in the universities and schools? In the resulting discussions of academic freedom (*Freiheit der Wissenschaft*), the minister of education made the most telling point. He called them *Katheder-Unsozialisten*. The socialist leader Franz Mehring warmly concurred.[20] Mehring optimistically took the hatred of the *Verein* as showing that capitalism was on its deathbed: these worthy professors were like doctors approaching the patient with scalpels to extract the tumors, but the patient, delirious, waves them away and more speedily sinks into ruin.

It is important to distinguish the anti-Manchester free trade aspect of the *Verein* from its state-intervention side. Engel is all for internal state intervention in the setting up of self-help institutions, but he often remained on the side of free trade— which led to his final ill-managed downfall in his spat with official protectionist grain policies. If we are to look for examples of self-help, we should think of Engel not as founding father of *Kathedersozialismus*, but as institutional intervener in the strife between capital and labor. A lecture of his shows him at his own optimistic worst: he thought that the day was about to dawn for worker-controlled factories.[21]

His lecture (to a distinguished audience including the crown prince) had an inadvertently inauspicious opening. Engel begins by quoting Napoleon III of France, who in 1844, while a prisoner of the Second Republic, had written "On the Extinction of Pauperism." After quoting Napoleon on worker-controlled factories, Engel writes, "23 years have passed since those words were written. Napoleon has since become Emperor of France. All the world now agrees that he is one of the greatest of statesman, although in the year 1844 he was taken for an idealist." Only four years after Engel's lecture Napoleon was back in prison again, this time in Germany.

Engel printed his lecture in the *Arbeiterfreund*, organ of the *Central-Verein für das Wohl der arbeitenden Klasse*. The names bear their reform-from-above nature on their faces. Engel characteristically begins by stating that "the political and economic world is taken up with strident claims about the imbalance between capital and labor in industry." He announces that we must not be swayed by polemics, but should instead attend to "positive facts." These lead to an alternative point of view, indicated by the subtitle of his lecture, the English words "Industrial Partnership." That was the name then given to arrangements by which the work force would share in the profits of their employer. The idea originated in France, in the early 1840s, not with Napoleon III but with Edmé-Jean Leclaire, once quite well known as "the father of profit-sharing." It became fashionable among English liberals. Engel dwells on one great success story then attracting popular attention. Between 1865 and 1867—when Engel lectured—a scheme of profit-sharing flourished at the Yorkshire collieries of Henry Briggs, Sons & Co. It had been deliberately introduced in order to get the men to leave their union, and for two years it worked well.

Describing the shining example of the Briggs mine, Engel ends with these glowing words: "Thus the social problem [of Capital and Labor] is a problem no more ... the translation of this solution into practical life has begun." He urges that the North German parliament follow the British lead in its next session, and "make a place for *Arbeitergesellschaft* within the institutions of a general German code of civil law." Alas, Briggs (like most other profit sharers) fared little better than Napoleon III. In 1873 the bottom fell out of the coal market, leaving the workers with virtually no income. They realized their need for solidarity, rejoined the union, and struck the mine.

5   The Value of a Human Being

Subsistence wages play as great a part in the reform economics of Ernst Engel as they do in the radical economics of Karl Marx. According to Engel, a worker must be paid the true cost of "living." This is not merely the cost of surviving and being able-bodied during the productive period of a working person's life. The true cost is what is required to maintain the person from a healthy childhood to a decent old age. Moreover—and what I am about to say now is not a caricature—each body is a little fund of capital. So much must be invested in maintenance between birth and 16 years of age, say, when the person becomes productive. This is entitled to interest at the going rate of 4%. A "subsistence" wage for a laborer is thus not just enough to stay alive, but enough, distributed over a productive lifetime, to cover the initial capital costs and to provide security in retirement.

Reflections such as these add fuel to the demand for employer contributions to sickness benefits, pension funds, and the like. They are the arguments of the welfare state, and it is too often forgotten, outside Germany, that Germany is the home of the welfare state. It was partly created by works such as Engel's "The Price of Labor" (1866) or "The Value of a Human Being" (1883).

Engel addresses these questions: Who is producer? Who is consumer? He argues for the importance of service industries as part of production. He laughs at those who say that only the production of material goods counts as production—for then the barber who shaves your beard is not a producer, but becomes a producer when he turns your hair into a wig.[22] He is keen that all cultural contributors, teachers and preachers, shall count as producers. Likewise consumption includes consumption of spiritual goods as well as material ones. (A football match is a material production, because, like a spa, it comes under the heading of health care, but a night at the opera is cultural, and a morning at church is ethical consumption.)

It is in terms like these that Engel formulates what was called *Engel's Law*: the prosperity of a group varies inversely as the proportion of group consumption needed for subsistence. The greater the part of the family budget spent on food clothing, shelter, heat, light, safety, and health care, the less prosperous is the family. To a philosopher this seems to be less a law than a definition, but it persists in economics textbooks to this day. It also has a little moral uplift on the side, for the larger part of the balance after subsistence is classified as *geistige Bildung*, so that the more prosperous the family, the more (by definition) it devotes to *g.B.* (Education, Church, Theater, Art, Scholarship, Literature, etc., etc.).

Engel repeatedly argued that the household budget is a much ignored tool of economics.[23] Here his model was Ferdinand Leplay, the great student of the European worker. Leplay was vociferously opposed to the number-crunching of the statistician-sociologists who patterned themselves on Quetelet. He despised averages, and wanted representativeness instead. In his first great book,[24] he picked 26 typical families spread across Europe, from nomads of the Steppes to a cutler in Sheffield. He described in great detail every item of consumption for the year, not just "food" but kinds, from cabbages to curds; not just "heat" but kinds, coal or wood. It is precisely this sort of information (together with material presented by Ducpetiaux to the 1855 International Statistical Congress in Brussels) that enabled Engel to state his law, and to define his measure of prosperity.

There is a problem about cross-cultural comparisons of prosperity, and about comparisons of a society at different epochs in its existence. Families in one time and place may differ in size from those in another. So we need a standard measure of the "consumption-need" of a family. It should be a standard such as the ohm, amp, or volt of electricity: let it be named after the great man of statistics. Call this standard the Quet, says Engel, loyally.

Engel assumes that adult males past 25 have equal subsistence needs and that adult females after 20 have equal needs, a little less than that of males. The basic unit is the infant. The subsistence needs of an infant shall be defined at 1 Quet. For immature people, the need is $[1 + \text{age}/10]$ Quets. Thus the needs of an adult male are 3.5 Quets, while those of a female are 3 Quets. (Farr had suggested a smaller discrepancy of $13:12$.)

Suppose you are told that 2,490 American families spend $289,000 on meat, while 104 German ones spend $8,170. This means little, says Engel, until we standardize, and see that the American budget allows $12.30 per Quet for meat per year, while the German one allows $7.79. In terms of such a measure he is able to produce his final work, a comparison of Belgian prosperity over a forty-year period.[25]

Engel's economics enables him to deduce that pension plans and the like—the social net—are part of workers' rights. His conclusions about the distribution of surplus value are less clear. As I make him out, the worker does not have a right to surplus value (where the surplus value is the "true" surplus value after the cost of living the whole life is deducted as a cost). It is, however, in the interests of society to create a prosperous populace, that is (according to Engel's law, or definition), a populace that has a fair amount of disposable income after subsistence.

Engel is ever an optimist. He is the world's greatest optimist about that bone of contention, the Malthusian question. In the nineteenth century, as today, there was much debate as to whether the geometric growth of population would necessarily overtake the arithmetically growing rate of production of bare necessities. As European prosperity increased after 1860, Malthusians declined. But in 1860 one could have ample ground for Malthusian skepticism. Pick up the statistical report of a *Kreis* more prosperous than many, Koblenz. A note of alarm is sounded. In 1834 the population was 55,258; by 1858 it had grown by 7,831 souls. But the horse population had declined by 12, and the number of cattle had diminished by 497.[26] Did not such statistics teach that the population/food ratio was getting worse before our very eyes?

Engel would have none of that kind of talk. There is no upper bound to population density.[27] Some of his arguments are a little odd. He is much impressed by some American writers who urge a sort of conservation of animal and vegetable stuff by recycling. No molecule, animal, vegetable, or mineral is ever lost, and "so" the land can support any number of people, if we treat it right. To increase the density of population, we need merely recycle faster.

6 Naive Vital Statistics

Engel was very good at collecting numbers. His forms for self-reporting by heads of households became widely used in many countries. But we now look back with hesitation on some of the numbers that he collected, and on the uses that he made of them. Vital statistics are a case in point.

The first political arithmetic has to do with birth and death. The immediate practical benefit of such data is in the formation of tables for life insurance and annuities. The history of that topic is a curious one, and the longtime lack of good tables presents a problem for the historian of statistics. By 1860, however, the principles were perfectly well understood. Engel's senior opposite number in London, William Farr, was no great mathematician, but he did know how to prepare a life table. Engel did not. This may be connected with the fact that the German states were, comparatively, latecomers to the insurance business, but by the time Engel took the helm in Berlin, two generations of insurance companies of all kinds had been around. Engel himself had run a specialized insurance company for the mode of insurance that he himself had invented. Hence Engel's rather incompetent use of vital statistics is rather glaring.

There are two kinds of oddity, one in the data collected, and the other in the use of the data. Life data were primarily collected in half-decades, so that one knew the number of people who died between 35 and 40, but not the distribution within that age group. This is a very substantial curtailment of information. In Holland and Sweden this information had long been available for half-years, not half-decades, and so one could construct quite delicate curves corresponding to mortality experience.

As for use of the data, it arises from real conceptual confusion. This does not originate with Engel, for it is expressed by Dieterici. It is not I who call the issue conceptual but Dieterici, in a paper *Über den Begriff der mittleren Lebensdauer* ... (1859). Dieterici thinks that if you want to know the average age of a living population, you can be satisfied with the average age at death. That was effectively Dieterici's last paper. One of Engel's first tasks on replacing Dieterici was to continue these reflections. In an extensive study,[28] he addresses the question of the mean age of the population at the time of the census, and considers the time span 1816–1860. He regards the mean age of the living as being identical to the mean age at death of the population over this extended time period. This is logically unsound. For a counterexample, think of so-called baby-booms. Ten years after a baby-boom—even in times of horri'ving infant mortality—the average age of the population will be quite different from the average age of the dying. Of course there could be empirical considerations that precluded counterexamples such as these, and made Engel's analyses reliable, by accident, as it were. But Engel does not seem even to realize the problem. Nor is it the case that the Prussian Bureau got much better at such questions during Engel's administration. The improvement were made not in his office but next door, in the bureau for Berlin City, directed by Richard Boeckh. His data about Berlin were as good as any in Europe, and his technical expertise at unpacking data was substantial.

The insensitivity of Engel's Bureau to these questions is the more curious in the light of the important theoretical advances made at that time. The work of Lexis is to my mind paramount,[29] but Knapp—a fellow founder of the *Kathedersozia-listen*—uses graphical methods for population statistics without any conceptual difficulty whatsoever.[30]

7   One or Two Newspapers?

Engel was mathematically unsophisticated, and he seems to have had no theory of statistical inference whatsoever. An entertaining story of his well illustrates this. He told it a quarter-century after the events that made the story.[31]

In the five years preceding 1848 Saxony had a strikingly reactionary government. There were essentially no newspapers, even though Leipzig and Dresden were Germany's publishing centers. The quality of censorship is indicated by the fact that the firm of Brockhaus, which did a lively trade in publication for Hungarian readers, was forbidden to publish *anything* in Hungarian, for there were no censors in Leipzig who could read Hungarian! But after the Parisian events of February 1848, there were popular uprisings, and the reactionary ministers were replaced by liberals. The presses rolled; newspapers appeared everywhere. All these events were shortlived. The 1849 Dresden uprising, fired by the great anarchist Bakunin, forced the royal family to leave, and there was temporary revolutionary success, soon to be put down with Prussian aid.

Engel writes of a tiny incident in these stirring times. Teubner, the Dresden publishers, started a newspaper in 1846, called the *Dresdener Tageblatt*, with a right-wing editor. After the press was unshackled in March 1848, every publisher started a newspaper. Brockhaus, for example, had its *Leipziger allgemeine Zeitung*, which became the *Deutsche Allgemeine*. Teubner dropped its unsellable reactionary *Tageblatt* on 31 March. On 1 April 1848 it commenced the *Dresdener Journal*, edited by Bidermann, the liberal professor. The *Tageblatt* editor was understandably miffed, and sued Teubner, claiming that the *Journal* was simply the same newspaper as the *Tageblatt*, and hence it was *his* newspaper, even if it had been renamed.

Expert testimony was summoned, and who better to ask than Saxony's energetic Engel? When are two newspapers identical? We would recognize this as a difficult question for statistical analysis even today: how many features must A and B share, how many differences must they exhibit, in order to us to say that they are, or are not, the same Z? For example, if we find many shards in different archaeological sites, how many features of design must they share, how many differences must they lack, in order for us to say they are or are not from pottery of the same period of civilization? It is noteworthy that the courts of those days recognized the question about newspaper identity to be a statistical one.

We learn much about Engel's style of work from his report to the court. When he was consulted, the *Journal* had been running for a year. He had some 12 months of daily issues for the *Journal*, and 21 months of issues of the *Tageblatt*, before its demise. Engel studied the last six months of the Tageblatt and the first six months of the *Journal*.

His justification for this "sampling" is curious. He says that he is using 36.3% of the total issues available to him (that is, he is using 4 quarterly periods out of 11 available). He compares this favorably with the much admired statistical survey of French agriculture, which studied only 25% of farming land in France, and inferred that the rest would be "analogous." He remarks that chemical analysis may rely on a sample of only 1/10,000 of the substance being analyzed. He does not mention what today we find crucial, that the population being sampled by the French statisticians is vastly larger than Engel's 1,000 odd daily issues of a newspaper. Nor does he notice that the French workers can sample different strata of their population. Of course Engel has no technique for comparing the reliability of his sample to that of the agricultural survey; what is indicative is that it does not even occur to him that in sampling, percentages of a population, without mention of population size, are almost meaningless.

Engel goes to town on the last 183 issues of the *Tageblatt* and the first 183 of the *Journal*. These must be compared in terms of their "external" and "internal" properties. There are only three relevant external properties: title, format and size, and times of publication. The titles are different, but the papers are the same size and appear at the same time of the day. Externally they agree on 2 out of 3 features. It is, however, the internal properties that are telling.

First we must divide the articles in the newspaper into suitable clasifications. Engel finds 10 major categories, which by the time he had made his subdivisions, come to 84 subcategories. The 10 major headings are:

1. official notices,
2. news of the royal family,
3. articles about the material circumstances of the people,
4. cultural conditions of the people,

5. moral ditto,
6. social ditto,
7. municipal affairs,
8. political and administrative news,
9. general news, biography, curiosities,
10. pure scientific news.

The 84th entry, a subentry under 10, is *Belletristik*. Classes (4) and (8) have the most subdivisions.

Engel now had a grand total of 366 issues containing 7,017 articles classified in 84 columns. Then come two further schemes of classification, based on the form and the content of these 7,017 articles. As for their form, they may either be leading articles (from the editors), or correspondents' reports, or *feuilletons*, i.e., feature articles or essays. As for the content, articles are sorted according to their "geographical forum," namely, the region in which the described events occur. Then they are classified according to their "tendency": liberal, liberal-conservative, reactionary, or indifferent. Engel notes that there is a problem about this. He is finishing his report in 1852, when authoritarian government has been restored, but he is classifying articles written in a period of revolutionary fervor. What was commonplace in mid-1848 will seem radical in 1852, but this must not be allowed to warp our classification.

Engel is now in a position to deluge the court with digests of information in prose, tables, and graphs. Typical example: the *Tageblatt* relied on correspondents for 38.36% of its articles, while 77.69% of the *Journal* pieces comes from correspondents; 50.85% of the *Tageblatt* was in the form of feuilletons, but only 16.51% of the *Journal*. The contrasts in political tendency are somewhat less striking, especially considering the provenance of the two successive dailies. Engel's graphical presentation is elegant to behold, if not so easy to follow. We are to take in, visually, the contrasts in distribution. We "see" that regularly the differences are so great that we cannot suppose that we are dealing with the "same" periodical. Engel also checks out the possibility that the *Tageblatt* has been gradually transforming itself into the *Journal*, by plotting monthly distributions according to tendency over the whole one-year period. This graphically refutes the hypothesis of gradual transformation from right-wing to left-wing tendency.

Engel concludes in a sentence: these are two distinct newspapers. Despite the endless calculations, there is no attempt at quantifying the inference itself. One is supposed simply to notice the difference between the two organs.

The report provides us with a rare glimpse at how Engel worked. Never loath to count, Engel summarizes his division of labor as follows:

1. orientation and familiarization with the case: 8 hours;
2. design of systems of classification: 12 hours;
3. characterization of the 7,017 articles according to Engel's scheme: 98 hours;
4. entering the classified articles into the tables: 320 hours;
5. numerical summaries of these results: 130 hours;

6. using these numbers to establish various perspectives on the the data: 97 hours;
7. graphical representations: 53 hours;
8. writing and editing the final report: 60 hours.

He spent 778 hours on this project. The editor of the *Tageblatt* did not win his case.

8   Philosophical Remarks

What have we here? Vignettes of a bureaucrat, and of a bureaucracy, surely. But what do they tell us about number, about statistics, about probability? Let us briefly fit the tale into a larger picture.

At the beginning of the nineteenth century Europe created an avalanche of printed numbers. It began to conceive of itself in numerical terms: for the first time the world was constituted by measurements or countings. People, too, were reconstituted in this process, for the counters had to count people according to kinds, according to categories, and they had to invent the kinds into which people sort themselves. The enumerators did not simply take established categories and count: they invented kinds of people, and thereafter people fell into those very classifications, made themselves into people who "fit," and so changed their very conceptions of themselves. The avalanche of numbers had that effect, but in Western Europe it had another one. It generated numbers that made statisticians imagine that there are purely probabilistic laws of nature. It became possible to think in terms of autonomous statistical laws.

The probabilistic revolution needed two elements. One was the vast amassing and publication of numerical data. The other was the perception that these data are lawlike. We find both elements well instantiated in the official statistics of France, especially those emanating from the reports of Paris and the Seine department after 1820, and those from the Ministry of Justice. They provide the path upon which Quetelet strides. It is important that the avalanche of numbers is not enough for this to be an available path. The proof of this is in (to continue my rocky metaphor) the morraine of cluttered numbers spewed out by the industry of Engel and his peers. There is no conception of lawlikeness in these numbers, and no taming of chance. Chance, for Engel, is what it is for Hume, a mere empty word. But why is there no taming of chance? My thesis is that it has to do with the way numbers are conceived, as descriptions of an organic and endlessly complex world that shall not be allowed to decompose into atoms that are then subject to statistical law. Had atomistic Hume had nineteenth-century numberings before him, he would have attached a real sense to "chance." Holistic Engel has the numbers without the atoms, which leaves him with the same stance about chance as Hume.

The story of Engel's Bureau is a story of absence. What is absent confirms what is important in the French tradition of the taming of chance. But, it will be asked, was it not German mathematicians, from Lexis to Bortkiewicz, who so rapidly advanced our understanding of probability in the years following Engel's retirement from the Bureau? Yes, but precisely by taking up the mathematical problem of the

French, the one previously addressed by Poisson and Bienaymé. Heyde and Seneta[32] provide a historical sketch of their problem: How is it possible to achieve statistical regularity out of individual irregularity? How can a statistically homogeneous set arise from the behavior of individuals who are heterogeneous? For Engel that is not a question, because mass phenomena are *sui generis*, the properties of wholes in themselves, and not constituted from atomistic parts. Only when the French questions are taken up do the German mathematicians provide a mathematical theory of *Massenerscheinung*, which leads on to von Mises and the *Kollektiv*. We might say, with only a semblance of paradox, that there can be no probabilistic "collectives" in a statistics that conceives the state in an essentially collectivist fashion.

Notes*

1. Ernst Engel, "Die Sparcassen in Preussen als Glieder in der Kette der auf dem Prinzip der Selbsthilfe aufgebauten Anstalten,"*ZKPSB*, 1 (1861), 85–108.

2. W. Saenger, "Das Preussische Statistische Landesamt 1805–1934," *Allgemeines Statistisches Archiv*, 24 (1935), 445–460.

3. M. Norton Wise, "Social Statistics in a *Gemeinschaft*: The Idea of Statistical Causality as Developed by Wilhelm Wundt and Karl Lamprecht" (Zentrum für interdisziplinäre Forschung, typescript, 1983).

4. Ian Hacking, *The Taming of Change* (Cambridge: Cambridge University Press, 1986).

5. See Richard Boeckh, *Die geschichtliche Entwicklung der amtlichen Statistik des Preussischen Staates: Eine Festgabe für den internationalen statistischen Congress in Berlin* (Berlin: 1863), and "Verzeichnis der von der königlichen Regierung auf dem laufenden erhaltenen statistischen Nachrichten," *ZKPSB*, 3 (1863), 287–308, for summaries of these events.

6. Ernst Engel, "Die Aufgaben des Zählwerks im Jahre 1880," and "Die Aufgaben des Zählwerks im deutschen Reiche am Ende des Jahres 1880. Unter besonderer Berücksichtigung preussischer Verhältnisse," *ZKPSB*, 19 (1879), 367–376, 1–70.

7. "Verzeichnis," *ZKPSB*, 3 (1863), 287–308.

8. Bruno Hildebrand, "Das statistische Seminar in Jena," *Jahrbücher für Nationalökonomie und Statistik*, 6 (1866), 77–78.

9. See, for example, Ernst Engel, *Das Zeitalter des Dampfes in technisch-statistischer Beleuchtung* (Berlin: 1880).

10. Ernst Engel, "Beiträge zur Statistik des Krieges von 1870–71," *ZKPSB*, 12 (1872), 1–320.

11. Ernst Engel, "Die wahren Verluste der königlich preussischen Armee im Kriege des Jahres 1866," *ZKPSB*, 7 (1867), 157–167.

12. Ernst Engel, "Die Methoden der Volkszählung, mit besonderer Berücksichtigung der im preussischen Staate angewandten," *ZKPSB*, 1 (1861), 149–212.

13. Ernst Engel, "Die Volkszählung, ihre Stellung zur Wissenschaft und ihre Aufgabe in der Geschichte," *ZKPSB*, 2 (1862), 25–31, on p. 26.

14. Ian Hacking, "Nineteenth-Century Cracks in the Concept of Determinism," *Journal of the History of Ideas*, 44 (1983), 455–475.

* *Abbreviation: ZKPSB: Zeitschrift des Königlich preussischen statistischen Bureaus.*

15. Engel, "Volkszählung," p. 27 (note 13).

16. A. L. Schlözer, *Theorie der Statistik* (Göttingen: 1804), vol. 1: *Einleitung*.

17. Ernst Engel, *Denkschrift über Wesen und Nutzen der Hypotheken-Versicherung und über die Rätlichkeit der Begründung einer Hypotheken- und Rückversicherungs-Anstalt im Königreiche Sachsen* (Dresden: 1856); *Die Hypotheken-Versicherung als Mittel zur Verbesserung der Lage des Grundcredits* (Dresden: 1858); *Beleuchtung der Bedenken gegen die Hypotheken-Versicherung und gegen die Errichtung einer Hypotheken-Versicherungs-Gesellschaft* (Dresden: 1858).

18. Ernst Engel, "Die Wohnungsnoth. Ein Vortrag auf der Eisenacher Conferenz," *ZKPSB*, 12 (1872), 379–402.

19. Emil Blenck, "Zum Gedächtniss an Ernst Engel. Ein Lebensbild," *ZKPSB*, 36 (1896), 231–238.

20. Franz Mehring, "Die Hetze gegen den Kathedersozialismus," *Die Neue Zeit*, 15 (2) (1896), 225–228.

21. Ernst Engel, "Der Arbeitsvertrag und die Arbeitsgesellschaft," *Der Arbeiterfreund* (1867), 371–376, 394.

22. Ernst Engel, "Das Gesetz der Dichtigkeit," *Zeitschrift des Statistischen Bureaus des Königlich Sächsischen Ministeriums des Innern*, 3 (1857), 153–182, reprinted in Engel, *Die Lebenskosten belgischer Arbeiter-Familien früher und jetzt* (Dresden: 1895).

23. Ernst Engel, *Das Rechnungsbuch der Hausfrau und seine Bedeutung im Wirtschaftsleben der Nation* (Berlin: 1882).

24. Pierre Guillaume Frédéric Leplay, *Les ouvriers européens* (Paris: 1855).

25. Engel, *Lebenskosten* (note 22).

26. *Statistische Übersicht und Mittheilungen über die Verwaltung des Kreises Coblenz* (Coblenz: 1860), p. 11.

27. Engel, "Gesetz" (note 22), pp. 180–182.

28. Ernst Engel, "Die Sterblichkeit und die Lebenserwartung im preussischen Staate und besonders in Berlin," *ZKPSB*, 1 (1861) 321–353; 2(1862), 50–69.

29. W. Lexis, *Einleitung in die Theorie der Bevölkerungsstatistik* (Strasburg: 1875).

30. G. F. Knapp, *Über die Ermittlung der Sterblichkeit aus den Aufzeichnungen der Bevölkerungs-Statistik* (Leipzig: 1868).

31. Ernst Engel, "Die Statistik im Civilprocess; eine Reminiscenz aus dem Leben der Presse im Jahre 1848; mit 6 Tafeln graphischer Darstellungen," *ZKPSB*, 13 (1873), 43–62.

32. C. C. Heyde and E. Seneta, *I. J. Bienaymé: Statistical Theory Anticipated* (New York, Heidelberg, and Berlin: Springer, 1977).

19 How Do Sums Count? On the Cultural Origins of Statistical Causality

M. Norton Wise

The concept of indeterministic statistical causation is new in the twentieth century and remains obscure. To illuminate the context of its original gestation two examples are here developed. Familiar from quantum mechanics, the idea had a prior and wider currency around the turn of the century. In central Europe, particularly Germany, statistical causation acquired much of its meaning from the political-social ideal termed Gemeinschaft *and from the idea prominent in psychology of a qualitative causation distinct from quantitative causation.* Gemeinschaft *connoted both essential unity in a whole society and essential diversity in its constituent individuals, while "qualitative causation" determined the behavior of a complex system without determining the behavior of its parts, and did not reduce to causes acting on parts. Both concepts inform the examples discussed.*

Case one describes Wilhelm Wundt's attempt to make psychology the foundation discipline of the human sciences in the way physics served the natural sciences. With qualitative causality replacing quantitative, the new psychology would trace the evolutionary growth of psychical energy in individuals and in spiritual communities as physics traced the conservation of physical energy in nature. From Karl Lamprecht's development of these ideas for a causal—thus scientific— cultural history emerged the first explicit statement of "statistical causality."

Case two explores the psychological philosophy with which Harald Høffding sought to unify, rather than differentiate, the human and natural sciences. Attributing qualitative and quantitative aspects to both realms, he emphasized the dialectic relationships of whole and parts, continuity and disconuity, rationality and irrationality. As shown in an extended analogy, Niels Bohr employed the same dialectics in formulating his approach to quantum mechanics, with its characteristic stationary states, discontinuous transitions, and statistical causation.

If we ask for the conditions under which the idea of indeterministic statistical causation entered modern thought, three generalizations appear valid. It arose in the period between 1870 and 1920; in association *always* with a belief in unanalyzable holism; and *usually* in central Europe. Taking those conditions as given I offer here two case studies intended to illuminate the discourse within which the new concept took shape. Those cases exemplify several further generalizations about the central European context whose full justification will require a lengthier study. First, statistical causation emerged among intellectuals who held a 'moderate liberal' political position, self-consciously opposed to both Anglo-French liberalism and to traditional conservatism. Second, they sought to bridge the gap between the natural sciences and the human sciences, typically through psychology. Finally, statistical causation attained legitimacy as a concept through the prior legitimacy of ideas about qualitative causation as opposed to quantitative. Leaving the latter two claims for the examples, I begin with a characterization of moderate liberalism and its significance for social statistics.

1 Moderate Liberalism and the Critique of Social Statistics in Germany

However minimal the classical liberalism of natural rights, natural law, and social contracts may have been in Germany before the revolution of 1848, even that minimum seems to have disappeared afterwards as the vast majority of liberals made their peace with Prussian power.[1] The process of accommodation reinforced traditional 'moderate' ideals, not only in the dominant National Liberal Party formed in 1867 but also in the more radical Liberal Peoples Party (*Freisinnige Volkspartei*). Moderate liberals defined themselves in opposition to what they regarded as the mechanical summing doctrine of the social contract. They espoused instead organic unity in a historically evolved national state. So pervasive was their denunciation of sums, aggregates, and averages as representative of the unified state that I shall represent their position in a single phrase: *German individuals do not sum.*

A standard reference for moderate liberal theory prior to the 1848 revolution was F. C. Dahlmann's *Politics, Reduced to the Ground and Measure of Given Conditions* (1835). Dahlmann led the Casino Party dominant in the Frankfurt Parliament. Like all moderates he sought a middle way between personal freedom and the prerogatives of the state. His solution was organic unity: "For the state is not merely something common among men, not merely something detached; it is at the same time something grown-together, a bodily and spiritually united personality." Citing Aristotle, he took the family as the original form of the state and made the state as natural as man himself. That view entailed denial of both natural rights and the man-made state based on natural law: "Neither are men by nature free and equal, nor is the state to be conceived as an artificial institution that was preceded by a nonpolitical state of nature."[2]

Dahlmann's liberalism consisted essentially in his insistence on the inviolability of constitutional law and on the right to representation of all the people in the making and changing of laws. His constitutional state was monarchical and the representation of the people corporate, a representation of "all the estates or branches of the people." Such corporate representation presumed essential diversity: "To persist, the form of government of a large state must be built out of materials that are not uniform but diversified and that are as little as possible artificial and as much as possible really existing."[3]

At every level of social organization—estates, political parties, labor unions, the family, and the nation—moderate liberals saw unity in the whole and diversity in the parts as equally essential aspects of the organic system. The prerogatives of the state did not derive from the will of the people, nor did the rights of the people derive from monarchical dispensation; rather in the just society state power and personal freedom legitimated each other mutually in a harmonious system of interrelationships. Organicists claimed both greater unity and greater individual freedom in their ideal state than any social contract could support, with its aggregation of isolated individuals. Thus the attack on the summing doctrine took two forms: the whole is greater than the sum of the parts; and the parts, being of essentially different kinds, cannot be averaged to isolate common features.

Many liberals to the left of Dahlmann took a far stronger position on personal freedom than he, advocating even the rights of man, but rarely did they have in mind anything like the self-defining and self-interested individual of classical theory. The reductionist physiologist Rudolph Virchow, for example, a leader of the radical liberals and an advocate of natural rights, absorbed during the 1850s the organic ideal. In "Atoms and Individuals" (1859) he denied that individuals summed like independent atoms and emphasized instead their individuality within the unity of the nation, like cells in an organism: "What is an organism? A society of living cells, a tiny well-ordered state, with all the accessories—high officials and underlings, servants and masters, the great and the small."[4] Ian Hacking has described the similar view of Salomon Neumann, who published in Virchow's journal of physiology, advocated the rights of man, and also believed in organic unity.

Heinrich von Treitschke, famous historian and propagandist of National Liberalism, gave classic expression to the right-wing form of organic unity: "As long as there have been states," he said in 1855, "... politics have ever been questions of power and not of law [Recht]. What is called constitutional law [Staatsrecht] is in my view only a euphemism; it is simply the statistical calculation and grouping of the political relationships of power existing in a state .... A positive constitutional law is good then when it makes dominant by law those social forces which possess most political capacity." Treitschke did not mean by a "statistical calculation" of power a sum of quantities of power lodged in independent individuals; the priority of the nation ruled his entire political philosophy. If Treitschke could nevertheless talk of a statistical calculation, he meant that "the state is an institution that is rooted completely in the nature of every man." Individual contributions of power derived not from individuals themselves but from the "relationships of power" in which they participated.[5]

A social-political ideal so ubiquitous as that of organic unity in diversity cannot have gone without a name. It was Gemeinschaft. As German intellectuals sought increasingly to differentiate their own national heritage from that of the French and British, especially around 1870 and after, they opposed Gemeinschaft (community) to Gesellschaft (society) as a whole to an aggregate. But the broad political connotations of Gemeinschaft by no means exhausted its significance, for it permeated the culture of central Europe as a structure of discourse embedded in the most various institutions and practices. In this sense it was not primarily a political doctrine at all, although the political connotations were never far to seek. Neither was it a concept that meant the same thing to all who used it; left liberals gave priority to individual diversity, while the right emphasized the unitary state. 'Gemeinschaft', like freedom and democracy, served as part of a way-of-talking about social formations that structured to some degree the way everyone talked, but that gained its meaning largely from the context and motives for its use. Thus no one will confuse the rather conservative and antiindividualistic rendering of Wilhelm Wundt, who begins my first case study, with the liberal-socialist intent of Ferdinand Tönnies in his classic Gesellschaft und Gemeinschaft (1887). Tönnies dedicated his second edition (1912) to the Danish philosopher Harald Høffding, who begins my second case. From the two perspectives emerged very different concrete concepts of statistical causality, both rooted in the same way-of-talking about ideal social life. Tönnies put it as follows: "*Gemeinschaft* is enduring and true living together, *Gesellschaft* only fleeting and pretended. It is therefore appropriate that *Gemeinschaft* should be understood as a living organism, *Gesellschaft* as a mechanical aggregate and artifact."[6]

Before turning to specific cases I must discuss briefly the question,What could it mean to count individual actions in a *Gemeinschaft*? Attempts to provide an answer brought a fundamental transformation in the interpretation of social statistics in Germany and in central Europe generally. Because the papers in this volume of Ian Hacking and Ted Porter supply extensive evidence for this claim, I shall here draw out only crucial aspects of it, relying on the widely cited work of Wilhelm Lexis.

The center for statistical thought in Germany was the *Verein für Sozialpolitik*, most of whose members seem to have moved on the leftward side of national liberalism. Dissatisfied with the social policy of the National Liberal Party and with the doctrinaire opposition of the radicals as well, they sought a policy 'above politics' that would unify society. The *Gemeinschaft* dominated their thinking. Sharing also the popular fascination with Darwinian evolution, they stressed the evolutionary historical character of the social organism and the variation and diversity of its parts. Their organic ideal, however, did not admit of mechanistic reduction. They rejected any idea of natural laws for the functioning or for the evolution of society in the abstract and sought instead empirical regularities in real historical cultures. Social statistics could play a limited role in their endeavor, but they rejected Quetelet's infamous attempt to found a social physics on the statistical study of the 'mean man' and the laws of his behavior.

Wilhelm Lexis, physicist turned economist and founding member of the *Verein*, brought this critique to a sharp focus in his *Massenerscheinungen* of 1877. What, he asked, was the meaning of the statistical regularities popularized by Quetelet? Two unattractive interpretations presented themselves: the mechanistic idea associated with Quetelet of natural law as a force acting in every person, despite his deviations from the mean man; or a mystical spirit standing above the society and controlling its average behavior. Lexis distinguished these possibilities from more agreeable alternatives on the basis of a simple comparison with a game of chance in which one draws a certain number of balls from an urn containing a mixture of black and white balls. A long series of such trials will yield a normal distribution for the ratio of black to white balls that is centered on the actual ratio in the urn. Focusing attention for the first time in the history of statistics on the meaning of the width or "dispersion" in the distribution, he argued that a width narrower than that expected for the urn drawings would indicate a force behind the scenes directing the result. A wider distribution would indicate more variation than drawings from a *single* urn with a fixed ratio of black and white balls would allow. But random drawings from a *variety* of urns with different ratios, in analogy to social groups living under different circumstances or changing over time, would produce the wider distribution.[7]

Despite some ambiguity, Lexis considered himself to have successfully eliminated the possibility of hidden forces, whether mechanical or mystical. Societies might occasionally exist in a steady state of dynamic equilibrium, in which case social statistics would exhibit normal dispersion and "the freedom of the individual would unite itself with the conditions of existence of the whole." But above-normal dispersion, the concomitant of "evolution," was the rule: "Human society is always busy trying, out of its own force and with its own responsibility, to change the basis of the state [of things], which, incidentally, even though it remained the same, would not represent a coercive law for the individual, but only conditions of his action."[8]

Lexis's critique offered strong support for the ideal of individual diversity within the evolving *Gemeinschaft*. Moreover, it began to suggest how one might conceive the interaction between cultural whole and individual such that statistical regularities arose. Because individuals were essentially diverse, social relations could not determine the action of an individual even approximately; they provided only "conditions of his action." Lexis put the point in a modern-sounding form: "The inductive inference from observed phenomena to unobserved phenomena, which in the domain of natural science attains an empirical certainty, leads in the inexhaustible richness of the life of man only to a greater or lesser degree of probability." In a footnote he quoted Gustav Rümelin: "Is it the case that the interpenetration of all psychical forces will perhaps always and everywhere evade scientific determination; is it the case that psychical forces differ from the physical and physiological precisely in that the latter possess an eternally unchangeable measure of capacity [*Leistungsfähigkeit*] while the former, for all the persistence of their basic form, are, with respect to their degree of strength, subject to a gradual inner transformation?"

The query suggests that probabilities for human action might be unanalyzable in principle, because psychical energy is not subject to a conservation law but undergoes transformations of its own independent of external influence or physiological change. That Lexis had not come to so definite an opinion is apparent from another remark in the text: "Every individual event stands to be sure in a strictly closed chain of causality, which one could also demonstrate in each given case, but the available systems of causes are so numerous and manifold that the coincidence of the individual cases appears to us only as chance." The ambiguity manifest in these remarks could be reproduced many times from other authors. in their very ambiguity, however, lay the opening, perhaps even the demand, for a new concept of probability and statistical causality. The new ideas seem to have emerged largely from considerations of psychology (or psychophysics) like those in Lexis's footnote.[9]

2   Wundt and Lamprecht: From Qualitative Causality in Psychology to Statistical Causality in History

Wilhelm Wundt literally educated a generation of experimental psychologists, directing 186 doctorates at his Leipzig institute from 1875 to 1919. His views on psychical causality were known to everyone who followed psychology. They are of interest here because he began his career thinking that "from statistical surveys [like

Quetelet's], more psychology can be learned than from all philosophies except Aristotle" (1862), yet by the 1890s he had rejected social statistics as he adopted "qualitative causality." I discuss only his later views.[10]

Wundt regarded the soul as a *Hilfsbegriff* (auxiliary concept) in psychology having the same status as matter in natural science. Our experience contained neither concept directly. Both were abstractions that assisted in explaining empirical facts causally. To reify either into a substantial entity was to create a "mythological-metaphysical" concept. For Wundt, "Natural science and psychology deal not with different contents of experience but with one and the same experience from different standpoints." Natural science regarded that experience in its outward, objective aspect, mediated by concepts, while psychology regarded it in its inner, subjective, and immediate aspect. The two touched each other at every point, but the *Hilfsbegriffe* of psychology, such as worth and purpose, could in no way be subsumed under those of nature, such as force and energy.[11]

In early works Wundt had espoused psychophysical parallelism, but he came to believe the parallel incomplete. Any given physiological condition might correspond to numerous psychological perceptions. He argued for an "independent psychical causality" complementing physical causality. That idea provided the organizing thread of Wundt's philosophy of psychology.[12]

Wundt enunciated three principles of psychical causality, referring essentially to three activities of the mind: perception, apperception, and willing. The direct "objects" of perception—actually *"processes* that change from one moment to another"—were complex psychical entities (*Gebilde*), which, by abstraction, could be analyzed into elements: pure sensations and simple feelings. According to Wundt's central "principle of creative synthesis," however, the whole was greater than the sum of its parts: "Psychical entities stand in definite causal relations to the elements composing them, but at the same time they possess *new* properties, which are not contained in the individual elements. In this sense all psychical entities are products of a creative synthesis [*schöpferische Synthese*]." Complementing this synthetic principle of perception, the "principle of psychical relations" governed apperceptive analysis into interdefining parts: "The partial psychical contents receive their meaning through the relations in which they stand to other partial contents." The "principle of psychical contrasts," finally, referred to the ordering in opposites of feelings associated with willing: e.g., desire and listlessness, excitement and calm. The oppositions led to mutual strengthening, so that "contrast" acted as an agent of "creative synthesis."[13]

Assuming continuity between successive psychical states, Wundt derived from these principles his laws of psychical development. The most important concerned growth and purpose. In contrast to the succession of physical states, which exhibited quantitative causation and conservation of energy, psychical states did not obey quantitative relations. The capacity of one state in relation to the next depended rather on its subjective value to the individual, on its quality: "The physical energy principle has for its content the *external, quantitative side of reality*, while the psychical principle contains the *internal, qualitative* side."[14] Qualitative energy, considered temporally in the light of "creative synthesis," implied a "principle of the growth of psychical energy," for each new state brought with it new qualities or values not present in the preceding. Thus, where physical energy values posited conservation, psychical energies posited growth.

A corollary to the principle of creative synthesis limited every causal explanation in psychology to retrospective analysis, in contrast to the predictive character of physical explanation. "For the character of the 'creative' lies precisely in the fact that only after the effect or product is given can we take account of its inner connection with its components or factors." Wundt called this teleological principle "heterogony of purpose." It described how the result of a synthesis, seen as success or failure in a purpose, acted back on the components to change their relations and thus to redirect the synthesis.[15]

These ideas suggest why Wundt thought his new psychology could serve as a foundational science for the *Geisteswissenschaften* in the way analytical mechanics served the *Naturwissenschaften*. While the causal laws of mechanics described force and energy, his principles would bring "purpose" and "value" under causal control. But Wundt's psychology did not stop at individuals; it extended also to their participation in a *geistige Gemeinschaft*, a community of language, myths, and morals. For his *Völkerpsychologie* Wundt carried his principles of psychical causality into the holist social-political ideal of moderate liberalism. One of the most important ethical tasks of the nineteenth century, he wrote, "consists in the *conquest of individualism*, in the foundation of a moral *Weltanschauung* that recognizes the value of the individual personality without thereby giving up the independent value of the moral community." In that this *Volksgemeinschaft* exhibited spiritual products present in the individual only as "tracelike inclinations" (*spurenweise Anlagen*), it was appropriate to attribute to the whole a consciousness (*Gesamtbewusstsein*) a will (*Gesamtwillen*), a spirit (*Gesamtgeist*), and a personality (*Gesamtpersönlichkeit*).[16]

Wundt carefully distinguished these *Hilfsbegriffe* of the *Geisteswissenschaften* from the mythological notion of a spirit of the community existing apart from the individuals, which he labeled "vulgar spiritualism." But he denounced as well the vulgar materialism of the "doctrine of the social contract, according to which the spiritual community was not supposed to be anything at all original and natural, but rather was to be traced back to an arbitrary union of a sum of individuals."[17] This political attack on the reductive summations of classical liberalism stood inseparably with Wundt's attempt to win for the holistic *Geisteswissenschaften* the status of natural science.

As a theory of the individual in a *Gemeinschaft*, Wundt's psychology had no place for statistics. As a logical matter, he noted purely empirical probabilities could never have the meaning of probabilities for the outcome of a particular case because without a priori or hypothetical knowledge one could not know that the determining causes were the same in each case. One could only assume that the totality of conditions was constant. The inductive inference of probabilities, therefore, was "always only valid as a conclusion *from many cases to many cases*." Without further discussion Wundt dismissed the mean-man: "For statistics employs large numbers, not in order to eliminate the more or less extensive deviations of individual observations, each of which already contains the entire sought-after law, but because ... the law is without any validity except for phenomena in the mass."[18] That statement asserts that although collective behavior is lawlike, individual behavior is lawless, not as a mere matter of accidental deviation from the mean but because statistical law is without *any* validity for the individual. The statement does not imply "vulgar spiritualism," but "creative synthesis": The spiritual forces of the totality have their origin only in the individual and can produce a spiritual life of the whole only in that they act back on the individual."[19] Wundt, even more clearly than Lexis, imagined that statistical stability derived from a heterogeneous variety of individuals whose actions were nevertheless functionally integrated in the total personality of the *Gemeinschaft*. The behavior of the whole, therefore, could not be regarded as a sum, first because the individuals were interdependent and second because the whole, as a qualitative unity, realized purposes that were not present in each individual except as *spurenweise Anlagen*.

Although statistics possessed little value for psychology, Wundt suggested, it might still find appropriate application in a causal account of history. In history, indeed, Wundt had already found an enthusiastic supporter, his Leipzig colleague Karl Lamprecht. Through social statistics Lamprecht attempted to escalate Wundt's concept of psychical causality into a vast scheme for laws of historical development. But from history, preeminent discipline among the *Geisteswissenschaften*, came the most vehement opposition to all supposed "laws" of society. Lamprecht's unsuccessful attempt to disarm that opposition reveals the challenge faced by all those who wished to pursue causal science as participants in a *Gemeinschaft*.

The obstacles appear in uncompromising form in J. G. Droysen's 1862 review of H. T. Buckle's famous *History of Civilization in England*. Droysen blasted Buckle for dilettantism and triviality, his sin being the claim that history could be considered a science only if it employed the methods and goals of natural science. For Buckle, simplifying Quetelet, that meant discovering and applying causal laws of social phenomena by analogy to mechanics. His ideal society would develop freely under natural laws of individual action. Social statistics would discover those laws by averaging out deviations. Such a reduction of individuals to anonymous atoms and the concomitant implication that statistical results represented the "sum of history" violated Droysen's most cherished views of the freedom of the moral man in a moral society:[20]

History concerns not merely the material [*Stoff*] on which history works. Additional to matter is form; and in these *forms* history has a restless, self-propagating life. For these forms are the moral communities [*Gemeinsamkeiten*] in which we bodily and spiritually become what we are .... In the community [*Gemeinschaft*] of the family, of the state, of the people, etc., the individual has raised himself above the narrow closet of his ephemeral ego in order ... to think and to act out of the ego of the family, of the people, of the state. And in this elevation of undisturbed participation in the action of moral might, according to its type and duty, not in the unlimited and unbound independence of the individual, lies the true essence of freedom. It is nothing without moral might; without that, freedom is immoral, a mere locomobile.

Lamprecht shared Droysen's combined disciplinary and political ideal of the *Geisteswissenschaften*, but with Buckle he also admired the power of statistical and of causal analysis. He had somehow to make Droysen cohere with Buckle, and he saw an opening within Droysen's own historiography. If the individual only realized his own personality through that of the whole, then the personality of the whole had also to be seen as active in history. "The ideas that powerful personalities push forward are nothing other than the tendencies of the psychical *Gesamtorganismus* of an age and of an historically delimited part of humanity." Individualist history required a complementary collectivist history, or history of culture: "For *Kulturgeschichte* is history of the life of the soul of human communities."[21] The psychical attributes of the *Gemeinschaft* no more connoted "supernatural hocuspocus" to Lamprecht than to Wundt. He sought through Wundt's psychology to place the history of the social totality midway between idealist history and natural science, drawing on the prestige of both. His essential conceptual step may be seen as assimilating the idea of a species to that of a spiritual community. Natural science aimed at knowledge of the species, the general as abstracted from the particular, which made analysis and synthesis under causal laws possible. One might treat a culture in the same way if its individual members could be regarded as of the same type, a species, and if the culture could be regarded as made up of these identical members. Strictly held, that scheme contradicted Lamprecht's concept of a cultural unity as consisting of "a majority of people" who had in common a "*Gesamtgefühl, Gesamtvorstellung, Gesamtwille*," which was "not identical with the sum of the individual factors." Nevertheless, this psychical totality, being "of fully immanent character," existed only through its individual members rather than in a separate spiritual realm. On that ground Lamprecht argued that "it is permissible to regard the persons who constitute this entity as identical." "With respect to this entity they are exemplars of a species; and a historical representation of the social entity is justified in operating with its totality as a species concept."[22]

The *Gemeinschaft* was a species. Not all individual members, however, exhibited the attributes of the totality. In what sense then were they all members of a species? Lexis and Wundt, among many others, had raised that problem implicitly. Lamprecht developed their answer in an explicit form: his *Gemeinschaft* was inherently a statistical concept, but cleansed of the egregious notion of an aggregate.

Before discussing that issue directly, it will be useful to consider further the sort of history that Lamprecht advocated and partly produced. He wished to write world history as the natural evolution of human societies considered as biological organisms, an "evolution of species" as defined above. He did not, however, wish "to repeat in the area of history Darwin's error in natural history. Darwin believed ... that one could explain the evolution of organisms from the effect of chemical-physical laws alone, in virtue of the somewhat mystical, intercalated concepts of adaptation and inheritance. He did not believe in the immanent power [*Potenz*] of evolution, as Goethe prophetically envisioned it. Today, however, there exists among many people scarcely a doubt that Darwin's mechanistic attempt at explanation does not suffice ...."[23] Historical evolution, Lamprecht argued, should be regarded as the evolution of social personalities under the action of forces (as conditions, *Zustände*) immanent in the forms of social organization themselves. Such forces of self-organization he labeled "social-psychical factors," the most original being economic organization, speech, and artistic expression. Other factors, such as customs, morals, and law, developed at a higher level.[24]

In pursuit of his ideal of *Kulturgeschichte* Lamprecht produced a monumental *Deutsche Geschichte* in sixteen volumes (1891–1909). It aimed at a universal history of the evolution of man based on the German case. Dividing his social-psychical factors into two kinds, Lamprecht constructed two parallel series of epochs of human evolution, a series for "material culture," based on economic organization, and a series for "spiritual culture," based on values and modes of thought. This psychophysical parallelism led to considerable misunderstanding on the part of his critics, who suspected another materialistic causal scheme. In Lamprecht's own conception psychical factors held primacy. He spoke, for example, of the task of interrelating the series as one of "psychologizing the economic stages."[25]

An essential feature of Lamprecht's scheme was the *Dominante*, the dominant type of individual in a particular epoch. In the economic series, for example, that type might at one time be capitalist. But coexisting with it were minority types such as feudal and socialist, the *Dominante* of other eras, some declining and others incipient.[26] One could not suppose that the behavior of the dominant type determined by itself either the character of a given era or the course of historical evolution. The personality and development of the culture depended on dynamic interaction of the very different elements within it.

It will now be clear why Wundt's scheme of psychical causality could provide the theoretical basis of Lamprecht's historiography. "Creative synthesis" described both the relation of individual personalities to the communal personality and the relation of already synthetic, social-psychical factors to the total culture: "The sum of all social-psychical factors forms in itself, in every period, a unity." In each of these relations the elements could be separated from the whole only retrospectively by abstraction. Temporal progression of society was again creative and followed Wundt's law of growth of qualitative psychical energy, for "historical life must move with constantly increasing psychical intensity." Each new synthesis acted back on its own elements to change their relation, leading to newly defined elements and a new synthesis. Thus any culture evolved progressively in a dynamic interaction of social relations.[27]

To recover the problem of statistics, we may summarize Lamprecht's "scientific" history as consisting of two parts: (1) empirical regularities in the progressive stages of evolution of a social species, a *Gemeinschaft*, and (2) qualitative causal laws explaining the structure of the *Gemeinschaft* and its dynamic character. The two parts together constituted the "biology of humanity" for they exhibited a form of "biological causality," a causality that was not necessary in a strict sense. "There exists no law that this acorn must become an oak tree ... but under normal conditions [*Verhältnisse*] acorn and oak tree behave in that way.... It is exactly the same with the demonstrated periods of social-psychical evolution: they need not ensue for a particular *Volk*; but if the *Volk* develops normally they do ensue." Similarly, a given individual need not exhibit the behavior characteristic of the stage of development of the culture, although most individuals would.[28]

To the question, then, whether his series of historical periods could be justified as a natural law of development applied to identifiable objects and established by induction, Lamprecht answered, no! It could be justified as a near-law, a "statistical regularity" applying to "statistical quantities" and established by "statistical induction." To complete that set of conceptual reorientations he coined the term "statistical causality," explaining, "Statistical causality is a conditioned causality; that is, it enters with necessary regularity [*Regelmässigkeit*] but not with absolute constancy [*Stetigkeit*]."[29]

Significantly, Lamprecht did not actually carry out the process of statistical induction, claiming essentially that history had done that job for him and that the results could be seen through comparison of different cultures. Behind this claim stood the idea that typical cultural products, as products of the whole culture, gave a statistical representation of it. Therefore, comparison of qualitative products— style in art, economic structure, etc.—might be a better method of forming inductive generalizations of the statistical kind than actual counting of individual behavior, though the latter would provide "an especially exact-appearing application of that comparative method."[30] Concomitantly, in elaborating his view of statistical causality, Lamprecht noted that "according to empirical observation one has necessarily to distinguish between regularly [statistically] and singularly [necessarily] entering causal relations." That claim resolved the traditional antinomy between freedom and necessity, between idealist history and natural science. On the one hand, "Social psychical causality exists necessarily and under all conditions for all human processes in the mass," while on the other, "Almost the most important guiding principle of my investigation [has been] that there is a relative, practical freedom of the individual, which, enrolled in the loftiest necessity of historical existence, nevertheless is called upon to act uniquely."[31]

In Lamprecht's scheme of statistical causality cultural characteristics did not represent average behavior but something like highly probable behavior arising from the creative synthesis of diverse elements. One is tempted to think of the dynamic interactions between his dominant and subordinate types and between various social-psychical factors as interfering probability amplitudes. Lamprecht offers considerable support for such an interpretation. Occasionally, however, he argued that nonnecessity of statistical causality derived from the fact that it was of "compound" (*zusammengesetzt*) character and hid in itself "simple causal relations that we cannot, or cannot yet, completely ascertain." But even that usage was not unambiguous. Lamprecht suggested elsewhere that such compounding represented "the qualitative moment of every causal connection" even in the natural sciences; "One thinks, for example, of the chemical characteristics of compound (*zusammengesetzt*) bodies in relation to their components." Thus the indeterminacy of compound causes might derive from their qualitative union.[32]

Such confusing juxtapositions of merely complex causes with unanalyzable qualitative unities show that Lamprecht (like Lexis) had not arrived at a settled view of statistical causality. Equally clearly, however, he was struggling to attain a conception of historical causation in which the results of social statistics would appear as manifestations of a holistic psychological dynamics of the *Gemeinschaft* rather than as a mechanical sum over independent individuals:[33]

The older psychology ... was a psychology of the individual. It knew man only as an abstract individual; to it *das Volk* was only a mechanical aggregate of people standing for themselves; they stood among one another without any creatively acting contact that would produce something new. It was the psychology of rationalism, which had not yet recurred to Aristotle's idea that the whole comes before the parts and that the parts only exist through the whole, that at least parts and whole are equally original.... It did not recognize, therefore, the concept of the natural society, nor that of the nation as the most perfect form of all natural societies.... Nothing is in this regard more characteristic for it than the doctrine of the social contract, whereby even the state, the highest spiritual community, is supposed to be nothing original and natural but rather only an arbitrary aggregate of individual people whose concrete individual wills have been directed to the production of such a community.[33]

3 Høffding and Bohr: Psychology and Statistics in the Bohr Atom

Several authors have related Niels Bohr's philosophy of atomic physics to analogous ideas of William James and Søren Kierkegaard. I see no reason to doubt the analogies or Bohr's recognition of them. There is little, nevertheless, that Bohr could have learned from them that he did not also learn in the local context of Copenhagen, especially in the writings of Harald Høffding. As a close friend of Bohr's father; as a regular participant in a discussion group meeting in the Bohr household; and as Bohr's philosophy professor at the University of Copenhagen in 1903/04, Høffding's presence was immediate. More important, the entire structure of Bohr's philosophy follows closely the framework advocated by Høffding. Important aspects of this correlation have been emphasized recently by Jan Faye, who also surveys the little direct information that exists about Bohr's interaction with Høffding.[34] I shall carry the argument further to show that not only Bohr's mature ideas; but crucial aspects of the detailed development of those ideas, can best be understood in relation to Høffding's philosophy.

I hasten to add that by tracing the existence of a close analogy I do not imply the thesis that Høffding's ideas were the cause of Bohr's. I imply rather that Bohr participated in an environment that gave meaning and support to ideas that Høffding expressed in particularly lucid form. Bohr might well have imbibed those ideas from his father, who shared at least some of them, or from others in their circle, such as the physicist C. Christiansen another of Bohr's teachers. Lacking as yet more detailed information about these others, I take Høffding as a particularly relevant source and one that Bohr certainly exploited in formulating his distinctive approach to physical theory.

3.1 Høffding's Psychological Philosophy

Høffding partook deeply in the perception of social and psychological fragmentation common around the turn of the century. With others he located that fragmentation in the stresses of industrial society, in "the tendency of modern culture to isolate or to mechanize the single elements of life." Invoking Rousseau and Schiller, he sought to revive the idea of "the harmonious unfolding of the soul as a supreme end" in opposition to "the modern slavery to work." Within society he hoped that "the individual personalities can develop themselves independently and yet in reciprocal harmony, so that there may be a *social organism* [*soziale Lebenstotalität*] analogous to the individual organism." [35]

Positioning himself on the left wing of moderate liberalism, Høffding advocated social reform and the rights of man; but not as natural rights—rather as a symptom and an assumption of modern social and political evolution. His idealization of organic harmony constituted rather a progressive's call to action than a conservative's nostalgia, for Høffding rooted his thought in an existential dilemma. Life, like Being in general, involved a constant battle for existence between isolated elements and integrated totalities: "Wherever we find Being, we also find within it such a strife going on between elements and totalities. The great question is whether out of this strife the elements or the totalities (the solar systems, organisms, souls, human societies) will come off victorious." [36]

In introducing Høffding's *Problems of Philosophy* to American students in 1906, William James identified the Copenhagen professor's views with his own holistic pragmatism and called him "one of the wisest, as well as one of the most learned of living philosophers." [37] The *Problems* first appeared in 1902. They constituted, as James said, Høffding's "philosophical testament." The testament reappeared in 1910 in an "extended edition" titled *Human Thought, Its Forms and Its Tasks*. Between those editions Niels Bohr, aged from seventeen to twenty-five, attended Høffding's introductory lectures on psychology, logic, and history of philosophy in 1903/04. Presumably he encountered at least the following basics of his professor's view.

1. *The Antinomy of Continuity and Discontinuity.* Høffding identified four chief problems of philosophy: consciousness, knowledge, Being, and values. All four reduced to one essential problem, the relation between continuity and discontinuity. [38] In continuity he located stability and conservation, while in discontinuity he found the source of new content and progress. As equally essential perspectives, the two poles formed an antinomy of contrary, but not contradictory, tendencies in human thought and life. The antinomy balanced both conservative and progressive social views and Kantian rationalist and British empiricist philosophies.

In psychology, continuity represented the fundamental striving of personality for wholeness, both in time and over the simultaneous elements of consciousness. The great question of psychology became then, "*Does our conscious life form a totality, a continuum, a little world for itself, or is it only an aggregate, a sum of elements and fragments?*" Rejecting as atomistic and mechanical the British associationist tradition of Locke, Hume, and Mill, Høffding asserted "that the so-called psychical elements are always determined by the relations in which we find them," But he also rejected the attempt of James and Bergson to make the "immediate stream of spiritual life" the source of its own parts: [39] "Consciousness and personality can as little be explained as the products of previously given elements, as organic life can be explained as the product of unorganic elements. On the other hand, consciousness and personality, just like organic life, come into being through a perpetual synthesis of elements not originally begotten by themselves. It is this antinomy which makes the genius of life and of personality so great a riddle."[40]

2. *Continuity as the (Unrealizable) Ideal of Knowledge.* Despite his antinomies, Høffding held continuity to be the ideal of knowledge as of personality. Rational understanding of concepts required their continuous connection through the ground-connsequence relation. By analogy (the Kantian analogy of experience) rational understanding of phenomena, whether physical or psychical, required their continuous causal connection. Discontinuity implied irrationality. Such discontinuities seemed apparent in unconscious intervals (e.g., dreamless sleep) and in qualitative differences both between separate states of consciousness and between the elements of any single state.[41] The psychologist had then to seek continuity in the face of discontinuity, reducing the latter to continuity where possible and carefully locating the irrational remainders where not. Høffding identified three irreducible sources of irrationality, in the relations of quality to quantity, time to causality, and subject to object. All are significant for interpreting Bohr's quantum mechanics.

Concerning quality and quantity, physiological reductionists had argued that because qualitative discontinuities in the psychical realm could not be eliminated, continuity could only be achieved by replacing psychical states by corresponding physical (physiological) states, to which the quantitative relations of causality and conservation of energy applied. Strictly, continuity required an idealized principle of causality applied to a literally continuous series of states. Thereby the relation A causes B would become an identity $A = B$ in a process of continuous evolution. But even supposing physiological continuity, Høffding objected, quantitative identity would not remove qualitative difference. Psychology had therefore to be treated independently.[42]

To relieve the problem of quantitative discontinuity between psychical states, Høffding postulated a "*potential psychical energy*" to fill unconscious intervals. If that idea seemed obscure, so did its physical analogue: "Words like 'inclination' [*Anlage*], 'possibility,' 'capacity,' 'disposition,' 'trace,' 'tendency,' have the same meaning as the word potential energy."[43] Just as physicists used potential energy to conserve the sum of potential and actual (kinetic) energy, so psychologists would (ideally) conserve potential and actual consciousness:[44]

The construction of the concept of the unconscious arises from the same need as the construction of the concept of potential energy in physics. One wants to maintain continuity to avoid an origin out of nothing. And in both realms the hole is plugged by an X. If one asks more closely what sort of X this may be, the physicist will be inclined to conceive his X as related to or analogous to phenomena of motion. With the same right the psychologist will conceive his X, the potential consciousness, as an analogue to [actual] consciousness, as a psychical differential.

Høffding generated his notion of psychical energy not by analogy to the mechanical view of nature, the "mechanical mythology," but by analogy to the electromagnetic view, popular around the turn of the century. Taking his cue from Maxwell and Hertz, as opposed to Boltzmann, he noted that the mechanical view gratuitously regarded extension and motion of matter as pure quantities, rather than qualities to be explained.[45] In addition, mechanism treated matter as an inert, static element separate from the dynamic element, motion or energy. The electromagnetic view, on the other hand, put electrons in the place of inert atoms and regarded them as forms of motion, "standing waves," in the imponderable ether. As to the ether itself, it could only be defined as "that which moves." By analogy, "We know the soul (just as the ether) only as acting. What we here know is just psychical energy, no passive 'essence'. The essence or the nature of the soul is activity."[46]

In thus showing how psychical energy might be regarded as the content of the soul, Høffding did not suppose that energy content determined the development of psychical processes. Rather he distinguished psychical energy from organization and purpose as content from form and as quantity from quality: "Equivalence is not the only viewpoint with respect to energy. That shows itself psychologically in that a state of the soul is not only described through the comprising energy that it controls, but also is characterized by the qualitative state of the psychical elements (thoughts, interests, purposes) that constitute the center of gravity, the real self."[47] Thus quantity and quality remained distinct categories for the description of the soul, as for all other realms of Being, with quality constituting irreducible individuality.

Quality appeared in another form when Høffding considered time in relation to causality. Temporal directionality, he argued, could not be got rid of no matter how successful one might be in reducing the causal series to a continuum through conservation of energy. "If A and B are equivalent, it is to a certain extent irrelevant whether I go from A to B or B to A. And yet, in every individual case it is either A that goes over into B or B into A."[48] The problem had been widely debated by physicists since the 1870s as the problem of irreversibility in the second law of thermodynamics, the law of increasing entropy in all processes of nature. Boltzmann's mechanistically conceived reduction to probability considerations was received in central Europe with skepticism at best. Max Planck sought an escape through continuum mechanics. The positivists Ernst Mach and Richard Avenarius regarded time as merely the order of experience, which might have been reversed. Energeticists treated the entropy law as derivative from the spatial distribution of energy. Their organizer, Wilhelm Ostwald, correlated the conservation and entropy laws with quantity and quality (intensity) of energy, respectively, and in the quantity-quality distinction found the opportunity to explain not only physics and chemistry but biological evolution, psychology, values, and sociology.[49]

Aware of all these approaches and borrowing heavily from them, Høffding nevertheless asserted temporality as an independent fact. While the entropy law provided a clear example, he placed his emphasis on the reality of history and of quality. "Even if the equivalence relation were carried out everywhere, there would always remain a historical element in our knowledge that is given in the direction in which the transition from one event to another occurs. There can be causal equivalence without value equivalence." Høffding's various expressions of this central problem reduce to the single question, To what degree does the specification of the energy of a given state determine its transition to a succeeding state? "And here we come upon the fact that every relation of equivalence—as, in general, every causal relation—first becomes effective in fact when certain conditions, especially forces of release, are present." By "forces of release" Høffding apparently referred to the trigger mechanisms for releasing large amounts of energy so often deployed as an analogy for the freedom of the will to direct material processes. But Høffding believed in no such direct interaction of mind and matter. The need for forces of release was to him inherent in all states of Being, whether psychical or physical. It represented totality and individuality in those states.[50]

If totality and individuality could control the release of energy, they might also control whether that release was continuous or discontinuous. "The old principle that nature is always in conformity with itself ... and makes no jumps, seems to be refuted by experience."[51] Høffding cited radioactive decay of radium as a phenomenon that seemed to violate conservation of energy. The new Darwinism, furthermore, emphasized the sudden origin of new species from mutations, and William James claimed that new insertions into psychical development could occur that were not derivable from previous processes. Such jumps represented the ultimate violation of continuous rational causation.

Irrationality emerged for Høffding in yet a third relation, that of the subjective and objective elements of cognition. He rejected attempts either to locate discontinuities of our knowledge in the knowing subject alone, with continuity in the objective world, or to locate them in the objective world alone. Contending that subjective and objective determine each other in an infinite series, he concluded that the principles of knowledge can never be fixed once and for all. "Thought must constantly be set to work afresh to find predicates for the determination of being, because the springs which feed the stream of thought are inexhaustible."[52] That view underlay the essentially critical cast of his philosophy.

3. *Critical Monism and the Symbolic-Dynamical View of Truth.* In keeping with his dialectical view of consciousness and knowledge, Høffding found in Being itself the "battle of the world views" between materialism and idealism. He regarded mind and matter as equally original aspects of our experience and looked for their unity in a common source. Recognizing, however, that our experience might not exhaust Being, he treated knowledge as a set of partial views attained by analogy to the basic type-phenomena of our experience. These type-phenomena included mind, matter, causation, and evolution.

Differentiating his own monism from the Spinozistic varieties of Leibniz, Fechner, and Wundt (which gave priority to mind), Høffding assumed complete parallelism, and ultimate identity, between mind and matter. Causation supplied the principle for attaining continuity in both realms. Most characteristic of his type-phenomena, however, was evolution, which expressed the reality of time. Time meant directedness and striving, striving to attain unity. Thus the struggle between totalities and elements entered as the very essence of time and evolution. But if evolution were a type-phenomenon, then Being itself involved a "battle for existence" and ought to be regarded as a becoming. The eternal incompleteness of knowledge reflected the incompleteness of Being. One could only take a critical perspective toward knowledge, for it could only be tentative. Høffding therefore labeled his philosophy *Critical Monism*: "It finds ... in Being a force struggling towards unification, which, by progressive evolution, overcomes the sporadic and hostile elements."[53]

Critical monism laid the basis for what Høffding called the *dynamical and symbolic notion of truth*. All principles of knowledge, including especially causation, were to be regarded as "intellectual habits" that enabled consciousness to realize its aim of unification of experience. Adopting the economic theory of Avenarius and Mach, Høffding recognized parsimony and utility as the criteria of validity of these principles. They were tools. "The significance of principles is that they may lead us to reach a rational understanding of our work. Their truth consists in their *valid application*; and this consists in their *working value*."[54]

Because the truth of a principle rested on its unifying power, its capacity to develop psychical energy, Høffding labeled his doctrine 'dynamical'. That term also connoted change over time, as opposed to Kant's static principles. Accordingly, principles and concepts were merely symbols of Being, albeit symbols adapted both to phenomena and to the demands of thought and thus not arbitrary symbols. Calling on modern physical science for support (Maxwell and Hertz and the electromagnetic view of nature), Høffding spread the positivist message: "The problem [of truth] reduces itself to finding a group of symbols which can be employed with entire *consistency*, and from which conclusions can be drawn that will be confirmed by new experiences which can themselves be again expressed by the same group of symbols. But by this method we never get rid of the possibility that another set of symbols might have expressed the actual experiences as well or better, and furnished equally verifiable deductions."[55]

In different situations different sets of symbols might be necessary. Thus the problem of truth required finding the conditions under which a given set of symbols could be applied. Høffding even suggested that Being "conceals simultaneous discords in itself, which make it impossible to construct a harmonious whole. ... philosophers have been too sure that Being in itself was a closed and constant totality, and that it was only our wills and minds that had to battle incessantly to exist and to attain harmony." Neils Bohr would propound the same doctrine of symbols, consistency, and discord in his attempt to understand the atom. He would surely have agreed with Høffding's quotation from Lessing: "It requires one to think gymnastically, not dogmatically."[56]

3.2 Reading Bohr from Høffding's Perspective

Niels Bohr is regarded by many as a heroic visionary of twentieth-century physics, by some as an obfuscating mystic.[57] All agree, however, that quantum mechanics emerged in 1925 under his direct tutelage. Here I discuss his earlier work, aiming primarily at his statistical viewpoint. The discussion divides into two components characteristic of his entire program: stationary states and transitions; mechanical transformability and correspondence. I examine Bohr's use of these ideas by drawing an analogy—as exact as possible—to Høffding's psychological philosophy.

1. *Stationary States and Transitions—the Problem of Individuality.* Høffding conceived psychical energy and causality by analogy to the electromagnetic view of nature. He hoped thereby to attain a unified view of psychical states and psychical transitions analogous to that between electrons and the electromagnetic field. We saw that he did not regard this unity of state and transition energies as determining either the direction or the distance of the transitions. But he went further; he questioned whether the dynamics of the field could ever completely explain the "static" elements within it, the unmoved motions constituting matter, molecules, and electrons.[58] Similarly, he doubted whether knowledge of the laws of psychical development could explain the static element in psychical states. This whole range of problems involved for Høffding the nature of individuality. He expected individuality to appear in both physical and psychical realms and at every level of organization, "There is here an entire ladder of increasing complications, from simple change of position, through physical and chemical processes, up to the organic, psychical, and social phenomena."[59]

Readers of Bohr's post-1925 interpretive essays will recall the similar series of levels through which he hoped to climb by generalizing his principle of complementarity. Like Høffding, he did not expect lower rungs of the ladder to explain higher ones—quantum mechanics would not explain life—rather the rungs formed a continuing analogy, a series of individualities. Bohr recognized in the atom the same three features that Høffding had remarked of individuality generally: (1) stability, (2) jumps, (3) self-directedness. I shall discuss primarily (1) and (2) in this section, leaving (3) for the next.

Concerning the stability of stationary states Bohr used such expressions in his later essays as "the essential non-analyzability of atomic stability in mechanical terms" and "the feature of individuality symbolized by the quantum of action." With respect to discontinuous transitions between stationary states he spoke similarly of "an individual process" and "the peculiar feature of indivisibility, or 'individuality', characterizing the elementary processes." In biology he found an analogy to "the self-preservation and self generation of individuals" in contrast to "the subdivision necessary for any physical analysis." That distinction legitimated the concept of purpose in any biology that would take "due regard to the characteristics of life in a way analogous to the recognition of the quantum of action in ... atomic physics."[60] As has often been observed, this biological analogy parallels closely the usage of Bohr's father, as well as Høffding's usage. But how and when did Bohr begin to structure his physics through the notion of individuality?

Essential clues to an answer appear in a retrospective lecture "On the Interaction between Light and Matter" delivered at Copenhagen in 1920. Bohr began from the "electromagnetic picture of the world," which he said exhibited "an inner harmony which is hardly attained by any other edifice within natural science." But several times he referred to that picture as simply "the electromagnetic theory": e.g.,

"According to the electromagnetic theory, [Newton's fundamental laws of mechanics] appear as special examples of the application of [Maxwell's] fundamental laws of electrodynamics." Thus, if the electromagnetic world view represented Bohr's own early presupposition, we should expect to see him sliding easily between electromagnetic theory and mechanics. Other factors point in the same direction. The beauty of the electromagnetic picture inhered for Bohr in "the detailed and natural manner in which it accounts ... for the interaction between light and matter." That meant that it provided a natural distinction between two aspects of a single system: "We may designate as matter the electric particles in the atoms and as light the force effect that can spread from one atom to another without the transfer of any electric charges." Bohr's language indicates that he regarded the distinction between matter and radiation, like that between mechanics and electromagnetism, as conventional.[61]

Despite its attractive unity Bohr pointed to two deficiencies in the electromagnetic view. First, it did not account for the essential stability of matter as represented in the fixed properties of the chemical elements. Second, it left unexplained Planck's empirically correct formula for black-body radiation, with its implication that in a large collection of atomic systems, regarded as simple harmonic vibrators, the energy states of the vibrating systems would have to be quantized, and could undergo only discrete transitions when emitting or absorbing radiation. The problems of stability and radiation are well-known as two that Bohr sought to solve with his atomic theory. The possibility, however, that he may originally have considered them as problems for the electromagnetic world view has not, to my knowledge, been remarked. That perspective would suggest that Bohr saw the problems in the same light as Høffding had: What is the relation between the dynamic and static elements in the matter-ether system?

In a lecture in 1919 in Leiden Bohr referred these problems not to the electromagnetic world view but to "the ordinary laws of mechanics and electro-magnetism," his standard phrase. But a puzzling feature appears here and in several later reports. After describing Planck's quantized energy states Bohr remarked that the application of the quantum hypothesis to heat radiation, like most other applications, had been "of a statistical nature, the phenomena under consideration depending not on the motion of a single system but on the behaviour of a great number of similar systems." Lecturing in Berlin in 1920 he repeated the point: "So far as statistical equilibrium is concerned only certain distinctive states of the oscillator are to be taken into consideration." These statements suggest that Bohr himself originally believed that the statistical calculation, made by summing over independent oscillators, reproduced phenomena that actually depended on correlated behavior among the atoms. Such a belief would have been natural according to the electromagnetic view in which atoms and field formed a single continuous system. He went on to emphasize the "great importance" of Einstein's having drawn attention to the fact "that the socalled photo-electric effect offered a possibility of an application of the theory to a problem of an essentially non-statistical nature," i.e., to a single atom. Later, calling this discovery "Einstein's great original contribution," Bohr would express it in his own language as the " 'individuality' characterizing the

elementary processes."[62] Given his emphasis, it seems likely that Bohr was recording a 'Eureka' experience of his own. Examination of his work in 1911–1912
reinforces that view as well as the impression that he had looked toward modifications in the electromagnetic world view to solve the problems of heat radiation
and of the atom.

Bohr's doctoral dissertation of 1911 applied the Lorentz electron theory to
calculate electric, magnetic, and thermal properties of metals. In Lorentz's scheme
the electrons formed an ideal gas moving in the space between fixed metal molecules, with mechanical collisions occurring as between elastic spheres. Bohr's project was essentially critical (in Høffding's sense): "to examine to what extent the
results obtained in [Lorentz's theory] depend on the special assumptions made" and
thus to prepare the way for a new theory that would integrate more of experience.[63]
He would conclude that a radical break was necessary.

Generalizing Lorentz's model to include realistic electron-molecule interactions,
Bohr calculated "stationary" effects (like conductivities) and "non-stationary"
effects (like emission and absorption of heat radiation) for the electron gas as a
whole. The terms are analogous to the later stationary states and transitions in a
single Bohr atom. The statistical calculations gave adequate results for the stationary cases, but heat radiation remained intractable. Bohr concluded,[64]

It seems impossible to explain the law of heat radiation if one insists upon
retaining the fundamental assumptions underlying the electromagnetic theory
[citing here Einstein and Planck]. This is presumably due to the circumstance
that the electromagnetic theory is not in accordance with the real conditions and
can only give correct results when applied to a large number of electrons ... or
to determine the average motion of a single electron over comparatively long
intervals of time ... but cannot be used to examine the motion of a single
electron within short intervals of time.

Here the problem lay with electromagnetic theory as applied to the interaction of
electrons between themselves and with the radiation field.

Bohr's calculation had assumed, consistent with classical theory, that the interactions of the electrons among themselves and their interactions with the heat
radiation would not affect their average motions in comparison with the effect
of the molecules. The electron-molecule interactions, furthermore, were isolated
mechanical events that established temperature equilibrium between the molecules
and the electrons. This "dynamical statistical equilibrium" would apply "if the
forces between the metal molecules and the electrons are of the same type as the
forces considered in ordinary mechanics."[65] The last phrase apparently meant to
Bohr two-body forces, excluding higher-order or irreducible multibody forces of
organization. "Dynamical statistical equilibrium" was the basic assumption of the
kinetic theory of gases for equilibrium states. It was equivalent to the Maxwell-
Boltzmann distribution of velocities, to 'molecular disorder', and to mechanical
analyzability. If it applied to the electron gas, Bohr showed, Planck's formula
would not result. Here the problem lay in the applicability of mechanics to individual electron-molecule interactions.

Bohr soon elaborated his interpretation of the failure of this condition, reaching at the same time for a nonmechanical explanation of the second law of thermo-dynamics itself. Claiming that Boltzmann's various arguments for deriving the entropy law from mechanics and probabilities were vacuous, he wrote to his friend C. W. Oseen, "With regard to the entropy theorem ... I believe that it cannot be justified on the basis of considerations of probability, but only by considering the manner in which deviations from what is called molecular disorder are produced." Such deviations, he suggested, arose from "systematic correlations between the position and velocity coordinates of the electrons," and "It is just these correlations that give rise to all the phenomena of interest."[66] Although Bohr crossed out these remarks, they show that he believed the discontinuity of Planck's oscillator states, and the entropy law as well, derived from nonmechanical, holistic phenomena among collections of electrons. Juxtaposed with the preceding statements, they show that Bohr made no sharp distinction between the failure of electromagnetic theory and the failure of mechanics, although he associated the former primarily with multielectron or electron-field interactions and the latter with electron-molecule interactions. Nowhere did he suggest that electromagnetic theory failed for the radiation field itself.

Twice in 1912 Bohr drafted similar remarks, intended as rejoinders to papers by J. Stark and O. W. Richardson, but again he kept them to himself.[67] He vacillated too between ascribing the nonclassical deviations to multielectron and to electron-molecule interactions. His breakthrough in 1913 should thus partly be seen as a decision to fix individuality in the single atomic systems rather than in the electron gas. That move occurred in the context of work on Rutherford's nuclear atom and its association with discontinuous radioactive transitions on the one hand and chemical stability on the other. Nevertheless, Bohr always regarded Einstein's emphasis on the singular, nonstatistical nature of the photoelectric effect to have been the turning point that put the problems of electromagnetic interactions inside the individual atom.[68]

I emphasize that in committing himself to the atomic explanation of Planck's quantized energy levels, Bohr did not give up a statistical explanation and adopt a nonstatistical one. He had all along regarded the statistical result, obtained by summing over isolated systems, as epiphenomenal, a result that corresponded to an underlying holistic process. Having located that process in the atom, statistics remained epiphenomenal; its results had now to be attributed to the individuality of the atomic system.

2. *Mechanical Transformability and Correspondence—Causation and Choice in Individual Transitions.* Harald Høffding took the principles of causation and evolution as type-phenomena of Being, where causation implied a continuing relation of identity expressed as continuous conservation of energy, and evolution referred to the temporal direction of the equivalence relation. Neils Bohr in 1918 made principles comparable to these the mainstays of his heuristic program for understanding more fully the atomic structure that he had proposed in 1913. I shall jump then from the prehistory of his first famous series of papers to the conceptual

content of his second. In doing so I assume known two features of Bohr's quantum postulates: first, the stationary states were fixed by their periodicity properties, in such a way that for simply periodic orbits twice the mean kinetic energy divided by the orbital frequency was an integral number of quanta of action; second, the radiation frequency during a transition was not related to the orbital frequencies of initial and final states, but was equal to their energy difference divided by Planck's quantum of action.

Beginning with continuous equivalence, we recall that establishing such continuity was to Høffding identical to attaining rational understanding. For that reason he had postulated potential psychical energy to fill the gaps between conscious psychical states. To attain a similar end for transitions between stationary states of an atom Bohr adapted a principle introduced by Paul Ehrenfest, the principle of adiabatic invariance. In his initial presentation of the ideas in 1916 Bohr repeatedly stated his purpose as an examination of the consistency of the postulates of the quantum theory in their application to a variety of different kinds of problems. Consistency in integrating experience was of course just Høffding's test of the adequacy of type-phenomena in his dynamical-symbolic notion of truth. Continuous causation realized consistency. In similar fashion Bohr argued that "the possibility of a consistent theory" based on the quantum postulates derived from the applicability of adiabatic invariance.[69]

Ehrenfest's principle allowed one to transform continuously the quantized states of one periodic system into those of another by slowly varying the forces involved. A strict mechanical proof showed that, under slow enough variation, the periodicity of the one system would transform continuously into that of the other. Planck's quantum condition for harmonic vibrators, for example, would transform into Bohr's quantum condition for electron orbits. Continuous equivalence guaranteed consistency.

In his fully developed application of the invariance principle in his 1918 "Quantum Theory of Line Spectra," Bohr continued his concern with consistency between different applications of the theory. But he now emphasized as well the possibility of obtaining a rational determination of the energies of the stationary states.[70] That possibility arose because Bohr's new paper generalized his earlier treatment of simply periodic systems to conditionally periodic systems, characterized by more than one periodicity and thus more than one quantum number. By passing through a suitable cycle of adiabatic transformations, including a condition of so-called degeneracy in which one of the periodicities vanished and all of the states corresponding to its quantum number had the same energy, any stationary state of a system could be transformed into another state of the same system. That result overcame a "fundamental difficulty": "We possess no means of defining an energy difference between two states if there exists no possibility for a continuous mechanical connection between them."[71] The result held particular importance because at that time Bohr was beginning to insist that energy and momentum were not conserved in individual transitions.

To emphasize that the question was one of consistent application of terms like energy, defined only in classical mechanics, to processes that violated mechanics,

Bohr renamed Ehrenfest's principle the "principle of mechanical transformability." This new appellation had the additional advantage that it did not suggest a thermodynamic (and thus statistical) result, as 'adiabatic' did. The principle applied to each individual system.[72]

The requirement of slow variation during 'mechanical' transformations provided Bohr with a fresh insight into the holistic nature of stationary states. 'Slow' meant a small change in the force-field experienced by the electron during one cycle of its oscillations (compare Bohr's 1911 remarks):[73]

This may be regarded as an immediate consequence of the nature of the fixation of the stationary states in the quantum theory. In fact the answer to the question, whether a given state of a system is stationary, will not depend only on the motion of the particles at a given moment or on the field of force in the immediate neighborhood of their instantaneous position, but cannot be given before the particles have passed through a complete cycle of states, and so to speak have *got to know* the entire field of force of influence on the motion.

Bohr's anthropomorphic terminology should not be dismissed as merely metaphorical. If indeed he thought of the stationary states and transitions as something like 'individuality' within the electromagnetic view of nature, then the atom as a whole—electron and field, as individual *within* environment—had to be regarded as an integral unit, just as in 1911–1912 he had supposed that Planck's discrete oscillator states derived from collective, nonmechanical motions in an electron gas. The extra something maintaining stability in the stationary states was harmony (periodicity), and harmony required the electron to know its entire field.

The quotation immediately above continues, "If thus in the case of a periodic system of one degree of freedom . . . if [the field's] comparative variation within the time of a single period was not small, the particle would obviously have no means to get to know the nature of the variation of the field and to adjust its stationary motion to it. . . ." With periodicity destroyed the electron would undergo nonmechanical transitions (mutations) to establish a new harmony. Similarly, in the case of a degenerate state, if an external field were applied to remove the degeneracy by inducing a new periodicity in the motion, the adjustment would have to occur "in some unmechanical way."[74] The field could never be increased slowly enough from zero that its change during one period was small, because the frequency of the perturbed motion would be proportional to the intensity of the applied field.

The second main principle of Bohr's 1918 synthesis, and the one that reinterpreted classical statistics on a holistic foundation, was the correspondence principle. He introduced that label only in 1920, but in rudimentary form had been using the idea since 1913 as another form of the consistency and continuity argument. At high quantum numbers the results of the quantum theory had to agree with classical results, for in this region of long orbital periods and small differences in periods between states the classical calculations agreed with experiment. Furthermore, the quantum mechanical frequencies for radiation emitted during transitions approached the orbital frequencies.

The new content of this guiding principle in 1918 derived in part from Einstein's elegant paper of 1916, which obtained Planck's black-body formula from the mere assumption of certain probabilities for spontaneous and induced transitions between stationary states. If that could be done, then in the region of long wavelengths (high quantum numbers), where classical calculations agreed with Planck's formula, the quantum transition probabilities had to agree with the classical predictions of intensity based on electron orbits. Expressing the orbital motion as a sum of harmonic components, Bohr correlated the quantum-mechanical radiation frequencies with the frequencies to be expected classically from harmonic vibrations. He then correlated the transition probabilities with the amplitudes of the harmonic components.[75] In this correlation lay the reinterpretation of statistics.

By the classical theory all fundamental frequencies and overtones associated with a given orbital motion would be emitted or absorbed simultaneously; the corresponding quantum frequencies could be emitted or absorbed only in completely separate transitions between a given stationary state and the neighboring states. Thus a single classical orbit corresponded to a large collection of quantized atoms undergoing different transitions, and no criterion for choosing the transition of a given atom was possible. On the one hand, therefore, the observed radiation might be regarded as a statistical result—a sum over radiations from independent electrons—but on the other hand the entire statistical result had to be present in each electron as a potential, a set of possibilites and probabilities that were essentially unanalyzable. In his retrospective essays Bohr would stress this feature as one "of indivisibility, or 'individuality,' characterizing the elementary processes": "The very idea of stationary states is incompatible with any directive for the choice between such transitions and leaves room only for the notion of the relative probabilities of the individual transition processes. The only guide in estimating such probabilities was the so-called correspondence principle...."[76]

There is little in these statements that was not present in Bohr's conception of the correspondence principle in 1918. They also contain little that Høffding had not stated more generally with respect to individuality. Høffding regarded both jumps and their direction as irreducibly individual and qualitative. Drawing on evolution as a type-phenomenon, he made variation an essential feature of Being, not continuous variation but mutation. He also regarded general laws as only more or less applicable to individual cases. Especially in ethics and social thought wide variations from any norm were to be expected: "One and the same demand applied to different individuals may enjoin upon each of them an entirely different task." But as a real phenomenon of Being, diversity occurred in all sciences. "Here is a world," Høffding wrote in 1909, "which is almost new for science, which till now has mainly occupied itself with general laws and forms. But these are ultimately only means to understand the individual phenomena, in whose nature and history a manifold of laws and forms always cooperate."[77]

Bohr's correspondence principle served just the function of connecting general laws and forms (classical mechanics and electromagnetic theory) with individual phenomena. It is significant, then, that he commonly referred to it as a "formal analogy," a term that appears throughout his writings. His usage conforms with Høffding's specification of the use of type-phenomena for understanding Being. All such usages, he said, have two motives: a *formal* motive "connected with the demand for continuity, a demand in which both personality and science coincide"; and a practical motive, involving the use of an *analogy* drawn from one part of experience to attempt to understand all experience. Høffding's type-phenomena were formal analogies; so too in his view were the old atomic theory, and the mechanical and the electromagnetic conceptions of nature. [78]

As a formal analogy, the correspondence principle helped to make the frequencies and the probabilities of individual transitions conceivable in classical terms, by relating the probabilities to amplitudes of harmonic oscillations present in both initial and final stationary states. Already in 1916 Bohr had recognized, because of the dispersion of light at transition frequencies, that the transition mechanism showed "a close analogy to an ordinary electrodynamic vibrator." [79] But while the analogy made sense for spontaneous transitions, it did not for induced transitions, for it suggested an increased amplitude of any harmonic component having the frequency of the inducing radiation. Einstein had emphasized that the photoelectric effect involved not a gradual buildup of energy in an oscillating electron followed by its escape from the atom, but a sudden transition. He had shown, furthermore, that if energy and momentum were conserved in the photoelectric process, then the electromagnetic energy had to be taken up by the electron as a single localized light quantum, a photon. Either one had to give up continuous electromagnetic field energy, and along with it the wave picture of light, or give up conservation of energy and momentum. By 1919 Bohr regarded conservation as "quite out of [the] question," a view he shared with several others, among them C. G. Darwin in England.

In July or August of 1919 Bohr drafted a letter to Darwin (never sent) in which he wrote, "I am inclined to take the most radical or rather mystical views imaginable." He supposed that the frequency of the incident radiation was just "the key to the lock which controls the starting of the interatomic process." [80] That view implies that Bohr did not believe that radiation induced transitions by increasing the amplitude of a harmonic component in the orbital motion. What then did determine the occurrence of a transition? We should notice first that Høffding, with his "forces of release" and his correlation of discontinuous "jumps" with nonconservation of energy, would have found Bohr's ideas quite congenial. Such ideas were fundamental to Høffding's notion of individuality. They did not, however, imply indeterminism of individual action: "Indeterminism, which teaches the existence of causeless acts of will, absolutely destroys the inner connection and the inner continuity of conscious life." The individual merely exhibited "indeterminateness," indeterminateness with respect to the characteristics common to its species. "What is inherited has therefore more or less indeterminateness, and the degree and direction in which it is developed is a question of individual experiences." Individuality determined behavior even though its qualitative nature was inaccessible to analysis: "Individuality appears in consequence an irrational whole, which admits of only approximate determination." [81]

Høffding also expressed himself clearly on the relation of statistical law to the individual: "... statistics show the influence of external conditions on human actions. But in respect of every single individual, the force of external conditions is always modified by the inner condition with which the individual confronts the external world."[82] He did not mean that the individual merely deviates more or less from type as a result of accidental circumstances.

Høffding's views on indeterminateness and statistics represent quite closely Bohr's views on atomic transitions induced by external radiation. The radiation did not cause transitions; it only conditioned them by resonating with internal harmonics (no resonance, no transition). The resonance process, furthermore, could not be analyzed exhaustively, because which resonance the electron would respond to remained completely indeterminate. Any definition of the electron's stationary state required knowledge of its periodicities, and knowledge of periodicities precluded exact knowledge of its motion in space and time. The periodicities thus inhered in the state as a whole. Statistics could decompose this whole in an artificial sense by treating it as a sum of many different electrons. That did not touch the individual process, however, except as a "formal analogy" between classical quantitative causation and the qualitative causation of atomic processes.

After 1918 Bohr regularly adopted the term 'conditions' for the effect of external radiation on the transition process, indicating a qualitative resonance between the radiation and the atomic motions with no quantitative exchange of energy. By 1922 he had begun to speak of "*latent*" oscillators connecting the stationary states and stimulated by the radiation field. These became the "virtual oscillators" of the famous Bohr-Kramers-Slater paper in 1924, but I shall not develop the idea further here.[83]

4 Conclusion

The Wundt-Lamprecht and Høffding-Bohr cases exemplify two very different ways in which a perceived qualitative dimension of reality was applied to the reinterpretation of statistical regularities. Wundt and Lamprecht regarded the qualitative as exclusively psychical and postulated a special psychical energy that grew by creative synthesis, thus allowing for evolution. Høffding and Bohr attributed the qualitative to both psychical and physical realms but restricted the concept of energy to the quantitative. Nevertheless, they recognized a qualitative determinant of change that they associated with evolution and that allowed for nonconservation of energy. In both cases 'qualitative' connoted unanalyzability and thus indeterminateness in principle, which opened the way for probabilistic interpretations of part-whole relations.

With respect to part-whole relations, especially those between the individual and the collective, the tendencies of the two diverge. While both emphasized unity in the *Gemeinschaft* and diversity among individuals, Wundt and Lamprecht saw the whole as maintaining its creative synthesis continuously. The unifying force and the attendant probabilities for individuals resided in the collective, even though they were expressed through the individuals. Høffding and Bohr did not regard the unity of the *Gemeinschaft* as continuous or as outside individuals, but as a dialectical unity deriving from the antinomy of continuity and discontinuity, whole and parts.

Concomitantly, they located the conditions of the collective in the individuals, who nevertheless expressed only those of its aspects that resonated with their own individuality. Each of them expressed in his own way the strife between totality and elements in the collective. Høffding characterized the difference in views when he accused Wundt of a "mystical dualism: the end lies in the 'universal will,' and the means in the 'individual will.' He has no space in his ethics for the tragic conflicts which may arise from the collision of the individual will with a historically formed universal will." Høffding understood Wundt as arguing for an antiindividualism that located the "motives," "tendencies," and "content" of the "spiritual individual" in his membership in society alone.[84]

Where Wundt and Lamprecht ascribed stable probabilities for diverse actions to the collective, Høffding and Bohr ascribed them to the individual. That the difference was not always easy to see is evident from Bohr's early work on electron theory, in which he vacillated between attributing statistical regularities to nonmechanical behavior in multielectron and single-atom interactions. After 1926 a similar problem arose with respect to wave-particle duality in quantum mechanics and its statistical interpretation. Most physicists have since accepted the view (on practical rather than psychological grounds) that the wave formalism applies only to collections of electrons and has no meaning for individual events. To Niels Bohr, however, the wave always characterized the probabilistic behavior of individual systems. Drawing his familiar analogy with individual consciousness, he associated the wave with holistic "sentiments," "feelings," and "instincts," and the particle with analytic "thought."[85] True to his original commitments, he found in these two sides of consciousness the antinomy of whole and parts characteristic of all aspects of experience, which required complementary viewpoints for full expression.

Notes

1. Leonard Krieger, *The German Idea of Freedom: History of a Political Tradition* (Chicago and London: University of Chicago Press, 1957), pp. 341–457.

2. F. C. Dahlmann, *Die Politik, auf den Grund und das Mass der gegebenen Zustände zurückgeführt*, 2nd ed. (Leipzig: 1847), p. 4; *Geschichte der französischen Revolution, bis auf die Stiftung der Republik*, 3rd ed. (Berlin: 1864), p. 223.

3. F. C. Dahlmann, *Die Politik* (note 2), pp. 9, 84.

4. Rudolph Virchow, "Atome und Individuen" (1859), in *Vier Reden über Leben und Kranksein* (Berlin: 1862), reprinted in *Disease, Life, and Man: Selected Essays by Rudolph Virchow*, trans. L. J. Rather (Stanford: Stanford University Press, 1958), pp. 120–141, on p. 130.

5. Letters of 25 March and 22 November 1855 in Max Cornicelius, ed., *Heinrich von Treitschkes Briefe* (Leipzig: 1912), I, pp. 295–296, 101; quoted in Krieger, *German Idea of Freedom*, pp. 365–366.

6. Ferdinand Tönnies, *Gemeinschaft und Gesellschaft: Grundbegriffe der reinen Soziologie*, 3rd ed. (Berlin: Curtius, 1920), p. 4.

7. Wilhelm Lexis, *Zur Theorie der Massenerscheinungen in der menschlichen Gesellschaft* (Freiburg: Wagner, 1877), p. 22.

8. W. Lexis, *Zur Theorie* (note 7), p. 92.

9. W. Lexis, *Zur Theorie* (note 7), pp. 7, 4ff.; cf. p.19.

10. Miles A. Tonker, "Wundt's Doctorate Students and Their Theses 1875–1920," and Carl F. Graumann, "Experiment, Statistics, History: Wundt's First Program of Psychology," in *Wundt Studies: A Centennial Collection*, ed, W. G. Bringmann and R. D. Twency (Toronto: Hogrefe, 1980), pp. 269, 38. Quotation by Graumann from *Beiträge zur Theorie der Sinneswahrnehmung* (Leipzig: 1862), p.xxv.

11. Wilhelm Wundt, "Der Begriff der Seele," *Grundriss der Psychologie*, 11th ed. (Leipzig: Kröner, 1913), para. 22.

12. W. Wundt, "Der Begriff" (note 11), p. 397.

13. W. Wundt "Der Begriff" (note 11), para. 23; also his *Logik: Eine Untersuchung der Principien der Erkenntnis und der Methoden wissenschaftlicher Forschung*, 2 vols., 2nd ed. (Enke: Stuttgart, 1895), II, pt. 2, pp. 268ff. and generally 241–290.

14. *Logik*, II (note 13), pt. 2, pp. 274–277; quotation p. 277.

15. W. Wundt, *Logik*, II (note 13), p. 280.

16. "Die Entwicklung geistiger Gemeinschaften," *Grundriss*, para.21, summarizes Wundt's *Völkerpsychologie*; also *Logik*, II, pt. 2, pp. 231–240; 291–296. His political philosophy is best expressed in "Ueber den Zusammenhang der Philosophie mit der Zeitgeschichte" (1889) and "Ueber das Verhältnis des Einzelnen zur Gemeinschaft" (1891), in *Reden und Aufsätze*, 2nd ed. (Leipzig: Kröner, 1914); quotation p. 54. See also *Grundriss*, p. 365. Wundt's politics, like his psychology, became increasingly holistic and conservative after he left Heidelberg for Leipzig in 1875. There he had been active in the liberal workers movement in the sense of Schulze-Delitzsch emphasizing self-help. See Gustav A. Ungerer, "Wilhelm Wundt als Psychologe und Politiker," *Psychologische Rundschau*, 31 (1980), 99–110.

17. *Grundriss* (note 11), p. 385.

18. *Logik*, I (note 13), pp. 442, 445; II, pt. 2, p. 474.

19. *Reden* (note 16), p. 54.

20. J. G. Droysen, "Erhebung der Geschichte zum Rang einer Wissenschaft," *Historische Zeitschrift*, 2 (1863), in *Historik: Historisch-kritische Ausgabe*, ed. Peter Leyh (Stuttgart: Frommann, 1977), p. 465; quoted by Karl Lamprecht, "Was ist Kulturgeschichte? Beitrag zu einer empirischen Historik," *Deutsche Zeitschrift für Geschichtswissenschaft*, N.F. 1 (1896/97), 75–150; reprinted in *Ausgewählte Schriften zur Wirtschafts- und Kulturgeschichte und zur Theorie der Geschichtswissenschaft* (Aalen: Scientia Verlag, 1974), pp. 257–327 (p. 287).

21. "Kulturgeschichte" (note 20), p. 291; "Die Kernpunkte der geschichtswissenschaftlichen Erörterungen der Gegenwart," *Zeitschrift für Socialwissenschaft* (1899); *Schriften*, p. 507. For a lucid summary of Lamprecht's historiography see Karl II. Metz, "Historisches 'Verstehen' und Sozialpsychologie: Karl Lamprecht und seine Wissenschaft der Geschichte," *Saeculum*, 33 (1982), 95–104.

22. "Kulturgeschichte" (note 20), pp. 263–266.

23. "Ueber den Begriff der Geschichte und über historische und psychologische Gesetze," *Annalen der Naturphilosophie*, 2 (1903); *Schriften*, p. 588.

24. "Kulturgeschichte" (note 20), pp. 295–297, for discussion of the idea of forces as *Zustände*.

25. "Biopsychologische Probleme," *Ann. d. Naturphilosophie*, 3 (1904); *Schriften*, pp. 597–603.

26. I thank Karl Metz for this point. See his "Historisches 'Verstehen'" (note 21), pp. 98–99.

27. "Kulturgeschichte" (note 20), pp. 308, 276, 296, 315.

28. "Ueber die Entwicklungsstufen der deutschen Geschichtswissenschaft," *Zeitschrift für Kulturgeschichte*, N.F. 5/6 (1897/98); *Schriften* (note 20), p. 456; "Die Entwicklung der deutschen Geschichtswissenschaft vornehmlich seit Herder" (1889 lecture), *Schriften* (note 20), p. 495; "Kulturgeschichte" (note 20), p. 318.

29. "Kulturgeschichte" (note 20), pp. 315–318; "Individualität, Idee und sozialpsychische Kraft in der Geschichte," *Jahrbücher für Nationalökonomie und Statistik*, 3F. 13 (1897); *Schriften*, p. 332.

30. "Kulturgeschichte" (note 20), p. 315. Cf. Wundt's discussion of the mean man, *Logik* (note 13), II, pt. 2, p.475, which refers to Lexis, whom Lamprecht also cites here.

31. "Individualität" (note 29), pp. 337, 338; "Kernpunkte" (note 21), p. 502n1.

32. "Individualität" (note 29), p. 332; "Kulturgeschichte" (note 20), p. 275n4.

33. "Kulturgeschichte" (note 20), p. 259ff.

34. Jan Faye, "The Influence of Harald Høffding's Philosophy on Niels Bohr's Interpretation of Quantum Mechanics," *Danish Yearbook of Philosophy*, 16 (1979), 37–72.

35. Harald Høffding, *The Problems of Philosophy*, trans. G. M. Fisher (New York: MacMillan, 1906), pp. 162–163.

36. Høffding, *Problems* (note 35), pp.164, 150.

37. Høffding, *Problems*, p. v.

38. Høffding, *Problems*, p. 8.

39. Høffding, *Problems*, pp. 14, 16; *Der menschliche Gedanke, seine Formen und seine Aufgaben* (Leipzig: Reisland, 1911), pp. 8–9.

40. Høffding, *Problems* (note 35), p. 19.

41. Høffding, *Problems*, pp. 63–65 (p. 26).

42. Høffding, *Problems*, pp. 28–32, 36–39.

43. Høffding, *Problems*, p. 44; Høffding, *Gedanke* (note 39), pp. 1–38, treats "psychical energy" generally (quotation p. 32).

44. Harald Høffding, rev. of F. Jodl, *Lehrbuch der Psychologie*, in *Vierteljahrsschrift für wissenschaftliche Philosophie*, 22 (1898), 220.

45. Høffding, *Gedanke* (note 39), pp. 304ff.; *Problems* (note 35), pp. 91, 88. In *Gedanke*, p. 365n, Høffding cited Paul Langevin (*Physics of the Electron*, 1904) and C. Christiansen, Bohr's physics professor (Programmschrift der Kopenhagener Universität, 1905) as exponents of the electromagnetic view.

46. Høffding, *Gedanke* (note 39), pp. 37–38.

47. Høffding, *Gedanke* (note 39), pp. 238ff.; see also pp. 38, 300–301, 314, and Høffding, *Problems* (note 35), p. 88.

48. Høffding, *Gedanke* (note 39), p. 314.

49. Høffding, *Problems* (note 35), pp. 98–100 (p. 93n40). On Boltzmann and Planck see Thomas S. Kuhn, *Black Body Theory and the Quantum Discontinuity 1894–1912* (Oxford and New York: Clarendon Press and Oxford University Press, 1978), esp. chapters 1 and 2.

50. Høffding, *Gedanke* (note 39), pp. 235, 315; Høffding, *Problems* (note 35), p. 105. On trigger mechanisms see Theodore M. Porter, "A Statistical Survey of Gases: Maxwell's Social Physics," *Historical Studies in the Physical Sciences*, 12 (1981), 77–116, esp. pp. 100–114.

51. Høffding, *Gedanke* (note 39), p. 316.

52. Høffding, *Problems* (note 35), p. 113.

53. Høffding, *Problems* (note 35), p. 136. Generally on type-phenomena see pp. 116–152; on evolution, "The Influence of the Conception of Evolution on Modern Philosophy," in *Darwin and Modern Science*, ed. A. C. Seward (Cambridge: Cambridge University Press, 1909), pp. 446–464.

54. Høffding, *Problems* (note 35), pp. 77, 71, 81.

55. Høffding, *Problems*, p. 90, and *Gedanke* (note 39), p. 305.

56. Høffding, *Problems*, pp. 120, 128.

57. For a provocative presentation of the latter view, see J. L. Heilbron, "The Earliest Missionaries of the Copenhagen Spirit," *Review d'histoire des sciences*, 38 (1985), 195–230.

58. Høffding, *Problems* (note 35), pp. 91–94.

59. Høffding, *Gedanke* (note 39), p. 316.

60. Niels Bohr, *Atomic Physics and Human Knowledge* (New York: Wiley, 1958), pp. 9, 8, 6, 34, 10.

61. "On the Interaction between Light and Matter," in *Niels Bohr: Collected Works*, gen. ed. L. Rosenfeld (vols. I (1972), III (1976), IV (1977) ed. J. R. Nielsen; vol. II (1981) ed. U. Hoyer) (Amsterdam, New York, and Oxford: North Holland), III, pp. 230–232.

62. Bohr, "On the Program of the Newer Atomic Physics," *Coll. Works* (note 61), III, pp. 206, 209; "On the Series Spectra of the Elements," III, pp. 243–244; Bohr, *Atomic Physics* (note 60), pp. 33–34.

63. Bohr, *Studies on the Electron Theory of Metals*, trans. J. R. Nielsen, *Coll. Works* (Note 61), I, p. 393; "Lecture on the Electron Theory of Metals," at Cambridge Philosophical Society, 1911, *Coll. Works*, I, p. 414; "Statements at Thesis Defense," *Coll. Works*, I, p. 97. My discussion of Bohr's electron theory supplements that in John L. Heilbron and Thomas S, Kuhn, "The Genesis of the Bohr Atom," *Hist. Stud. Phys. Sci.*, 1 (1969), 211–290 (on electron theory, pp. 213–237).

64. Bohr, *Electron Theory* (note 63), p. 378. See Bohr, "Electron Theory" (note 63), p. 419, for a similar conclusion regarding magnetism.

65. Bohr, "Electron Theory" (note 63), p. 413; also Bohr, *Electron Theory* (note 63), p. 300.

66. Bohr to Oseen, 1 Dec. 1911, *Coll. Works* (note 61), Vol. I, pp. 430–431.

67. Bohr, *Coll. Works* (note 61), I, pp. 438, 444. Also Bohr to Harald Bohr, 28 May 1912, I, p. 553.

68. Bohr, "Rutherford Memorandum" (1912), *Coll. Works* (note 61), I, p. 137; "On the Constitution of Atoms and Molecules," *Philosophical Magazine*, 26 (1913), 5, in *Coll. Works*, I, p. 165; "On the Application of the Quantum Theory to Periodic Systems" (1916, unpublished), *Coll. Works* (note 61), II, p. 434. For a detailed discussion of the complex evolution of the Bohr atom see Heilbron and Kuhn, "Genesis" (note 63), pp. 238–290.

69. Bohr, "Application of the Quantum Theory" (note 68), pp. 433, 435ff., 444, 452, 460; cf. Høffding, *Problems* (note 35), pp. 90–94.

70. Bohr, "On the Quantum Theory of Line Spectra," *Dan. Vid., Selsk. Skrifter, naturvid. -mat. Afd.*, 1918–1922, 8. *Raekke, IV. 1, 1–3. Coll. Works*, III, pp. 4, 87, 75, 105, 109n. I omit discussion of the associated problem of "rational determination" of a priori probabilities of the states (pp. 75, 92).

71. Bohr, "Line Spectra" (note 70), pp. 75–90ff.

72, Bohr, "Line Spectra," p. 74; Bohr to Ehrenfest, 18 May 1918, pp. 11–12.

73. Bohr, "Line Spectra," pp. 87ff. (my emphasis).

74. Bohr, "Line Spectra," p. 89.

75. Albert Einstein, "Zur Quantentheorie der Strahlung," *Verhandlungen der Deutschen Physikalischen Gesellschaft*, 18 (1916), 318, and *Physikalische Zeitschrift*, 18 (1917), 121. Bohr, "Line Spectra" (note 70), pp. 70–73, 80–82, 98–102.

76. Bohr, "Line Spectra" (note 70), p. 81; Bohr, *Atomic Physics* (note 60), pp. 34, 35.

77. Høffding, *Problems* (note 35), p. 171; "Influence of . . . Evolution," p. 458.

78. Høffding, *Problems* (note 35), pp. 117, 120–122.

79. Bohr, "Application of the Quantum Theory" (note 68), p. 449.

80. Bohr to Darwin, 1919 (undated), discussed by Martin J. Klein, "The First Phase of the Bohr-Einstein Dialogue," *Hist. Stud. Phys. Sci.*, 2 (1970), 1–39, on pp. 20–21.

81. Harald Høffding, *Outlines of Psychology*, trans. M. E. Lowndes (London: MacMillan, 1896), pp. 346, 352ff., 353.

82. Høffding, *Outlines* (note 81), p. 351.

83. Bohr, "On the Application of the Quantum Theory to Atomic Structure. Part I: The Fundamental Postulates," *Proceedings of the Cambridge Philosophical Society, Supplement* (Cambridge: Cambridge University Press, 1924), p. 36, from *Zeitschrift für Physik*, 13 (1923); *Coll. Works* (note 61), III; pp. 458–499, on p. 493. N. Bohr, H. A. Kramers, and J. C. Slater, "The Quantum Theory of Radiation," *Phil. Mag.*, 47 (1924), 785–802, and *Zs. f. Phys.*, 24 (1924), 69–87.

84. Harald Høffding, *Modern Philosophers: Lectures Delivered at the University of Copenhagen during the Autumn of 1902*, trans. A. C. Mason (London: MacMillan, 1915), pp. 35, 28. See also his *Brief History of Modern Philosophy*, trans. C. F. Sanders (New York: MacMillan, 1935), pp. 280–284.

85. Bohr, *Atomic Physics* (note 60), pp. 21, 27, 52, 93.

Name Index for Volumes 1 and 2

Page numbers in boldface refer to volume 2.

Subject Index for Volumes 1 and 2

Page numbers in boldface refer to volume 2.

Printed in the United States
by Baker & Taylor Publisher Services